METHODS FOR THE STUDY OF MARINE BENTHOS

Third Edition

METHODS FOR THE STUDY OF MARINE BENTHOS

Third Edition

Edited by

Anastasios Eleftheriou* and Alasdair McIntyre†

**Department of Biology, University of Crete, Greece and Hellenic Centre for Marine Research, Crete, Greece*
†School of Biological Sciences, University of Aberdeen, UK

Blackwell
Science

Editorial Offices:
Blackwell Science Ltd, 9600 Garsington Road, Oxford OX4 2DQ, UK
Tel: +44 (0) 1865 776868
Blackwell Publishing Inc., 350 Main Street, Malden, MA 02148-5020, USA
Tel: +1 781 388 8250
Blackwell Science Asia Pty Ltd, 550 Swanston Street, Carlton, Victoria 3053, Australia
Tel: +61 (0)3 8359 1011

First edition published 1971
Second edition published 1984
Third edition published 2005 by Blackwell Science Ltd

Library of Congress Cataloging-in-Publication Data
Methods for the study of marine benthos / edited by Alasdair McIntyre and Anastasiois Eleftheriou. – 3rd ed.
p. cm
Second edition edited by N.A. Holme and A.D. McIntyre.
Includes bibliographical references and index.
ISBN-13: 978-0-632-05488-6 (hardback : alk. paper)
ISBN-10: 0-632-05488-3 (hardback : alk. paper)
1. Dredging (Biology) 2. Benthos–Research–Methodology. 3. Marine biology–Methodology. I. McIntyre, A. D. II. Eleftheriou, Anastasiois.

QH91.57.D7M47 2005
578.77′7′0723–dc22
2004027447

ISBN 0-632-05488-3

A catalogue record for this title is available from the British Library

Set in 11/13pt Times
by TechBooks, New Delhi, India
Printed and bound in India
by Replika Press Pvt. Ltd, Kundli

The publisher's policy is to use permanent paper from mills that operate a sustainable forestry policy, and which has been manufactured from pulp processed using acid-free and elementary chlorine-free practices. Furthermore, the publisher ensures that the text paper and cover board used have met acceptable environmental accreditation standards.

For further information on
Blackwell Publishing, visit our website:
www.blackwellpublishing.com

Sand Ripples in Shallow Water – Cover photograph by Igor Ruza.

Contents

Contributors

Antony J. Bale, Plymouth Marine Laboratory, Prospect Place, West Hoe, Plymouth PL1 3DH, UK

Brian J. Bett, DEEPSEAS Group, George Deacon Division, Southampton Oceanography Centre, Southampton SO18 1UH, UK

Maura G. Chapman, Centre for Research on Ecological Impacts of Coastal Cities, Marine Ecology Laboratories, A11, University of Sydney, NSW 2006, Australia

Anastasios Eleftheriou, Department of Biology, University of Crete, Vassilika Vouton, P.O. Box 2208, 71409 Heraklion, Crete, Greece and Hellenic Centre for Marine Research, P.O. Box 2214, Heraklion 71003, Crete, Greece

John D. Gage, Deep-Sea Benthic Group, Scottish Association for Marine Science, Oban PA37 1QA, UK

Carlo Heip, Netherlands Institute of Ecology, Centre for Estuarine and Marine Ecology, Korringaweg 7, P.O. Box 140, 4400 AC Yerseke, The Netherlands

Peter M.J. Herman, Netherlands Institute of Ecology, Centre for Estuarine and Marine Ecology, Korringaweg 7, P.O. Box 140, 4400 AC Yerseke, The Netherlands

Andrew J. Kenny, Centre for Environment, Fisheries and Aquaculture Science, Burnham Laboratory, Remembrance Avenue, Burnham-on-Crouch, Essex CMO 8HA, UK

Alasdair D. McIntyre, School of Biological Sciences, University of Aberdeen, Aberdeen AB24 2TZ, UK

Tom Moens, Department of Biology (Marine Biology Section), University of Gent, K.L. Ledeganckstratt 35, 9000 Gent, Belgium

Derek C. Moore, Fisheries Research Services, Marine Laboratory, P.O. Box 101, 375 Victoria Road, Aberdeen AB11 9DB, UK

Colin Munro, 1 Orchard Cottages, Coombe Barton, Shobrook, Crediton, Devon EX17 IB5 UK

Heye Rumohr, Institut fur Meereskunde, Dusternbrooker Weg 20, D-24105 Kiel, Germany

Chris J. Smith, Hellenic Centre for Marine Research, Crete Centre, P.O. Box 2214, Heraklion 71003, Crete, Greece

Paul J. Somerfield, Plymouth Marine Laboratory, Prospect Place, West Hoe, Plymouth PL1 3DH, UK

Antony J. Underwood, Centre for Research on Ecological Impacts of Coastal Cities, Marine Ecology Laboratories, A11, University of Sydney, NSW 2006, Australia

Jaap van der Meer, Department of Marine Ecology and Evolution, Royal Netherlands Institute for Sea Research, P.O. Box 59, 1790 AB Den Burg, Texel, The Netherlands

Dick van Oevelen, Netherlands Institute of Ecology, Centre for Estuarine and Marine Ecology, Korringaweg 7, P.O. Box 140, 4400 AC Yerseke, The Netherlands

Richard M. Warwick, Plymouth Marine Laboratory, Prospect Place, West Hoe, Plymouth PL1 3DH, UK

Dedication

We would like to dedicate this volume to the memory of Norman Holme, the prime mover of the first edition and a dedicated scientist whose contribution to marine science continues to be much appreciated.

Preface to the Third Edition

The International Biological Programme (IPB), a world-wide plan of coordinated research on the biological basis of productivity and human welfare, covered terrestrial and aquatic environments, and ran for almost a decade from 1964. Recognising the need for guidance in methodology, the IBP arranged for the publication of a series of handbooks on techniques with the aim of achieving comparability of results all over the world. One of these handbooks dealt with the study of marine benthos. It was targeted partly towards newcomers to the field, partly towards isolated workers who had no ready access to the literature, and partly towards those outside the topic whose work led them into benthos studies.

Initially published in 1971, it was substantially revised in the second edition twelve years later to take account of new approaches and interests. Inevitably, even that edition has been overtaken by events, including advances in technology and the demand for increasingly detailed information on bottom-living communities. This demand stems from several sources, not least the requirement around the world for environmental impact statements in the context of major developments in the marine environment. Pollution assessments also continue to call for knowledge of bottom-living communities, while the new focus on ecosystem management of fisheries will depend on a much greater understanding of the inter-relationships between fish and their benthic prey.

The third edition of the handbook has aims similar to those of the original and retains the same layout with the same number of chapters and most of the previous topics. However, there are several notable changes: the chapter on ship position fixing has been dropped and considerations of phytobenthos previously contained within a single chapter have been appropriately redistributed. In further re-arrangements, underwater photography and video are considered in two different chapters on imaging techniques and diving, while a new chapter on the deep sea has emerged from the previous chapter on macrobenthic techniques. All remaining chapters have been substantially revised and expanded. It should be noted that all but three contributors are new and therefore their treatment of the different topics reflects this fresh approach.

During the re-writing of the handbook care has been taken to keep to a minimum any significant overlapping in the descriptions of gear equipment and methods to be used for different purposes or different biota, by cross-referencing or by means of a very brief outline of a commonly used gear or method. However, where a specific sampler or technique was deemed essential to the work to be carried out, this information has been reiterated. Such is the case in the chapter on macrofauna techniques and the benthic deep-sea sampling, and on the acoustic techniques used for determination of the seabed characteristics, shared by the chapter on imaging techniques and the chapter on seabed sediment studies.

Over the last decade, the evolution of benthic methodology has been steady rather than uniform, as is clearly reflected in the new edition. Many facets of benthic research have been accompanied by significant technological research in acoustic techniques, in deep-sea technology, diving techniques, ideas and methods in planning benthic surveys and energy flow studies. Yet in contrast, Chapter 5 on macrofauna techniques has retained much of the previous content, which is indicative of a slowdown in technological progress in this field and perhaps of a scarcity of ideas in technical and methodological issues, with research relying on established methods. One can hardly come to the conclusion that the existing methodology concerning sampling of an extremely complex environment has reached any degree of perfection. It cannot be stressed enough, however, that many of the tools described in this handbook are complementary and should be used in parallel where appropriate.

There is a range of urgent and unresolved priorities concerning the identification and the linking of benthic patterns and processes and the development of suitable techniques for such linkages. Biogeochemical fluxes across the water–sediment interface, the role of bacteria in the benthic environment, the dynamics of recruitment, the linking of biodiversity to ecosystem and carbon flow and its role in ecological stability are but a few of the priority areas in benthic ecological research which need to be accompanied by new technological development.

The last decade has been pivotal in recognising that benthic research in the future should be conducted in collaboration with all the other disciplines. Moreover, having established that there is a benthic–pelagic coupling, it has been recognised that the underlying mechanisms and interactions, particularly between diversity and ecosystem functions, are indeed complex and can be understood only through a multi-disciplinary research network. However, it is now accepted that the only way to understand the functioning of marine ecosystems is to formulate reasonable conceptual models based upon information concerning exchanges and interactions of the ecosystem components resulting from the collaboration between the sub-disciplines of benthic ecology and other disciplines. Further developments on field studies to test relationships are also required, including manipulative experiments.

Recent global concerns about the declining health of the sea and the depletion of marine resources and the biodiversity of marine life caused by anthropogenic influences and climatic changes highlight the importance of the benthos and the benthic environment. These ecosystems have been shown to be a sensitive index of such alterations and changes and as such demand long-term and effective monitoring.

Anastasios Eleftheriou, *Heraklion*
Alasdair D. McIntyre, *Aberdeen*
Editors

Preface to the Second Edition

During the 12 years since the first edition of this handbook was prepared for the International Biological Programme (IBP) there have been not only considerable advances in techniques for sampling, sorting and interpretation of data from programmes of benthic studies, but also a radical change in outlook, which is reflected in the objectives to which such programmes are directed. Except in a few instances (notably the exploration of hydrothermal vents in the deep sea), sampling programmes are now directed less towards exploration of new habitats, and more towards a fuller understanding of the interrelations between organisms, both within the benthic community and with the overlying waters. With this has been linked a need to study fluctuations in benthic communities under altered environmental conditions, leading to projects described as 'ecologial impact studies' or 'environmental monitoring programmes'.

This edition attempts to take into account such changing needs and there has been some rearrangement of text to cover the considerable amount of material published in the past 12 years. In some instances our task has been lightened by the appearance of a work summarizing or reviewing advances in a particular field, but elsewhere we have shortened or eliminated sections which no longer appeared immediately relevant in order to make way for new material.

The objective of the handbook remains broadly the same as outlined in the preface to the first edition – that is, to be a general introduction to the subject and to indicate the range of possibilities, rather than to enter into precise details, which are obtainable in the literature cited.

We are aware that among some workers – notably those studying the Baltic Sea – significant advances have been made in the intercalibration of sampling, sorting and identification techniques. Nevertheless, we have refrained from laying down more than general guidelines as we are aware of the very wide range of conditions under which work is carried out worldwide in relation to sea state, depth of water, type of sediment, research ships and gear available, and conditions for processing samples and data, both ashore and afloat.

In this new edition some of the chapters (3, 8 and 9) have been revised by their original authors, others have involved the collaboration of additional contributors (1, 2, 6 and 7) and some (4 and 5) have been written afresh.

We would like to acknowledge the continuing support and advice of colleagues in our different laboratories in the preparation of this new edition. In particular the editors would like to thank A. Varley and library staff at the MBA, G.W. Battin for redrawing some of the diagrams, R.L. Barrett for assistance with the editing and M. Rapson and S. Marriott for preparation of the typescript. The illustration for the cover was prepared by A. Rice and A. Eleftheriou. Acknowledgement of sources for diagrams and photographs is given at appropriate points in the text.

Norman A. Holme, *Plymouth*
Alasdair D. McIntyre, *Aberdeen*

Preface to the First Edition

This handbook has been written to meet the needs of three kinds of workers: the newcomer to the field, the isolated worker without access to large libraries or the advice of colleagues, and those in related disciplines who may for some reason require to collect or study biological samples from the seabed. Because of the very varied nature of the seabed in different regions and because of the differing requirements of individual workers, we have from the first avoided laying down definite rules and procedures to be adopted. However, there is a right and a wrong way to set about things, and the handbook had to contain more than a set of platitudes. The best general advice that can be given is for a preliminary survey to be made to see the range of possibilities for study. Before a large-scale survey at sea is made, for example, some preliminary sampling, perhaps by dredge, should be made to show the types and size range of the animals and plants to be studied, the nature of the deposits, and the topography of the seabed. Only then will it be possible to decide what techniques to adopt in the main survey.

This is an international handbook, but at the same time it is an English-language edition written by workers in one part of the world. We appreciate that we may not have been able to give adequate coverage to some techniques which may have been successfully adopted in other countries, and similarly the list of gear suppliers in the Appendix may seem to be somewhat restricted geographically. Such limitations may be remedied in a later edition, should this appear, and meanwhile it may be possible to have French and Spanish translations of this handbook produced through the assistance of FAO.

Acknowledgements to those who have kindly given permission for reproduction of figures are given in the text. Our thanks are due to the staff of the drawing offices at Plymouth and Aberdeen for redrawing some of the figures, and to Mr A. Eleftheriou of Aberdeen who provided the cover drawing. Helpful discussions with colleagues in our different laboratories, and with other workers at the Areachon meetings, have contributed much to this Handbook, and special mention must be made of the assistance given by library staff at our laboratories and at the National Institute of Oceanography, Wormley, Surrey, both in the preparation of this handbook and of the Bibliography. We would also like to thank Mrs W.D.S.

Kennedy and Mrs H. Readings for assistance in the editing of the handbook, and Mr A. Varley for compiling the index.

Norman A. Holme, *Marine Biological Laboratory, Plymouth, England*
Alasdair D. McIntyre, *Marine Laboratory, Victoria Road, Torry, Aberdeen, Scotland*
July 1970

Chapter 1
Design and Analysis in Benthic Surveys

A.J. Underwood and M.G. Chapman

1.1 Introduction

Since the previous edition of this book, there have been numerous advances in soul-searching (Peters, 1991), methods of analysis (Clarke, 1993), commentaries on logic (Underwood, 1990) and need for better understanding of environmental impacts (Schmitt & Osenberg, 1996; Sparks, 2000). As a result, it is not possible in a general summary to review comprehensively even the new material, let alone everything relevant to the topic of improved benthic sampling. Suffice it to say that quantified sampling, particularly to test applied and logically structured hypotheses about patterns and processes in marine habitats, is of increasing importance to underpin understanding of natural processes, to predict changes in response to environmental influences, to help with management and conservation of diversity, natural resources, systems and functions. Taking care in the acquisition of quantitative information should therefore be of paramount importance to all marine biologists and ecologists.

The previous edition of this book was published in 1984. During that year, Hurlbert (1984) also published a devastating critique of the failure of many published studies to demonstrate a valid basis for reaching conclusions because the samples analysed were inappropriately or not at all replicated. His study was confined to the *published* studies, in better journals, refereed and subject to independent scrutiny. The overall situation, taking into account rejected papers, less intensively scrutinised journals and the flood of unreviewed grey literature, was clearly much worse. Nor does it seem, by practical inspection of more recent literature and through reviewing manuscripts and applications for grants, that the situation, though now better, has improved substantially.

This chapter therefore presents a general overview of issues. It is not, nor could it be, a 'cook-book' of procedures that might work. Rather, it is a serious attempt to consider fundamental issues of replication, in space and time, the nature of variables examined, issues about designing comparative sampling programmes and so forth.

These topics are considered against a general background of logical structure. The issue is a simple one – if the aims and objectives of any study are not clearly identified at the outset, the least damaging outcome will be wastage of time, money and resources. The worst outcome will be a complete lack of valid information on which to build understanding, predictive capacity and managerial/conservatory decision-making. Where aims are vague, designs of sampling are usually (if not always) inadequate, data do not match necessary assumptions, analyses are invalid and conclusions suspect.

In contrast, where aims and purposes are logical, coherent and explicit, it is usually possible to design a robust, effective, efficient and satisfactory programme of sampling to allow aims to be achieved with minimal uncertainty. This seems such common sense that it does not need stating – but common sense indicates that the world is flat and that the sun rotates round the earth. Common sense, therefore, is not enough.

The starting point for studies needing quantitative sampling is that objectives are clear, the variables to be measured are defined and the sorts of patterns anticipated in the data are clearly identified (as testable hypotheses). As much information as possible has been collected and understood about the processes operating and their scales in space and time, about the biological interactions in assemblages and their responses to environmental variables. The constraints of money, time and equipment are all understood. In other words, the professional components of scientific work are all in place. Under these circumstances, it should be possible to design sampling to achieve minimal probabilities of error in analyses.

As a result, the focus here is on general issues and procedures to provide help and guidance with setting objectives, formulating hypotheses and designing sampling. This will serve as an aide-memoire for contemplating issues of logic when dealing with spatial and temporal variability in biological systems. It is intended to provide an introduction into the broader literature which has made so many advances in methodologies for dealing with the problems of biological complexity in the real world.

1.2 Variability in benthic populations

Surveys must be designed to take into account the fact that benthic animals and plants are extremely patchy in distribution and abundance. Patchiness is caused by processes external to the assemblage, particularly disturbances and recruitment, in addition to processes operating within the existing assemblage. Although anthropogenic disturbances are often very severe (e.g. large-scale trawling or dredging can cause extreme changes in benthic assemblages; Hall & Harding, 1997; Lindegarth et al., 2000), natural disturbances are common and important contributors to spatial variability of populations. These vary from small-scale disturbances, e.g. being overturned by waves affects the assemblage on a boulder (Sousa, 1979) and potentially the assemblage in the sediment below it, to large-scale processes, such as erosion of nearshore sediments (Shanks & Wright, 1986) and destruction of assemblages in response to large storms (Underwood, 1998).

The most important contribution to patchiness is, however, probably due to unpredictable and variable patterns of recruitment (Underwood & Denley, 1984). Both settlement itself and post-settlement mortality typically vary at a hierarchy of spatial scales (Caffey, 1985; Gaines & Bertness, 1992). Patterns at larger scales tend to be more predictable because most species are confined to particular habitats within a biogeographic range (Brown & Gibson, 1983). At small scales within patches of habitat, however, there is considerable variability in recruitment (Keough, 1998), caused by local environmental variation (e.g. topographic features of habitat or localised water currents) or by the existing assemblage itself (e.g. gregarious settlement in response to conspecific adults or consumption of larvae by large numbers of sessile species). The arrival in a site of larvae competent to settle is itself influenced by a multitude of external processes, many of which act in the water column well away from the site of settlement. The localised processes, both physical and biological, that influence recruitment, are interactive, so recruitment is extremely variable in space and time.

In addition, numerous interactions within the assemblages themselves continually alter patterns of abundances and these effects, too, occur at a range of spatial scales. For example, predation may decrease abundances at the scale of a shore or habitat, but within that habitat, may eliminate species from certain patches but leave other patches alone. Feeding by eagle rays or shore birds can create extreme small-scale patchiness in abundances of their prey, although these effects are complicated by environmental factors, such as currents and movement of sediment (reviewed by Thrush, 1999). Even in areas with heavy predation, the prey may settle in particular microhabitats, where they can grow large enough to escape predation (Dayton, 1971).

Competition, either for space among sessile animals or plants or for food among mobile animals, also contributes greatly to patchiness of assemblages. Therefore, overgrowth or dislodgment of one species by another (Keough, 1984) causes very patchy assemblages of sponges, ascidians and other colonial animals in subtidal habitats and of barnacles and various types of algae in intertidal habitats. Although superior competitors, either for space or food, may eliminate inferior species, the relative strength of interspecific competition may be balanced with that of intraspecific competition, thus ensuring that neither species is eliminated, but that both persist in very variable and patchy numbers.

Processes such as these are better understood for benthic assemblages living on hard surfaces because, first, the patterns are often readily visible and, second, the processes are relatively easily investigated experimentally. They are, however, also important for assemblages in soft sediments, where recruitment may be equally

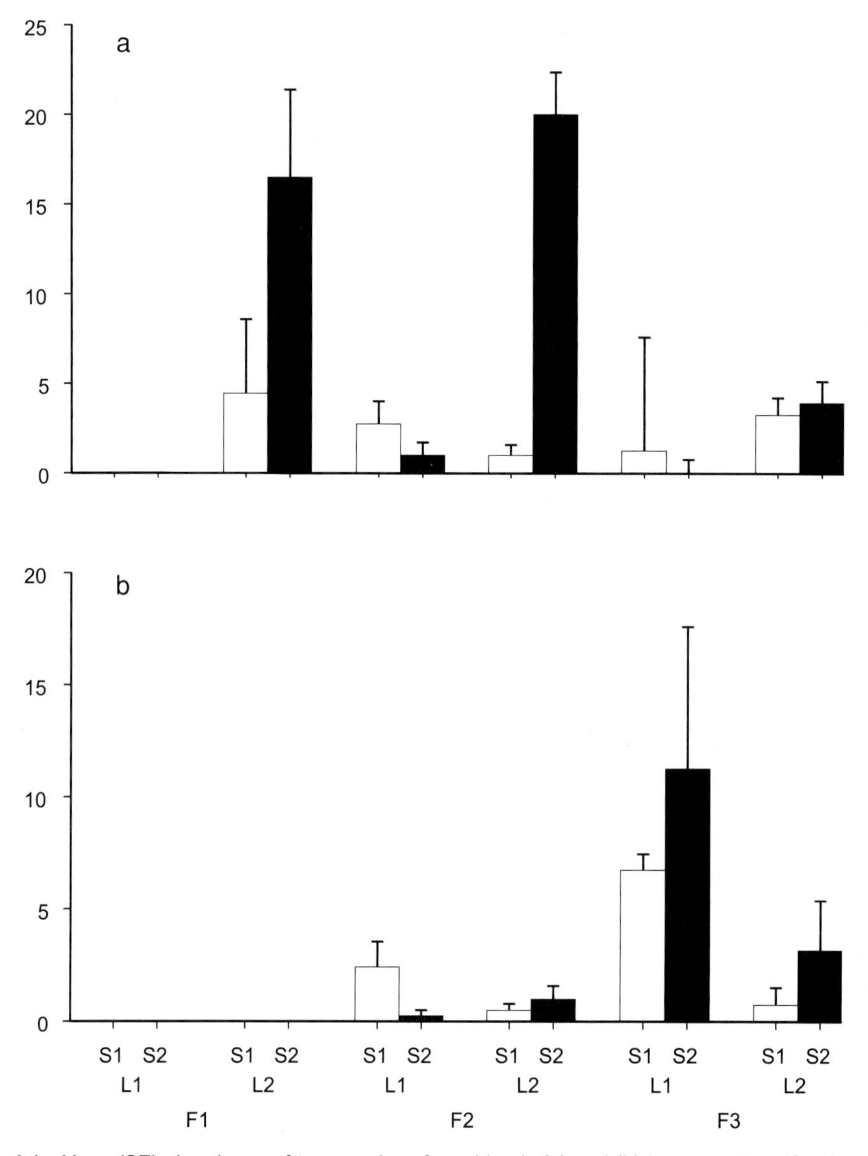

Fig. 1.1 Mean (SE) abundance of two species of amphipods (a) and (b) between sites (S1, S2; 10s of metres apart) in each of two locations (L1, L2; 100 m apart) in each of three mangrove forests (F1, F2, F3; km apart). Note that each species shows significant variation at each spatial scale, but these differ between the two species. Patterns of variation at the scale of sites and locations also vary from one mangrove forest to another.

variable (Skilleter, 1992) and local disturbances, competition and predation alter local abundances, causing very patchy distributions at a hierarchy of spatial and temporal scales (Morrisey et al., 1992a, b; Fig. 1.1). To estimate abundances of benthic animals and plants accurately, measures must be made at the range of spatial scales relevant for the species, assemblage or process under consideration.

1.3 Appropriate replication

Appropriate spatial replication

Whatever the hypothesis being tested and the ultimate use of the data from sampling, spatial replication is a mandatory component of any benthic study. The large variability in numbers and varieties of benthic species from place to place at many spatial scales (see above) creates fundamental problems for determining at what scales replication is necessary.

Consider a relatively simple problem of determining the influence of the type of sediment on the numbers and types of benthic species. To keep the example very simple, suppose that a particular species of polychaete is predicted to be more abundant where sediments are coarse than where fine sediments form a major proportion of the total. The hypothesis to be tested is therefore that abundance will, on average, be greater in an area of coarse sediment than in a corresponding area of fine sediment. Further, suppose that patches of the different types of sediment are about 800 m in diameter. Finally, as is virtually inevitable, imagine that numbers of worms per m^2 of habitat may vary substantially at scales of tens of metres and at scales of hundreds of metres, even in the same type of sediment. Consider a sampling scheme in which 10 box cores are sampled in one patch of coarse sediment and 10 cores are taken from a patch of finer sediment some 2 km away. The mean number of worms is greater in the cores from the coarser sediment – consistent with the hypothesis. This does *not* allow a valid interpretation that the difference is associated with the difference in sediment (as predicted). Such an interpretation or influence is confounded by the alternative that numbers are different simply because the two areas are 2 km apart. The alternative models explaining why numbers of worms vary are that the sediments differ or that natural variation from place to place, regardless of sediment, causes the numbers to be different. The only way to separate these two models is to predict (from the first model) that differences from one type of sediment to the other are greater than expected from natural variation in either type of sediment.

The study therefore requires replication of sites of each type of sediment to estimate natural variation that is not associated with the type of sediment. It is sometimes argued that such a study as described is replicated because there were several cores in each patch. This is an example of what Hurlbert (1984) called 'pseudoreplication' – the replicate units in each sample are at the wrong scale and are estimating variability at 10s of metres, not at 100s of metres. A sensible design would solve the problem by sampling several patches of coarse sediment spaced, say, 2 km apart and several patches of finer sediment at similar spacing, about 2 km from the nearest area of coarse sediment sampled. The two types of patch should be *interspersed*, i.e. chosen to be higgledy-piggledy on a map, so that no systematic trend or gradient makes them different for reasons other than the type of sediment (see Underwood (2000) for examples and illustrations of this issue).

Using such a design, the variation of scales of tens of metres within a patch and at hundreds of metres from patch to patch of the same type of sediment can be estimated. A hierarchical, or spatially nested analysis of the data can then be used to test the hypothesis that the difference, on average, between the two types of sediment is greater than the natural spatial variation from patch to patch of the same type. Examples of such analysis for benthic infauna are given in Green and Hobson (1970) and Morrisey et al. (1992a, b) and the form and structure of the analyses were described in detail in Underwood (1997a).

The lengthy description of a very simple case is made necessary by the large number of unreplicated studies, with logically invalid conclusions, that still keep appearing in (or are submitted to and rejected from) marine ecological journals. Note that better sampling does not replace good biological knowledge. So, if the variation in numbers of polychaetes from site to site with finer sediments is known from previous studies to be of the order of 10s to 100s of worms per m^2 and there are 1500 more worms per m^2 in the single coarser site sampled than in the single finer site in a particular study, it can be argued that this size of difference is much greater than natural variation among sites with finer sediments. Therefore, the conclusion that more worms are found in coarser sediments is valid.

This argument will only be correct if there has been adequate previous study to demonstrate the general validity (from area to area of the world and from time to time) of the notion that numbers of worms vary naturally by tens to hundreds per m^2. It is not usually the case that there is sufficient information of this type available. It is, in fact, often the case that previous studies were not replicated at appropriate spatial scales to reach such general conclusions. If such an argument is used, it is essential to provide the details of the previous studies that might justify it. Often, it is more efficient (and scientifically, because it does not depend on inductive inferences, it is always more valid) to use appropriate replication in any new study.

One further example of problems of logical conclusions from sampling at only one or two spatial scales will be considered. Suppose that dredging is being done at several places in an estuary, to keep channels open for shipping. The sediments in shipping channels are contaminated by heavy metals. Sampling is required to detect any impact on benthic infauna in areas around the sites being dredged (not in the sites being dredged – they are not just impacted, the habitat is actually removed). The anxiety is that fine sediments with associated heavy metals may wash from the dredged sites to areas up to a hundred metres away. A replicated study can be designed (as in Fig. 1.2), with several dredged areas and several controls being sampled. In each area, replicated patches of sediment are sampled. Natural spatial variation from patch to patch and area to area is estimated and analysis can reveal any systematic difference between sites near dredging and control sites.

Suppose, however, that the movement of contaminated sediment, if accidentally released from a dredged area, is actually much larger than anticipated. Contaminated fines may now be dispersed over the whole estuary, thus deleteriously affecting all of

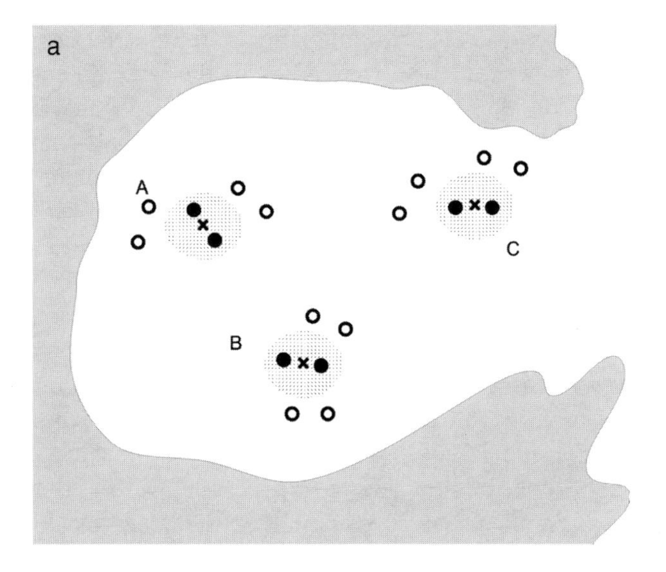

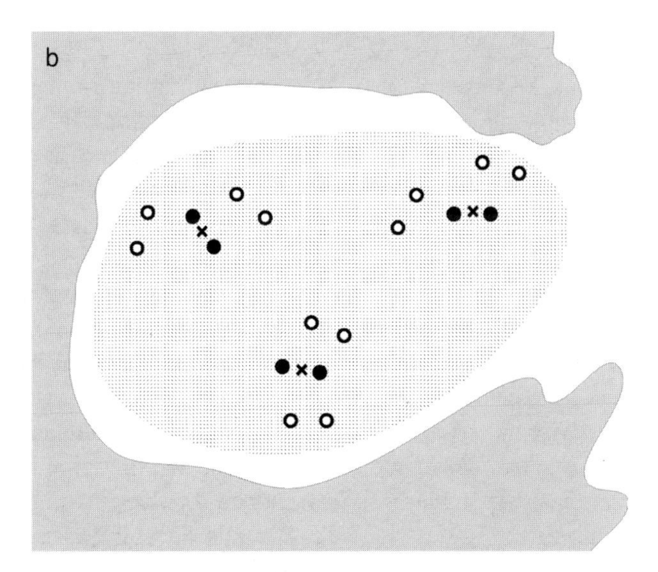

Fig. 1.2 Sampling to detect an impact due to escape of fine sediments from dredging at X in areas A, B and C in an estuary. In each area, there are sampling sites (•) within the distance predicted that sediment would disperse if accidentally released (i.e. over the stippled areas in (a)). Control sites (○) are outside the predicted area of impact and at similar distances apart as the potentially impacted sites. In (b) is the actual, much larger (stippled) area impacted; no impact would be detected because all the controls are also affected.

the control sites, in addition to the sites immediately adjacent to dredging. Changes in benthic fauna due to heavy metals will now occur over all the sites sampled, but there will be no apparent impact because the control sites and the sites next to dredging will not differ in analyses of the data. It will appear that during the course of the study there has been an estuary-wide change in fauna not associated specifically with dredging. This sort of situation requires sampling at much larger scales, best of all in other estuaries where dredging is not occurring.

Designs of this type and procedures for analysing the data to detect impacts in such situations were discussed in detail by Green (1979) and Underwood (1992, 1994). The moral of this example is clear – when in doubt about the relevant spatial scale, use a design that can detect changes or differences at one or more of several of the possible scales.

Appropriate temporal replication

Many hypotheses concern temporal changes, e.g. seasonal patterns of variation or potential changes in populations due to disturbances. One of the major problems with many such studies is the confounding of temporal and spatial variability (Underwood, 1993). For example, to measure seasonal patterns of abundance, it is common to collect a sample, using a set number of replicates, each month (or each season) in one or more places. Tests of temporal pattern then typically compare the abundances from one month (or season) to another to the variability among replicates in the different seasons (using procedures such as analyses of variance). The problem with such comparisons is that variation among seasons is indeed a measure of temporal variation, but variation within a sample is calculated from measures of spatial variation, because the replicates taken at any particular time are all at the same time, even though scattered in space. In such a design, seasonal (or other temporal) patterns are not contrasted against temporal variation within each season, but against spatial variation.

To test for seasonal (or other *a priori* selected scales of temporal variation), temporal variation among the factors of interest must be compared to temporal variation within each factor of interest. In other words, temporal variation among seasons must be compared to the magnitudes of variation that occur in each season. To measure such variability, it is essential to collect samples at a number of times within each season. With two or more scales of temporal sampling, seasonal or other long-term trends can be identified against background noise. Where there is no measure of shorter-term temporal variation and such variation is large, quite spurious seasonal (or other temporal) patterns will be seen in the data (Fig. 1.3).

Different scales of temporal sampling are extremely important for identifying environmental impacts. Disturbances to the environment may either be short-lived (pulse disturbances) or persist for long periods of time (press disturbances) (Bender et al., 1984). The responses of organisms to either type of disturbance may be relatively short-term (i.e. a pulse response), for example, abundances may rapidly increase, but soon drop to normal levels, irrespective of whether the disturbance persists or ceases. Alternatively, populations may show long-term responses (i.e. press responses) to continuing disturbances (because the disturbance continues to exert an effect) or to pulse disturbances (because the disturbance, although ended long ago, caused some long-term change to some other environmental or biological

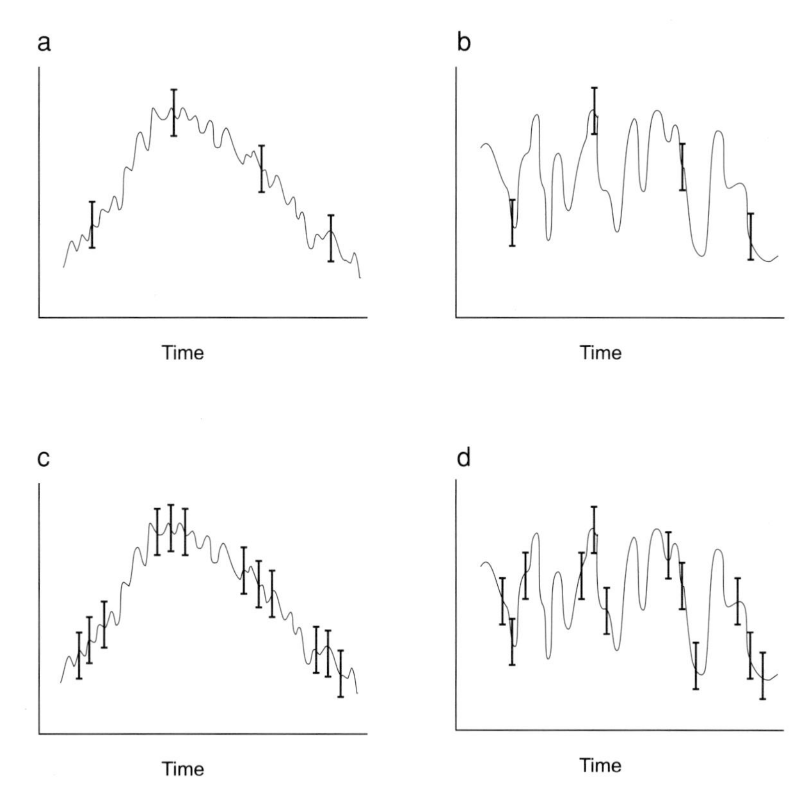

Fig. 1.3 With only a single sample at a series of time intervals a long time apart (e.g. each season), an apparent seasonal pattern could be identified in the variable being measured whether (a) there is, indeed, a long-term seasonal trend or (b) there is considerable short-term variability, but no long-term trend. Short-term temporal sampling is needed within each season. This provides the correct form of within-season replication to measure seasonal changes and identifies (c) long-term trends from (d) background 'noise'.

variable). The experimental designs needed to distinguish among pulse and press responses to pulse or press disturbances have been thoroughly described in by Underwood (1991) and Glasby and Underwood (1996). They all need a sampling design that can measure temporal variation at different temporal scales, measured at the spatial scales relevant to the pulse and press responses.

For examining most environmental impacts and many other ecological hypotheses, the temporal scales of change are not known and can seldom be predicted. The only way to identify important temporal change is to use a hierarchical temporal sampling scheme. Such designs (Underwood, 1991) not only identify whether any change indicates a pulse or press response, but also quantify temporal variances at a number of scales. Impacts that change variances may be more common than and as important as those that change means (Underwood, 1991). If an impact causes much greater fluctuations in abundance (e.g. by altering water currents so that recruitment is more unpredictable), the average abundance of a species may not change over long periods of time, but the species may be very rare or very common

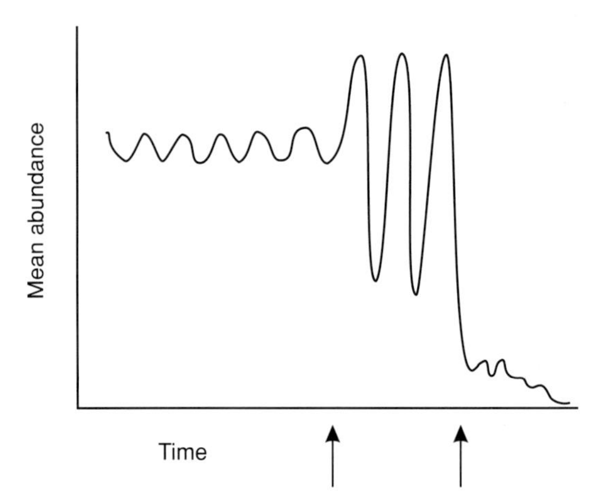

Fig. 1.4 Environmental impacts (at the times indicated by arrows) that affect the variability in abundances of a species, rather than the mean abundance, can cause increased vulnerability to further disturbance. If a subsequent impact occurs during a phase when abundances are very small, the species may easily be driven to local extinction.

at different times, depending on recent recruitment (Fig. 1.4). Such populations may be far more vulnerable to cumulative, but relatively minor impacts causing local extinctions, if a subsequent impact occurs during a period when abundance is small. Sampling designs that measure changes in mean abundance at different temporal scales, in addition to changes in temporal variances of abundance, are most likely to be successful in identifying environmental impacts and natural ecological variations.

1.4 Size of sampling unit

Because abundances typically vary at a range of spatial scales, the size of the sampling unit selected to sample populations is very important in identifying patterns of abundance. For example, consider a population of polychaetes or other small benthic animals that typically aggregate in clusters, about 10 cm in diameter, with the clusters spaced about 20–30 cm apart (Fig. 1.5a). Therefore, some patches of sediment have very large numbers of animals, whereas other patches have very few or no animals. Sampling with very large cores or quadrats, for example, 50 cm × 50 cm, will suggest that the animals are very regularly spaced throughout the site. Each quadrat is likely to sample one cluster, with perhaps a few individuals from adjacent clusters, giving very similar measures of abundance in each replicate (Fig. 1.5b). This size of sampling unit is too large to measure the spatial pattern because the ecological processes causing these patterns are operating at scales of 30 cm or smaller.

Sampling with a quadrat smaller than 30 cm, e.g. 15 cm (Fig. 1.5c) provides a more accurate picture of the spatial variability of these organisms. These units will sample some clusters, giving very large measure of abundance and some spaces

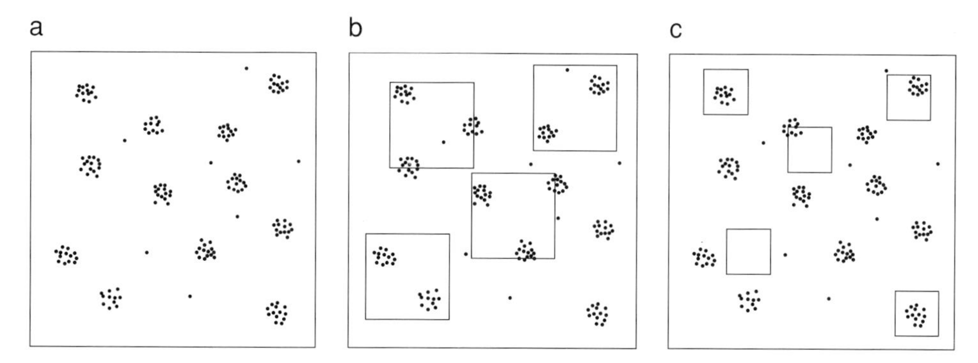

Fig. 1.5 (a) Small benthic animals frequently aggregate into clusters separated from one another. (b) Sampling with units that are much larger than the clusters tends to produce data that suggest a very regular distribution because each unit samples a similar number of individuals. (c) Sampling with smaller units shows the very clustered spatial pattern, with some units sampling clusters and others sampling the bare space among the clusters.

among the clusters, giving very small measures. The average of the replicates should be the same as when large quadrats are used (when scaled to the same area), as long as each set of quadrats representatively samples the population. The variance will, however, be much larger because it will more accurately represent the true pattern of dispersion, thus identifying the scale at which the ecological processes causing these patterns are operating.

In many cases, particularly for organisms in sediments or for very small or cryptic organisms, one cannot observe the patterns of dispersion prior to sampling, so it is not always easy to determine the appropriate size of a sampling unit. In this case, a pilot study in one site at the start of the experiment, in which a range of sizes of sampling units are used, may identify such patterns and save a lot of effort later. Alternatively, some good knowledge of the natural history of the species being investigated, or other similar species, or a critical evaluation of the sizes of sampling units others have used, with the reasons why and the patterns identified, may help.

When one is quantifying patterns of abundance of a number of species in the same habitat, different sized sampling units can be used for the different species simultaneously. Therefore, for intertidal and subtidal rocky shores in Australia, a 50 cm × 50 cm quadrat has been shown to provide independent samples that accurately quantify patterns of dispersion and abundance for a number of the larger gastropods and barnacles (Underwood, 1981). There are many smaller species that show most variability at the scale of a few cm, e.g. littorinid snails (Underwood, 1981). To count these in a 50 cm × 50 cm quadrat is not only time-consuming (there can be many thousands in a quadrat of that size), but it would also not measure variation at the scale that is most important to these small animals. It is most useful therefore to count these and similar small species in replicate smaller subquadrats, scattered representatively over the area of the large quadrat. Similarly, there may be large animals, such as sea urchins or large whelks that may be seldom sampled in a

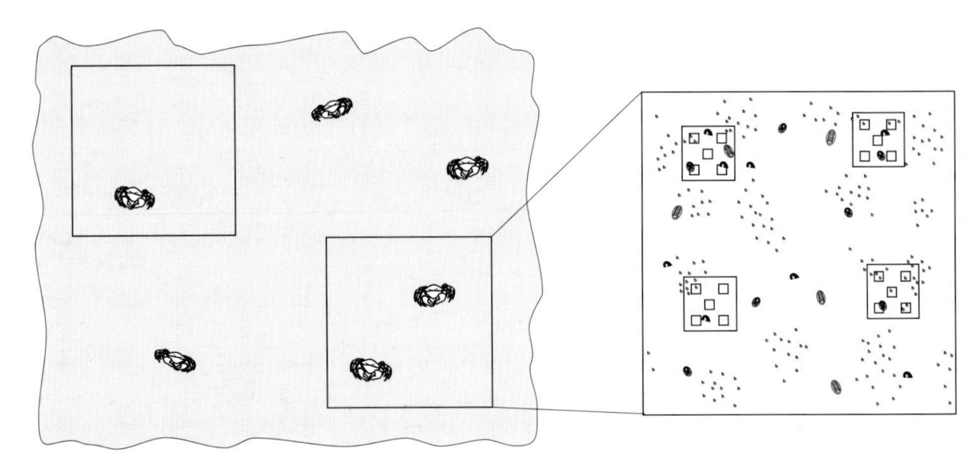

Fig. 1.6 When sampling a suite of species of different sizes and abundances, the most time-effective design is to sample large animals in large quadrats (or other sampling units) scattered over the site, smaller animals in a number of subquadrats, nested inside each of the large quadrats and very small or numerous animals in a set of even smaller areas scattered inside each subquadrat. Such sampling measures variation of the small animals over the same extent of the site as the large animals and allows the variation of the small animals to be partitioned among the different spatial scales.

50 cm × 50 cm quadrat, without a very large number of quadrats being used. Therefore, these animals may be sampled in large units, say 3 m × 3 m, within which one may sample some 50 cm × 50 cm quadrats, within each of which one may sample some 5 cm × 5 cm subquadrats (Fig. 1.6). Each of these scales is relevant to the species being measured and no one scale is appropriate for all species.

Another problem that can arise from the use of inappropriate or only one size of sampling unit occurs in attempts to measure patterns of association between abundances of two species. Consider two species of amphipods living in shallow, subtidal sediments. When sampled with large quadrats, abundances of the two species may be very strongly positively correlated (Fig. 1.7a, b), i.e. they increase in abundance together. When sampled with small quadrats, abundances may be very strongly negatively correlated (Fig. 1.7c, d), i.e. where there are many of Species A, there are few of Species B and vice versa. These two different patterns could arise because the two species are responding similarly to some environmental variables, but in opposite directions to others. Therefore, there may be a large-scale gradient across the site, e.g. water depth, and both species are more common in shallower than in deeper water. Simultaneously, there may be small ripples in the sediment across the study site, with one species living upcurrent and the other downcurrent of each ripple. This would separate the animals at a scale smaller than 25 cm, thus the negative correlation picked up with the smaller sampling unit (Fig. 1.7e).

Use of only one of the units would lead to the conclusion that the two species were either positively *or* negatively correlated. In fact, they are negatively *and* positively correlated. The different sized units also direct attention to the scale of

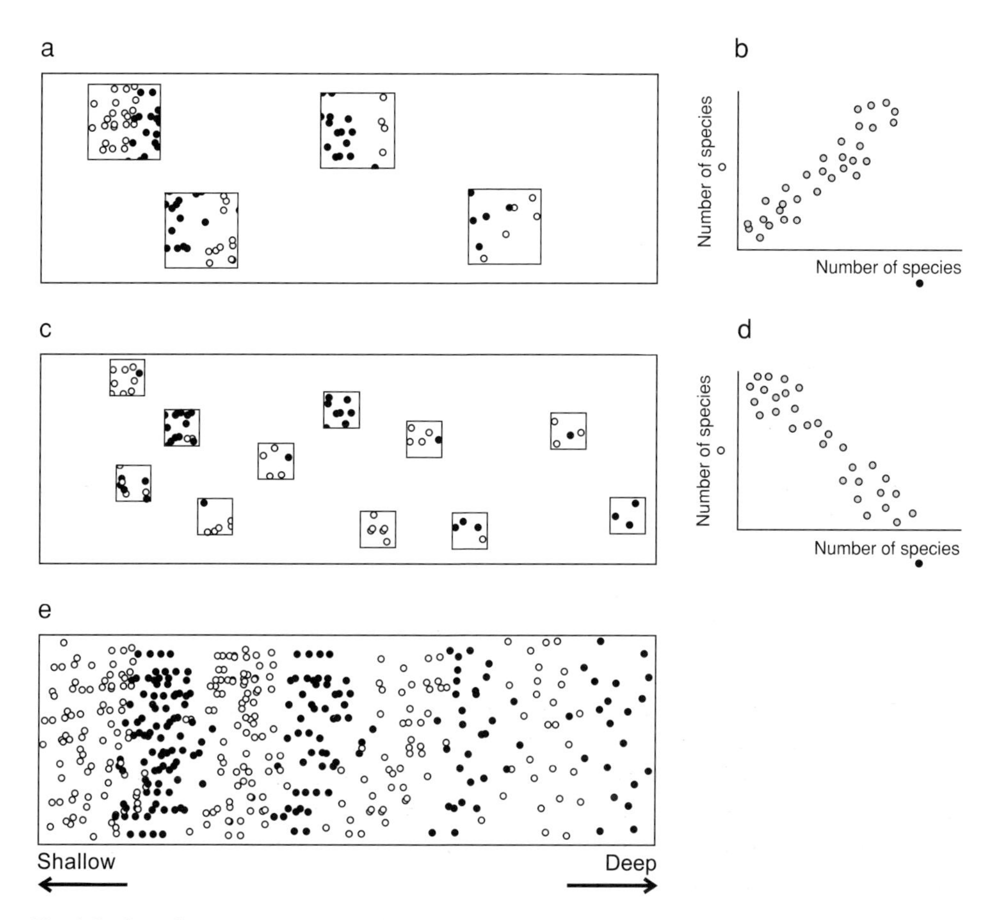

Fig. 1.7 Sampling two species using a large sampling unit (a) shows a positive relationship between the species (b), whereas a smaller unit (c) shows a negative relationship (d). (e) This occurs because the two species are responding to small-scale environmental variables that tend to separate them, while each of them increases in the same way across a broad environmental gradient. Note that the actual pattern in (e) is not visible or known when sampling most types of benthic organisms, so correlations at one scale may not be a good, nor complete description of relationships between species.

the important ecological processes. Therefore, it is necessary to consider factors that can cause positive correlation at the scale of the entire site, perhaps over tens of metres, in addition to factors that cause negative correlation at the scale of cm within the entire site. The opposite pattern could equally be found. For example, species may be positively correlated at a small spatial scale, for example, if they use the same microhabitat. At a large spatial scale, e.g. across a depth gradient, they may be negatively correlated, with one species more common in shallow water and one more common in deeper water.

It is often very difficult to identify individuals of many sessile animals and plants. They are therefore usually sampled *in situ* by estimating the percentage of space that they occupy. This can be done by photographing patches of habitat and determining cover from the photographs using a grid of dots or planimetry, or directly in the field. Cover can be estimated from a grid of points, with the species under each point being scored (Underwood, 1981), or by dividing a quadrat into a series of smaller grids and estimating cover in each of these on a 0–4 scale (Benedetti-Cecchi et al., 1996). Comparing different methods of sampling cover of sessile organisms in the field, Foster et al. (1991) and Benedetti-Cecchi et al. (1996) have shown that estimates of cover vary according to the method used, and a pilot study may be important to determine which method gives greater accuracy and precision.

It is clear that the size of the sampling unit used will strongly influence the spatial patterns that are identified and, hence, the understanding of ecological processes. In their extensive review on this topic, Andrew and Mapstone (1987) strongly emphasised the need to consider carefully the sizes (and numbers) of sampling units used in a study to optimise any sampling design, rather than automatically use methods that are reported in the literature, without any idea of their suitability.

1.5 Independence in sampling

Many analytical procedures require that data are independently measured from place to place, time to time and replicate to replicate. This is a particularly important assumption for many univariate procedures (regressions, analysis of variance, chi-squared tests). The issues are not so stark for some multivariate analyses and for many tests where data are permuted to generate distributions of test statistics (Clarke, 1993).

Correlations in space occur among data where there are influences of density in one area on what happens in surrounding areas. For example, for the numbers of the two species illustrated earlier (see Section 1.4), there would be positive correlations between data points taken on the peaks of ripples (where numbers of one species are relatively large). There would be negative correlation between any pair of sample units where one is in a trough (where numbers are small) and one is on a peak (where numbers are large).

Many biological processes cause correlations in spatial data. For example, if some species is scattered at random across an area of habitat, there will be no pattern of correlation between the numbers in pairs of sample units, wherever they are placed. If, in contrast, numbers are non-random because predatory fish eat the animals and the activity of fish is concentrated in a few sites, say where there are rocks, there are now patches relatively devoid of the prey animals. Two or more sampling units (cores, grabs, quadrats) landing in one of the areas where predators are active will have small numbers of prey and there will thus be some positive correlation in numbers found in sample units where numbers are small.

Non-independence through time (more commonly called serial correlation) can be a much larger problem. Analyses of differences in numbers of animals (numbers of species, whatever variable) are fraught with difficulties because the numbers in any given area tend to be correlated with future numbers. Even after trends due to seasonal cycles or mortality through time have been identified in analyses, there is often a tendency for the data at one time to be related to those at the next time, or a few subsequent times of sampling.

There are two general issues about serial correlation. The first is to determine whether it is present in any given set of data and the second is to determine what can be done about it. A general test for serial correlation is the Durbin–Watson test (Durbin & Watson, 1951), which uses residuals from whatever trend or time-course has been fitted to the data. If e_i represents the residual at time i and there are t times of sampling:

$$d = \frac{\sum_{i=1}^{t} (e_i - e_{i-1})^2}{\sum_{i=1}^{t} e_i^2}$$

is compared with tabulated values when there is no serial correlation. If positive correlation is present, differences between adjacent times of sampling are smaller than expected by chance and d will be smaller than in uncorrelated series of data.

If serial correlation is present and many times of sampling (or distances for spatial analyses) are available, time-series analytical procedures will help. Procedures are described in detail in Box and Jenkins (1976), Diggle (1990) and Cliff and Ord (1973). One possibility is to repeat the Durbin–Watson test at different temporal (or spatial) lags. Thus, compare e_i with e_{i-2}, e_{i-3}, etc., to identify an interval at which there is no longer any noticeable correlation. Data from such distances or times apart are therefore independent and could be analysed by traditional procedures. This must, however, result in a loss of data and a considerable waste of effort to collect data that are subsequently not useable.

What this means in practice for sampling biota is that, usually, it is not possible to finish up with various temporal or spatial samples. Therefore, it is crucial to think in advance about the biological and environmental processes operating that will cause positive or negative correlations in the data. Then, it is often possible to ensure that sampling is done at large enough distances or times to ensure that the correlations do not turn up in the sampled data. Procedures such as hierarchical sampling schemes will also help us to allow relevant analysis and some forms of fractal analysis may offer alternative approaches (Leduc et al., 1994).

1.6 Multivariate measures of assemblages

Examination of many types of models about natural ecological processes, such as predation or competition, or about changes in response to environmental disturbance or management, requires measures of assemblages of species. Differences in assemblages from place to place, or changes through time are complex because several variables can change simultaneously. These include the species or taxa found in each sample, their relative abundances and their distribution among replicates within each sample, i.e. their spatial variation. Each of these is equally important in understanding natural ecological processes or changes in response to disturbances. Analytical tools that can identify changes in assemblages need therefore to take all of these factors into consideration.

Several procedures attempt to reduce such multivariate data into univariate measures that summarise aspects of the entire data set, e.g. species richness, evenness, dominance, etc. None of these measures, however, deals simultaneously with the entire variability of the data. More useful procedures examine components of a multivariate set of data simultaneously, thus measuring the magnitude of differences between samples in composition of species, abundances and spatial pattern. These procedures are well described elsewhere (Clarke, 1993) and will be introduced only briefly here.

They generally work on the same principle (illustrated in Fig. 1.8). Abundances (biomasses or equivalent measures) are measured in a suite of taxa (Fig. 1.8a) in each of a matrix of samples, each with a number of replications. A similarity (or dissimilarity) matrix is then calculated for all pairs of replicates across all of the samples in the matrix (Fig. 1.8b). A number of measures of similarity can be used, but the Bray–Curtis coefficient of similarity:

$$S_{jk} = 100\left[1 - \frac{\sum_{i=1}^{p} |y_{ij} - y_{ik}|}{\sum_{i=1}^{p} (y_{ij} + y_{ik})}\right]$$

where p is the number of species and j and k represent any two sample units, is generally considered most suitable for ecological data which tend to have many zero values and where abundances tend to be over-dispersed among replicates (Clarke, 1993). The Bray–Curtis measure is also not affected by species that are absent from both of the sample units being compared. When two replicates are identical, the similarity measure is 100% (dissimilarity 0%) and when they have no species in common, similarity is 0% (and dissimilarity 100%).

These measures of (dis)similarity can then be compared within and among samples, assuming replication in each sample, using analyses such as ANOSIM (Clarke, 1993) or NPMANOVA (Anderson, 2001). These test the null hypothesis that the average magnitude in these measures between samples is not greater than that within samples. These tests have still not been extended to consider many complex designs and extreme care is needed in interpreting any significant result because, as described above, differences between samples are affected by the species present, their abundances and their occurrences across replicates. These tests do not allow identification of the contribution that each of these is making to any overall patterns of differences.

As with univariate analyses, one must also consider transforming the data prior to analysis. It is common to see the data transformed to $X^{0.5}$ or $X^{0.25}$, in an attempt to minimise the effects of very common species and to allow rare species to contribute more to any patterns of difference. Transformations of different severity (e.g. $X^{0.5}$

a

Data	s1	s2	s3	s4	s5	s6	s7	s8
Species 1	6	12	0	21	0	0	19	11
Species 2	1	0	0	0	2	0	0	0
Species 3	15	3	6	0	0	0	0	0
Species 4	2	1	1	0	1	0	0	0
Species 5	10	20	3	3	3	53	0	0
Species 6	54	55	23	31	79	87	99	56
Species 7	0	12	10	11	3	24	2	0
Species 8	0	13	23	0	0	15	0	0
Species 9	0	1	0	3	0	0	0	3
Species 10	0	5	0	0	0	5	1	0
Species 11	1	0	1	1	0	0	1	0
Species 12	0	0	1	0	3	0	0	0
Species 13	0	0	2	0	0	0	0	0
Species 14	0	0	0	3	0	4	0	0
Species 15	3	6	5	9	0	0	5	0
Species 16	6	0	1	0	0	0	0	0
Species 17	0	0	0	0	0	0	0	3
Species 18	18	1	10	1	3	2	0	3
Species 19	47	34	34	47	64	33	38	44
Species 20	2	1	1	0	0	0	0	0

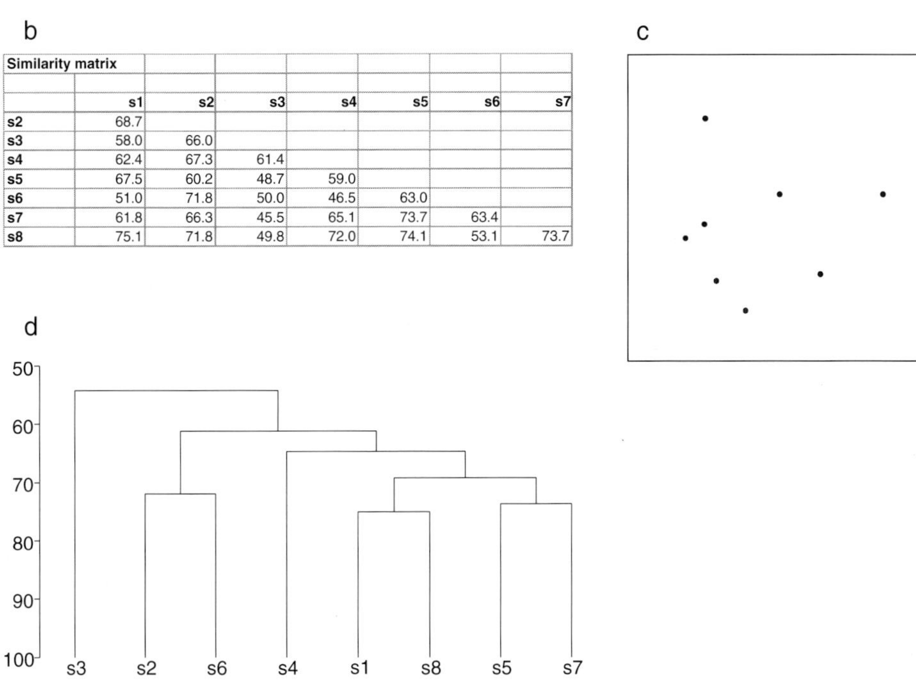

b

Similarity matrix	s1	s2	s3	s4	s5	s6	s7
s2	68.7						
s3	58.0	66.0					
s4	62.4	67.3	61.4				
s5	67.5	60.2	48.7	59.0			
s6	51.0	71.8	50.0	46.5	63.0		
s7	61.8	66.3	45.5	65.1	73.7	63.4	
s8	75.1	71.8	49.8	72.0	74.1	53.1	73.7

Fig. 1.8 Many useful multivariate analytical procedures convert (a) a matrix of samples (in this case the columns are abundances of a suite of taxa – the rows in the table), or (b) other variables into a matrix of similarities or dissimilarities. This summarises differences across all variables between each pair of samples into a single distance, or measure of difference between the pair. This information can be presented graphically in many ways. (c) nMDS plots illustrate similarity among samples in two- or three-dimensional space, using the ranked distances. Points closer together on the plot are more similar and vice versa. (d) Clustering attempts to identify 'natural' groups of samples by identifying those which are more similar to each other.

is not as severe a transform as $\ln(X)$) affect the patterns found (Olsgard et al., 1997) and, thus, the interpretation of any analysis. Another consideration is the level of taxonomic resolution to use. Frequently, very similar results are obtained when organisms are sorted to species, or to genera, or families, or larger groups (Gray et al., 1988; Clarke, 1993). Spending time at the start of any project examining the effects of different transformations or levels of taxonomic resolution on the results obtained may well save considerable effort later. It is, however, very important to consider these issues carefully, especially with respect to the model and hypothesis being tested.

Measures of (dis)similarity can also be presented graphically, e.g. in non-metric multidimensional scaling (nMDS) plots. Such plots attempt to place all replicates in two- or three-dimensional space, maintaining the relative distances (or measures of dissimilarity) among the replicates (Fig. 1.8c). Replicates that plot closely together contain similar assemblages and replicates that plot further apart contain more dissimilar assemblages. Such plots are particularly useful for illustrating differences among samples and are generally used in conjunction with analyses, such as ANOSIM.

In addition, samples can be subjected to cluster analyses, of which there are a range of different methods (Clifford & Stephenson, 1975). These attempt to find natural groups of samples, rather than to examine differences between predetermined groups. These groupings are often displayed in a dendrogram (Fig. 1.8d) against a scale of (dis)similarity (e.g. Bray–Curtis similarity), which identifies the degree of similarity that separates samples into different groups.

1.7 Transformations and scales of measurement

There are a number of reasons why data may need to be transformed prior to analysis. Some parametric and non-parametric analytical procedures (analysis of variance, Kruskall–Wallis tests of ranks) have underlying assumptions, for example, that the data are normally distributed, or that variances among samples are not heterogeneous (i.e. are of the same magnitude). If these assumptions are not met, there is increased possibility of Type I errors. Although different analyses have different levels of sensitivity to making Type I errors, depending on the particular assumption being violated (Underwood, 1997a), it may be necessary to transform data prior to analysis to meet the assumptions of the analysis (Winer et al., 1991).

Three types of ecological processes cause variances to vary excessively among samples. Counts of randomly dispersed animals or plants in quadrats, cores or other sampling units will be distributed approximately as Poisson distributions. This occurs when there are no ecological processes operating to cause the individuals to aggregate or spread out. In a Poisson distribution, the variance equals the mean. Therefore, when comparing samples in which means differ, it is also inevitable that

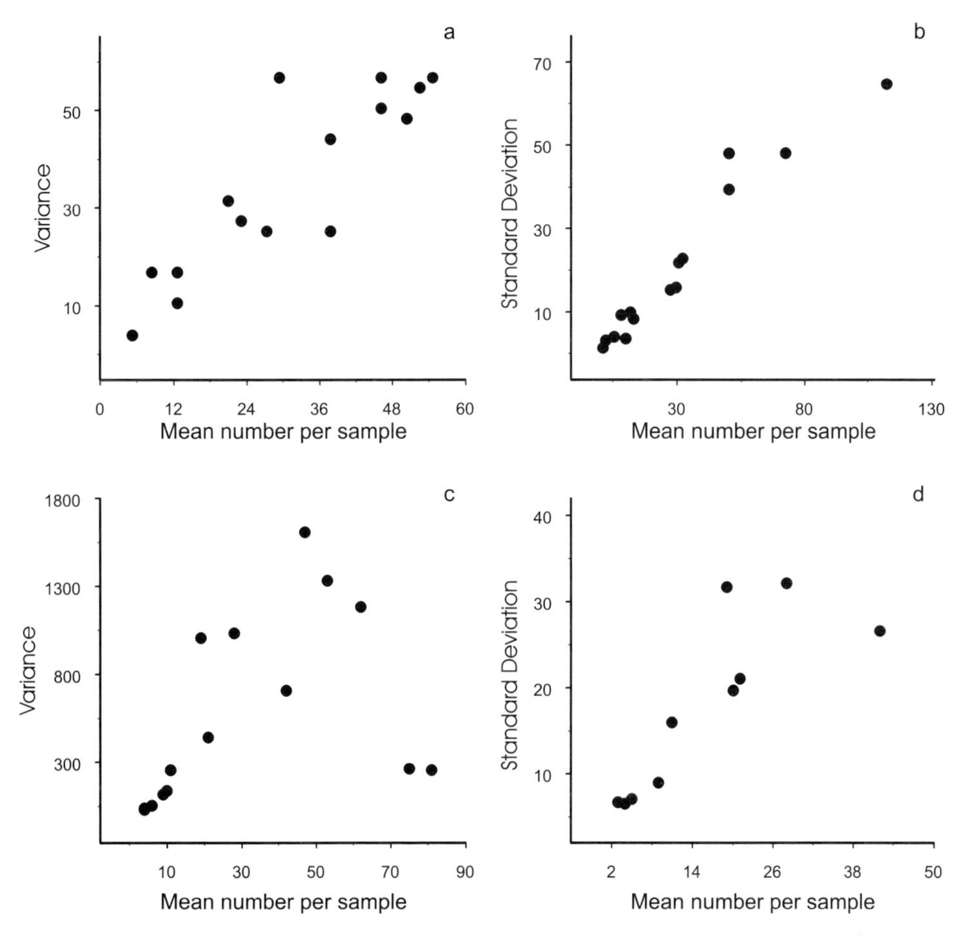

Fig. 1.9 (a) Linear relationship between means and variances in randomly distributed species; (b) linear relationship between standard deviations and means in log-normally distributed data; (c) relationship between variances and means of proportional (or percentage) data; (d) relationship between standard deviations and means between constrained sets of proportional (or percentage) data. (a) and (b) are numbers of microgastropods per cobble; (c) and (d) are percentage covers of unoccupied space underneath intertidal boulders.

variances will differ (Fig. 1.9a). Comparing samples with very large differences among means is likely to give significant values for heterogeneity of variances (e.g. using Cochran's test). In situations such as this, where the variance–mean plot is roughly linear, the data should be transformed using $\sqrt{X+1}$ to eliminate these problems.

Many ecological processes cause variances to increase exponentially with increases in means; the data are log-normally distributed. For example, sizes of organisms caused by differing rates of growth, distances moved during dispersal, the concentration of enzymes in animals of different sizes. It is common for most members of a population to grow at relatively similar rates or disperse relatively similar distances, whereas a few grow extremely fast or disperse very long distances. These will cause log-normal distributions in the data. Similarly, small animals may have large concentrations of enzymes, but large animals have small concentrations. If the concentration is scaled to the size of the animals, measures will be very large for small animals. In cases such as these, the standard deviation, not the variance, is approximately linearly related to the mean (Fig. 1.9b). Such data will be more likely to meet assumptions of normality and homogeneity of variances if they are log transformed prior to analysis (to whichever base seems most appropriate). Unfortunately, if there are zeros in the data, a value (e.g. 0.1 or 1, depending on the range of the data themselves) needs to be added to each datum prior to transformation. Adding these numbers needs careful consideration because it can alter the relative relationships among means, especially in cases where the data are very variable, e.g. lots of zeros or very small numbers in some samples.

Finally, if data are percentages (or proportions), they may be distributed as binomial distributions. In such a case, variances may be small for samples where means are near the limits of the distribution (e.g. 0% or 100%), but relatively large when means are towards the middle of the distribution (e.g. 50%). In such cases, an arc-sine transformation may remove heterogeneity of variances (Fig. 1.9c). When the data do not span most of the range of 0–100%, they will often appear to be distributed in a lognormal distribution, so a transformation to logarithms may be more appropriate (Fig. 1.9d).

In some cases, even if variances are not heterogeneous and, therefore, transformation is not necessary to meet assumptions of some statistical test, it may still be appropriate to transform data prior to analysis. Therefore, if the ecological processes influencing the patterns being measured are multiplicative, rather than simply additive, it is appropriate to transform all data to logarithms when testing for differences among mean values. This would ensure that doubling the mean from 10 to 20 from one site to another would give the same measure of difference as doubling from a mean of 100–200. This may be very important, for example, for testing predictions about effects of pollutants on populations of animals across a number of sites, where the natural density of animals differs greatly from site to site. If the pollutant is hypothesised to decrease densities similarly across a range of natural environmental conditions (and therefore a range of natural densities), transforming the densities to logarithms will allow for such a test because the test will not be confounded by variation in natural densities. More insights into the use of transformations in ecological sampling can be found in Elliott (1977).

In any case, whatever transformation is used for whatever reason, it must be noted that transforming data does change the scale over which the data are distributed and relationships between means and variances. They therefore also change the test of any hypothesis. It is essential that the relationships between the hypothesis being tested, the analytical procedures being used and the scale in which the data are analysed be clearly maintained if one is to interpret logically the results of any statistical tests in terms of ecological processes.

1.8 Data-checking and quality control

Rigorous control of the quality of data is an essential requirement for any research project. This does not only mean collecting relevant, independent, representative data in the field or laboratory for appropriate tests of hypotheses of interest. It also includes very careful translation of the data into the format necessary for analysis. These days, most data are stored in databases or spreadsheets on computers. It is essential that as much quality control goes into checking that the data in the spreadsheet match those collected in the laboratory or field as went into collecting the data in the first place. One infallible, but time-consuming way to ensure that the data match is to make sure that after each set of data is entered into the computer, it is also checked by two people, at least one of whom is independent of the collection of the data. This picks up many errors that arise from illegible handwriting, tiredness or the sheer boredom associated with entering large amounts of data into databases or spreadsheets.

If it is not possible to develop such foolproof procedures, there are several possible short-cuts. For example, if one has a large set of univariate data, such as the sizes of urchins, the diversity of amphipods or the numbers of polychaetes, from a number of sites, it is often useful to simply calculate the standard deviation (SD) for each site. If data have been omitted, or very aberrant values have been incorrectly entered, they will cause a very small or a very large measure of SD in a site, relative to other sites (Table 1.1). Therefore, a very unusual SD may indicate an error either in collection of the data in the first place, or in transcribing data into the computer. This

Table 1.1 Calculations of standard deviations from counts of small snails in 15 samples. Note the very large SD for sample 5 in column 2, due to a value of 233 being entered instead of 23. When this was corrected (column 3), sample 5 still had a large SD, but this was correct. It was due to naturally large abundances and patchiness in this sample.

Sample	SD (not checked)	SD (checked)
1	0.00	0.00
2	0.45	0.45
3	4.92	4.92
4	0.00	0.00
5	101.04	15.96
6	8.26	8.26
7	7.46	7.46
8	7.60	7.60
9	2.59	2.59
10	7.14	7.14
11	5.45	5.45
12	5.02	5.02
13	4.04	4.04
14	3.42	3.42
15	3.65	3.65

procedure is quick and easy to use as a form of data-checking, but will not indicate relatively small errors, only those that cause substantially different estimates of standard deviations in samples.

When data from a number of species (or variables) have been collected, another form of quality control is to calculate distance measures among the different replicates (e.g. Bray–Curtis dissimilarities for species, or Euclidean distances for abiotic variables). If these are plotted as an nMDS (non-metric multidimesional scaling) plot in two or three dimensions, the plot will quickly show outliers, i.e. replicates that have abnormal values for one or more species/variables (Fig. 1.10a). Available procedures to identify which species/variables most contribute to measures of dissimilarity (Clarke, 1993) focus on those data that should be checked.

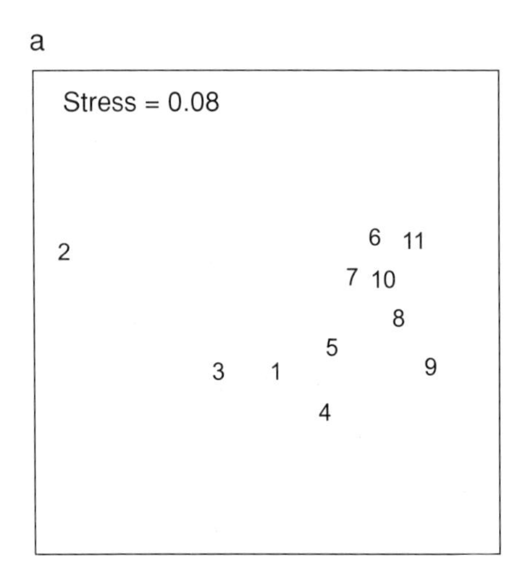

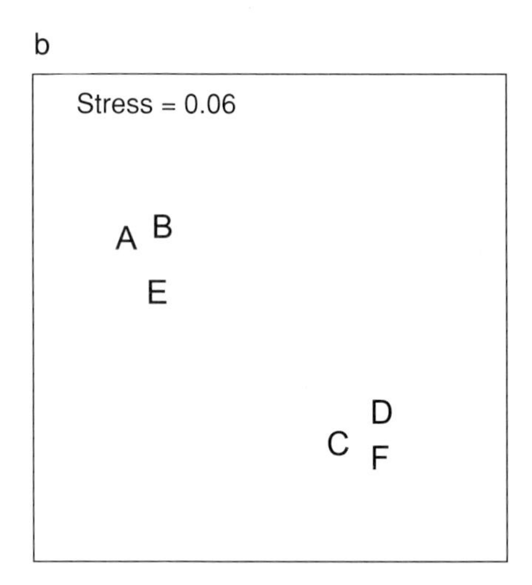

Fig. 1.10 (a) Replicate two in this sample of macrobenthos from a mangrove forest forms an outlier in the nMDS plot because the data for 3 of the 37 taxa in the set of data were omitted from this sample (and were therefore erroneously given a value of 0). When this was corrected, the replicates grouped into a single cluster. (b) If the points from a number of replicates in a single sample cluster into two or more distinct groups, it often suggests that the replicates are coming from two distinct strata and the sampling design should be stratified.

Of course, not all samples with excessively large or small SDs or which form outliers in nMDS plots are necessarily in error. They may simply reflect large natural variability. Similar procedures can also be used to determine whether the sampling strategy is appropriate or not. For example, if replicates cluster into two distinct groups (Fig. 1.10b), it suggests that one might be dealing with two habitats or sets of environmental conditions, each with a different set of taxa. A stratified sampling strategy would probably be better in such a case, with, for example, A, B and E treated as replicates within one stratum and C, D and F as replicates in another (Fig. 1.10b) (see Chapman & Underwood, 1999). Similarly, if a species were very aggregated in some sites, but more randomly distributed in others, the SDs would tend to fall into two distinct groups. This too suggests that sampling should not be done across all sites as if they were equal, but sites should be stratified into two groups (see also Section 1.11).

Visual examination of patterns of means, variances, correlation and multivariate data is as important as the statistical tools used to analyse the data. It can provide important information about errors in data and about patterns of variability within and among samples. It should therefore be considered an important precursor to any analysis, rather than looking at the data after statistical differences have been identified.

1.9 Detecting environmental impacts as statistical interactions

Much benthic sampling is done to detect or measure the size of environmental impacts. It is therefore worth considering the sorts of designs of sampling that are best able to do this. Of course, many situations exist in which optical designs are not possible because of the lack of time before a disturbance causing an impact or lack of resources to achieve adequate spatial replication. Nevertheless, if there is clarity of thought about what is really necessary to achieve the best chance of reliably and unambiguously detecting impacts, there will be the maximal likelihood that the best outcome will emerge when the design must be less than optimal. By understanding what is needed, greater understanding can be achieved about what is not available and the consequences for interpretation.

The main features of sampling designs necessary to detect impacts have been described in detail by Green (1979) and Underwood (1994, 2000). In an ideal design, there should be data from before to after a disturbance that might cause impacts, so that, if an impact occurs, it can be demonstrated that it appeared after the disturbance purported to have caused it. There must be proper temporal replication before and after the disturbance to provide reliable estimates of average conditions (Bernstein & Zalinski, 1983; Stewart-Oaten et al., 1986) and to estimate temporal variance, which might itself be altered by an environmental disturbance (Underwood, 1991).

There must be replicated, undisturbed controls, to demonstrate that an impact, if it occurs, is associated with the disturbed area and is not a general phenomenon happening in some habitat and not due to that disturbance (Green, 1979). 'Undisturbed' in this context means subject to any other influence or process, except the particular disturbance under investigation (Underwood, 2000). Control sites should be replicated to prevent confounding of interpretations (Underwood, 1992 and see 'Appropriate spatial replication' section above).

Finally, as convincingly revealed by Green (1979), an environmental impact must always be found as an ecological interaction. An impact is a change from before to after a disturbance (planned or accidental) which is not the same in the disturbed area as in undisturbed controls. Therefore, there is a lack of consistency in the temporal pattern of change of the variables being measured in the disturbed area and the patterns of change in the controls (Fig. 1.11). Accordingly, to analyse impacts, it is necessary to design sampling to provide data that can be analysed to detect and interpret statistical interactions.

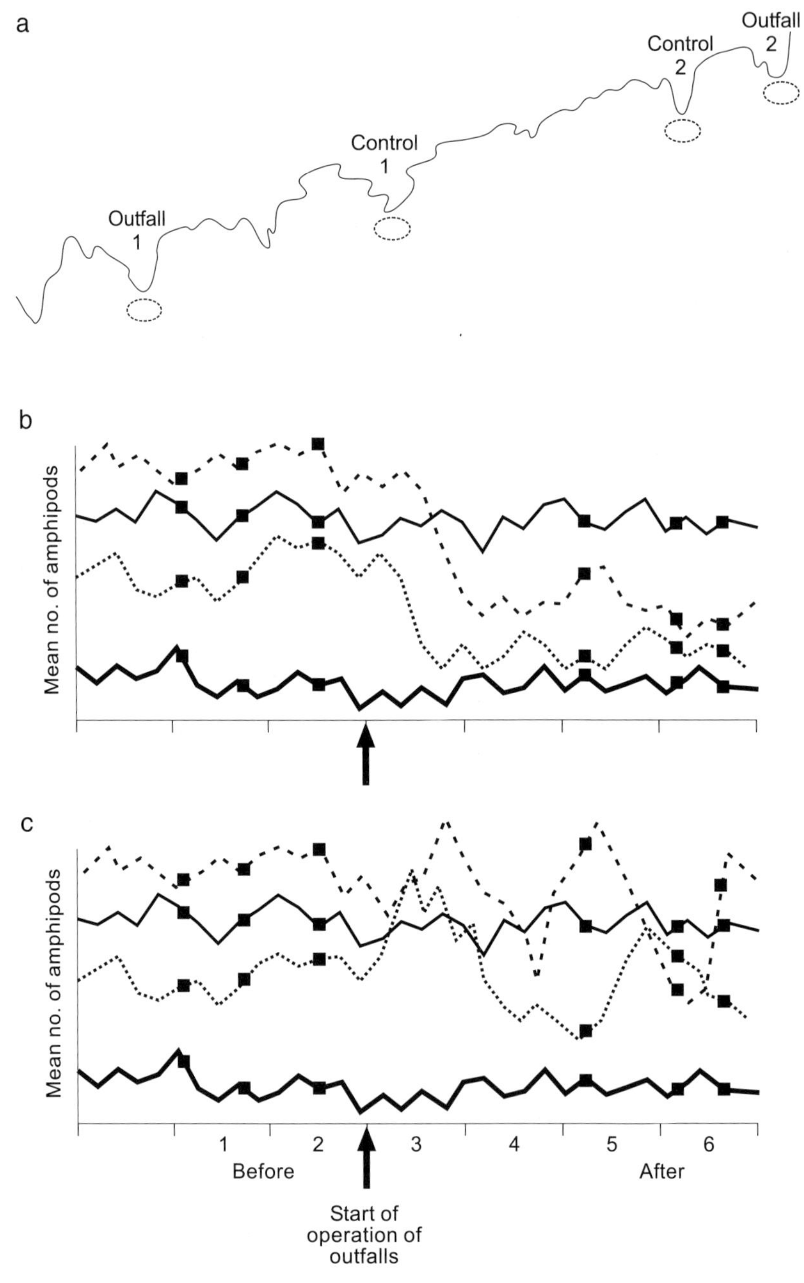

Fig. 1.11 (a) Sampling to detect impacts from construction of coastal sewage outfalls. Two outfalls (---, . . .) and two similar control sites (—, **—**) are sampled three times (■) before and again after the outfalls begin to discharge. (b) Data that would indicate a large-scale, consistent 'press' impact. (c) Data that would identify shorter term fluctuating impacts.

Such an ideal design is illustrated in Fig.1.11 and the analysis is in Tables 1.2 and 1.3. The situation concerns construction of sewage outfalls along a coastline. They are to be situated on rocky headlands with fast currents, deep water and coarse offshore sediments. Two outfalls were examined, with two corresponding controls with similar features of habitat (rocky headlands with fast currents, deep water

Table 1.2 Framework for analysis of hypothetical sampling to detect impacts due to construction of sewage outfalls (illustrated in Fig. 1.11). *O* indicates the comparison of mean numbers of an amphipod between outfall and control locations and is a fixed factor (see Winer et al., 1991; Underwood, 1997a). There are two (randomly chosen) Locations of each type, indicated by *L*(*O*). Sampling Before and After the outfalls were built represents a fixed factor; there are three randomly chosen Times of sampling Before and again After the outfalls start operation. $n = 5$ replicate cores are sampled at each time in each location.

Source of variation		Degrees of freedom	Test
Outfalls versus Controls	= O	1	Irrelevant
Before versus After	= B	1	Irrelevant
O × B		1	Indicates large-scale, long-term impact
Locations (Outfall or Control)	= L(O)	2	Irrelevant
B × L(O)		2	Indicates different impacts at the two outfalls
Times of sampling (Before or After)	= T(B)	4	Irrelevant
O × T(B)		4	Indicates possible fluctuating impacts
L(O) × T(B)		8	Indicates possible fluctuating impacts that differ between outfalls
Residual		96	
Total		119	

Table 1.3 Framework for an asymmetrical (beyond BACI) analysis of sampling to detect impacts due to construction of a single sewage outfall. There is now no replication of outfall locations, but all other details are as in Table 1.2.

Source of variation		Degrees of freedom	Test
Outfalls versus Controls	= O	1	Irrelevant
Before versus After	= B	1	Irrelevant
O × B		1	Indicates large-scale, long-term impact
Locations (Control)	= L(C)	1	Irrelevant
B × L(C)		2	Irrelevant
Times of sampling (Before or After)	= T(B)	4	Irrelevant
O × T(B)		4	Indicates possible fluctuating impacts
L(C) × T(B)		4	Irrelevant
Residual		72	
Total		89	

and coarse sediments). Any coast-wide change in benthic fauna that is not due to discharge of sewage will affect controls and the outfall locations. An impact will cause a different change where there are outfalls from where there are no outfalls (hence an interaction in the spatial difference between outfall and control locations from before to after the outfalls are commissioned).

Data are collected three times (essentially chosen at random) over two years before the outfalls begin to discharge sewage and then three times over a period of two years, starting a couple of years after the outfalls are commissioned (Fig. 1.11). At each time of sampling, some number, say five, replicate areas of sediment would be taken in each location and the animals of some species of interest (perhaps an amphipod) collected in each area. From such data, the analysis in Table 1.2 can be completed. A large and consistent press impact (Bender et al., 1984; Underwood, 1989, 1991) would cause an interaction identified as $O \times B$ in Table 1.2. A shorter-term, more fluctuating response (perhaps a pulse response; Bender et al., 1984) would appear as an interaction $O \times T(B)$ in Table 1.2. Where an impact has a different effect on fauna in the two locations with outfalls, there will be an interaction $B \times L(O)$ or $L(O) \times T(B)$ depending on the rates of change caused by the sewage. Impacts causing changes in temporal variance rather than changes in mean abundance of animals (Fig. 1.11b) can also be analysed (Underwood, 1991, 1994).

Of course, in most cases, there is only one planned disturbance (in the present example, only one outfall would be planned). Consequently, there is a loss of replication of the disturbed ('outfall') location and correspondingly less certainty about any impact. There must, however, still be replication of control locations, ensuring that extreme forms of confounding do not happen. Such asymmetrical designs are still analysable and interpretable (Table 1.3) and have been called 'beyond BACI' designs (Underwood, 1992) to distinguish them from earlier, spatially unreplicated and confounded BACI designs (Stewart-Oten et al., 1986). The acronym BACI is for 'before-after/control-impact' sampling, where there is one (and only one) disturbed (called 'Impact') location and only one control. The disadvantages and logical problems with BACI designs are well known (Underwood, 1992; Smith et al., 1993).

In many situations, there can be no sampling before a disturbance, because it is accidental (such as oil spills) or the impacts were unforeseeable and only start to appear after the disturbance has started. Such situations can still be analysed (Glasby & Underwood, 1996; Chapman et al., 1995), but, again, there is a loss of certainty about the cause of the impact. Without earlier data there is less certainty that the impact started after the disturbance.

There is now a substantial body of reviews of literature on sampling and analysis of environmental impacts starting from Green (1979) and, more recently, Spellerberg (1991), the papers in Schmitt and Osenberg (1996) and Sparks (2000). There are still many problems in analyses of complex samples for multivariate measures, but interactions caused by impacts can, in the simplest cases, be analysed by permutation procedures (Anderson, 2001).

1.10 Precautionary principles and errors in interpretations

Precaution has become a guiding goal for decision-making in a framework of ecological sustainability (see Cameron & Abouchard, 1991; Wynne, 1992). Despite different views by lawyers, managers and scientists on what precautionary principles mean (Dovers & Handmer, 1995), there is consensus that precautions can be defined in terms of environmental decisions. Given the uncertainty about the scientific information underpinning managerial decisions (and, although often forgotten, the uncertainty in all the architectural, economic, engineering, legal, political and social information), it is important to remember that mistakes will occur in interpretation of analyses of quantitative ecological data. In statistical terms, there are two major types of mistakes (Table 1.4). Type I errors occur when a null hypothesis is rejected by a statistical procedure, but is, in fact, true. Type II errors occur when a null hypothesis is false (some alternative is true), but is retained. In environmental decision-making, the hypothesis is that there will be an impact due to some disturbance. For some other ecological studies, hypotheses are predictions such as:

(1) Based on the observation that faunal assemblages vary with depth and composition of sediments vary with depth, it can be proposed that differences in composition of sediments cause the difference in assemblages. Therefore, a relevant hypothesis is that, for any particular composition of sediments, similar fauna will be found at all depths. Also, to any depth, the fauna in a particular type of sediment will be similar to those in the same type of sediment at other depths.
(2) Given previous observations of fewer amphipods in areas where rays feed, the model can be proposed that predation or disturbance by the rays decreases the number of amphipods. This leads to the hypothesis that the areas where rays are experimentally prevented from entering will develop larger numbers of amphipods than in the corresponding control areas.

Sampling is then done to test the hypothesis (hypotheses) and data are analysed statistically. A statistical procedure includes choosing a probability (α) of Type I

Table 1.4 Two types of error in conclusions from statistical tests.

	As a result of sampling, the Null Hypothesis is:	
	REJECTED You believe there has been an environmental impact	RETAINED You believe there has been no environmental impact
Unknown to you, the Null Hypothesis is:		
TRUE	TYPE I ERROR With probability α	CORRECT DECISION
FALSE	CORRECT DECISION	TYPE II ERROR With probability β

error – the probability that the data collected cause rejection of the null hypothesis as the outcome of the test. So, an impact is found where there is not one, different fauna seem to be present in similar sediments whereas in reality there is no difference, rays seem to make a difference, but in reality numbers of amphipods are influenced by something else.

Type II errors are the opposite situation – no impact is found, even though there is one; fauna actually differ, but no difference is found; numbers of amphipods are not significantly different in the presence and absence of rays, even though rays are, in fact, important predators. The probability of making Type II errors (β) is, for most studies, not known.

In environmental studies, failure to find an impact (a Type II error) is much more serious than spuriously claiming to find one (Type I). Where an impact is believed to be present, more work will usually be done (to check, to determine its extent, to test further hypotheses about its consequences). Almost certainly, the mistake will be found. In contrast, where a real impact is missed, no particular follow-up will occur and environmental degradation will continue. Precautionary principles therefore require that errors should be of Type I and not of Type II (Gray, 1990, 1996; Peterman & M'Gonigle, 1992; Underwood, 1997b).

There is no space here to demonstrate how to reduce the probability of Type II errors. Details are given in Green (1989), Peterman and M'Gonigle (1992), Underwood (1997b). The probability of not making a Type II error, i.e. of correctly rejecting a full hypothesis (finding an impact) when it is false (there is an impact), is $(1 - \beta)$ and is called the power of a test (see Table 1.4). Power is increased where

- α, the probability of Type I error is increased;
- the intrinsic variance or 'noise', in the measurements is small;
- the size of effect is large, i.e. large differences are expected if the null hypothesis is false;
- the sizes of samples are large.

Of these, the variance is a property of the system being measured (see Section 1.11). α and the sizes of samples (numbers of replicate sites, times of sampling, sample units in each site) are chosen by the experimenter. Thus, they should be chosen to achieve large power, i.e. great capacity to reject null hypotheses (to find differences where they exist).

The key to power analysis is the effect size. To calculate power of any sampling and analysis requires that the amount of difference among times, places, habitats, experimental treatments, etc., be specified in advance. Thus, the hypothesis that rays influence amphipods is devoid of real information. What is required is a statement, based on available knowledge, that, if rays decrease numbers of amphipods, removing them from some experimental areas will cause an increase of X amphipods per unit area of habitat. Where the original observations provoking the study were that fewer amphipods are present where rays feed (see above), X is known from these observations (it is the observed difference hypothetically caused by rays).

For assessment of environmental impacts, effect sizes are more difficult to establish. A good yardstick is that the effect sizes (how much difference in densities of the species affected or how much change in any other relevant variable is expected, if there really is an impact) should be based on what responses would be triggered by the discovery of that size of impact. So, suppose that a reduction of 10% in numbers of crustaceans in some mud-flat near an industrial outfall would be an impact, but would not cause any change in regulation or management. In contrast, a reduction of 60% would trigger immediate regulatory responses. It is therefore very important to be able to detect a 60% difference, but of limited or no value to be able to find much smaller impacts.

A sensible effect-size would be somewhere around 50% change and the power of the study should be made large (say, >90%) in order to have a good chance of finding the impact if it occurs (Green, 1989; Underwood, 1997a). Where resources (money, time, equipment [i.e. money, money, money]) do not then allow adequate replication to achieve large power, α (the probability of Type I error) should be increased to achieve reductions in β (the probability of Type II error). Mapstone (1995) has described in detail how to trade off the lack of resources (increasing β) with increasing the chance of mistakenly finding impacts (increasing α) for assessments of environmental impacts.

1.11 Precision and size of samples

As indicated above (see Section 1.10), the power of tests is a function of the sizes of samples used. In general, the precision of an estimate from some sample is increased as a function of the size of sample. A typical measure of precision is the standard error, which is (sample variance/size of sample)$^{1/2}$ and clearly decreases (so precision increases) as size of sample increases. Wherever possible (and it is always desirable) a maximal acceptable imprecision should be specified. It is possible to estimate the variance of the variable being measured and thereby to calculate how many replicates should be included in a sample to achieve the necessary precision.

There are, however, several features of design of sampling that help increase precision of estimates of abundances of organisms. First of all is stratification. Wherever it is possible to make a 'map' of abundances – from previous studies in the literature or from pilot studies, stratification of sampling will often substantially reduce imprecision. As a simple example, suppose that it is generally known that a particular species of sea urchin is generally more abundant (per box-core) in areas of very coarse sediment than where sediments are finer. Suppose that in your study area, there is about 25% of the seafloor composed of coarse sediments, in several large patches. The remaining areas are finer sediments. You have sufficient funds to take a total of 16 cores to estimate the average numbers of urchins per m^2 of sea floor, as part of an ongoing study of disturbances due to trawling. Suppose you

Table 1.5 Sampling urchins in an estuary with two types of sediments. In (a), $n = 16$ cores are taken at random over the whole area; some are in coarse, some are in finer sediments. In (b), $n = 4$ cores are taken in the areas of coarse sediment and $n = 12$ in areas of finer sediment. Stratification substantially increases the precision of the estimate of mean number per core.

(a) Numbers of urchins in 16 cores
19, 100, 77, 13, 1, 15, 20, 17, 90, 77, 8, 22, 8, 78, 14, 29
Mean = 36.8; Variance = 1169; SE = 8.5

(b) Numbers of urchins
Coarse sediment: $n = 4$
102, 80, 100, 95
Mean = 94.3; Variance = 98.9; SE = 5.0

Finer sediment: $n = 12$
8, 10, 5, 6, 26, 23, 25, 26, 17, 8, 1, 27
Mean = 15.2; Variance = 95.8; SE = 2.8

Combined sample:
Mean = 34.9; SE = 2.5

now take the samples at random positions across the area, resulting in the data in Table 1.5a. This gives an estimate of mean abundance of 36.8, with SE = 8.5.

If, instead, you stratified the sampling, i.e. you took 25% (or 4 of your 16 samples) in the areas of coarse sediment and the remainder in the other habitat, you would have the data in Table 1.5b. Now, the mean number of urchins is estimated separately in each habitat, with separate estimates of imprecision and then combined for the whole area to give a mean abundance of 35.6, with SE = 2.6. This method is much more precise because it removes the variation among replicate cores that is due to the systematic differences between the two types of habitat. Of course, it also gives explicit information about the densities of urchins in each of the two habitats, which may or may not be useful (depending on the actual hypotheses being tested).

Combining the samples from each habitat as done here is only valid if the variances in numbers of urchins in the two areas are measured independently. Alternative methods, in particular, how to stratify samples so that the number of replicates taken in each stratum is proportional to the variance in the stratum, have been described in detail (Cochran & Cox, 1957; Cox, 1958) and other issues in marine ecology were reviewed by Andrew and Mapstone (1987).

A different method of expressing imprecision is to calculate the confidence interval for some mean estimated from a sample. A 95% confidence interval indicates a range of values that have a 95% probability of including the true mean being estimated. If the confidence interval is small, the unknown mean has been estimated fairly precisely.

A 95% confidence interval for a sample mean $\bar{X}X$ from a sample of size n is:

$$\bar{X} \pm t_{0.05,(n-1)df} * \sqrt{\text{Sample variance}/n}$$

As the size of sample (n) increases, the confidence interval will decrease because the standard error (i.e. $(\text{variance}/n)^{1/2}$) is smaller and because there are more degrees of freedom ($(n - 1)$ increases), giving a smaller value of t.

In any study of differences in means through time (seasons, before/after events, etc.) or across space (habitats, patches, depths, etc.), it makes sense to combine samples, so that the estimate of variance around each mean comes from a combined estimate of variance, with more degrees of freedom than from each sample alone. Thus, putting the samples together as in an analysis of variance improves the precision of sampled estimates of means, as illustrated in Table 1.6. For this to be valid, the variances of the various samples should be similar and this can be tested by various procedures. The example in Table 1.6 considers samples of $n = 5$ cores from each of the three habitats. The combined confidence intervals are much smaller than the individual ones. To achieve the same precision (i.e. the same small confidence interval) for each habitat separately would have required samples of $n = 7$ for each habitat, i.e. 1.4 times more work.

Table 1.6 Sampling an amphipod in three habitats to demonstrate precision of combined samples. Data (hypothetical) are numbers in $n = 5$ cores in each habitat. Variances were homogeneous (Levene's test, F with 2, 12 *df*, = 1.18, $P > 0.30$).

		Habitat		
		1	2	3
Replicate	1	10	28	13
	2	14	31	9
	3	3	33	3
	4	5	26	0
	5	2	19	1
Mean		6.8	27.4	5.2
Variance		25.7	29.3	31.2
SE		2.3	2.4	2.5
95% CI (4 *df*)		6.3	6.7	6.9
Combined variance		28.7	28.7	28.7
Combined SE		2.4	2.4	2.4
95% CI (12 *df*)		5.2	5.2	5.2

1.12 Gradients and hierarchies in sampling

There has been debate about appropriate sampling designs to use for analyses of influences along an environmental or other gradient. For example, when testing a hypothesis about the influence of discharge of freshwater from an estuary, it is appropriate to sample along a gradient from the mouth of the estuary. Similarly and more obviously, sampling down a depth gradient requires sites to be along the gradient. In other cases, the situation is not so clear-cut and whether or not sampling along a gradient is the best option depends entirely on the hypothesis being tested.

Furthermore, even if a gradient is to be sampled, the spacing of samples is dependent on the precise issue under examination.

To provide an illustration of the issues, consider a simple case of discharges of contaminants from an outfall on the coast (see Underwood (2000) for more details). In all three situations considered here, it is proposed that pollution due to the contaminants will be revealed by differences in the assemblages of animals in sediments near to as opposed to far away from the outfall. In the first case, the design of the outfall included modelling of the probable dispersal of contaminants and their dilution away from the pipe (Fig. 1.12a). The requirement of the ecological sampling is to test the hypothesis that concentrations of contaminants in the animals in sediments do, indeed, conform to the modelled pattern of distribution. In this case, samples must be taken along the gradient, at fairly regular intervals, to have the greatest chance of detecting departures from the prediction. In the second case (Fig. 1.12b), the issue of concern is not the actual spatial pattern of contamination, but the distance from the outfall at which concentrations of contaminants fall below some critical value. For example, if the outfall is discharging pesticides in a run-off from housing, there may be, for human safety, a maximal allowable concentration of contaminants in scallops on the seafloor. It is therefore important to focus attention on the limit within which it is not safe to harvest scallops for human consumption. In this case, sampling effort should be entirely focussed on the areas where modelling indicates that concentrations of contaminants are below the critical threshold. The resources and effort should be used to identify the 'safe' boundary rather than the actual spatial gradient of contamination.

Mostly, however, there are major disconnections between gradients of contamination and actual biological responses, i.e. pollution (Phillips, 1978; Spellerberg, 1991; Raimondi & Reed, 1996; Keough & Black, 1996). Such disconnections are due to the inertia of biological systems, i.e. animals or assemblages not being affected by that contaminant, so that there is no response despite chemical signals. Or, biological responses can occur when concentrations of contaminants are minimal, or undetectable, because of bio-accumulation in individual animals. Finally, populations may be unaffected by pollution because of widespread dispersal of their larvae, maintaining populations from elsewhere (Underwood & Peterson, 1988). In such cases, analyses along gradients are not likely to be the best idea, because there is no clear indication of the course or extent of the gradient. Sampling at sites near the outfall and comparisons with sites well away, which are control sites because they are far enough away to be unaffected by the outfall, is an efficient alternative (Fig. 1.13a). As discussed above (see 'Appropriate spatial replication'), uncertainty about the scale of the potential pollution requires sampling at more than the spatial scale.

Another design that may be useful concerns tests of hypotheses about differences in assemblages due to some installation or intrusion into a habitat. For example, it may be proposed that fauna in sediments near rocky reefs are different from those further away. Reasons for differences may be physical (waves, water-flow and

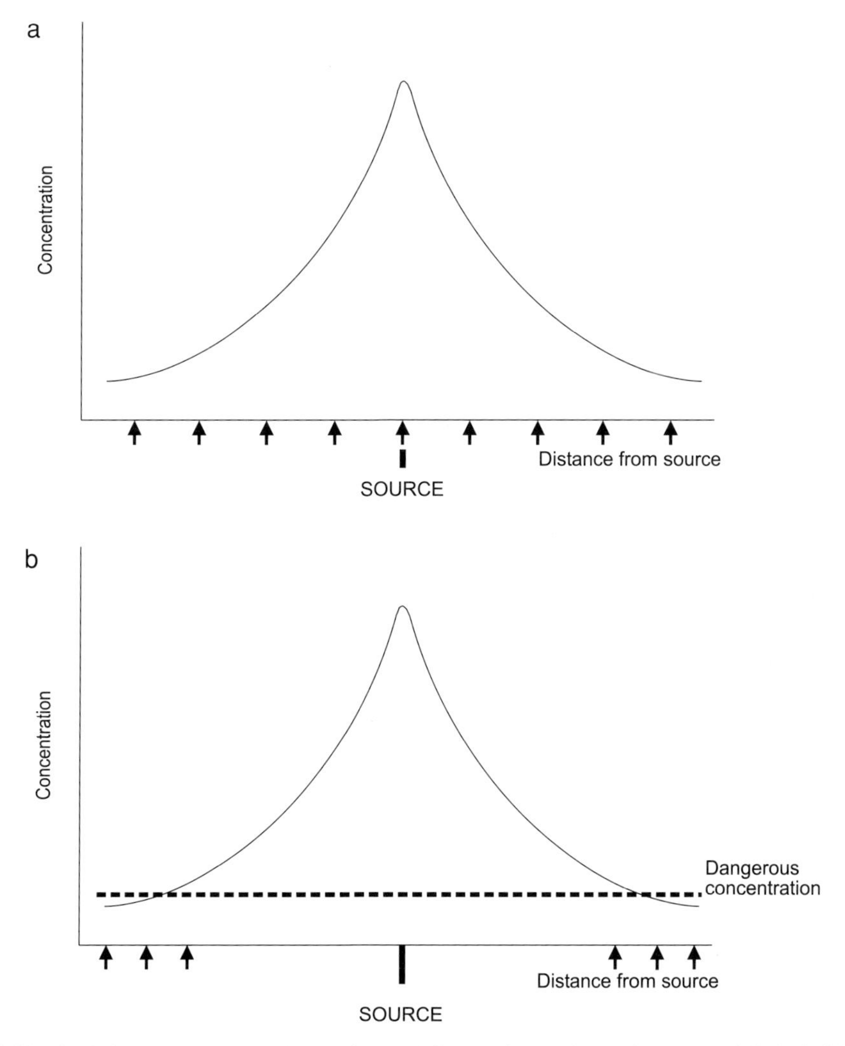

Fig. 1.12 Sampling relevant to gradients of some influence (e.g. pollution from an outfall pipe): (a) the issue is the spatial pattern of decreasing impact based on predicted spatial dilution of contaminants, so sampling along the gradient away from the outfall is necessary; (b) the major concern is risk to health due to pollution being over a defined 'safe' limit in animals at some distance from the outfall. Sampling is therefore concentrated around the areas where this safe limit is supposed to occur.

sediments may all be affected by the presence of a reef) or biological (predatory fish are associated with reefs and therefore eat animals close to a reef, but not those further away). In this case, sampling is needed to test the hypothesis that assemblages close to a reef are different from those further away. Samples are taken near a reef and at some distance (say, 10 m) from it (Fig. 1.13b).

The hypothesis is that any difference is greater than normal spatial differences, so a further set of samples is taken at 20 m from the reef. If the reef influences the

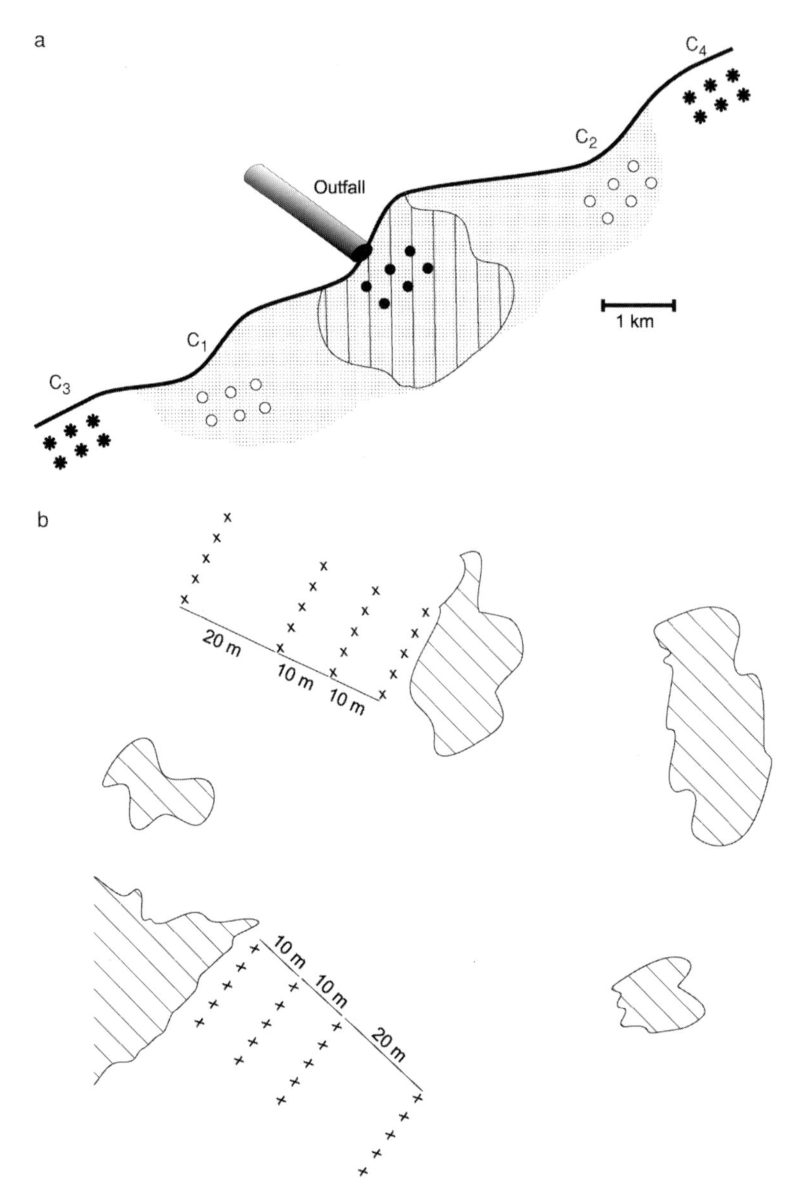

Fig. 1.13 (a) Sampling to handle uncertainty about the scale of an impact due to an outfall. It has been proposed that any impact would only extend over a small distance (shaded), so samples are taken from that area and from controls on either side, outside the area expected to be affected. The area influenced may, however, be much larger (stippled), so other control samples are taken at much greater distances from the outfall. (b) Sampling to test the hypothesis that assemblages in sediments are influenced by the presence of a rocky reef (reefs are shaded). For two replicate reefs, $n = 5$ cores are taken next to the reef (0 m) and at 10, 20 and 40 m from it. The reef will cause samples at 0–10 to differ more than 10–20, or 0–20 to differ more than 20–40 (i.e. more than similar distances where there are no reefs; for more details, see text).

fauna, the difference in multivariate measures of the assemblage from the samples at the reef and at 10 m will not be the same as the natural variation at that scale, i.e. the difference from 10 m to 20 m. If the scale of potential influences is not known in advance, other distances can be sampled. Thus, a further sample at 40 m allows a comparison of the differences between samples at the reef and at 20 m with the natural variation between 20 and 40 m (Fig. 1.13b). A detailed example of the use of this design was described by Kelaher et al. (1998) for infauna in sediments near boardwalks in mangrove forests.

The point to be understood is that sampling designs must be modified for each particular situation and the general principles to guide designs must be flexible frameworks (Green, 1993), responsive to the needs of the particular hypotheses being tested (Underwood, 1997a).

1.13 Combining results from different places or times

Sometimes, a particular study is quite small and, on its own, cannot be particularly revealing about a process under investigation. It is therefore important to think about mechanisms for combining outcomes of such studies in a meta-analysis, i.e. a combination of results of similar tests in different places and times and conditions, of the same hypothesis. General issues about ecological meta-analyses were pioneered by Gurevitch *et al.* (1992).

Here, some procedures are introduced. First, in many experimental studies, the samples are quite small and the outcomes are not clear, because the tests are not powerful enough to provide unambiguous interpretations. There are two general procedures available for combining results of several small tests. Suppose the outcomes of experimental treatments (say, mean numbers of some polychaete at five different depths) are available from several small studies (perhaps six different cruises have collected such data, but each only had two or three samples from each depth). The means from the five depths can be ranked from smallest to largest for each study and summed. This gives the frequency out of six studies of each depth having the smallest, 2nd, 3rd, 4th or largest number of worms. Such data can be analysed by a test devised by Anderson (1959) to determine non-random patterns in the rankings (see Underwood & Chapman (1992) for a detailed example).

Second, there may be a test of some hypothesis in several studies. Each gives a probability of the data being likely to be due to chance errors in sampling, but only a few (or none) are significant. The probabilities from these small tests can be combined using Fisher's (1935) test:

$$C = -2\sum_{i=1}^{k} \log_e(P_i)$$

where k is the number of tests and P_i is the probability from each test ($i = 1, \ldots, k$). C is distributed as χ^2 with $2k$ degrees of freedom if there are, in fact, no differences in the set of tests. Thus, large values of C would cause rejection of the null hypotheses over a set of tests. An example is given in Table 1.7.

Table 1.7 Example (using hypothetical data) of meta-analysis of numbers of worms per core ($n = 5$ cores per sample) amongst seagrass and in empty patches in seagrass in an estuary. The proportion of all seagrass made up by each habitat is given and is the weighting for the Stouffer–Liptak procedure (for all other details, see text).

Habitat	Proportion (= weight)	Mean (S.E.) no. of worms Seagrass	Bare patches	$t_{1\text{-tail, } 8\,df}$	P	Z_i
Dense *Posidonia*	0.40	62 (10.1)	51 (9.7)	0.79	0.227	0.749
Sparse *Posidonia*	0.30	15 (2.7)	11 (2.9)	1.01	0.170	0.954
Zostera	0.13	23 (3.0)	18 (3.2)	1.13	0.146	1.054
Halophila	0.17	15 (1.5)	9 (1.9)	2.49	0.019	2.075

Fisher's combinatorial test, $C = 18.28$; Stouffer–Liptak test, $Z_w = 1.98$

Finally, in some situations, several small tests have been done, but the meta-analysis requires them to be weighted, for example, to reflect different amounts of habitat across a study area. Suppose, for example, it is proposed that numbers of species of crustaceans are greater in a certain estuary in areas with seagrasses than in open, sandy sediments in patches in the seagrass beds. Three species of seagrass form beds in the estuary and one of them (*Posidonia*) occurs in densely or sparsely covered beds. Thus, there are four habitats of interest. In the estuary, there are different areas of each type of habitat.

Samples of $n = 5$ replicate cores were taken amongst seagrass and $n = 5$ cores were taken in bare patches within the bed. For each such pair of samples, a one-tailed t-test was done. There were significantly more crustaceans amongst seagrass in only one of the four tests (Table 1.7), but the probabilities were quite small in several tests. To test the hypothesis of a difference between seagrass and bare areas over the whole estuary, the tests were combined by Fisher's (1935) method (above), giving $C = 18.28$, with 8 df. This was significant at $P < 0.02$, indicating greater numbers of worms amongst seagrasses across the whole set of data. This test does not take into account the different areas occupied by the different habitats. So, a Stouffer–Liptak procedure (Folks, 1984) was used, which weights the outcomes of tests so that large habitats count for more than do smaller ones in a test of the estuary as a whole. The test statistic is:

$$Z_W = \sum_{i=1}^{k} W_i \cdot Z_i \bigg/ \sqrt{\sum_{i=1}^{k} W_i{}^2}$$

where Z_i is the standard normal score for P_i, the probability in the test and W_i, is the weighting for the ith habitat. The weighting is the proportional area of the estuary occupied by that habitat, so that widespread, common habitats count for more in the outcome. The probability of getting a value as large or larger than Z_W can be found from the standard normal distribution. In this case, $Z_W = 1.98$, with $P < 0.03$. There is a major difference between areas with seagrass and empty areas across all habitats. A more detailed example from marine ecological data may be found in McDonald et al. (1993).

Thus, even where resources do not allow large amounts of replication and adequate power, it is possible to combine the outcomes of repeated or similar studies, provided they are designed in logically comparable ways and have specific, well-identified hypotheses so that the procedures for combining them are valid.

1.14 Conclusions

It should be clear from the above that there are important and close connections between biological knowledge about systems, logical development of ideas, models and hypotheses, design of sampling, analysis of data and interpretations of the information. Of course, like everything else in science (and life!), there will be continuing disagreements about the role and purpose of hypothesis-testing in marine research (see Stewart-Oaten, 1996; Suter, 1996). This can be a rich debate or a sterile one. Here, since it would be a distraction, it will receive no mention. Suffice it to say that those who favour risk analysis in environmental assessments and who favour estimation of magnitudes of differences among sites, times, habitats, conditions, rather than structured tests of formally explicit hypotheses would surely all agree about the needs for care in developing sampling programmes? Without advance thought about appropriate scales of replication, it is no more possible to put valid confidence intervals around estimates of mean numbers of animals (or any other parameter being estimated) than it is to do a valid statistical test of some hypotheses about a parameter. Estimators and other describers of the world have the same issues for design as do experimentalists testing hypotheses to advance scientific understanding.

The issues are similar, whatever problem is being investigated. Biological systems are variable in space and time at many scales, because of many interacting processes. It is the responsibility of biologists and ecologists to understand the consequences of such variation and the ensuing interactions and non-independent patterns so that sampling can be planned to take all of the issues into account. Where 'ideal' sampling is not possible, what is uncertain in the data can be considered before any conclusions are reached. This will prevent unplanned difficulties preventing any value in interpretations. If major problems are inevitable with a certain design, alter the design or investigate more preliminary hypotheses in order to unravel the problems.

This introductory assessment of some of the issues is selective and seriously incomplete – there are many other issues. Its purpose is to identify the sorts of issues that should be considered when planning any programme of sampling in marine benthic systems, whatever the issue of concern in the study. Understanding the themes discussed should, at least, provide warnings about pitfalls and some of the vocabulary needed to translate a particular study into meaningful questions to ask statistical advisers.

The problems of pollution, fragmentation and destruction of habitat, over-harvesting of resources, restoration of degraded habitats, conservation of biodiversity, the control of introduced species, global warming and rises in sea level, etc., are vast and urgent. Never has there been such a need for good scientific understanding and advice about what to do and how, where and where. This science deserves the best scientific practice, so improving logic, design, analysis and interpretation of studies is an urgent and ongoing task for marine scientists. Complacency and lack of professionalism will continue to undermine the role of science and will continue to slow down the implementation of solutions to current urgent problems. Getting designs of sampling right entails getting problems identified and understood, besides providing links to valid analysis and interpretation. Improved sampling designs are the keys to improved scientific contributions to social needs.

Acknowledgements

Preparation of this paper was supported by funds from the Australian Research Council through the Centre for Research on Ecological Impacts of Coastal Cities. We are grateful to M. Button and V. Mathews for help with preparation of the figures.

References

Anderson, M.J. (2001) A new method for non-parametric multivariate analysis of variance. *Austral Ecology*, **26**, 32–46.

Anderson, R.L. (1959) Use of contingency tables in the analysis of consumer preference studies. *Biometrics*, **15**, 582–590.

Andrew, N.L. & Mapstone, B.D. (1987) Sampling and the description of spatial pattern in marine ecology. *Annual Review of Oceanography and Marine Biology*, **25**, 39–90.

Bender, E.A., Case, T.J. & Gilpin, M.E. (1984) Perturbation experiments in community ecology: theory and practice. *Ecology*, **65**, 1–13.

Benedetti-Cecchi, L., Airoldi, L., Abbiati, M. & Cinelli, F. (1996) Estimating the abundance of benthic invertebrates: a comparison of procedures and variability between observers. *Marine Ecology Progress Series*, **138**, 93–101.

Bernstein, B.B. & Zalinski, J. (1983) An optimum sampling design and power tests for environmental biologists. *Journal of Environmental Management*, **16**, 335–343.

Box, G.E.P. & Jenkins, G.M. (1976) *Time Series Analysis: Forecasting and Control*. Holden Day Inc., San Francisco.

Brown, J.H. & Gibson, A.C. (1983) *Biogeography*. C.V. Mosby Co., St Louis.

Caffey, H.M. (1985) Spatial and temporal variation in settlement and recruitment of intertidal barnacles. *Ecological Monographs*, **55**, 313–332.
Cameron, J. & Abouchard, J. (1991) The precautionary principle: a fundamental principle of law and policy for the protection of the global environment. *Comparative Law Review*, **14**, 1–27.
Chapman, M.G. & Underwood, A.J. (1999) Ecological patterns in multivariate assemblages: information and interpretation of negative values in ANOSIM tests. *Marine Ecology Progress Series*, **180**, 257–265.
Chapman, M.G., Underwood, A.J. & Skilleter, G.A. (1995) Variability at different spatial scales between a subtidal assemblage exposed to the discharge of sewage and two control assemblages. *Journal of Experimental Marine Biology and Ecology*, **189**, 103–122.
Clarke, K.R. (1993) Non-parametric multivariate analyses of changes in community structure. *Australian Journal of Ecology*, **18**, 117–143.
Cliff, A.D. & Ord, J.K. (1973) *Spatial Autocorrelation*. Pion Ltd., London.
Clifford, D.H.T. & Stephenson, W. (1975) *An Introduction to Numerical Classification*. Academic Press, New York.
Cochran, W.G. & Cox, G. (1957) *Experimental Designs, Second Edition*. John Wiley & Sons, New York.
Cox, G. (1958) *The Planning of Experiments*. John Wiley & Sons, New York.
Dayton, P.K. (1971) Competition, disturbance and community organization: the provision and subsequent utilization of space in a rocky intertidal community. *Ecological Monographs*, **41**, 351–389.
Diggle, P.J. (1990) *Time Series: A Biostatistical Introduction*. Clarendon Press, Oxford.
Dovers, S.R. & Handmer, J.W. (1995) Ignorance, the precautionary principle, and sustainability. *Ambio*, **24**, 92–97.
Durbin, J. & Watson, G.S. (1951) Testing for serial correlation in least-squares regression. *Biometrika*, **38**, 159–178.
Elliott, J.M. (1977) *Methods for the Analyses of Samples of Benthic Invertebrates*. Freshwater Biological Association Scientific Publication No. 25, Reading, MA.
Fisher, R.A. (1935) *The Design of Experiments*. Oliver and Boyd, Edinburgh.
Folks, J.R. (1984) Combination of independent tests. In: *Handbook of Statistics 4. Nonparametric Methods* (Eds P.R. Krishnaiah & P.K. Sen), pp. 72–94. North-Holland, New York.
Foster, M.S., Harrold, C. & Hardin, D.D. (1991) Point versus photo quadrat estimates of the cover of sessile organisms. *Journal of Experimental Marine Biology and Ecology*, **146**, 193–204.
Gaines, S.D. & Bertness, M.D. (1992) Dispersal of juveniles and variable recruitment in sessile marine species. *Nature*, **360**, 579–580.
Glasby, T.M. & Underwood, A.J. (1996) Sampling to differentiate between pulse and press perturbations. *Environmental Monitoring and Assessment*, **42**, 241–252.
Gray, J.S. (1990) Statistics and the precautionary principle. *Marine Pollution Bulletin*, **21**, 174–176.
Gray, J.S. (1996) Environmental science and a precautionary approach revisited. *Marine Pollution Bulletin*, **32**, 532–534.
Gray, J.S., Aschan, M., Carr, M.R., Clarke, K.R., Green, R.H., Pearson, T.H., Rosenberg, R. & Warwick, R.M. (1988) Analysis of community attributes of the benthic macrofauna of Frierfjord/Langesundfjord and in a mesocosm experiment. *Marine Ecology Progress Series*, **46**, 151–165.

Green, R.H. (1979) *Sampling Design and Statistical Methods for Environmental Biologists*. John Wiley & Sons, Chichester.
Green, R.H. (1989) Power analysis and practical strategies for environmental monitoring. *Environmental Research*, **50**, 195–205.
Green, R.H. (1993) Application of repeated measures designs in environmental impact and monitoring studies. *Australian Journal of Ecology*, **18**, 81–98.
Green, R.H. & Hobson, K.D. (1970) Spatial and temporal structure in a temperate intertidal community, with special emphasis on *Gemma gemma* (Pelecypoda: Mollusca). *Ecology*, **51**, 999–1011.
Gurevitch, J., Morrow, L.L., Wallace, A. & Walsh, J.S. (1992) A meta-analysis of competition in field experiments. *American Naturalist*, **140**, 539–572.
Hall, S.J. & Harding, M.J. (1997) Physical disturbance and marine benthic communities: the effects of mechanical harvesting of cockles on non-target benthic infauna. *Journal of Applied Ecology*, **34**, 497–517.
Hurlbert, S.J. (1984) Pseudoreplication and the design of ecological field experiments. *Ecological Monographs*, **54**, 187–211.
Kelaher, B.P., Chapman, M.G. & Underwood, A.J. (1998) Changes in benthic assemblages near boardwalks in temperate urban mangrove forests. *Journal of Experimental Marine Biology and Ecology*, **228**, 291–307.
Keough, M.J. (1984) Dynamics of the epifauna of the bivalve *Pinna bicolor:* interactions among recruitment, predation, and competition. *Ecology*, **65**, 677–688.
Keough, M.J. (1998) Responses of settling invertebrate larvae to the presence of established recruits. *Journal of Experimental Marine Biology and Ecology*, **231**, 1–19.
Keough, M.J. & Black, K.P. (1996) Predicting the scale of marine impacts: understanding planktonic links between populations. In: *Detecting Ecological Impacts: Concepts and Applications in Coastal Habitats*. (Eds R.J. Schmitt & C.W. Osenberg), pp. 199–234. Academic Press, San Diego.
Leduc, A., Prairie, Y.T. & Bergeron, Y. (1994) Fractal dimension estimates of a fragmented landscape: sources of variability. *Landscape Ecology*, **9**, 279–286.
Lindegarth, M., Valentinsson, D., Hansson, M. & Ulmestrand, M. (2000) Interpreting large-scale experiments on effects of trawling on benthic fauna: an empirical test of the potential effects of spatial confounding in experiments without replicated control and trawled areas. *Journal of Experimental Marine Biology and Ecology*, **245**, 155–169.
Mapstone, B.D. (1995) Scalable decision rules for environmental impact studies: effect size, Type I, and Type II errors. *Ecological Applications*, **5**, 401–410.
McDonald, L.L., Erickson, W.P. & Strickland, M.D. (1993) Survey design, statistical analysis, and basis for statistical inference in coastal habitat injury assessment: Exxon Valdez oil spill. In: *Exxon Valdez Oil Spill: Fate and Effects in Alaskan Waters*. (Eds P.G. Wells, J.N. Butler & J.S. Hughes), pp. 296–311. American Society for Testing and Materials, Philadelphia.
Morrisey, D.J., Howitt, L., Underwood, A.J. & Stark, J.S. (1992a) Spatial variation in soft-sediment benthos. *Marine Ecology Progress Series*, **81**, 197–204.
Morrisey, D.J., Underwood, A.J., Howitt, L. & Stark, J.S. (1992b) Temporal variation in soft-sediment benthos. *Journal of Experimental Marine Biology and Ecology*, **164**, 233–245.
Olsgard, F., Somerfield, P.J. & Carr, M.R. (1997) Relationships between taxonomic resolution and data transformations in analyses of a macrobenthic community along an established pollution gradient. *Marine Ecology Progress Series*, **149**, 173–181.

Peterman, R.M. & M'Gonigle, A.B. (1992) Statistical power analysis and the precautionary principle. *Marine Pollution Bulletin*, **24**, 231–234.

Peters, R.H. (1991) *A Critique for Ecology*. Cambridge University Press, Cambridge.

Phillips, D.J.H. (1978) The use of biological indicator organisms to quantitate organochlorine pollutants in aquatic environments: a review. *Environmental Pollution*, **16**, 167–229.

Raimondi, P.T. & Reed, D.C. (1996) Determining the spatial extent of ecological impacts caused by local anthropogenic disturbances in coastal marine habitats. In: *Detecting Ecological Impacts: Concepts and Applications in Coastal Habitats* (Eds R.J. Schmitt & C.W. Osenberg), pp. 179–198. Academic Press, San Diego.

Schmitt, R.J. & Osenberg, C.W. (Eds) (1996) *Detecting Ecological Impacts: Concepts and Applications in Coastal Habitats*. Academic Press, San Diego.

Shanks, A.L. & Wright, W.G. (1986) Adding teeth to wave action: the destructive effects of wave-borne rocks on intertidal organisms. *Oecologia*, **69**, 420–428.

Skilleter, G.A. (1992) Recruitment of cerithiid gastropods (*Rhinoclavis* spp.) in sediments at One Tree Reef, Great Barrier Reef. *Journal of Experimental Marine Biology and Ecology*, **156**, 1–21.

Smith, E.P., Orvos, D.R. & Cairns, J. (1993) Impact assessment using the before-after-control-impact (BACI) model: concerns and comments. *Canadian Journal of Fisheries and Aquatic Science*, **50**, 627–637.

Sousa, W. (1979) Disturbance in marine intertidal boulder fields: the nonequilibrium maintenance of species diversity. *Ecology*, **60**, 1225–1239.

Sparks, T. (Ed) (2000) *Statistics in Ecotoxicology*. John Wiley & Sons, Chichester.

Spellerberg, I.F. (1991) *Monitoring Ecological Change*. Cambridge University Press, Cambridge.

Stewart-Oaten, A. (1996) Goals in environmental monitoring. In: *Detecting Ecological Impacts: Concepts and Applications in Coastal Habitats*. (Eds R.J. Schmitt & C.W. Osenberg), pp. 17–28. Academic Press, San Diego.

Stewart-Oaten, A., Murdoch, W.M. & Parker, K.R. (1986) Environmental impact assessment: 'pseudoreplication' in time? *Ecology*, **67**, 929–940.

Suter, G.W. (1996) Abuse of hypothesis testing statistics in ecological risk assessment. *Human and Ecological Risk Assessment*, **2**, 331–347.

Thrush, S.F. (1999) Complex role of predators structuring soft-sediment macrobenthic communities: implications of changes in spatial scale for experimental studies. *Australian Journal of Ecology*, **24**, 344–354.

Underwood, A.J. (1981) Structure of a rocky intertidal community in New South Wales: patterns of vertical distribution and seasonal changes. *Journal of Experimental Marine Biology and Ecology*, **51**, 57–85.

Underwood, A.J. (1989) The analysis of stress in natural populations. *Biological Journal of the Linnean Society*, **37**, 51–78.

Underwood, A.J. (1990) Experiments in ecology and management: their logics, functions and interpretations. *Australian Journal of Ecology*, **15**, 365–389.

Underwood, A.J. (1991) Beyond BACI: experimental designs for detecting human environmental impacts on temporal variations in natural populations. *Australian Journal of Marine and Freshwater Research*, **42**, 569–587.

Underwood, A.J. (1992) Beyond BACI: the detection of environmental impact on populations in the real, but variable, world. *Journal of Experimental Marine Biology and Ecology*, **161**, 145–178.

Underwood, A.J. (1993) The mechanics of spatially replicated sampling programmes to detect environmental impacts in a variable world. *Australian Journal of Ecology*, **18**, 99–116.

Underwood, A.J. (1994) On beyond BACI: sampling designs that might reliably detect environmental disturbances. *Ecological Applications*, **4**, 3–15.

Underwood, A.J. (1997a) *Experiments in Ecology: Their Logical Design and Interpretation Using Analysis of Variance*. Cambridge University Press, Cambridge.

Underwood, A.J. (1997b) Environmental decision-making and the precautionary principle: what does this mean in environmental sampling practice? *Landscape and Urban Planning*, **37**, 137–146.

Underwood, A.J. (1998) Grazing and disturbance: an experimental analysis of patchiness in recovery from a severe storm by the intertidal alga *Hormosira banksii* on rocky shores in New South Wales. *Journal of Experimental Marine Biology and Ecology*, **231**, 291–306.

Underwood, A.J. (2000) Trying to detect impacts in marine habitats: comparisons with suitable reference areas. In: *Statistics in Ecotoxicology* (Ed. T. Sparks), pp. 279–308. John Wiley & Sons, Chichester.

Underwood, A.J. & Chapman, M.G. (1992) Experiments on topographic influences on density and dispersion of *Littorina unifasciata* in New South Wales. In: *Proceedings of the Third International Symposium on Littorinid Biology*. (Eds J. Grahame, P.J. Mill & D.G. Reid), pp. 181–195. The Malacological Society of London, London.

Underwood, A.J. & Denley, E.J. (1984) Paradigms, explanations and generalizations in models for the structure of intertidal communities on rocky shores. In: *Ecological Communities: Conceptual Issues and the Evidence*. (Eds D.R. Strong, D. Simberloff, L.G. Abele & A. Thistle), pp. 151–180. Princeton University Press, New Jersey.

Underwood, A.J. & Peterson, C.H. (1988) Towards an ecological framework for investigating pollution. *Marine Ecology Progress Series*, **46**, 227–234.

Winer, B.J., Brown, D.R. & Michels, K.M. (1991) *Statistical Principles in Experimental Design, Third Edition*. McGraw-Hill, New York.

Wynne, B. (1992) Uncertainty and environmental learning: reconceiving science and policy in the preventative paradigm. *Global Environmental Change*, **2**, 111–127.

Chapter 2
Sediment Analysis and Seabed Characterisation

A.J. Bale and A.J. Kenny

2.1 Introduction

This chapter describes practical methods for characterising sediments physically and compositionally. It does this on two levels: the first level is that determined by point sampling with cores and grabs, the second level introduces relatively new acoustic (remote sensing) methods that characterise bed sediment over large spatial scales. Both approaches complement each other; point sampling provides direct information for comparison with biological data in samples as well as 'ground truth' for remotely sensed methods. Remotely sensed surveys put 'point sample' information into perspective in relation to the broad spatial distribution of sediment properties, benthic habitats and their patchiness. This chapter is not intended as a thorough grounding in sedimentology or acoustic surveying but aims to provide basic, practical guidance compatible with biological studies along with references to sources of more detailed information. It should allow workers to define recent, surface sediment substrates associated with biological studies and, possibly, to infer ecological information from a knowledge of the sediment properties in a particular environment.

The sediment composition at a given site depends on the source material, whether of geological or biological origin or both. It depends crucially on the energy (waves, tides, currents) of the environment and the balance between erosion and accumulation. For a discussion of physical processes which govern sediment transport and accumulation see Dyer (1979) and Soulsby (1997). A key factor affecting sediment properties is the biota itself. Various components of the biota can destabilise sediments through burrowing and feeding, whilst others act to stabilise and modify the substrate through root development, construction of worm tubes, faecal pelletisation, and secretion of EPS (extracellular polymeric substances) (Paterson, 1997; Decho, 1990). Seasonal and climatic changes in the balance between these processes can significantly influence sediment erosion and accumulation processes in estuaries (Widdows et al., 2000). Both physical and biological aggregations of grains modify the size distribution and properties of the sediment. Furthermore, particularly in silt-rich sediments, the adsorption of organic molecules at particle surfaces and electrostatic, inter-particle attractions mean that these sediments behave cohesively rather than as individual grains. In classical sediment petrology (where most of the practical techniques for sediment grain size analysis were developed) the analyst generally seeks to remove organic material from samples in order to describe the population of primary mineral grains within the sediment. However, there is now a growing appreciation that for biological and environmental sedimentology, the *in situ,* natural properties of the sediment are more relevant for interpreting the physical behaviour of the sediment and animal–sediment relationships than those derived from the petrology-based analysis of mineral grain sizes and properties.

Sampling and storage

The design of sampling schemes to define a sediment/biotic environment depends very much on the questions that are to be addressed by the study, and by the nature and scale of the patchiness of the particular substrate, as well as practical and economic considerations (these aspects are covered more fully in Chapter 1). In practical terms, sampling techniques and apparatus must be consistent with the purpose for which the samples are required. To take an extreme example, there is no point in attempting to quantify silt content in samples obtained from an open grab system where unknown and variable quantities of fines have been washed out of the sample during collection. Useful information on sampling equipment is given by Kramer et al. (1994). For information on the practical efficiency of various sampling devices, readers are referred to Chapter 5 and to Blomqvist (1991), Somerfield and Clarke (1997) and Rumohr (1999). Several studies point to the efficiency of coring systems such as the Craib corer (Craib, 1965) and the Barnett–Watson multi-corer (Barnett et al., 1984) (see Chapter 6) that maintain the integrity of the sample, along with its surface 'fluff' and overlying water. The key principle of both of these systems is that they employ a bed frame and hydraulically damped core penetration to minimise disturbance of the mobile surface layer (fluff) which, if lost due to bow wave effects, can bias surface sediment composition (Bale & Morris, 1998). However, these corers are really only suitable for silt-rich (cohesive) sediments and do not work well in compacted sands or sediments containing gravel or stones. Furthermore, core samples from these instruments are generally too small (50–100 mm diameter) to sample macrofauna adequately. The Reineck box corer (square cores of 0.5 m width and 0.7 m depth) takes representative samples (good retention of fines) which are large enough for macrofaunal sampling. Unfortunately, coring equipment of this size requires a substantial vessel with winches and a gantry capable of hauling the 2–3 tonne suction load as the corer is pulled out of the sediment. If such a vessel is not available then compromises will have to be made and the choice of sampler will have to take account of the capabilities of the vessel. None of the coring systems performs well in the presence of stones and gravel and in such cases, large mechanical or gravity-operated grabs may be the only option. Clearly operators should acquaint themselves with the limitations of each device and be aware of the variations in sample integrity that can be introduced through instrument wear and tear and by changes in weather conditions and sea state. Murdoch and Azcue (1995) have produced a comprehensive manual of the practical aspects of sediment sampling.

Video cameras in suitable underwater housings attached to the frame of sampling equipment can sometimes provide valuable insight into the operation of the sampling device as well as allowing the nature of the bed to be assessed. Another approach in relatively shallow waters is to employ divers (see Chapter 4) to collect sediment samples with hand corers. On intertidal sediments, samples can be collected by hand using plastic core tubes or by manual excavation, and in these cases there can be a high degree of confidence in the integrity of the sample as long as appropriate care is taken.

If sediment samples are to be used for analyses of chemical or biological constituents then preservation must be considered to minimise the bacterial degradation of organic material. For work on redox-sensitive chemical species, core samples may need to be processed and stabilised under nitrogen in a glove box. In unpreserved samples, the death of animals and subsequent degradation of organic material can quickly alter sediment parameters such as organic content and redox potential. Conversely, many common biological preservatives may compromise subsequent chemical analyses. There are several options for sample preservation which fall into the categories of poisons (e.g. mercuric chloride, sodium azide, chloroform, antibiotics) or fixatives (e.g. formaldehyde, glutaraldehyde) and the selection would depend on the purpose for which the sediment was subsequently required. The relative merits of various preservatives are reviewed by Knauer and Asper (1989). In some cases, representative sub-samples for granulometric and chemical analyses should be removed from the main sample before treatment with preservatives, and stored differently. Preservation may be achieved by freezing at minus 20°C or lower, in which case glass containers should not be used. Preliminary experiments to assess the influence of containers and storage conditions on the substances of interest are strongly recommended for analyses of several sediment constituents.

Efficient labelling of samples is essential as this is the only factor relating the subsequent analyses to a specific sampling site. During sample collection, most workers adopt a sequential numbering system added to the container by an indelible marking pen or a similar tool. However, most pens quickly become inefficient during inclement weather and external labels can also become defaced by abrasion during transport. It is therefore a very good practice to have a secondary label, on waterproof paper, placed within the container. Log sheets should be used to record all relevant data at the time of sampling, i.e. position, weather/sea conditions, visual observations of sediment character and the sample identification. Buller and McManus (1979) advise that each new batch of samples received in a laboratory should be checked for numbering consistency against the log sheets to ascertain that all the samples expected are present. It is much easier to rectify any anomalies such as misnumbered or unreadable labels or to identify missing samples at this early stage rather than after a number of samples have been removed for analysis.

In summarising this section on samples, two aspects should be highlighted. First, make use of the best statistical advice available for determining sampling patterns and sample numbers (see Chapter 1). Second, try to make sure, where possible, that sample containers, storage conditions and preservation techniques are compatible with all the proposed determinations.

2.2 Particle (grain) size analysis

Grain size analyses define the sedimentary environment and give an insight into the physical regime. Although a wide range of techniques is available for grain size analysis (McCave & Syvitski, 1991; Lloyd, 1985), there are fewer techniques that are appropriate for aquatic sediment samples (Kramer et al., 1994). Grain size analysis attempts to define the dimensions of a particle, or population of particles, using a single parameter and, except for spherical particles, this is always a compromise since cylinders, rectangles, pyramids, plates and irregular shapes all require more than one dimension to define their size. Additionally, various analytical methods measure different characteristics of particles in attempting to define their size. For example, with microscopy the long and short axes of a plan view are averaged to give a diameter. The vertical dimension is assumed to be similar, though it is probably less than the cross section because the particle will tend to rest in its most stable attitude. Electro-sensing devices measure the true particle volume irrespective of the shape of the particle and assume the particle to be spherical in assigning a dimension. This value is the effective spherical diameter (ESD). Sieves measure the smallest section of a particle since elongated particles can pass through a mesh lengthwise. For sand, sphericity is a reasonable assumption but for silt-sized material containing clay platelets and irregular biogenic particles and debris, this assumption is less valid. There is a range of definitions for particles which attempt to accommodate shape factors (Allen, 1990) but for most purposes the assumption of sphericity and the use of the ESD is the only practical option. The diameter, or ESD, is the standard definition of the size of a particle.

For sediment with significant silt content, particle size analysis is operationally dependent and consistent sample preparation and handling are required. Sand grains, however, are simpler to analyse and good size distribution information can be obtained easily with a set of certified test sieves. Although electro-optical methods can now measure sand size particles up to 2 mm diameter, the characterisation of larger particles will generally require the use of sieves. For many purposes, sieves are the most practical option for describing gravel and sand sizes down to 63 μm. Sieves are arranged sequentially in a tower, largest apertures uppermost, and used with a mechanical shaker. Sieve analyses are 'low tech', and the equipment is inexpensive, and results are readily comparable between laboratories. On the debit side, sieving is labour-intensive and slow; sieves also need to be well maintained by washing, brushing and drying between uses to clean the meshes and avoid corrosion. Mesh size and condition should be checked periodically using a suitable microscope and calibrated eyepiece graticule. Before progressing to the methods employed for sands and silts, it is worth studying the concept of grade scales and the conventions used in sedimentology.

Grade scales

The object of particle size analysis is to characterise the sediment as a frequency distribution of grain sizes. In this process grain size is the independent variable in a continuous distribution, i.e. natural particles have potentially an infinite number of sizes ranging from the sub-micron to many millimetres. To be able to define such a distribution and to generate descriptive statistics, an arbitrary set of finite intervals is generally employed to convert the continuous distribution to a discrete series. The Udden/Wentworth scale (Wentworth, 1922) which combines numerical intervals with rational definitions (fine sand, coarse sand, etc., see Table 2.1) has been almost universally adopted amongst marine ecologists and geologists.

Table 2.1 The Udden/Wentworth grade scale based on powers of 2 and $\sqrt{2}$ and equivalent phi values.

Broad description	Description	2 scale (mm)	$\sqrt{2}$ scale (mm)	ϕ (phi)
		256		−8
	Cobble	128		−7
		64		−6
Gravel		32		−5
	Pebble	16		−4
		8		−3
	Granule	4		−2
	Very coarse sand	2		−1
			1.41	−0.5
	Coarse sand	1		0
			0.71	0.5
	Medium sand	0.5		1.0
Sand			0.351	1.5
	Fine sand	0.25		2.0
			0.177	2.5
	Very fine sand	0.125		3.0
			0.088	3.5
		0.062		4.0
			0.044	4.5
		0.031		5.0
			0.022	5.5
		0.0156		6.0
Silt			0.011	6.5
		0.0078		7.0
			0.0055	7.5
		0.0039		8.0
Clay		<0.0039	<0.0039	

The scale is geometric, based on a dimension of 1 mm and a ratio of 2 and has the advantage of increasing the resolution of information with decreasing particle size. If required, the resolution can be increased using a ratio of root 2 or the 4th root of 2 (Table 2.1). For further information on grade scales, workers should consult a standard sedimentology text such as in Lindholm (1987). A logarithmic transformation of the Wentworth scale gives the phi (ϕ) notation (Krumbein, 1938):

$$\phi = -\frac{\log_{10}(\text{dia.mm})}{\log_{10} 2}$$

The phi notation stemmed from the need for a graphical mechanism for data manipulation in the days before electronic calculation. However, in the words of Lindholm (1987), 'the phi scale is almost meaningless to most biologists, archaeologists and engineers who report grain sizes, as measured, in metric units.' He goes on to suggest that there are now good arguments for abandoning the phi notation, particularly since the calculation of measures of central tendency (median, skewness, sorting etc.) graphically has been superseded by computerised methods. Nevertheless, as environmental scientists, we need to be aware of the phi notation because it appears in sedimentology literature and because of the degree to which it is embedded in the methods of calculating sediment distribution parameters. The use of phi scales is discussed by Buller and McManus (1979) and considered further in this chapter within the section on data presentation.

Dry sieving clean sands

Clean sands, e.g. with less than 5% silt and negligible organic matter, are easy to deal with in the dry state and are simple to analyse by sieving. The analysis requires a stack of Wentworth grade sieves within the range 2000–2062 μm. The stack should be closed at the bottom with a pan and covered at the top with a lid. A mechanical shaker is essential when several samples have to be analysed and each sample should be shaken for a fixed interval – typically 15 minutes. A pre-weighed sample, e.g. 25 g of oven-dried sand, is introduced to the top sieve. After 15 minutes of agitation, the weight of the material retained by each sieve, and in the bottom pan (<63 μm), must be determined (Fig. 2.1). Characterising larger particles such as gravels and pebbles can also be done with appropriate sieve screens, but prohibitively large samples are often required. For materials in this range and for larger pebbles and boulders, it is often adequate to estimate sizes visually.

Rapid partial analysis of silt (or mud) content

For many purposes, including benthic ecology, sedimentology and evaluation of chemical contaminant data, a single determination of the silt content (sometimes termed mud content) is an extremely good indicator of the overall character of a sediment sample and has a predictive capability for determining mechanical

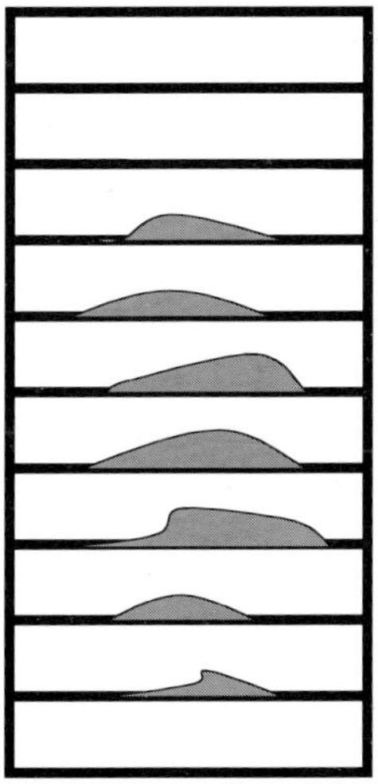

0.5 phi sieve stack	mesh size mm	wt %	cumulative % coarser	cumulative % finer
	1.00	0	0	100
	0.71	0	0	100
	0.50	4	4	96
	0.351	11	15	85
	0.25	18	33	67
	0.177	30	63	37
	0.125	21	84	16
	0.088	10	94	6
	0.062	6	100	0
	pan	0	100	0

Fig. 2.1 Schematic and table showing the results obtained from sieving a sand sample with a series of Wentworth interval sieves.

properties (Flemming & Delafontaine, 2000). This single determination greatly reduces the work compared with a complete spectral analysis of size. The objective is to determine the silt fraction as a percentage of the total sediment weight by sieving at 63 μm. This method is fairly robust in that as long as all the silt is washed out of the sand remaining on the sieve, the results are largely independent of the sample preparation and the sieving operation; i.e. it does not matter how much the silt/clay structures are degraded as the objective is merely to remove them from the sand to determine the sample weight loss. If, however, it is important to maintain the integrity of the silt/clay fraction, then this process can also be undertaken on undried (natural) sediment.

Method for silt content – dry sediment

(1) Take an accurately weighed sample of oven-dried sediment of about 25 g.
(2) Place the sample with about 250 ml tap water in a beaker and add 10 ml of 6.2 g/l sodium hexametaphosphate $(NaPO_3)_6$ to aid dispersion of clay particles. Break up the sediment with a glass rod and then stir mechanically for 10–15 minutes. Allow the sediment to soak overnight and then stir again for an additional 10–15 minutes.
(3) Wash the dispersed sediment suspension from the beaker on to a 63 μm sieve. This can be done manually with the sieve partially immersed in a white basin containing clean water so that the sediment is submerged. The sample is then sieved by 'puddling' so that the fines fall through to the basin. Replace the water in the basin at intervals and continue sieving and washing the sediment

until no further fines are washed out. Alternatively, the sample can be sieved using a mechanical shaker connected to a water supply. In this apparatus the lid of the sieve has transparent viewing windows and is fitted with a water inlet generating a spray, and the bottom tray has a drain outlet that can either be led to a large receiver or to waste. During the sieving operation water is introduced at mains water tap pressure at intervals until the effluent from the bottom pan runs clear and the sand in the sieve is clean.

(4) Finally, transfer the sieve and contents to an oven and dry rapidly at 100°C. Successive weights should be checked to determine the time required to achieve a constant weight, i.e. no further weight loss. At this stage the weight of sand remaining in the sieve can be determined and subtracted from the original dry weight of the sediment to allow a percentage silt to be calculated:

$$\text{Silt content (\%)} = \frac{(\text{wt sample} - \text{wt sand}) \times 100}{\text{wt sample}}$$

Some workers like to carry out further dry sieving of the dried sand over white paper to check for any further silt loss as a result of drying. If required, the sand fraction can be graded through the sand sieve series. In some environments, the sand fraction may include large organic debris (leaves, twigs). This might be evaluated by further sieving at, e.g. 2 mm or, alternatively, evaluated as weight loss on combustion (see later).

Method for silt content of wet sediment

A variation on the previous method which allows water content to be determined as well as avoiding initial drying of the sample is to sieve a representative portion of the original wet sediment. This may be important if, for example, part of the silt fraction is required for further analysis. To determine silt content in this way, approximately 50–60 g of the wet sediment should be weighed accurately and then dispersed in 250 ml of tap water with sodium hexametaphosphate. The method is identical to that used for oven-dried sediment but the silt fraction must be quantified. This will require a large receiver, e.g. a 10 l plastic measuring cylinder, to retain the silt and washings. The dry weight of sand is determined as before but the weight of silt and clay must also be determined and this is achieved using gravimetry.

Method for gravimetry

Gravimetry is an essential method for anyone engaged in sediment work; it is also used for the determination of suspended solids' concentration and the calibration of optical suspended load meters.

(1) Measure and record the total volume of the silt suspension collected.
(2) Thoroughly mix the suspension by shaking and/or mechanical stirring so that a representative volume of sub-sample can be withdrawn for filtration. Ideally,

the volume chosen should provide the maximum load on the filter without significant clogging. The volume filtered must be recorded.

(3) Transfer the sub-sample onto a pre-weighed GF/C or GF/F glass fibre filter in a vacuum filtration unit. For maximum accuracy, filters need to be pre-washed by filtering 50–100 ml of particle-free, distilled water and then dried to constant weight. Weights need to be determined to 5 decimal places of grams.

(4) When the sample has been filtered, add two 50 ml volumes of distilled water to wash salt and sodium hexametaphosphate from the sediment. When this has been drawn through, remove the funnel from the filter unit but leave the filter on the sinter under vacuum and use a distilled water wash-bottle to rinse salt water from the rim of the filter.

(5) Using flat tip forceps, remove the loaded filter from the sinter while still under vacuum so that excess water is drawn out. Then, dry to constant weight and determine the filter load by difference:

$$\text{Silt load (g l}^{-1}) = \frac{(\text{wt filter 2} - \text{wt filter 1) (g)} \times 1000}{\text{volume filtered (ml)}}$$

(6) To determine the total weight of silt in the original sample, multiply the silt load by the volume of silt suspension.

(7) Calculate the silt content as a percentage of the weight of sand plus silt/clay:

$$\text{Silt content (\%)} = \frac{\text{wt silt} \times 100}{(\text{wt sand} + \text{wt silt})}$$

Method for water content

From the data obtained for the wet silt content, the water content can be calculated as a percentage of the original wet weight by subtracting the total weight of sand plus silt/clay from the original sample wet weight:

$$\text{Water content (\%)} = \frac{(\text{wt wet sample} - (\text{wt silt} + \text{wt sand})) \times 100}{(\text{wt wet sample})}$$

Analysis of the silt–clay fraction

No single analytical method covers the entire size range of sediment particles. Sieving tends to be efficient for particles down to 63 μm; below this, alternative methods are required. Although sets of electro-formed ‘micro-sieves’ are available (e.g. Fritch, Veco) with apertures down to 3 and 7 μm, respectively, they are not suitable for routine analyses of natural sediments which often contain colloidal organic material and mucous substances that rapidly clog small pores. Similarly, filtration membranes can be obtained with a range of pore sizes (0.1–10 μm) (e.g. Millipore™, Nuclepore™) that could be used to screen particles for research purposes but, as with micro sieves, they are not a practical (or economic) option for routine analysis of sediment samples.

There are a number of options available if grading of the silt–clay fraction is required. Until the early 1980s, the only practical method of sizing silt/clays was the Andreasen pipette technique. This involved determining sedimentation rates by allowing dispersed particles to settle through a known distance/time, employing Stoke's law and assuming a constant particle density. This technique had numerous drawbacks being very time-consuming, extremely operationally dependent, and rather unsuitable for natural sediments that may contain a range of materials with different densities. The problem of variable densities applies equally to instruments that measure the decrease of X-ray density during sedimentation (e.g. Sedigraph) and to sedimentation balance instruments. Additionally, Stokes settling assumes spherical particles and a significant proportion of particles in this fraction are non-spherical biogenic debris or clay plates that settle differently. Since electro-resistive and optical analytical methods are now widely available, no further consideration will be given to sedimentation-based sizing techniques. Readers who require information on this technique are referred to earlier editions of this handbook.

Aggregation status

Before looking at the contemporary analytical methods for silt size particles, it is worth revisiting the sample 'handling' dilemma. In natural silts, primary mineral grains are the exception rather than the norm. In practice, a high percentage of the sediment will have been digested and evacuated as faecal pellets many times over by deposit feeders and the like. Additionally, surface sediment particles will have been deposited primarily as particle aggregations (flocs) resulting from particle–particle collisions in the water column. All these structures will have different degrees of resistance to breakage; flocs break at the slightest disturbance and faecal pellets can be initially quite resistant to breakage but become more fragile with time. Thus, in effect, the size distribution determined for a silt fraction generally reflects the degree of mechanical and chemical dispersion applied to the sample. This process of disturbance would have started from the moment the sample was collected and the extent of disruption would vary with the operational procedures in use. There are really only two options, neither of which is ideal. One option is to apply the methods that completely disperse every structural element so that the size distribution of primary particles is measured. This would need experimentation to optimise but might include, for example, chemical 'digestion' of organic material with hydrogen peroxide followed by ultrasonic dispersion and the use of sodium hexametaphosphate dispersant as described previously. The other option is to adopt a carefully standardised procedure so that a minimal but consistent degree of disruption is applied to each sample. This approach probably best reflects the true nature of the sediment silt properties *in situ* at the time of sampling.

Electro-resistance particle counters

Developed in the 1940s for blood cell counting, electro-resistance counters are one of the most efficient particle sizing techniques available because they can both count and size particles in a given sample volume. Examples of these instruments are manufactured by Beckman-Coulter and Micromeritics. The method is based on measuring changes in electrical resistance across a small aperture in a glass barrier separating two electrodes immersed in an electrolyte (Fig. 2.2). The increase in resistance between the two electrodes when a particle passes through the orifice, when calibrated, is a measure of the particle volume. Reduced pressure generated by a mercury manometer is employed to draw samples through the orifice from the stirred beaker and the volume of suspension drawn through the aperture is accurately measured to allow the system to count and size particles on a per volume basis. Several thousand particles per second are individually counted and sized with great accuracy and the method is independent of particle shape, colour and density.

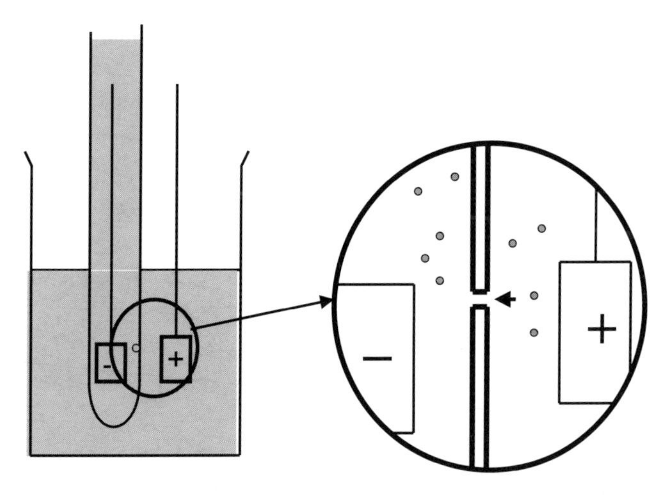

Fig. 2.2 Schematic diagram of the sensing zone in an electroresistive particle sizer.

Although almost unique in the ability to both count and size particles automatically, the electro-resistance method has some limitations. First, this method detects the volume of a particle irrespective of its shape and assigns each particle the diameter of a spherical particle with the equivalent volume (the ESD). Thus cylindrical particles are given a dimension which is shorter than the long axis and greater than the actual cross section. It should, however, be remembered that similar problems also apply to sieves in that cylindrical particles can pass through a mesh lengthwise. Most importantly, the suspension of particles has to be sufficiently diluted so that only individual particles pass through the aperture. An individual aperture is only sensitive over a limited size range of 2% to 60% of the aperture diameter. The maximum theoretical size range that can be measured is 0.4–1200 μm but this

Table 2.2 Particle size ranges that can be measured with each Coulter aperture.

Nominal aperture diameter (μm)	Nominal particle diameter range (μm)
15	0.4–9
20	0.5–12
30	0.6–18
50	1.0–30
70	1.4–42
100	2.0–60
140	2.8–84
200	4.0–120
280	5.6–168
400	8.0–240
560	11–336
1000	20–400
2000	40–1200

can only be achieved by several sequential analyses of each sample using different apertures (see Table 2.2).

In practice, it is usually possible to find a single aperture that covers the required size range, e.g. for examining silt samples screened at 63 μm, a 100 μm aperture allows measurements in the range 2–60 μm. The presence of particles in the sample that are larger than the aperture will lead to blockages so all samples will need to be pre-screened in some way. Work with the smallest particles (aperture of 50 μm and less) is extremely demanding in terms of minimising background counts in the suspending medium and in reducing electrical noise to an absolute minimum. Although the Coulter Multisizer^{TM} can size grains up to 1200 μm and should, in theory, be capable of sizing sand, in practice, the problems of maintaining heavy particles in suspension make sand difficult to analyse. An upper limit of 100 μm is perhaps reasonable, though the possibility exists of increasing this with the addition of glycerol to the suspending electrolyte to increase viscosity and density. Additionally, with large grain sizes, the large sample volumes required to achieve statistically significant counts also become a limiting factor. Detailed operating procedures and methods are provided in the manuals for particular instruments.

The Coulter Multisizer^{TM} can resolve a particle population into 256 size intervals. Using spreadsheets, electroresistive data can easily be recalculated into groups to approximate Wentworth grade intervals. However, it must be remembered that sieve data generate mass in intervals and the electroresistive techniques generate volume distributions. For sediments, volumes could be converted crudely to mass by the assumption of a typical density.

Laser diffraction particle sizing

Particle sizing by laser diffraction has been developed over the last 30 years by a number of instrument manufacturers (Malvern, Cilas, Beckman-Coulter). The method relies on the measurement of near forward angle light scattering (FALS),

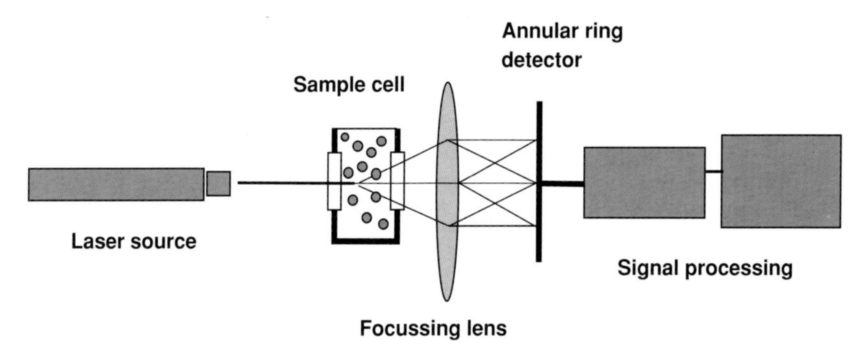

Fig. 2.3 Schematic diagram of a laser diffraction instrument based on the Malvern Instruments apparatus.

also called low angle light scattering (LALS), which is generated when a particle suspension is illuminated with a laser source (Fig. 2.3). The systems then employ Mie and/or Fraunhofer theory to deconstruct the radially symmetrical diffraction pattern into a size distribution. Early instruments operated over restricted size ranges but current instruments can now handle large size ranges, e.g. 0.05–3500 μm (Malvern Instruments) and 0.04–2000 μm (Beckman-Coulter) in one measurement. But, as mentioned in relation to the electro-resistance method, the problems of keeping heavy sand grains suspended in the optical sensing zone in representative numbers may cause errors. This is because in a sand/silt sample, sand comprises relatively low numbers, but a significant proportion of the mass or volume. The large dilutions required for laser diffraction quickly introduce errors in the sampling of sand grains causing a large bias in the mass distribution. For this reason, experience suggests that for samples comprising sand and silt, pre-screening at 63 μm with a sieve is an advantage from the sample-handling viewpoint. Laser diffraction can then measure the silt material at low dilution. With the appropriate sample handling systems, laser diffraction systems can also measure the size distribution of sands up to about 1 mm diameter though not if silt is present. It is relatively easy, using either spreadsheets or the instrument manufacturers' software, to recombine the data into one size distribution.

Sub-sampling of sediments

Laser diffraction, like the electroresistance technique, requires dilute suspensions, otherwise the laser beam is completely obscured or multiple scattering (particle–particle scattering) occurs. A major consideration, therefore, is the preparation of representative sub-samples. Sub-sampling sand is relatively easy through standard cone and quartering techniques or using sample splitters. If the sediment has been wet-sieved at 63 μm, the silt fraction will be in suspension and a representative volume may need to be taken for further dilution before analysis. Settled material needs to be uniformly suspended but care needs to be taken when stirring suspensions to make sure that the mixing is gently turbulent and not laminar, otherwise centrifugal effects can lead to the grading of material within the container. One answer to this could be that the container holding the sample is fitted with baffles or vanes. Some instruments are linked to sample handling systems that allow the optimum dilution of a suspension to be obtained before it is presented to the diffraction instrument. Replicates can be analysed to test reproducibility but care needs to be exercised to ensure that the samples are not consistently biased as well. Wet mud samples that are not screened by wet sieving should be thoroughly mixed before sub-samples are taken for analysis.

Microscopic examination and characterisation

There are a number of microscopic approaches to particle characterisation that can be suitably applied to sediment analysis. Microscopy can be used either as an adjunct to sample preparation prior to other methods of analysis (i.e. to assess the degree of aggregation, or the nature of particles) or as a direct method of counting and sizing.

With an eyepiece graticule calibrated against a stage micrometer, sand size particles can be observed easily at 100× or 200× magnification (obviously a calibration is required for each magnification). Ensuring that samples are representative is an important factor. Dry sand grains are effectively small marbles and are often difficult to mount for microscopic examination. One approach is to take a small piece of transparent adhesive tape (Sellotape™) e.g. 15 × 25 mm, and use forceps to pass the tape through the sample so that a single layer of particles adheres to the sticky side. The tape can then be placed on a slide, particles upwards, with a cover slip laid over the top to keep the tape flat. Silt size particles are best viewed wet using a cavity slide and require higher magnification to resolve. If allowed to dry, samples of silt and clay can aggregate and individual particles become difficult to resolve. Filtration of small particles on polycarbonate membranes (e.g. Whatman's Cyclopore™ & Millipore's Isopore™ range) is another method of presentation for microscopy, these membranes being sufficiently transparent to light for microscopy and inert to most stains.

Whilst microscopy is an invaluable aid in determining the nature of samples, sizing of particles by this method is time-consuming because of the number of particles that need to be counted to achieve reliable information. Experiments by Kennedy and Mazullo (1991) suggested that measuring 50–60 particles from a well-sorted sample with a mean diameter in the region of 100 μm could define the mean within 10%. However, in multi-modal samples or where information on the distribution is required, more particles need to be counted: possibly 200–300 or even as much as 1000 for some samples.

Image analysis

Automation of optical microscopy can be achieved, either partially or completely, using image analysis software packages combined with digital imagery (Kennedy &

Mazullo, 1991; Soille, 1999). Being automated, the task of characterising sufficient particles to allow the size distribution to be determined with confidence is easier and the results more objective. Bernard et al. (1999) employed scanning electron microscopy (SEM) and a motorised sample stage with image analysis to automatically accumulate particle size characteristics. Parallel elemental composition gathered for each particle using an X-ray fluorescence facility allowed relationships between the size and composition to be explored.

Particle stains

There are various stains and fluorophores which can be employed in conjunction with microscopy to allow a wide range of specific components within the sediment mixture to be identified. For example, Periodic and Schiff (PAS) (Whitlach 1981) and Rose Bengal allow organic components to be distinguished from mineral grains. A stain specific to ligneous plants can allow terrestrial material to be distinguished from marine plants (Pocklington & Hardstaff, 1974). Bacteria can be visualised using acridine orange (Daley, 1975) or DAPI (Porter & Feig, 1980) and epifluorescence microscopy.

Presentation and analysis of grain size data

The results of sediment analyses can be portrayed in a number of ways. In the simplest form, sediment can be assigned a verbal description based on the Wentworth grade scale, e.g. 'coarse sand' or, 'pebbles and greater'. On the basis of analytical information, this can be taken a stage further and given a numerical value using criteria relevant to the study, e.g. 85% (by weight) > 1 mm (coarse sand and larger). At the other extreme, sediment size and compositional properties can be incorporated with environmental and biological information using multivariate statistical methods such as multi-dimensional scaling (MDS) to describe and differentiate between environments (Warwick et al., 1991). Falling within this spectrum, there are a number of approaches to the presentation of sediment properties; a selection is discussed below.

Ternary diagrams

Ternary diagrams allow a sediment sample to be characterised with respect to three variables, for example, the percentages of sand, silt and clay determined by sieving. Equally, the three variables could be pebbles, gravel and sand. Using triangular graph paper, it is possible to present the percentage content of three variables as a single point. In practice, each of the three vertices of the triangular graph is labelled for one of the three variables. The distance from each base to its corresponding vertex is taken as 100%. The location of a sample with 60% sand, 30% silt and 10% clay would be as shown in Fig. 2.4a. When a number of sediments are plotted on the same graph, it is often possible to classify the sediments into groups that

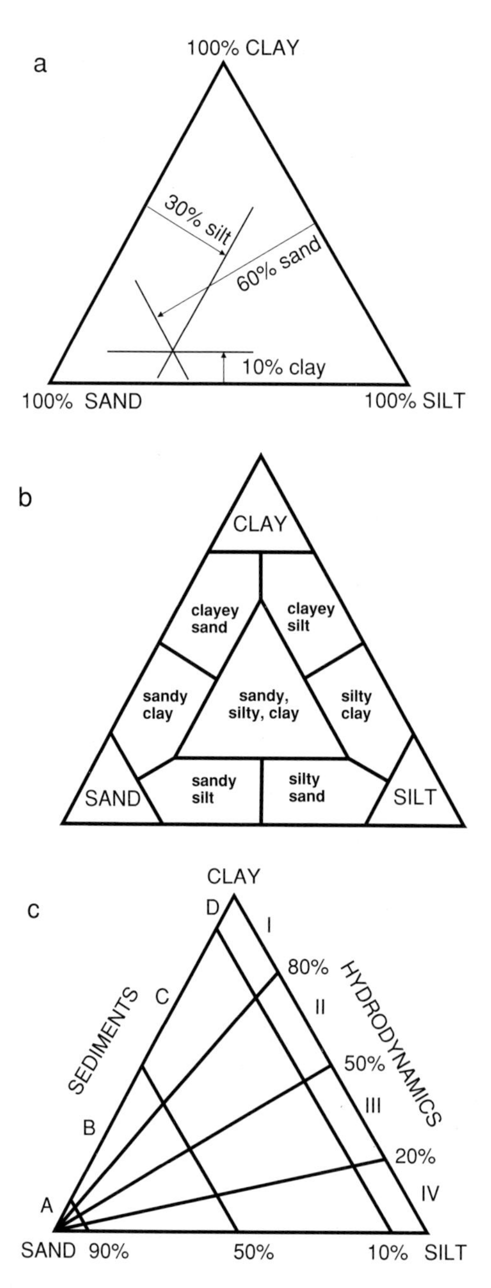

Fig. 2.4 Examples of ternary diagrams for plotting three variables describing sediments. (a) Showing how a sample with a 60, 30 and 10% (sand, silt, clay) composition is located within the plot. (b) Example of classification after Shephard (1954). (c) Classification including an interpretation of the hydrodynamic environment after Pejrup (1988).

have regional ecological significance. The triangular graph can then locate field samples into 'classes' for descriptive purposes. Figure 2.4b shows an example of an arbitrary but intuitive classification scheme by Shepard (1954). Pejrup (1988) has proposed a revised set of classes (Fig. 2.4c) where sediments are split into four classes (A–D) depending on their sand–silt composition. The corresponding silt–clay gradient is split into four equal classes (I–IV) indicative of hydrodynamic conditions. This concept is further expanded by Flemming (2000).

Histograms

Histograms are by far the most visually illustrative method of portraying grain size distributions and are employed as a matter of routine by most of the instrumentation manufacturers (e.g. Coulter, Malvern). They do not, however, lend themselves to further statistical analysis, except, perhaps, that the mode, or modes, (most common sizes) can be observed. Typically, size intervals used by instrumentation manufacturers, and those of the Wentworth grade sieves, employ geometric rather than arithmetic intervals to increase resolution in the smaller sizes. Since the cubic relationship between diameter and mass (or volume) skews mass–frequency relationships towards the largest particles, the use of logarithmic scales tends to transform sediment distributions into bell-shaped distributions which can be easily compared. Conventionally, size information is plotted on the x-axis and frequency (wt%) on the y-axis. Historically, phi intervals were plotted increasing to the right so that the true (millimetre) size decreased to the right but this somewhat anomalous convention is currently being questioned. McCave and Syvitski (1991) promote the plotting of log size in μm or mm increasing from left to right on the lower side of the diagram and phi parameter (if required) increasing to the left along the top (see Fig. 2.5). This convention will be followed from here.

Cumulative frequency curves and distribution statistics

The use of cumulative frequency curves has been described in almost every sedimentology text of the last few decades. This form of presentation, effectively stacking each bar of the histogram vertically upwards by the amount of the previous bar, has virtually no presentational value in that even quite significant differences in a histographical presentation are difficult to resolve visually from a cumulative plot. The principal advantage of this construction is that it allows percentile values to be extracted (see Fig. 2.6) in order that statistical parameters such as coefficients of sorting and measures of skewness and kurtosis (Lindholm, 1987; Buller & McManus, 1979) can be calculated.

One factor that needs to be considered here, and it stems from the convention to plot histograms with particle size increasing left to right, is that, to achieve a conventional (increasing from left to right) sigmoid, frequency-distribution, the plot needs to be cumulative 'per cent finer' rather than the 'per cent coarser' which is

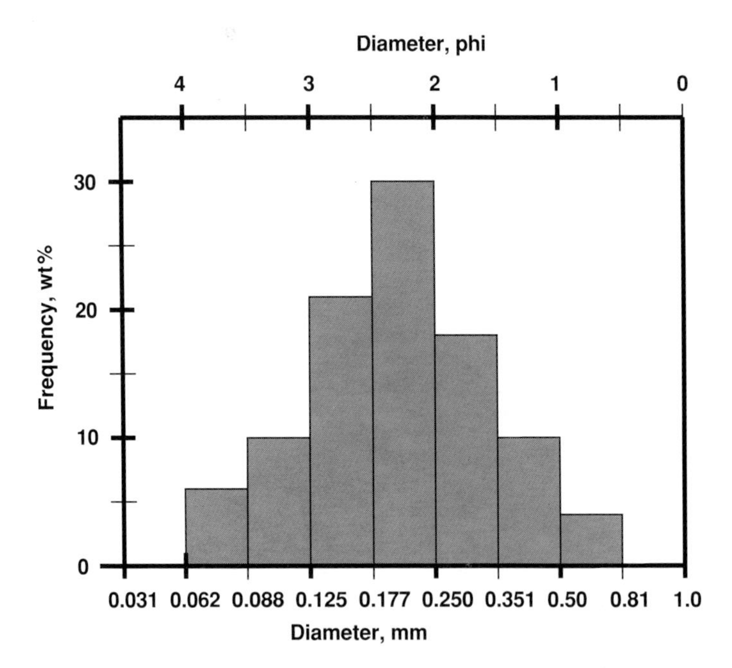

Fig. 2.5 A histogram, frequency–weight plot of the data from Fig. 2.1 employing both grain diameter in mm on a logarithmic scale and phi units on the *x*-axes.

invariably presented in older textbooks. A per cent finer plot is the inverse of the per cent coarser and is the reverse of the way in which, for example, sieve data are collected (see Fig. 2.1).

Although the graphical construction of the cumulative frequency curve allows manual extraction of percentile values, modern data processing routines using computing technology can automate the fitting of a curve to the data and the subsequent calculation of the statistical parameters. The standard, sigmoid, cumulative-frequency curve is well fitted by the expression:

$$y = a/(1 + b\mathrm{e}^{-cx})$$

Using a spreadsheet system such as Microsoft® Excel, for example, a set of particle size data pairs (x = size, y = cumulative per cent finer) can be fitted to the above expression using Solver® to refine the fitting parameters (a, b and c) iteratively and minimise the least-squares error. Once the best fit has been derived, x values equivalent to specific percentiles can be extracted automatically and inserted into the chosen statistical parameters (Buller & McManus, 1979).

Alternatively, software such at the US Geological Survey's SEDSIZE program, which is freely available on the Internet (http://water.usgs.gov/software/sedsize.html) will perform these calculations. Although DOS-based and 'unfriendly' by Windows™ standards, this is a robust piece of code. The program takes size distribution data derived from sieves or other analytical sources and

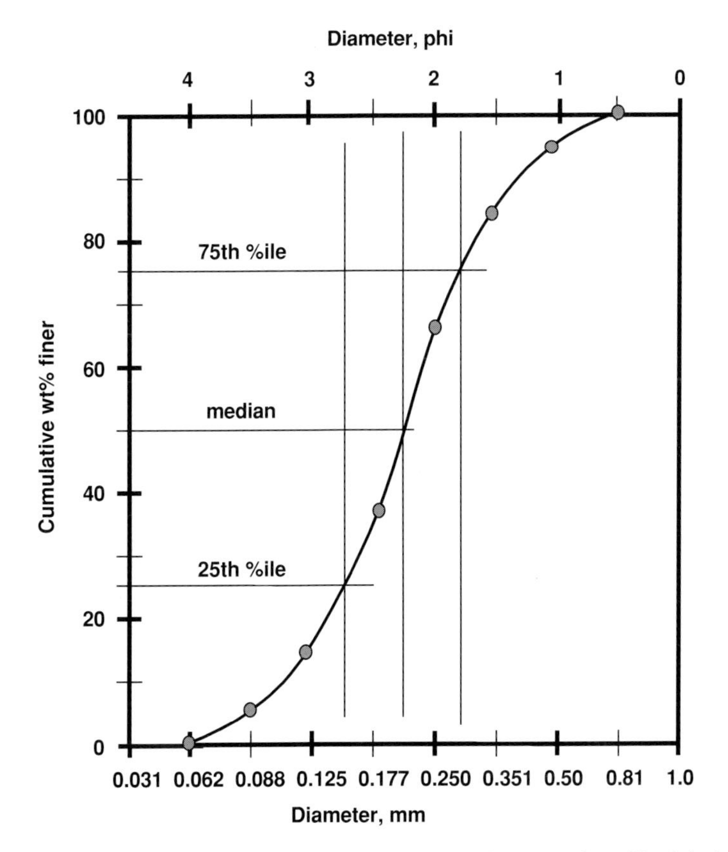

Fig. 2.6 A cumulative, 'per cent finer', frequency–weight plot of the data from Fig. 2.1 showing how percentile values are extracted.

outputs conventional statistical parameters based on millimetre (Trask, 1930) or phi notation (Krumbein,1938; Folk, 1966); it will also give percentages in the Wentworth grades of sand, silt and clay.

Statistical parameterisation of particle size data

The descriptive parameters that have been developed to describe features of sediment size distributions are analogous to the mean and standard deviation employed with the normal distribution in conventional statistics. The parameters fall into the following categories:

- measures of central tendency (median, mean, mode)
- measures of scatter about a central value (dispersion, deviation, sorting)
- measures of the degree of asymmetry (skewness)
- measures of the degree of peakedness (kurtosis)

Methods for calculating a number of descriptive statistics are given in Table 2.3. Clearly, values derived by one equation, e.g. for dispersion, are not comparable with

Table 2.3 Examples of statistical measures used to define grain size distributions. Adapted from Buller and McManus (1979) with the permission of Cambridge University Press.

System	Median	Mean	Dispersion/sorting	Skewness	Kurtosis
Metric	$Md = P_{50}$	$M = \frac{P_{25} + P_{75}}{2}$	$QD_a = \frac{P_{25} - P_{75}}{2}$	$Sk_a = \frac{P_{25} + P_{75} - 2Md}{2}$	$Kq_a = \frac{P_{25} - P_{75}}{2(P_{10} - P_{90})}$
			$So = (P_{25}/P_{75})^{1/2}$	$Sk = (P_{25} P_{75})/(Md)^2$	
Phi	$Md = \varphi_{50}$	$M_\varphi = \frac{\varphi_{84} + \varphi_{16}}{2}$	$\sigma_\varphi = \frac{\varphi_{16} - \varphi_{84}}{2}$	$\alpha_\varphi = \frac{M_\varphi - Md_\varphi}{\sigma_\varphi}$	$\beta_\varphi = \frac{1/2(\varphi_{95} - \varphi_5) - \sigma_\varphi}{\sigma_\varphi}$
		$M_Z = \frac{\varphi_{84} + \varphi_{50} + \varphi_{16}}{3}$	$\sigma_I = \frac{\varphi_{16} - \varphi_{84}}{4} + \frac{\varphi_5 - \varphi_{95}}{6.6}$	$Sk_I = \frac{\varphi_{84} + \varphi_{16} - 2\varphi_{50}}{2(\varphi_{16} - \varphi_{84})} + \frac{\varphi_{95} + \varphi_5 - 2\varphi_{50}}{2(\varphi_5 - \varphi_{95})}$	$K_G = \frac{\varphi_{95} - \varphi_5}{2.44(\varphi_{25} - \varphi_{75})}$

Md, median; *M*, mean; QD_a, arithmetic quartile deviation; *So*, Trask sorting coefficient; Sk_a, arithmetic quartile skewness; *Sk*, geometric quartile skewness; Kq_a, arithmetic quartile kurtosis (Trask, 1930; Krumbein & Pettijohn, 1938; Krumbein, 1939); M_φ, phi mean diameter; σ_φ, phi deviation measure; α_φ, phi skewness measure; β_φ, phi kurtosis measure (Inman, 1952); M_z, mean size; σ_I, inclusive graphic standard deviation; Sk_I, inclusive graphic skewness; K_G, inclusive graphic kurtosis (Folk & Ward, 1957). NB: These equations have been adapted to take account of the fact that plotting cumulative frequency against size in mm increasing from left to right (McCave & Syvitski, 1991) requires a per cent finer plot, rather than per cent coarser, in order to generate a 'conventional' sigmoid curve increasing from left to right.

values calculated using another expression for dispersion and certainly, except for median values, metric and phi-based parameters are not interchangeable. Buller and McManus (1979) provide a thorough overview. These methods are limited when dealing with multi-modal distributions that arise when a number of processes, possibly biological and physical, are affecting a sediment. It may be that the principal modes are the only usable descriptors in these circumstances.

Another approach to the calculation of statistical parameters is based on the calculation of moments which lends itself to automated, computerised methods, but requires information over the full size spectra (i.e. closed distributions). In sediment analysis, the information in the 'tails' typically has the greatest uncertainty, or is lacking altogether. This method is described by Lindholm (1987) and the limitations are reviewed by Buller and McManus (1979).

GIS mapping and correlation of parameters

The spatial mapping of sediment properties on a geographical basis, particularly employing GIS (geographical information systems) to overlay numerous parameters, (e.g. sediment characteristics, biological information, chemical information, bathymetry and hydrographic data), in a coordinated spatial reference, can be very informative (DeMers, 1997; Burrough & McDonnell, 1998). Distributions mapped in this way often provide good visual relationships between sediment characteristics and physical or biological factors such as bathymetry and currents in deep water or wave action, desiccation and algal distributions on intertidal sediments. With spatially referenced data storage and visualisation techniques, remotely sensed data (see Chapter 3) can be incorporated and used to inform process models that draw on a number of components within the GIS to generate further insight (Gittings, 1999).

2.3 Sediment characterisation

Bulk and dry density, water content, porosity and permeability

Measurements of the mass physical properties (e.g. bulk and dry density) and the porosity of sediment, i.e. in submerged sediment, the space between the grains that is occupied by water, are closely inter-related. Porosity is affected by compaction and grain size and is one of the factors measured *in situ* by geophysical methods of sediment characterisation (Dunn et al., 1980). The mass physical properties of a sediment are related to its mechanical strength and behaviour. For example, critical erosion shear stress is related to sediment dry density (Delo, 1988). Systematic relationships between bulk density, water content and sediment composition often reflect environmental conditions.

Care must be taken when examining sediment properties as a function of core depth. Bird and Duarte (1989) and Flemming and Delafontaine (2000) show how vertical distributions of properties of sediments such as particulate organic carbon (POC) and bacterial numbers can be misinterpreted when related to sediment mass rather than volume. This is because the bulk density of sediment (mass per unit volume) changes with compaction (water content) and composition (sand/silt ratio). These authors recommend that parameters associated with sediments, such as contaminants, should be quoted as per unit volume of sediment rather than per unit mass of sediment.

Flemming and Delfontaine (2000) give the following definitions:

Wet bulk density (BD_w) reflects the relationship between the total water-saturated mass (M_t) and the volume of the water-saturated sample (V_t) and is given by:

$$\mathrm{BD_w} = M_t/V_t$$

Dry bulk density (BD_d) is the relationship between the mass of dry solids (M_s) and the volume of the water-saturated sample (V_t) is calculated as:

$$\mathrm{BD_d} = M_s/V_t$$

Water content (absolute water content, W_a) is given as the mass of water (M_w) in a sample as a percentage of the original wet mass (M_t) of sediment and is the preferred unit, thus:

$$W_a = (M_w/M_t) * 100$$

Thus absolute water content never exceeds 100%, unlike relative water content (W_r, given by $(M_w/M_s) * 100$) which can result in water content of several hundred per cent (which is illogical). These values can all be measured directly using gravimetric means. Absolute water content (W_a) has been shown to be a universal master variable by which differences between sediments from various environments can be normalised (Flemming & Delafontaine, 2000). For example, the relationship between water content and dry bulk density from a wide range of environments is described by a single regression with a high correlation (Fig. 2.7).

Porosity (ϕ), also termed voids ratio, is directly related to water content in that it represents the fractional pore space in a sediment but is presented in terms of volume:

$$\phi = V_p/V_b$$

where V_p is the volume of pores and V_b is the bulk volume of the sediment. The values are dimensionless and generally quoted as decimals but may be multiplied by 100 to give percentages. Determination of porosity requires a measurement of the volume of water in a sediment sample and the volume of the dry sediment. Just as with water content, both of these can be measured gravimetrically. The weight (and

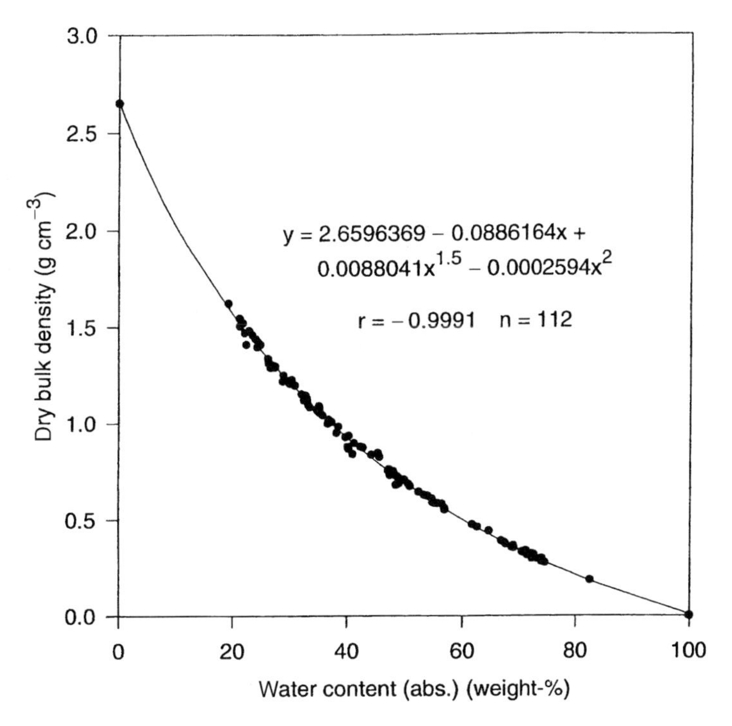

Fig. 2.7 Dry bulk density as a function of water content (W_a) for pooled data from the Wadden Sea, Griefswalder Bodden and Bay of Mont Saint Michel. Reprinted from Flemming and Delafontaine (2000), with permission from Elsevier Science.

thus volume) of water in a sediment can be determined by subtracting the weight of washed (to remove salt) and dried (to constant weight) sediment from the weight of the wet sediment. Washing can be achieved by repeated centrifuging, decanting and resuspension in distilled water. Dialysis in running water is an alternative method of removing the salt. For approximate results, washing can be omitted by calculating the weight contribution of salt from seawater density tables by assuming that the salinity of the interstitial water is the same as the overlying water. The volume of the dry sediment can be estimated from its weight by assuming a density of 2.65 but, for accurate work, the sediment volume of the sample needs to be determined from its actual density. The underlying method for this is the same as the laboratory determination of specific gravity of any material:

(1) The representative sample of wet sediment is placed in a tightly stoppered, pre-weighed, weighing jar and the weight of wet sediment determined.
(2) The sample is then transferred to a centrifuge tube which has been previously weighed in air and weighed suspended but completely immersed in water.
(3) Remove salt by adding distilled water, shaking, centrifuging and pouring away the overlying water. Repeat this two or three times.
(4) The sediment and tube should then be weighed suspended, immersed in water.

(5) Afterwards the tube is dried to constant weight and the dry weight of sediment determined.
(6) The difference between the dry and immersed weights of the sediment is equivalent to its volume (V_s) by displacement of water.
(7) The difference between the wet and the dry weight of sediment is the volume of interstitial water (V_p).

Thus:

$$\phi = V_p/(V_s + V_p)$$

Permeability is a concept taken from petrology, which is important for hydrocarbon yield/flow and to ground water percolation. Permeability is not necessarily related to porosity because the extent of interconnectivity between interstices is not known, nor are the sizes of throats or pores which link them known. For sediments, permeability is understood as the rate at which water under a constant head, h, passes through a cylindrical core of sediment of length L and cross section A. By applying a constant head of water to a core in a sealed apparatus, the coefficient of permeability, P, is given by:

$$P = QL/hAt$$

where Q is the volume of water flow in time t.

Organic matter content

Examination of the compositional elements of sediments will require some knowledge of analytical chemistry or the facility to draw on the services of an analytical chemist. However, since the parameters discussed below are indicators of the biology residing within, or on, the sediment system, they constitute valid topics for this chapter. The organic content of sediment can provide an insight into sediment cohesion, information on the potential nutritional value to deposit feeders, and on the oxygen demand within the sediment.

High temperature oxidation (elemental analysers)

A number of manufacturers (Perkin-Elmer, Thermo, Leco) produce elemental analysers (C, N, O, H, S) in which mg samples of oven-dried (90–100°C) sediment are oxidised at high temperature (1000–1300°C). This is currently the method of choice for determining carbon that is detected as CO_2. For standards and total carbon, the precision can be excellent. However, for sediments, the method is complicated by the difficulty in distinguishing between organic and inorganic carbon. There are two approaches: the first is to use thermal oxidation at a temperature that drives off all the organic material as CO_2 and then to measure the remaining (unaffected?) inorganic carbon with an elemental analyser and determine the organic carbon as the difference between total carbon and inorganic carbon. This is not straightforward because temperatures sufficiently high to oxidise refractory organic material are probably too high to prevent the decomposition of magnesium carbonate. The second (recommended) approach is to use a non-oxidising acid to dissolve the carbonates, ideally without solubilising or volatilising the organic material. Suitable candidates are hydrochloric (HCl, 2–6N), phosphoric (H_3PO_4, 1M) and sulfurous (SO_2, 5% w/v) acids. HCl is recommended by King et al. (1998) following an inter-comparison between 10 laboratories. The remaining organic fraction of the carbon can then be determined directly.

Loss on ignition (LOI)

A simple estimate of the organic content of sediments can be derived from the mass loss on ignition (LOI) in a muffle furnace. Pre-weighed, oven-dried (90–100°C) samples are typically ignited at 450°C for 6–8 hours (or to constant weight). Ignition temperatures ranging from 300°C to 600°C are found in the literature but it is generally considered that carbonates (principally shells) start to decompose at temperatures above 400°C (Weliky et al., 1983). The presence of carbonates can be assessed by looking for effervescence on the addition of dilute acid (e.g. 1 M HCl). Even for carbonate-free sediment, however, the LOI method is operationally dependent on temperature. Mook and Hoskin (1982) showed that for organic-free sediments with an appreciable clay content (83%), significant weight loss (10–20%) at temperatures over 400°C could be attributed to loss of 'structural water' within the clay matrices. For samples with lower clay contents, 'structural' or 'lattice' water may be insignificant and several workers have achieved very good relationships between loss on ignition values and POC (e.g. Craft et al., 1991; $R^2 = 0.990$). Typical factors relating POC to LOI fall between 0.5 and 0.2 (Table 2.4) though often with a large spread at a given site.

Table 2.4 Examples of conversion factors determined from particulate organic carbon (POC) and loss on ignition (LOI) measurements showing that the relationship is site-specific; POC = LOI multiplied by factor.

Reference	Factor	LOI values	Temperature, °C
Craft et al. (1991) Estuarine marsh	0.50	0–10%	450°C, 8 h
Byers et al. (1978) Sediment spiked with chitin compound	0.436	22.9%	475–500°C, 4–6 h
Wirth and Weisner (1988) North Sea sediments	0.24	0–10%	500°C
Morris, et al. (1982); Stephens et al. (1992) Turbid region of Tamar Estuary	0.29	8.2%	300°C, 2 h
Leong and Tanner (1999) Hong Kong, Tolo Harbour	0.197 and 0.165	12.3 and 12.6%	440°C
Sutherland (1998) Fluvial sediments (silt and clay), Hawaii	0.14	15–20%	450°C, 16 h

Wet oxidation

Wet oxidation using chromic acid–sulphuric acid mixtures and either titration or coulometric analyses of evolved CO_2 have been employed routinely and are straightforward, but they are time-consuming, use noxious reagents and require a reasonable analytical chemical skill. Wet oxidation methods also underestimate organic carbon by 20–30% in oxic samples (Byers et al., 1978; Leong & Tanner, 1999) and overestimate (140% of example) in anoxic sediments (Leong & Tanner, 1999) due to the presence of other reduced species (e.g. Fe, Mn) which 'consume' oxidant during the titration. This method is now superseded by the high temperature oxidation methods described above.

Chlorophyll

Chlorophyll is the key photosynthetic molecule and is measured as a proxy for photosynthetic biomass both in the water column and, more relevant to this chapter, on the surface of inter-tidal sediment or sub-tidal sediment within the photic zone. The microphytobenthic population provides much of the energy for epibenthic grazers and also influences the stability of the sediment (Paterson, 1989, 1997) through the production of mucopolysaccharides, also known as extracellular polymeric substances (EPS) (see Section 2.3).

There are three key stages in the determination of chlorophyll: (i) sampling and preservation, (ii) extraction of the chlorophyll from the cells and (iii) quantification of the chlorophyll in the extract. Sampling water column chlorophyll usually involves filtering a known volume of water through a GF/C or GF/F filter pad. Samples of sediment are usefully acquired using a small core of known dimensions which can be extruded so that a known depth (e.g. 2 mm) and area of sediment can be taken. Because corers tend to compact soft sediment, the wet weight of the sample should be determined. Water content and bulk density values from replicate samples can then be used to convert wet weight to dry weight or sediment volume. Samples for chlorophyll analysis should be kept in the dark and stored deep-frozen at the lowest temperature available. The chlorophyll molecule degrades rapidly to phaeopigments at ambient temperatures, especially in the presence of sunlight. Mantoura et al. (1997) recommend storage of samples in liquid nitrogen ($-196°C$) immediately after collection though for short periods (e.g. up to 4 weeks) $-80°C$ is sufficient. Various solvents can be used for the extraction of chlorophyll though acetone is the least toxic and is preferred (Wright et al., 1997). Samples should be disrupted for 15–25 seconds using an ultrasonic probe in 5 or 10 ml of 90% acetone in a plastic centrifuge tube, left for 10–20 minutes and then centrifuged at 3000g for 5 minutes to separate the particles from the solvent. Samples should be kept in the dark and as cold as possible during these operations and extracts stored in a freezer until analysis. In a given sample, a number of other chloropigments, carotenoids and chlorophyll degradation products will be present and interfere with chlorophyll when analysed spectrophotometrically. Sediments, in particular, are typically high in chlorophyll degradation products (phaeopigments) because of the accumulation of detrital plant material. The recommended method of analysis for chlorophyll is therefore to use high-performance liquid chromatography (HPLC) to separate and quantify chlorophyll *a* from the mixture of pigments. Spectrophotometric methods of quantifying chlorophyll in extracts are a more practical prospect for the sediment chemist but users must be aware that the results will be compromised by the presence of the other pigments. Cross-calibration of spectrophotometric measurements with some HPLC analyses may provide a means of assessing the reliability of these data. Various spectrophotometric equations employing up to three absorption measurements at specific wavelengths are detailed by Jeffrey and Welschmeyer (1997). Fluorometric analyses of extracts enable greater sensitivity at low chlorophyll concentrations but this is generally not needed for sediment pigments.

EPS carbohydrate

It is now widely accepted that the microphytobenthic population, especially epipelic diatoms, have a large impact on the stability of intertidal and shallow sediments through the exudation of mucopolysaccharide material that effectively binds surficial sediment particles together. Several workers (Black, 1997; Tolhurst et al., 2000) have shown that the critical erosion threshold of cohesive sediment can be dramatically reduced (virtually to the point of zero cohesion) by chemically removing organic material (peroxide digestion), or by poisoning the biota (mercuric chloride, copper sulfate). The erosion threshold can be increased again in direct proportion to EPS added to organic-free sediment (Tolhurst et al., 2000). Not surprisingly, colloidal EPS and chlorophyll are well correlated (Underwood & Smith, 1997, 1998). However, EPS is now considered a valuable measure of intertidal sediment character in its own right and the measurement of colloidal EPS, as carbohydrate, is not difficult. A quantity of wet sediment (e.g. 0.2–0.3 g) (for which the water content should be determined separately) or of dried, or freeze-dried sediment (e.g. 0.1–0.2 g), is weighed into a plastic centrifuge tube and 5 ml of ultra-pure water is added. The resultant slurry is agitated vigorously for 10 seconds using a vortex stirrer and then left to 'extract' for 15 minutes at 20°C. Following this, the slurry is centrifuged at 3000*g* and 1 ml of the supernatant analysed for carbohydrate using the phenol-sulphuric acid assay outlined by Underwood et al. (1995). This method measures total colloidal carbohydrate (which is all that most workers do) but in order to determine the true EPS, the polymeric material within an extract must be separated from the low molecular weight carbohydrate. This is achieved by precipitating the polymer fraction in 70% ethanol and decanting the low molecular weight carbohydrate. After a second wash with 70% ethanol, the polymer is 'resolubilised' in distilled water and assayed for carbohydrate as above. EPS is typically 20–40% of the total carbohydrate.

Temperature

The temperature of sediments would be closely related to that of the overlying water though significant deviations can be experienced on intertidal sediments which can heat or cool considerably depending on ambient weather and insolation. For submerged sediments, a 'high-tech' approach to measuring temperatures can be achieved with digital thermometer probes mounted on bottom-landing vehicles. For most practical purposes, a standard thermometer inserted into the centre of a grab or core sample immediately on recovery will provide a good indication of *in situ* temperature. For field work, a digital thermometer with a metal probe is more robust than a mercury-in-glass thermometer.

Eh and pH measurements

Indicators of acidity and redox balance (reduction–oxidation status) in sediments, pH and Eh, respectively, are geochemically inter-related. Microbial decomposition of organic material firstly consumes oxygen and then reduces nitrate to nitrite and ammonia. If oxygen or nitrate are not replaced through diffusion, continued oxidation of organic material results in the reduction of metal oxides such as Fe and Mn to an ionic form, and of sulfate to hydrogen sulfide giving rise to the familiar 'rotten eggs' smell of some anaerobic sediments. Microbial respiration of organic carbon produces CO_2 which, in solution, is weakly acidic. These processes of early sediment diagenesis are described fully by Berner (1971). Since the sedimentation of organic material tends to be associated with relatively quiescent environments, and the scope for oxygenation of sediments through physical disturbance or diffusion is consequently reduced, anaerobic conditions tend to be associated more with silts and clays than sands.

The pH of the interstitial liquids in cores can be measured using conventional glass electrodes in combination with Hg-$HgCl_2$ or Ag-$AgCl_2$ reference electrodes. A number of manufacturers make pH probes which are suitable for insertion into semi-solid materials. Redox potential is generally measured using an inert Pt electrode in conjunction with a standard hydrogen electrode or a reference electrode of known, fixed Eh. Suitably formed electrodes can be profiled through soft sediment cores using a rack and pinion system to achieve mm depth resolution (Mortimer et al., 1998). An important consideration when working with anaerobic or anoxic sediments is that reduced species will quickly oxidise in the presence of atmospheric oxygen. This can be controlled by storing and manipulating sediment samples in a nitrogen-filled glove box.

In situ *sediment characterisation methods*

As a result of the problems associated with sediment disturbance during sampling and the labour-intensive aspect of manual analyses, a growing number of devices are being developed for *in situ* measurements and observations. Brooke Ocean's, free fall cone penetrometer (FFCPT) can determine sediment structure and shear strength in the surface 2 m of soft sediments. Geotek's SAPPA (Sediment Acoustic and Physical Properties Apparatus) will measure geotechnical properties using a heavy, bottom-landing device that inserts acoustic probes 1.3 m into the sediment. Directly relevant to benthic biology, a number of *in situ* imaging systems allow surface sediments to be visualised under water so that the biota can be assessed and sediment grain sizes and even redox conditions estimated (Rumohr & Schomannn, 1992). Examples of these are Benthos' REMOTS™ and NOAA's SPI (Sediment Profiling Imagery). Both these systems employ a bottom-landing frame to insert a prism and mirror system slowly into the surface 20 cm of the sediment. A camera contained within an underwater housing then photographs the section of sediment exposed by the vertical face of the prism.

2.4 Remote acoustic sensing

Acoustic techniques can generate maps that reveal the geophysical characteristics of the seabed and allow inferences to be drawn regarding the geology and modern day (Holocene) sedimentary processes (see also Chapter 3). These maps can make an important contribution to studies of benthic communities, survey planning and sampling design as well as allowing the evaluation of natural and man-made disturbances of the seabed over large spatial scales. When used in conjunction with 'ground-truthing' methods (e.g. grabs, corers and underwater photography as described earlier), remote acoustic sensing enables the delineation of habitats (mainly physical attributes) and biotopes (habitat and community attributes). This section describes the principal techniques for remote acoustic sensing of seabed topography (bathymetry) and of the physical properties of surface sediments.

Background

Imaging of the seabed was revolutionised in the 1940s when, for the first time, relatively high frequency echo-sounders were positioned in such a way as to insonify a large swathe of seabed (Fish & Carr, 1990). The first, so-called sidescan sonars were rather crude, having low resolution, and could only reliably be used to detect large targets such as shipwrecks. However, over the last 30 years or so, rapid developments in acoustic electronics, largely in support of defence applications, have given rise to instruments that can be applied to environmental monitoring and research needs (Kenny et al., 2003; JNCC, 2001). The principal advances in the technology relate to the precise control of the phase and amplitude properties of the acoustic signal allowing high-resolution (almost photographic quality) images of the seabed to be obtained. Further developments in acoustic mapping technology have been associated with the increase in digital processing power and acoustic beam steering and focusing techniques which generate multiple beams of sound capable of the same high-resolution imaging, but at fast tow speeds (up to 15 knots).

In addition, software applications are continually being developed to provide greater data control and improved visualisation functions. Most systems now support *real-time* visualisation of sonar data as true (geo-corrected) sonograph mosaics.

In general, acoustic remote seabed mapping or sensing instruments fall into one of two generic types, namely (i) low grazing angle *swathe* systems (e.g. sidescan sonars; Fish & Carr, 1990; Newton & Stefanon, 1975; Kenny, 1998; Green & Cunningham, 1998) and (ii) *normal* incidence beam formation echo-sounders (single or multi-beam systems; Foster-Smith & Gilland, 1997; Magorrian, et al., 1995; Loncarevic et al., 1994; Hughes Clarke, 1998; Burns et al., 1985; Chivers et al., 1990; Greenstreet et al., 1996). Systems which process normal incidence beam data for habitat classification purposes are generally referred to as Acoustic Ground Discrimination Systems (AGDS).

The distinction between the two types of sonar is very important since they insonify the seabed in different ways and therefore the output requires different interpretation methods. The low grazing angle (broad-beam), swathe systems insonify, as the name implies, a wide swathe of seabed due to their low grazing angle but the beam is narrow in azimuth (that is, as you look down on the beam it is narrow). In order to achieve the low grazing angle the sonar transducer/receiver has to be close to the seabed and is generally towed rather than mounted on the ship's hull. The advantage of this configuration is that small objects protruding from the seabed cast large acoustic shadows. The acoustic geometry of the sonar footprint therefore makes 'swathe' sidescan systems most suitable for detecting small objects on the seabed including changes in bed roughness.

The echo-sounders, or AGDS, by contrast, have normal incidence beams to accurately quantify changes in bed level. To achieve good object detection the beam geometry must be narrow, which is in direct contrast to the sidescan sonar system. A schematic showing the beam geometry of a typical echo-sounder associated with an AGDS is shown in Fig. 2.8. However, it should be noted that this is a simplification since the actual sonar lobes have very complex shapes which are seldom exactly the same between soundings owing to the subtle changes in the properties of the water from one location to the next. Normal incidence (AGDS) systems fall into two main categories, single and multiple beam configurations. The main generic acoustic systems are described in more detail below, outlining their advantages and disadvantages for various seabed mapping applications.

Low grazing angle swathe systems (sidescan sonar)

Sidescan sonar is defined as an acoustic imaging device used to provide wide-area images of the seabed. The system typically consists of an underwater transducer connected via a cable to a shipboard recording device (Fig. 2.9). In basic operation, the sidescan sonar recorder charges capacitors in the tow fish through the cable. On command from the recorder, the stored power is discharged through the transducers, which in turn emit the acoustic signal. The emitting lobe of sonar energy (narrow in azimuth) has a beam geometry that insonifies a wide swathe of the seabed,

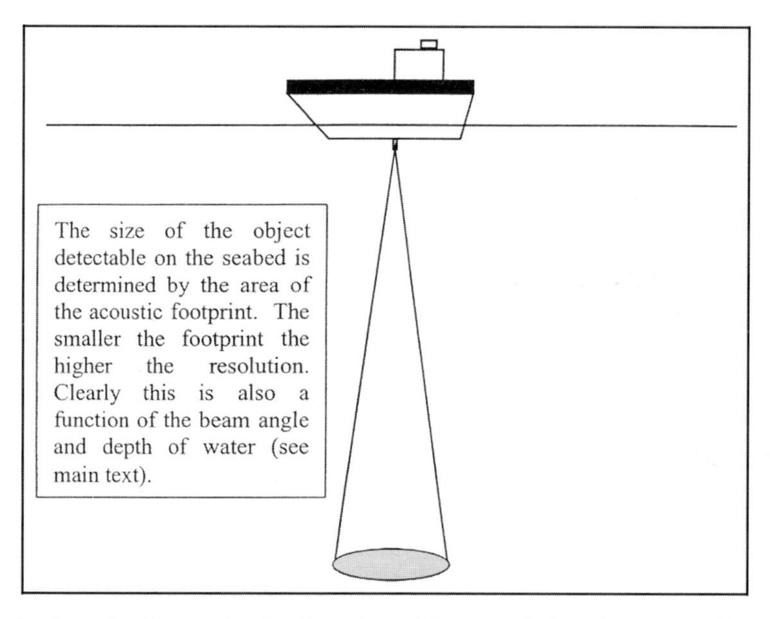

Fig. 2.8 A schematic diagram showing the nature of the acoustic footprint generated by a 'normal' incidence narrow-beam echo-sounder.

Fig. 2.9 A typical sidescan sonar towfish capable of operating at different frequencies, horizontal range and resolution.

particularly when operated at relatively low frequencies, e.g. <100 kHz. Then, over a very short period of time (from a few milliseconds up to 1 second), the returning echoes from the seafloor are received by the transducers, amplified on a time-varied gain curve and then transmitted up to the recording unit. The recorder further processes these signals and calculates the proper position for each signal in the final record (pixel by pixel) and displays these echoes one scan, or line, at a time on electro-sensitive paper and/or a VDU. A typical system design for sidescan sonar is shown schematically in Fig. 2.10.

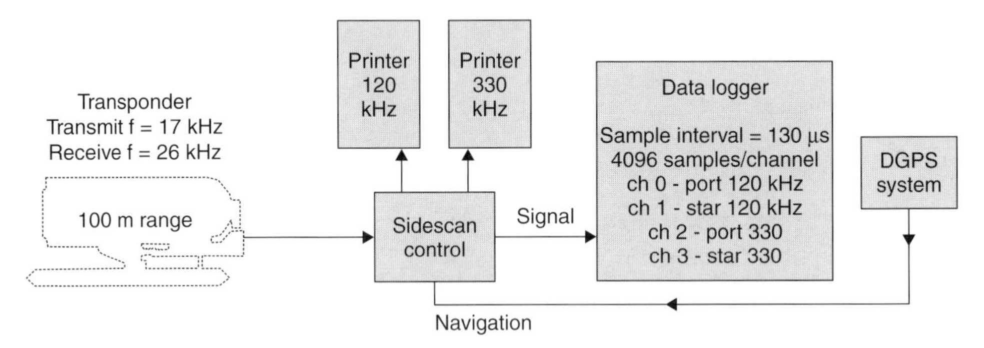

Fig. 2.10 A schematic diagram of a typical sidescan sonar system layout. Reprinted with permission of the Geological Survey of Canada.

Modern high (dual) frequency digital sidescan sonar devices can offer very high-resolution images of the seabed that can detect objects as small as tens of centimetres at a range of up to 100 m either side of the tow fish (total swathe width 200 m), although the precise accuracy will depend on a number of factors. For example, the horizontal range between the transducer and the seabed is affected by the frequency of the signal and the grazing angle of the signal to the bed, which is itself determined by the altitude of the transducer above the seafloor. Some typical limits associated with sidescan sonar are as follows: operating at 117 kHz under optimal seabed conditions and altitude above the seabed, a range of 300 m (total 600 m swathe) can be obtained. In general, resolution increases with decreasing range; for example, 0.1 m resolution is typically obtained with a range of 50 m (100 m swathe) whereas 'only' 0.3 m resolution is obtained at a range of 150 m.

In general, there is a trade-off between the area which can be mapped in a given time and the resolution or detectability of seabed features within the mapped area. For example, a sidescan sonar system operating at 500 kHz can potentially detect features measured in decimetres, but this can only be achieved along a narrow swathe of about 75 m per channel and therefore the typical area which can be mapped in an hour is relatively small. In contrast, the systems which operate at lower frequencies of around 50 kHz have much greater range and can be towed at faster speeds which allow a greater area of seabed to be mapped in a given time (see Table 2.5). The new generation of multi-beam and synthetic aperture sidescan sonars greatly reduce the trade-off between resolution and coverage by using beam steering and focusing techniques to generate several acoustic beams on each side of the sonar fish. Since each beam covers only a small part of the total swathe, these systems allow wide coverage of the seabed at fast tow speeds (10 to 15 knots) but with the resolution of a narrow swathe.

Table 2.5 The object resolution *versus* range for two sidescan sonar systems. Adapted from Kenny et al. (2003).

Range (m)	Spacing between soundings (m) @ 4 knots	120 kHz Sidescan 0.75° beam width (m)	330 kHz Sidescan 0.3° beam width (m)
25	0.07	0.33	0.13
50	0.13	0.65	0.26
100	0.26	1.30	0.52
200	0.52	2.60	1.00
500	1.30	6.50	n/a

It should be noted that the quality (or amplitude) of sidescan sonar data is potentially susceptible to variation when investigating the same area of seabed on successive occasions. For example, the grey-scale (signal amplitude) between swathes for the same area of seabed will be noticeably different if approached from different angles and at different states of the tide. This source of variation in signal amplitude for the same area or type of seabed causes problems when trying to classify the sonograph, since ground truth samples (grabs and underwater cameras) may reveal the seabed to be the same but the sonograph indicates differences. Sidescan does not normally produce bathymetric data. However, sidescan sonar provides information on sediment texture, topography and bedforms and the low grazing angle of the sidescan sonar beam over the seabed makes it ideal for object detection.

A basic understanding of how the sidescan record is generated is essential in order to understand how to interpret the record. Figure 2.11 summarises the relationship between the intensity of the returning echoes and the shape of the seabed (or objects). The returning echoes from one pulse are displayed on the recorder as one single line, with light and dark portions of that line representing strong or weak echoes relative to time. There are many variables such as waves, currents, temperature and salinity gradients, which will affect the sonar data.

Whilst there are efforts underway to make sidescan sonar interpretation an objective and semi-automated process, interpretation at present remains very much a qualitative analysis and requires experienced operators. As highlighted in Fig. 2.11, there are two important attributes of the seabed that will affect the intensity of grey-scale in the sonograph, namely:

- The material properties of the substrata. This will determine the intensity of the acoustic reflectivity of the seabed. For example, in most cases, rock and gravel are stronger reflectors than sand or mud and will therefore show up darker on the sonograph.
- The shape of the seafloor (or topography). Up slopes facing the towfish are stronger reflectors than down slopes.

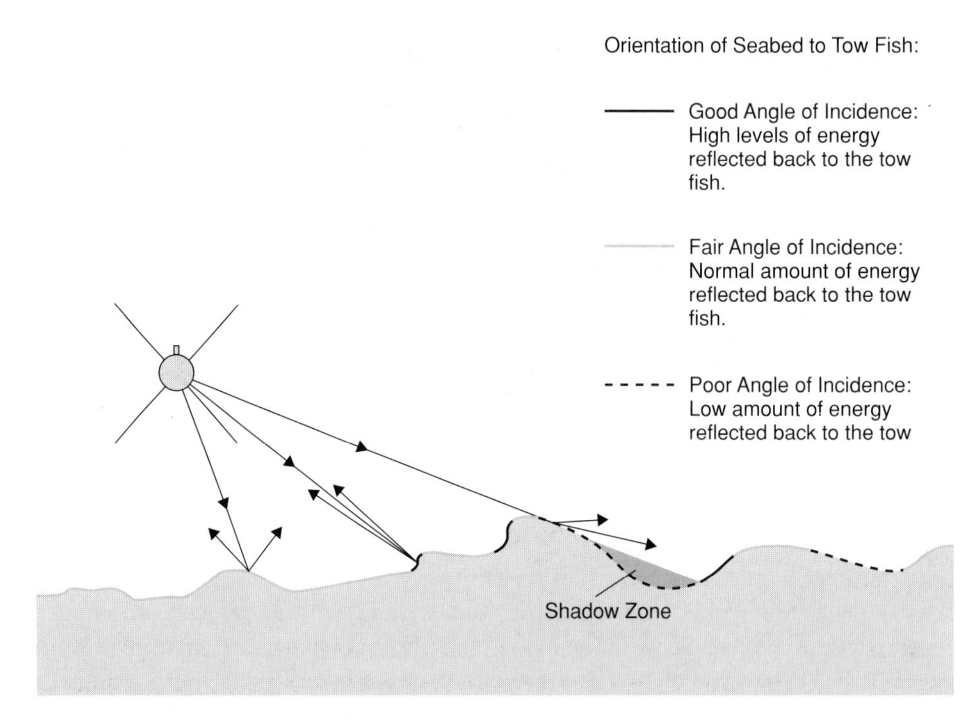

Fig. 2.11 A schematic diagram of sidescan sonar echoes. The key refers to areas of seabed with different reflective properties based upon their angle of incidence relative to the sono-fish.

Since material reflectors and topographical reflectors often produce similar results on the sonograph, the operator must interpret the image carefully in order to differentiate the actual composition of the seabed from noise. Shadows are the single most important feature of sidescan sonographs since they provide the three-dimensional quality to the two-dimensional image (Fig. 2.12a) and the interpreter relies on their position, shape and intensity to accurately interpret most sonar records.

The height of objects on the bed can be estimated from the sonar record using the following equation:

$$H_t = \frac{(L_s \times H_f)}{R}$$

where H_t is the height of the target (m), L_s is the length of shadow cast by the target (m), H_f is the height of the fish above the seabed (m) and R is the distance (m) along the hypotenuse between the towfish and the end of the shadow cast by the object.

During a sonar survey, navigational information from the survey vessel will be incorporated and stored in digital form with the acoustic information. Any other data that may enhance the interpretation such as field notes, bathymetry data and information on the distribution of sediment types (e.g. from Admiralty charts) should also be collated for comparison with the sidescan sonar information.

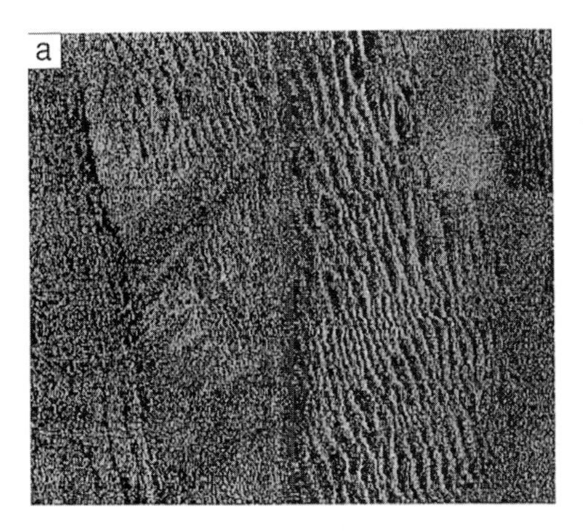

Fig. 2.12 (a) A sidescan sonograph (Datasonics SYS 1500 Digital Chirps™) of an area of seabed in the English Channel (UK) composed of gravel with a veneer of sand waves, which are clearly visible. The sonograph covers an area of about 400 m by 400 m. (b) A photograph of the seabed from the same area as the sonograph, although the photograph covers a much smaller area of about 0.1 m^2 (Brown et al., 2001).

An acoustic survey of an area will generally consist of a number of parallel swathes of information. Using proprietary software, the swathes are geographically located and geocorrected to compensate for the slant-range and layback of the sonar (see below). This process generates a mosaic of the seabed on which all the data are accurately positioned. Any features of particular interest which are identified can be enlarged and further enhanced, if required. Most sidescan sonar processing software will allow other information to be overlain to enhance the sidescan sonar images. It should also allow for annotation of the processed data so that objects and sediment types can be labelled and mapped out.

An estimate of towfish layback can be made by applying Pythagoras's theorem:

$$L = \sqrt{(C^2 - D_f^2)}$$

However, this does not take account of the catenary effect which lessens the layback, but this only becomes a problem for long cable deployments (>15 m). In the equation, L is the layback, C is the amount of in-water cable and D_f is the depth of the towfish.

Single beam echo-sounders (AGDS)

A typical echo-sounder system is made up of a transducer, transceiver and a computer processing system which integrates and controls all the separate components. The simplest echo-sounder is the single beam type which generates a short pulse of sound at a single frequency (usually between 30 kHz and 200 kHz) that travels through the water and rebounds off the seabed. The echo is detected by the transceiver which converts the mechanical energy into an electrical signal that is displayed on a screen or printed on paper. Normal incidence single-beam echo-sounders may be used to obtain a variety of information about the reflective characteristics of the seabed. The size of the transducer will determine the geometry of the sonar beam, particularly its beam width, which is important in defining the size of the acoustic footprint on the seabed. For example, an echo-sounder with a beam angle of 15°, operating in a water depth of 30 m, would insonify an area of seabed with a radius of about 7 m. Although the area insonified by the echo-sounder directly under the vessel tends to be circular in shape, the geometry is complex (having many acoustic side lobes) which, in practice, gives rise to a complex acoustic footprint. This will tend to limit the ability of the system to discriminate seabed characteristics consistently. For a summary description of AGDS, see Table 2.6. The extent to which sound is absorbed or reflected by the seabed, like sidescan sonar, largely depends upon the shape and composition of the seafloor, and these two parameters are often utilised by acoustic ground discrimination systems to classify the nature of the seabed (Burns et al., 1985).

Table 2.6 The three most commonly used AGDS systems to date. Adapted from Kenny et al. (2003).

System	Remarks
QTC – View™	Analysis of first return echo signals only using PCA analysis and presentation of acoustic properties of the seabed in *Q*-space
RoxAnn™	Uses the backscatter information from the first echo to characterise seabed roughness and the reflectance of the second echo via the sea surface to characterise hardness
EchoPlus™	Dual frequency digital signal processing system using first and second echo analysis technique, including compensation for changes in frequency, pulse length and power levels. Unprocessed base band signals are also obtained

RoxAnn™ is an AGDS that has been frequently used for environmental studies in the UK. The system uses electronic methodology to derive values for the tail part of the first return echo (E1) and the whole of the first multiple return echo (E2). While E2 is primarily a function of the gross reflectivity of the sediment, and therefore broadly equates to hardness, E1 is influenced by the small to meso-scale backscatter from the seabed and is therefore used to describe the 'roughness' of the seabed. By plotting E1 against E2 various acoustically defined seabed types can be discriminated (Chivers et al, 1990; Heald & Pace, 1996). With appropriate ground truth calibration, acoustic discrimination systems can be remarkably effective at showing where changes in seabed characteristics occur. However, great caution should be exercised in trying to directly compare readings taken on different surveys, as it is often difficult to be sure whether the echo-sounder is operating under the same conditions, especially when there may be intervals of months or years between the surveys. Also, the lack of swathe coverage of the bed results in the need to undertake extensive spatial interpolation to provide full-coverage maps of the seabed as highlighted by comparing sidescan sonar output with AGDS for the same survey tow (Fig. 2.13). To overcome the lack of swathe coverage of single beam systems multi-single transducer systems placed on booms and multi-beam forming echo-sounders have been developed.

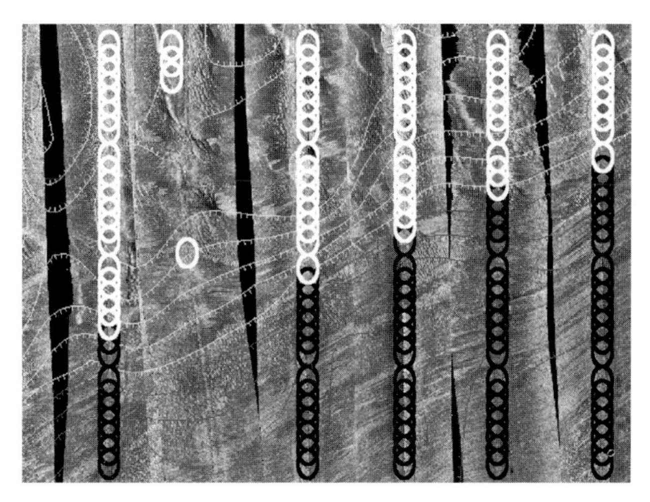

Fig. 2.13 A sidescan sonograph covering an area of about 1 km from left to right clearly showing a boundary in the nature of the seabed (running from lower left to upper right). The seabed in the upper section of the figure is composed of large sand waves and in the lower section the seabed is composed of gravel with sand veneers. In both cases the AGDS (QTC-View) has identified the same boundary by discriminating the two bed types as indicated by white and black circles, respectively (Brown *et al.*, 2000).

Multi-beam echo-sounders (MBES)

A major advantage of both multi- and single-beam systems over sidescan sonar is that they generate quantitative bathymetric data that are amenable to classification and image processing (see Fig. 2.14) but, unlike the sidescan sonars, the narrow beam width (which makes them ideal for quantitative analysis) makes them less useful for object detection, particularly when the objects are small <1 m (Brissette & Hughes Clarke, 1999). A typical high-resolution set-up of MBES would be a 1.5° beam width in 30 m of water providing a 0.8 m diameter footprint.

Two factors will control the potential bathymetric target resolution of a multi-beam echo-sounder, namely (i) the distance between soundings (both across and along tracks) and (ii), the size of the nadir footprint. Table 2.7 presents the results of two MBES systems, one with a 3.3° beam width and the other (higher resolution) with 1.5° beam width (Kenny et al., 2003). Both systems are compared operating under varying conditions of water depth and speed. It should be noted that the higher resolution system (Simrad EM3000^{TM}) is not appropriate for applications in deeper water (>400 m); indeed for detecting objects of about 1 m^2

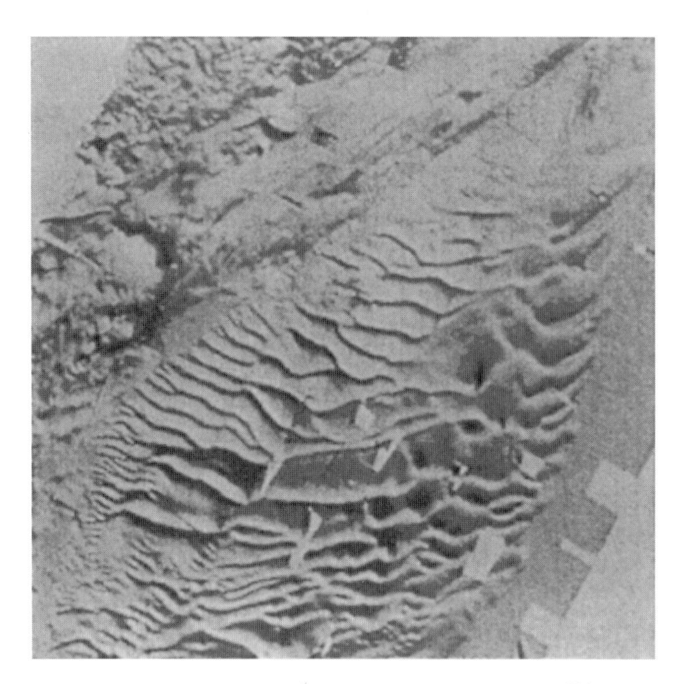

Fig. 2.14 Multi-beam bathymetric data generated using a Simrad EM1000™ system. The data were geocorrected, mosaiced and sun-illuminated and covered an area of about 3 km^2. Reprinted with permission of the Geological Survey of Canada, see also Loncarevic et al. (1994).

Table 2.7 The relative performance of two multi-beam echo-sounder systems. Adapted from Kenny et al. (2003).

Water depth	Spacing between soundings @ 12 knots	EM1000 3.3° beam			EM3000 1.5° beam width		
Metres	Metres	Footprint (m) nadir	Footprint (m) 30°	Footprint (m) 75°	Footprint (m) nadir	Footprint (m) 30°	Footprint (m) 75°
50	1.6	2.9	3.3	12	1.3	1.5	5.0
100	3.2	5.8	6.6	24	2.6	3.0	10.0
200	6.4	11.6	13.2	48	5.2	6.0	n/a
500	16	29	33	n/a	n/a	n/a	n/a
1000	32	58	66	n/a	n/a	n/a	n/a

the optimum operating conditions are survey speeds of up to 12 knots in 50 m of water.

Multi-beam systems differ significantly from multiple single beam transducer systems configured on booms, in their ability to beam form, thereby allowing them to cover an area theoretically up to 12 times the water depth in a single swathe. However, in practice, the effects of transducer roll, signal refraction and attenuation limit the area that MBES can cover, particularly if the survey is required to meet International Hydrographic Office bathymetric tolerance standards.

2.5 Conclusions

Clearly, the choice of acoustic system will depend on the objectives of the survey and the size of the area to be mapped. For example, for broadscale mapping at continental shelf scale (typically in 200 m of water), where relatively large geological features such as sand waves and reefs are of interest, multi-beam echo-sounders are often the preferred choice when quantitative bathymetric data are required in conjunction with object detection on the scale of tens of metres. However, for inshore areas in water depths of up to about 50 m, where there may be a need for identifying small (<10 m) habitat features, a combination of MBES and sidescan sonar may be preferable in order to ensure that both quantitative bathymetric data (1 m resolution) and qualitative high-resolution habitat relief data (10 cm resolution) are obtained.

Commercial suppliers – Disclaimer

Inclusion or exclusion of commercial suppliers in this chapter does not imply recommendation or endorsement by the authors, PML or CEFAS.

References

Allen, T.E. (1990) *Particle size measurement*. Chapman and Hall, London.

Bale, A.J. & Morris, A.W. (1998) Organic carbon in suspended particulate material in the North Sea: effect of mixing resuspended and background particles. *Continental Shelf Research*, **18**, 1333–1346.

Barnett, P.R.O., Watson, J. & Connelly, D. (1984) A multiple corer for taking virtually undisturbed samples from shelf, bathyal and abyssal sediments. *Oceanologica Acta*, **7**, 399–408.

Bernard. P., Eisma, D. & Van Grieken, G. (1999) Electron micro-probe analysis of suspended matter in the Angola Basin. *Journal of Sea Research*, **41**, 19–33.

Berner, R.A. (1971) *Principles of Chemical Sedimentology*. McGraw-Hill, New York.

Bird, D.F. & Duarte, C.M. (1989) Bacteria-organic matter relationship in sediments: a case of spurious correlation. *Canadian Journal of Fisheries and Aquatic Sciences*, **46**, 904–908.

Black, K.S. (1997) Microbiological factors contributing to erosion resistance in natural cohesive sediments. In: *Cohesive Sediments* (Eds. N. Burt, R. Parker & J. Watts), pp. 231–244. John Wiley & Sons, Chichester.

Blomqvist, S. (1991) Quantitative sampling of soft-bottom sediments: problems and solutions. *Marine Ecology*, **72**, 295–304.

Brissette, M.B. & Hughes Clarke, J.E. (1999) *Sidescan versus multibeam object detection and comparative analysis*. Ocean Mapping Group, University of New Brunswick, Canada. http://www.omg.unb.ca.

Brown, C.J., Cooper, K.M., Meadows, W.J., Limpenny, D.S. & Rees, H.L. (2000) An assessment of two acoustic survey techniques as a means of mapping seabed assemblages in the eastern English Channel. *ICES*, CM. 2000/T:02.

Brown, C.J., Hewer, A.J., Meadows, W.J., Limpenny, D.S., Cooper, K.M., Rees, H.L. & Vivian, C.M.G (2001) Mapping of gravel biotopes and an examination of the factors controlling the distribution, type and diversity of their biological communities, *Science Series Technical Report* 14, p. 43. Centre for Environment, Fisheries, Aquaculture Science, Lowestoft, UK.

Buller, A.T. & McManus, J. (1979) Sediment sampling and analysis. In: *Estuarine Hydrography and Sedimentation* (Ed. K.R. Dyer), pp. 87–130. Cambridge University Press, Cambridge.

Burns, D., Queen, C.B. & Chivers, R.C. (1985) An ultrasonic signal processor for use in underwater acoustics. *Ultrasonics*, **23**, 189–191.

Burrough, P.A. & McDonnell, R.A. (1998) *Principles of Geographical Information Systems*. 2nd Edition, xiii, p. 333. Oxford University Press, Oxford.

Byers, S.C., Mills, E.L. & Stewart, P.L. (1978) A comparison of methods of determining organic carbon in marine sediments, with suggestions for a standard method. *Hydrobiologica*, **58**, 43–47.

Chivers, R.C., Emerson, N. & Burns, D. (1990) New acoustic processing for underwater surveying. *Hydrographic Journal*, **42**, 8–17.

Craft, C.B., Seneca, E.D. & Broome, S.W. (1991) Loss on ignition and Kjeldahl digestion for estimating organic carbon and total nitrogen in estuarine marsh soils. *Estuaries*, **14**, 175–179.

Craib, J.S. (1965) A sampler for taking short, undisturbed cores. *Journal du Conseil Permanent International pour l'Exploration de la Mer*. **30**, 34–39.

Daley, R.J. (1975) Direct counts of aquatic bacteria by a modified epi-fluorescence technique. *Limnology and Oceanography*, **20**, 875–882.

Decho, A.W. (1990) Microbial exoploymer secretions in ocean environments: their role(s) in food webs and marine processes. *Oceanography and Marine Biology, Annual Review* (Ed. H. Barnes) **28**, 73–153.

Delo, E.A. (1988) *Estuarine Muds Manual*. Report SR 164. Hydraulics Research, Wallingford, UK.

DeMers, M.N. (1997) *Fundamentals of Geographic Information Systems*. xvii, p. 486. Wiley, Chichester.

Dunn, I.S., Anderson, L.R. & Kiefer, F.W. (1980) *Fundamentals of Geotechnical Analysis*. Wiley, New York.

Dyer, K.R. (1979) Estuaries and estuarine sedimentation. In: *Estuarine Hydrography and Sedimentation* (Ed. K.R. Dyer), pp.1–18. Cambridge University Press, Cambridge.

Fish, J.P. & Carr, A.H. (1990). *Sound Underwater Images: A Guide to the Generation and Interpretation of Side-scan Sonar Data*. American Underwater Search and Surveys Ltd, p. 189. Lower Cape Publishing, Orleans, MA.

Flemming, B.W. (2000) A revised textural classification of gravel-free muddy sediments on the basis of ternary diagrams. *Continental Shelf Research*, **20**, 1125–1138.

Flemming, B.W. & Delafontaine, M.T. (2000) Mass physical properties of muddy intertidal sediments: some applications, misapplications and non-applications. *Continental Shelf Research*, **20**, 1179–1197.

Folk, R.L. (1966) A review of grain-size parameters. *Sedimentology*, **6**, 73–93.

Folk, R.L & Ward, W.C. (1957) Barazos River bar: a study in the significance of grain size parameters. *Journal of Sedimentary Petrology*. **27**, 3–26.

Foster-Smith, R. & Gilland, P. (1997). Mapping the seabed. *CoastNET*, **2**, 1.
Gittings, B.M. (Ed.) (1999) *Integrating Information Infrastructures with Geographical Information Technology* (Innovations in GIS 6). Vol. xv, p. 280, Taylor & Francis, London.
Green, M. & Cunningham, D. (1998). Seabed imaging techniques.*Hydro International*, **2**, 4.
Greenstreet, S.P.R., Tuck, I.D., Grewar, N.G., Armstrong, E., Reid, D.G., & Wright, P.J. (1996) An assessment of the acoustic survey technique, RoxAnn, as a means of mapping sea-bed habitat. ICES, *Journal of Marine Science*, **54**, 939–959.
Heald, G.H. & Pace, N.G. (1996) Implications of a bi-static treatment for second echo from a normal incidence sonar. In: *Proceedings of the 3rd European Conference on Underwater Acoustics*, pp. 649–654.
Hughes Clarke, J.E. (1998) Detecting small seabed targets using high-frequency multibeam sonar. *Sea Technology*, **6**, 87–90.
Inman, D.I. (1952) Measures for describing the size distribution of sediments. *Journal of Sedimentary Petrology*, **22**, 125–145.
Jeffrey, S.W & Welschmeyer, N.A. (1997) Spectrophotometric and fluorometric equations in common use in oceanography. In: *Phytoplankton Pigments in Oceanography* (Eds. S.W. Jeffrey, R.F.C. Mantoura, & S.W. Wright), pp. 597–615. UNESCO, Paris.
JNCC (2001). *Natura 2000, Marine Monitoring Handbook*. (Ed. J. Davies et al.) UK Marine SACs Project, p. 405. Joint Nature Conservation Committee, Peterborough.
Kennedy, K & Mazzullo, J. (1991) Image analysis method of grain size measurement. In: *Principles, Methods, and Application of Particle Size Analysis* (Ed. J.P.M. Syvitski), pp. 76–87. Cambridge University Press, Cambridge.
Kenny, A.J. (1998) A biological and habitat assessment of the sea-bed off Hastings, southern England. *ICES*, CM., 1998/E:5. p.21.
Kenny, A.J., Cato, I., Desprez, M., Fader, G., Schuttenhelm & Side, J. (2003). An overview of seabed mapping technologies in the context of marine habitat classification. *ICES Journal of Marine Science*, **60**, 411–418.
King, P., Kennedy, H., Newton, P.P., Jickells, T.D., Brand, T., Calvert, S., Cauwet, G., Etcheber, H., Head, R., Khripounoff, A., Manighetti, B. & Miquel, J.C. (1998) Analysis of total and organic carbon and total nitrogen in settling oceanic particles and a marine sediment: an interlaboratory comparison. *Marine Chemistry*, **60**, 203–216.
Knauer, G. & Asper, V. (1989) *U.S. GOFS Working Group on Sediment Trap Technology and Sampling*. U.S. GOFS Planning Report No 10.
Kramer, K.J.M., Brockman, U.H. & Warwick, R.M. (1994) *Tidal Estuaries: Manual of Sampling and Analytical Procedures*. Balkema, Rotterdam.
Krumbein, W.C. (1938) Size frequency distributions of sediments and the normal phi curve. *Journal of Sedimentary Petrology*, **8**, 84–90.
Krumbein, W.C. (1939) Graphic presentation and statistical analysis of sedimentary data. In: *Recent Marine Sediments* (Ed. P.D. Trask), pp. 558–591. Thos. Murby, London.
Krumbein, W.C. & Pettijohn, F.J. (1938) *Manual of Sedimentary Petrology*. Plenum, New York.
Leong, L.S. & Tanner, P.A. (1999) Comparison of methods for determination of organic carbon in marine sediment. *Marine Pollution Bulletin*, **38**, 875–879.
Lindholm, R.C. (1987) *A Practical Approach to Sedimentology*. Allen and Unwin, London.

Lloyd, P.J. (Ed.) (1985) Particle size analysis 1985. In: *Proceedings of the 5th Particle Size Analysis Conference*, University of Bradford, 16–19th Sept 1985 Wiley-Interscience, Chichester.

Loncarevic, B.D., Courtney, R.C., Fader, G.B.J., Piper, D.J.W., Costello, G., Hughes Clarke, J. E. & Stea, R.R. (1994) Sonography of a glaciated continental shelf. *Geology*, **22**, 747–750.

Magorrian, B.H., Service, M. & Clarke, W. (1995) An acoustic bottom classification survey of Strangford Lough, Northern Ireland. *Journal of the Marine Biological Association of UK*, **75**, 987–992.

Mantoura, R.F.C., Wright, S.W., Jeffrey, S.W., Barlow, R.G. & Cummings, D.E. (1997) Filtration and storage of pigments from microalgae In: *Phytoplankton Pigments in Oceanography* (Eds. S.W. Jeffrey, R.F.C. Mantoura & S.W. Wright), pp.283–306. UNESCO, Paris.

McCave, I.N. & Syvitski, J.P.M. (1991) Principles and methods of geological particle size analysis. In: *Principles, Methods, and Application of Particle Size Analysis* (Ed. J.P.M. Syvitski), pp. 3–21. Cambridge University Press, Cambridge.

Mook, D.H. & Hoskin, C.H. (1982) Organic determinations by ignition; caution advised. *Estuarine, Coastal and Shelf Science*, **15**, 697–699.

Morris, A.W., Loring, D.H., Bale, A.J., Howland, R.J.M., Mantoura, R.F.C. & Woodward, E.M.S. (1982) Particle dynamics, particulate carbon and the oxygen minimum in an estuary. *Oceanologica Acta*, **5**, 349–353.

Mortimer, R.J.C, Krom, M.D., Watson, P.G., Frickers, P.E., Davey, J.T. & Clifton, R.J. (1998) Sediment-water exchange of nutrients in the intertidal zone of the Humber estuary, UK. *Marine Pollution Bulletin*, **37**, 261–279.

Murdoch, A. & Azcue, J.M. (1995) *Manual of Aquatic Sediment Sampling*, p. 224. Lewis, Boca Raton, FL.

Newton, R.S. & Stefanon, A. (1975) Application of side-scan sonar in marine biology. *Marine. Biology*, **31**, 287–291.

Paterson, D.M. (1989) Short-term changes in the erodibility of intertidal cohesive sediments related to the migratory behaviour of epipelic diatoms. *Limnology and Oceanography*, **34**, 223–234.

Paterson, D.M. (1997) Biological mediation of sediment erodibility: ecology and physical dynamics. In: *Cohesive Sediments* (Eds. N. Burt, R. Parker & J. Watts), pp. 215–230. John Wiley & Sons, Chichester.

Pejrup, M. (1988) The triangular diagram used for classification of estuarine sediments: a new approach. In: *Tide-influenced Sedimentary Environments and Facies* (Eds. P.L. de Boer, A. van Gelder & S.D. Nio), pp. 289–300. Reidel Publishing, Dortrecht.

Pocklington, R. & Hardstaff, W.R. (1974) A rapid semiquantitative screening procedure for lignin in marine sediments. *Journal du Conseil. Conseil International pour l'Exploration de la Mer*, **36**, 92–94.

Porter, K.G. & Fieg, Y.S. (1980) Use of DAPI for identifying and counting aquatic microflora. *Limnology and Oceanography*, **25**, 943–948.

Rumohr, H. (1999) Soft bottom macrofauna: collection, treatment, and quality assurance of samples. *ICES Techniques in Marine Environmental Sciences,* no 27. ICES, Denmark.

Rumohr. H. & Schomann, H. (1992) REMOTS sediment profiles around an exploratory drilling rig in the Southern North Sea. *Marine Ecology Progress Series*, **91**, 303–311.
Shepard, F.P. (1954) Nomenclature based on sand-silt-clay ratios. *Journal of Sedimentary Petrology*, **24**, 151–158.
Soille, P. (1999) *Morphological Image Analysis: Principles and Applications*. Springer, Berlin.
Somerfield, P.J. & Clarke, K.R. (1997) A comparison of some methods commonly used for the collection of sublittoral sediments and their associated fauna. *Marine Environmental Research*, **43**, 145–156.
Soulsby, R. (1997) *Dynamics of Marine Sands*. Thomas Telford, London.
Stephens, J.A., Uncles, R.J., Barton, M.L. & Fitzpatrick, F. (1992) Bulk properties of intertidal sediments in a muddy, macrotidal estuary. *Marine Geology*, **103**, 445–460.
Sutherland, R.A. (1998) Loss-on-ignition estimates of organic matter and relationships to organic carbon in fluvial beds. *Hydrobiologia*, **389**, 153–167.
Tolhurst, T., Paterson, D.M. & Gust, G. (2000) The influence of an extracellular polymeric substance (EPS) on cohesive sediment stability. Abstract: *INTERCOH 2000*, Delft, Sept 4–8, 2000.
Trask, P.D. (1930) Mechanical analysis of sediments by centrifuge. *Economic Geology*, **25**, 581–599.
Underwood, G.J.C., Paterson, D.M. & Parks, R.J. (1995) The measurement of microbial carbohydrate exopolymers from intertidal sediments. *Limnology and Oceanography*, **40**, 1243–1253.
Underwood, G.J.C. & Smith, D.J. (1997) Predicting diatom exopolymer concentrations in intertidal sediments from sediment chlorophyll *a*. *Microbial Ecology*, **35**, 116–125.
Underwood, G.J.C. & Smith, D.J. (1998) In situ measurements of exopolymer production by intertidal epipelic diatom-dominated biofilms in the Humber estuary. In: *Sedimentary Processes in the Intertidal Zone* (Eds. K.S. Black, D.M. Paterson & A. Cramp), pp. 125–134. Geological Society Special Publication, London.
Warwick, R.M., Goss-Custard, J.D., Kirby, R., George, C.L., Pope, N.D. & Rowden, A.A. (1991) Static and dynamic environmental factors determining the community structure of estuarine macrobenthos in SW Britain: why is the Severn Estuary different? *Journal of Applied Ecology*, **28**, 329–345.
Weliky, K., Suess, E., Ungerer, C.A., Muller, P.J. & Fischer, K. (1983) Problems with accurate carbon measurements in marine sediments and particulate matter in seawater: a new approach. *Limnology and Oceanography*, **28**, 1252–1259.
Wentworth, C.K. (1922) A scale of grade and class terms for clastic sediments. *Journal of Geology*, **30**, 377–392.
Whitlach, R.B. (1981) Animal-sediment relationships in inter-tidal marine benthic habitats: some determinants of deposit-feeding species diversity. *Journal of Experimental Marine Biology and Ecology*, **53**, 31–45.
Widdows, J., Brown, S., Brinsley, M.D., Salkeld P.N. & Elliott, M. (2000) Temporal changes in intertidal sediment erodibility: influence of biological and climatic factors. *Continental Shelf Research*, **20**, 1275–1289.
Wirth, H. & Wiesner, M.G. (1988) Sedimentary facies in the North Sea. In: *Biogeochemistry and Distribution of Suspended Matter in the North Sea and Implications for Fisheries*

Biology (Eds. S. Kemp, V. Dethlefse, G. Liebezeit & U. Harms), pp. 269–287. Mitt. Geol. –Palaont. Inst. Univ. Hamburg. Heft 65.

Wright, S.W., Jeffrey, S.W. & Mantoura, R.F.C. (1997) Evaluation of methods and solvents for pigment extraction. In: *Phytoplankton Pigments in Oceanography* (Eds. S.W. Jeffrey, R.F.C. Mantoura & S.W. Wright), pp. 261–282. UNESCO, Paris.

Chapter 3
Imaging Techniques

C.J. Smith and H. Rumohr

3.1 Introduction

The progress in our understanding of benthic ecology, especially in sublittoral habitats, has in the past necessitated the use of traditional sampling methods in which samples of sediment or fauna are removed for processing. These methodologies disturb, if they do not actually destroy the objects under investigation, give no idea of the representativeness of the sample and do not allow *sensu strictu* repeated sampling at the same spot.

Observational methods, however, are non-destructive and allow information from one identical object to be gathered over time. This approach has long been common in rocky shore ecology where researchers can reach their study sites without getting wet and can document their observations directly *in situ*. While sample material is rarely removed, visual impressions are recorded. Other non-destructive methods have included the observation of invertebrate behaviour in aquaria. Image recording in both these situations can be undertaken with non-specialist camera systems. However, when imaging devices are taken into an underwater environment, specialist systems and carriers must be deployed.

Imaging methodologies can be used in a variety of studies:

- Survey
- Observation and documentation (non-destructive and non-selective)
- Recording of behaviour/activities
- Identification and enumeration of features
- Evaluation and performance of samplers
- Location of samplers/samples and platforms

Common in-water imaging systems can be classified primarily into two types: acoustic and optical (Rumohr, 1995). Acoustic methods allow for wide-area (km range) low-resolution imaging (5+ m range) whilst optical methods allow smaller area (m range) high-resolution images (cm range). Table 3.1 shows the scales and resolutions of the most common imaging methods applied to benthic ecology (for more information see Kenny *et al.*, 2003). Additional information is also given in Table 3.1 in terms of comparative economics of capital cost and processing.

Table 3.1 Comparison of different imaging methods standardized to 1 hour ship time data gathering at 100 m depth. Cost of equipment is indicative as prices vary according to system complexity. Processing time is related to the 1 hour of data collected; more data may not require exact multiples of this.

Method	Area coverage (km^2)	Scale (m)	Resolution (m)	Equipment cost (Euro)	Processing time (hour)
Multi-beam	1	200	1.0	>50 000	8
Sidescan sonar	1	200	0.2	30 000	8
Bottom discrimination sonar	0.1	10	10–100	22 000	2
Laser line scan system	0.05	7	0.01	>75 000	8
Towed underwater TV	0.002	1	0.02	16 000	1–2
ROV	0.001	1–2	0.01	20–40 000	1–2
Still image camera	2*		0.01–0.001	10 000	2
Sediment profile imagery	2*	0.15	0.002	26 000	2

*This is the rough area that can be representatively covered in 1 hour.

Different methods have different uses and give different types of results. In an area survey it may be beneficial to use more than one methodology with a nested sampling approach that covers different resolutions and scales. All methods are relatively expensive, requiring technical knowledge to operate as compared to a traditional grab or dredge, but all are relatively fast methods with minimal processing unless high levels of detail and analysis are required. Images can provide a great deal of information, besides giving direct measurements of visible parameters; they give a feel for an area, an idea whether it might be useful to take further samples, the ability to document over time. On this subject we can only repeat that 'a picture is worth a thousand worms' (Solan et al. 2003).

Since the last review on imaging methods appeared (Holme, 1984) there have been two major technological breakthroughs in the field. The first was the application of charge-coupled device (CCD) cameras for underwater viewing and the second was the ongoing introduction of digital recording. CCD cameras have been in use for some time and are now the sensor of choice for newer digital recording devices. This chapter gives details of primary methods for scientific imaging, looking from wide-area imaging systems to smaller area higher resolution systems, an overview of which is given in Table 3.1. The imaging methods are then followed by sections on carrier platforms, special applications, laboratory imaging and analysis. An excellent review of scientific applications of imaging techniques to benthic sciences has been recently completed by Solan et al. (2003). Recent examples include the use of video for assessing patch dynamics (Parry et al., 2003) and invertebrate stocks (Smith & Papadopoulou, 2003), sediment profile imagery for anthropogenic impacts (Nilsson & Rosenberg 2000; Smith et al., 2003), scaled imaging with underwater video (Pilgrim et al., 2000) and acoustic methods for habitat mapping (Greenstreet et al., 1997).

3.2 Acoustic imaging

Acoustic methods and their use in seabed characterisation are dealt with in detail in Chapter 2, but some relevant aspects are summarised here. Acoustic mapping techniques are an essential part of the imaging approach in recording physical attributes, habitat (see Kenny et al. 2003) and community patterns of seafloor habitats at different spatial scales. Acoustic devices can be ranked according to their resolution capacity and the area covered. Historically, single-beam echo-sounders developed for depth measurement have been used to depict bottom structure as well as some sedimentological properties depending on the reflecting properties of the seafloor. For all acoustic methods the basic principles apply that the lower the frequency the longer the range, but the higher the frequency, the greater the resolution.

Acoustic ground discriminating systems (AGDS)

Single-beam echo-sounders (30 kHz to 3.5 MHz) may be used to obtain a variety of information about the reflective characteristics of the seabed. Such systems reveal the contour and depth of the seabed and indicate the thickness and structure of the sediment layers. They send a pulse of sound at a particular frequency that reflects from the seabed and the echo is then picked up by a transducer. The time between the sending and the return of the signal is traditionally used to measure the water depth. There are presently a number of ground discriminating systems where details of the echo are electrically processed to obtain information on bottom type. While the RoxAnn™ and Echoplus use an electronically gated tail part of the first return echo (E1 – roughness) and the whole of the first multiple return echo (E2 – hardness), the QTC system (Quester Tangent Corporation) analyses the first echo by means of principal component analysis. The resulting 64 parameters are used for ground discrimination. It should be noted that the majority of AGDSs are hull-mounted and are limited to use in shallow waters (<200 m depth) either by frequency or by increasing the size of beam footprint and therefore resolution. Recent developments apply the analysis of echo-sounder returns to include multi-beam systems, therefore enabling wider swathes to be analysed.

Sidescan sonar

Sidescan sonar is an acoustic imaging device (100–1000 kHz) used to provide wide-area, high-resolution pictures of the seabed. A geologists' tool for some time, it is now routinely used by benthic biologists. This acoustic method is used for charting seabed features (reefs, sand ripples, seagrass, gross changes in sediment type) and revealing special sediment structures of both biogenic and anthropogenic origin (e.g. feeding mounds or trawl tracks). The system typically consists of a towed underwater transducer (towfish/torpedo) connected via a fibre optic or coaxial cable to a shipboard control unit and recording device. The emitted narrow lobe of sonar energy has a beam geometry that insonifies a wide swathe of the seabed. The returning echoes from the seafloor are received by the transducers and, amplified on a time varied gain curve, transmitted to the recording unit. The recorder further processes these signals into either analogue or digital format, and then prints the echoes on paper or on a monitor screen, scan by scan over time, building up a waterfall image of the seabed.

Modern high-frequency sidescan sonar devices offer a high-resolution image of the seabed that can detect objects in the order of tens of centimetres at a range of up to 100 m on either side of the towfish. Sidescan sonar can produce, under optimal conditions, an almost photo-realistic picture of the seafloor. Over several georeferenced swathes, an overall resulting image, or mosaic, of the area can be built up forming an area map, where geological and sedimentological features, and even biological features (i.e. seagrass beds) are easily recognisable. A handbook for seafloor imagery analysis has been produced by Blondel and Murton (1997). Considerable increase in resolving capacity down to centimetres may be available in the future from the full use of synthetic aperture sonar (SAS) in seafloor imaging.

In the future, sidescan sonar may play a larger role in seabed discrimination. Sidescan processing software is now becoming available and will automatically classify features on the seabed; however, as in single-beam AGDS, ground truthing is necessary with higher resolving methods (e.g. sediment grabs, stills, video).

Swathe bathymetry

Swathe bathymetry through hull-mounted multi-beam or inferometric methodologies is a relatively new seabed mapping technology that produces high density geo-located depth measurements through digital processing techniques, and can be used to create impressive shaded-relief or colour topographic maps. A major advantage of multi-beam systems over sidescan sonar is that they generate quantitative bathymetric data, which could also be used for classification. They may also utilise backscatter data to form images similar to those of sidescan data, but with lower resolution, partly due to the variable height above the seabed of the hull-mounted sensors. The beam width makes them less useful for object detection when the objects are less than 1 m^2 and they require very accurate information on navigation, roll, pitch and sway and also calibration of sound velocity.

3.3 Video

Underwater video camera systems

Two broad classes of camera system are used underwater: daylight cameras and low light (intensified) cameras. The former are standard CCD (charge coupled device) monochrome and colour cameras, whilst the latter comprise intensified cameras

Table 3.2 Typical sensor choices for optical imaging – type of image sensor, limiting sensitivity in lux, typical horizontal range, indicative cost in kilo Euro (1000 m depth rating) and typical application.

Sensor	Limiting sensitivity (lux)	Limiting resolution (TVL)	Typical range (m)	Cost (KEuro)	Application
Mono CCD	0.1	570	5	3.5	Near field inspection and monitoring
Colour CCD	0.5	470	2–3	7	Standard survey
SIT	5×10^{-4}	600	20	17	Navigation, all round perception
ISIT	5×10^{-6}	525	40	30	Far field viewing and natural behaviour studies
Gen 1 ICCD	1×10^{-3}	480	23	14–17	Far field viewing and natural behaviour studies
Gen 2 ICCD	5×10^{-5}	450	23	19–22	Far field viewing and natural behaviour studies

and include both tube (monochrome SIT – silicon intensifier target) and CCD (ICCD – intensified CCD) cameras. The choice of camera is based principally on the sensitivity required by the user since underwater viewing range is related to light sensitivity of the camera (Table 3.2). The intensified systems are significantly more sensitive, but are also significantly more expensive. Most underwater cameras are CCD cameras, working in the daylight range with the addition of lights for work at night or at depth. Because of dissipation of light in water, the effective viewing range of a standard CCD camera will be a few metres, with better performance from a monochrome than from a colour camera. Intensified cameras are generally used for longer-range viewing and, with the same amount of light as a CCD camera, may have a viewing range of 10 m. They are also useful for research purposes where light may induce non-natural behaviour, for example, in fish behaviour studies, or in turbid conditions where additional lighting may make visibility worse through backscatter. Though the monochrome SIT camera is based on 35-year-old tube technology, in terms of price and resolution it is still one of the best systems available, its major drawback being that the tube is very easily damaged by pinpoint light sources, especially the sun. While monochrome cameras tend to have a high degree of resolution with sufficient lighting, colour cameras do have the advantage of presenting natural colours and therefore have more easily interpreted images. Sensitivity of cameras is also governed by spectral response, and SIT cameras are ideally suited for underwater applications as maximum sensitivity occurs in the blue/green region, coinciding with maximum transmission of water. CCD cameras have maximum response in the red light region where transmission in water is much poorer; however, they can be used with infrared lighting in low ambient light conditions over short ranges. This can be ideal for close range behavioural studies where the subjects are insensitive to infrared, both in the laboratory and field. The highest quality of recording uses 3-CCD cameras (broadcast quality standard) where light entering the camera lens is split through a prism into red, green and blue light with a single CCD sensor for each colour. Because light is split into constituent wavelengths, 3-CCD cameras are highly sensitive to light and require a high level of ambient lighting.

The use of stereo video with parallel-mounted, matched cameras has been driven by the need for underwater inspection in oil field applications and where a three-dimensional view is required with regard to manipulation. The left and right cameras can be observed through head mounted mini-monitors on the left and right eye, or through using polarising glasses and having two monitors overlapped by projection through a prism. A more recent development, the TV-Trackmeter system (Technomare, Italy), uses twin cameras and a computer processor. One single image is seen by the operator, but the operator can track objects or make 3D measurements in real time.

Lenses

Choice of lenses is another important feature of the camera depending on the application required. For general purposes, a wide-angle lens will cover more extensive areas at close range and have a good depth of field, but may have some distortion at the edges of view and pick up incidental light reflections near the surface. Optimum viewing angle for general underwater work is 60–70°. The use of zoom lenses underwater is becoming more common, especially in ROV systems and for close-up observation studies. At full zoom, they require a stable platform, better lighting and have a more limited minimum focal distance and depth of field than a conventional lens. In a self-contained system, the use of a zoom may not be necessary as this function can be undertaken by the carrier vehicle moving up to the subject.

Housings

Video cameras must be encased to protect them from seawater and pressure. Shallow water self-contained systems can be constructed of plastic/epoxy or metal with irregular shapes, as the pressures exerted in diving depths are within the strength tolerances of modern light materials. Deep systems are encased in cylindrical housings with a clear port at one end and electrical/video connection at the other. Plastic/epoxy resin housings can be used to depths of no more than 300 m. Metals including anodised aluminium, stainless steel or titanium are used in construction of all deeper water housings. Titanium, with the best strength to weight factor, is the most expensive material. Anodised aluminium is commonly used for small housings in the 600–1000 m range. The technology behind composite materials and ceramics continually improves, with the result that they may appear in deeper housings in the future.

The ported ends of the housings are simply sealed (commonly pop-on/pop-off, with a safety retainer) and rely on O-rings to prevent entry of water. Normally, two

O-rings are used for an endcap seal, an in-line seal and a face-plate seal. Flooding of housings is usually caused because of inattention to the O-ring, which is the least expensive part of the system.

Clear ports can be constructed of acrylic, glass or crystal. Acrylic is easily scratched, but because it has the same refractive index as water, light scratches are not seen underwater, but can be polished out of the face using a fine abrasive cleaner. For deeper applications, glass or crystal ports are used. The shape of the port is very important as a flat port will reduce the angle of view, make objects appear larger and produce some distortion at the edge of the picture. A domed port, although more difficult to produce, will restore the angle of view and real object size with little distortion.

Data transmission

Video data can either be stored immediately, or transmitted and stored. To store the data on-board the carrier vehicle, a video recorder is necessary, which will require a low volume, low power recorder. Hi-8 format or Digital Video fit these criteria (see 'Storage Media' section below).

Transmission to the surface is by umbilical cable, which will contain various wires to cover both power and control as well as data transmission. Data transmission can be either analogue through a simple coaxial cable (composite signal), or for maximum quality, digital through a fibre optic cable. Fibre optics require special end members for connection and digitisation of the original analogue signal and are therefore more costly options. However, they have the benefit of a very thin cross-section, light weight and broadband data transmission over long distances, with no signal attenuation over typical usage at sea. The maximum transmission distance without amplification for an analogue cable without significant loss of video signal is approximately 500 m, which can be increased up to 2000 m with amplification. The use of surface amplification may also amplify electrical 'noise' that has been picked up in the cable. Cables can have a diameter from a few millimetres up to several centimetres depending on their function and may be armoured with the inclusion of wire or kevlar sheaths. They may also be oil or tar-filled to prevent water ingress if the outer covering should be punctured. The use of armoured cables allows pulling load to be put on the cable and instrumentation, including cameras, can be deployed directly from the cable without the need for a separate steel wire or rope. The armoured cable will allow the instrumentation to be retrieved manually or by winch.

Wireless-through-water transmission of video is not currently possible at full-frame rate in real time (25 frames per second in Europe), because of limitations in speed of transmission and bandwidth (limitations imposed by low frequencies required). Where applications do not require real time it may be possible, whilst recording full video underwater, to transmit compressed still images (low quality) periodically to the surface, with updates in the range of one per minute. Research is underway to increase wireless ability using compressed data.

Once at the surface, images can be wireless-broadcast in a number of standard ways (radio transmitter or satellite) to show live images at any distance from the source. Live images have also been transmitted through the Internet for teaching purposes (low resolution) through cellular phone connection. Compression is required, but as technology improves (speed of computers and phone lines), the quality of transmitted images will continue to improve.

Format

Across the world there are three basic international colour TV standards for transmission, recording and playback.

- PAL (Phase Alternating Line): Western European format (Europe, Australia, East coast of South America). 4.43 MHz colour sub-carrier, 625 lines, 50 Hz, 25 frames per second.
- SECAM (Supérieure Elégante Couleur et l'Affaire Magnifique): French format (France Eastern Europe, Middle East, many parts of Africa). 4.43 MHz colour sub-carrier, 625 lines, 50 Hz, 25 frames per second.
- NTSC (National Television Standards Committee): American format (America and territories, Canada, Japan, Korea, West coast of South America). 3.58 MHz colour sub-carrier, 525 lines, 60 Hz. Modified NTSC with 4.43 MHz colour sub-carrier, 30 frames per second.

The camera, monitor and recording system must be compatible (i.e. have the same format). Video recorders and high quality monitors are often multi-format, and reasonably cheap systems can now be found to convert input to output across formats (e.g. Panasonic NV-1 recorder), although there will be some loss in the original quality.

Storage media

There are several standards for the recording of video images and comparative data are shown in Table 3.3. VHS using composite input (single coaxial) has been the most common standard. S-VHS has higher resolution but requires a signal input with two separate lines for chrominance (colour information) and luminance (brightness information). This system is widely used although the input may be composite as there are few S-VHS cameras available or used in underwater applications. The highest quality (broadcast quality) recording format is Betacam which has replaced U-matic as the large format 'broadcast quality' recording system. It is both bulky and expensive. Hi-8 is used where low volume and good quality are required, e.g. in diver-operated housings, although this format is now being replaced by digital format.

Digital recording is becoming the new standard format, because of the increasing resolution, constant quality (there is no loss of quality on copying), low power

Table 3.3 Typical recording medium choices for optical imaging – cassette recording type, resolution in TV lines, maximum tape length in minute, volume of cassette in relation to digital video (DV) cassette, Euro cost of high quality cassette and typical available signal inputs to recorder.

Medium	Resolution (TVL)	Maximum tape length (min)	Cassette volume	Cassette cost (Euro)	Signal input
VHS	260	240	4	4	Composite
S-VHS	>400	240	4	12	Composite, S-video
Hi-8	>400	90	1.5	10	Composite, S-video
U-matic HB	270 colour 370 mono	90	6	15	Composite
BetaCam SP	>600 colour	90	8	20	Composite, S-video, component
Digital (DV)	500	90	1	10	Digital, component, composite, S-VHS
DVD	500	120	1	4	Digital, composite, S-VHS

requirements, small size of cassette and the ease of importing into a computer. Currently, digital recorders use small tapes, but new technologies include mounted discs, PC card storage devices and CD-sized disks (Digital Video Disk – DVD). Digital recordings can be compressed on to a disk with a standard DVD disk able to hold approximately two hours of video material. At present, the machines for recording live video onto DVD are becoming much more common and cheaper. Digital disk media give the added advantage of being able to jump immediately to any part of the recording, and software is readily available for non-experts to edit videos on their desktop computers.

Power supply

Power supply is an important consideration for underwater video operations and particular attention should be paid to the stability of the voltage and frequency. Video cameras actually have small power requirements and work on 12 or fewer volts. A UPS (uninterruptible power supply – stabilised power supply) system is advisable where possible if only to power the surface recorder. For the underwater system, if the loading is too high for a UPS system, then a separate power supply is advisable. At sea, loading on a shared generator will vary with the ship's requirements and the mere act of turning on a cooker or heating system can cause onscreen and recording interference. Voltage irregularities may be smoothed in transformation, but frequency alterations have greater effects on recorders.

High voltages and currents are often encountered in underwater systems and extreme care should be taken to ensure that the system is correctly maintained, protected and earthed. Attention is drawn to the UK association of diving contractors code of practice for electricity underwater (AODC 1985).

Video monitors

For a system operator, a high quality 14- or 16-inch monitor will be optimal and it should be sited to reduce external light reflections, preferably in a darkened room. Control operations using video should not be overcrowded and it is recommended that a separate monitor be placed elsewhere for viewing by large numbers of personnel and at sea, especially for operations underway, a monitor should be placed on the bridge so that the captain may view the operations directly.

3.4 Photography

Large-area satellite or aerial photography is generally applicable in coastal fringes where features can be detected on the shore, intertidal areas, or coastal waters (phytal regions under ideal conditions with calm and clear water). Satellite images may have a resolution of 5–1000 m and aerial photographs 0.5–5 m. Both types of images are prone to atmospheric shading (clouds) and for both cases there is an absolute need for 'ground truth' information with data gained using traditional methods. Hyperspectral or mono-spectral (e.g. infrared) images can be used to discern various patterns such as vegetation, which may not be visible in normal visual images.

Underwater photography has been used since soon after the invention of the camera, but came into prominence with SCUBA diving and the proliferation of, first, the Nikonos series of cameras, housed land cameras and then smaller, cheaper marinised compact cameras. Most underwater cameras are in the standard 35 mm format, although some large format cameras (70 mm) are also used. Deeper specially housed cameras will normally take large film packs (200–500 frames) and deep sea systems that may be deployed in excess of 6 months, can have special films packs with up to 4000 frames. Data backs that print onto the film are used to identify the sequence, time date and other information on or between consecutive frames. An external flash is needed as a light source with a battery pack able to charge the flash for the required deployment duration. The flash is best placed to one side of the camera forming an angle of approximately 45° between flash, subject area and camera. This provides shadow, which is very useful for understanding structures seen in the image.

Still photographs are a necessary complement to any video recording since photographic resolution is much better than any video documentation (video loses resolution with increased processing). Underwater photography can supplement the semi-quantitative data from divers and, whether deployed by diver or remotely, allows meso-scale areas or transects to be recorded accurately. Furthermore, it permits life-like reproduction of fine structures and live positions of invertebrate fauna which are normally destroyed during sampling.

Underwater photography was, and is, much used in deep-sea investigations. In particular, the famous Edgerton cameras (BENTHOS Inc., USA) opened the door to a new field of research (Kidd & Huggett 1980; Foell & Pawson 1986).

Digital cameras

Digital cameras rely on CCD sensors, but encode and store the image in digital rather than analogue format. They have been available in the market for many years with BENTHOS having established the first underwater systems by adapting Kodak digital land cameras. Recent changes have included much smaller devices with much greater sensitivity, better image compression and storage capacity. Several manufacturers now offer underwater models. The relationship between digital video and digital still cameras has recently become blurred with both systems providing access to recording both digital and still images. Until now video has been primarily stored on tape and still images on solid-state media. Solid-state media can be used to store short lengths of video, but tape is, at the time of writing, still the best medium for long-play recording. Some video cameras offer both storage media and some still image cameras offer the ability to output composite or digital video for recording on a separate device. Transmission of still images to remote units remains a relatively slow process, but cameras are now offered in housings with electronic media storage giving the ability to record several thousand images. Images can be downloaded directly to a computer on the surface through special connector ports. With sensitive CCD sensors, separate flashguns are not always necessary and good quality normal underwater lighting can be used over short ranges. Unfortunately, there is no single standard for digital recording because of the never-ending increase in computer power, image compression software development and miniaturisation of components.

Illumination

Full colour light penetration is limited underwater with absorption of higher wavelengths (red and orange) within a distance of a few metres. Overall light penetration depends on local conditions and may be from metres in eutrophic sediment-rich waters to 100+ m in parts of the oligotrophic areas such as the Mediterranean. Artificial light must be used for optical imaging purposes at greater sea depths to bring out the 'natural' colours in colour images. Lighting systems normally consist of pressurised housings (at least around the bulb) and must be powered either from external sources (cable supply) or be self-contained with batteries. Self-contained systems are time-limited because of the high energy demand of the lights. Lights may have different types of reflectors and domes to produce different angles of diffusion. Halogen bulbs are the most common type for general underwater lighting in use and have the advantage of low-cost replaceable bulbs, rampable setting for brightness, but they produce a yellow/orange light that has limited penetration and hence range. Lights more commonly available today include high intensity discharge (HID) gas arc lamps which require a more substantial housing for the electrical ballast system, take 2–3 minutes to reach full operational brightness, but have a longer bulb life and produce a much more striking blue/white illumination (similar wattage produces 4–6 times as much light). Halogen lights burn at approximately 2500–3000 K, whereas HID lights burn at 4000–6000 K. The lower the colour temperature, the more red or yellow the light (more easily absorbed), the higher the temperature, the whiter the light and the more realistic the visualisation of colour and contrast.

For still film-based photographic systems, a dedicated flashgun is necessarily synchronised to the camera. However, modern digital camera systems can produce acceptable images on standard underwater lighting (especially HID lights).

Coverage and correct angle of illumination to the sensing head (photographic or video) is a key feature in producing good quality images. Offsetting the lighting produces greater shadow, highlighting parts of the image and giving an idea of 3D representation, although large shadows will hide parts of the image. Large offsets will also reduce the amount of light backscatter to the lens in the case of high water turbidity.

Calibration and measurement

Live or recorded video can be calibrated for measurements either prior to use or continuously online. Prior to use, a fixed grid with measuring bars can be placed in front of the camera and the image recorded. The calibration will hold as long as the view does not change. Thus, a sled-mounted camera with a fixed angle of view and fixed distance above the focus plane can be calibrated by filming a measured grid fixed to the seabed. The grid should be larger than the field of view so that any aberration towards the edge of the grid in the viewed image can be noted. In the playback of recorded material the grid can be copied on to a transparent slide and fixed on the screen of the monitor. All objects in the plane of the calibrations can be measured fairly accurately, but there will be inaccuracies in objects above or below the measuring plane. This method can be used effectively for density measurements of particular features. For continual scaling, a number of lasers can be used. Lasers produce focused beams and if two lasers are parallel mounted at a known distance apart, the viewed laser spots will remain parallel at the same distance apart if directed on to a perpendicular plane. This method gives the ability to make size measurements in the area of the projected laser beams. A third laser can be added for size and distance measurements. This laser forms the third side of a right-angled triangle for the projected beams, but it is mounted at a very slight offset angle. The distance between the third beam and the other two varies with the distance of the projection and by using simple trigonometry, distance is calculated as a function of the offset angle and the distance between the third and either the first or second beam. Images can be calibrated with measurements made by means of computer software.

3.5 Carrier platforms

Imaging sensors can be carried on a wide variety of carrier platforms involving both simple and complex, mobile and static platforms, depending on the work to be carried out. Figure 3.1 shows a cost duration profile for a number of different types of platform.

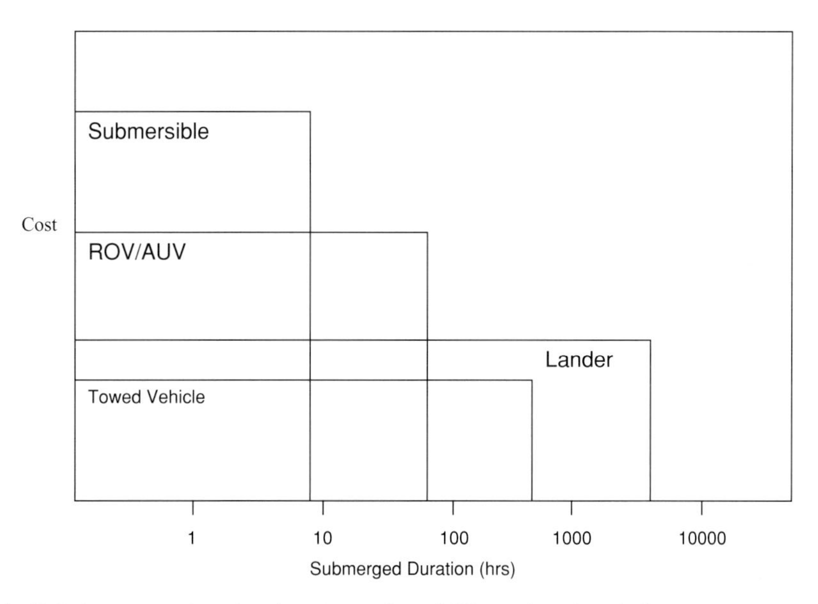

Fig. 3.1 Relative cost against duration of recording of different imaging platforms.

Diving

This area is covered in more detail in Chapter 4, but a few important points are highlighted here.

Divers were used as carrier platforms even before the introduction of SCUBA equipment. Because of their relative freedom of movement underwater, small self-contained units are the most popular, especially camcorders and photocameras (such as the popular Nikonos series). With recent advances in mixed gas technology and decompression issues, divers can now routinely go deeper and penetrate harsher environments. The relevance to imaging is that pressure housings must take this into consideration and whilst diving camera systems were once proofed to 30 m depth they must now be routinely proofed to 100 m depth. With a self-contained camera, the diver has direct control to select what to film and how best to carry this out, but should be aware of the special limitations and techniques that need to be applied underwater including lighting techniques, stability, positioning, patience and maintaining water clarity. It should be noted that with a low-light camera, more features might be seen on video than are visible to the diver in the water.

Drop frames

Camera systems can be mounted on remote frames, either self-contained or using a cable. Where a frame is suspended above the seabed, it will move with the motion of the surface vessel, making viewing the image unpleasant in rough weather conditions. The camera can be on a fixed mount or mounted on a pan and tilting motor allowing the camera view to be changed. Underwater lights may be fixed to the pan and tilt so that they can track the camera view. To prevent a suspended frame from rotating, a tail fin can be attached to keep the frame oriented into the general direction of movement or into the current. Cameras can also be mounted on to bottom sampling equipment for both directed sampling (the final drop of equipment is controlled over a particular area) and to assess the area from which the sample comes (e.g. information on representativity). Self-contained cameras are also deployed on landers where a system is free-deployed, and later retrieved by using an acoustic release (or time release), releasable ballast weights and sub-surface floatation. These systems can have deployment times ranging from hours to months and consequently require low volume, low power systems and for long-term deployments will usually work on time lapse, with photographs or video taken at timed intervals.

Specialized towed platforms

Towed camera sleds are in contact with the seabed so that the camera system can be maintained with a fixed view of the seabed. The camera view can therefore be easily calibrated for quantitative viewing. A vertical view gives good imaging over a horizontal plane. However, very low speeds must be maintained for acceptable resolution and there is little information gained concerning relief. Sled cameras are therefore often mounted to look obliquely forward, giving images with an expanding forward view, which is more difficult to quantify. The advantage of this view is that it gives some warning of obstacles in front of the camera and that there is better visual information in the third dimension of relief and height of an object above the seabed. Towed sleds are commonly used to 500 m depth with umbilicals and live viewing of images. Management of the tow cable/umbilical is very important as it can easily drag in front of the sled and should therefore be buoyed at intervals close to the sled with submersible buoys. Maintenance of the correct speed is also very important for a clear camera view. Too fast a speed will blur video pictures, while too slow a speed may allow the cable to drop in front of the sled or the towing vessel to lose steerage way. Shand and Priestley (1999) describe a sled system whose basic design is used by several institutions around Europe (Fig. 3.2).

Fig. 3.2 Schematic of 'Aberdeen-type' towed video sled.

Several specialised towed or drifting vehicles has been developed in-house by large research institutions and private companies, principally for deep-water research, for example, TOBI and SHRIMP (Southampton Oceanography Centre), SAR (IFREMER) and ARGO (WHOI). A brief review of these platforms is given in Wernli (2000). These vehicles may contain camera systems for recording the seabed features and may be independent, have umbilical transmission systems, or be acoustically or thin-umbilical controlled. Cameras are positioned to look downward and can be switched on and off by altitude switches. The altitude can be printed on or between photographic frames on still systems so as to quantify the camera view. The major problem of using these systems is to maintain the vehicle within a particular altitude range of the bottom, which must be done by constant alteration of the length of paid-out cable. The sensor systems must be well protected within the camera frame to reduce the damage caused by inevitable impact with the bottom (frame polishing events).

Some specialised towed vehicles with controllable surfaces can allow the vehicle to offset from the tow path in a vertical or lateral direction. The RCTV (remote controlled television) has four rotating cylinders that utilise the Magnus effect to provide a lift in the vertical or lateral plane, whereas the ROTV (remotely operated television) has a box kit structure with small wings on the leading edge to provide the same lift. A disadvantage of this type of vehicle is the requirement for forward movement; it cannot be stopped in mid-use to investigate a target item further. The vehicles can be loaded with a variety of sensors (cameras, lights, sonars and other sensors) and are used commercially to undertake pipeline, cable and route surveys. The main scientific role of these vehicles is in fisheries research, either in trawl behaviour or fish behaviour with respect to trawling. The vehicle is towed adjacent to a trawl and can manoeuver all around the trawl or even inside, for visual or acoustic inspection.

Remotely operated vehicles

Remotely operated vehicles (ROV) are tether-controlled and powered vehicles with independent motors and payloads comprising sensor (navigation and imaging) and optional equipment (manipulators, tool packages, scientific samplers, etc.). Throughout the world there are some 200 commercially available models ranging from simple eyeball ROVs in the water (1–2 kW power, 30–600 m depth) to workhorse ROVs used in the oil industry (50–100 kW power, 600–3000 m depth) and bottom crawlers for cable and pipeline burial. There are also a number of specifically developed or adapted ROVs for scientific research, some of the larger of which are shown in Table 3.4 with further details in Wernli (2000). Figure 3.3 shows several of these large systems. The sensor load that an ROV can carry depends on its size, power availability and the specification of the umbilical cable for transmission purposes. For small ROVs, cameras may be mounted within a

Table 3.4 Large science-orientated ROV systems in service.

ROV name	Depth rating (m)	Weight (kg)	Payload (kg)	Horse power	Operator	Country
Aglantha	2000	740	100	26	University of Bergen	Norway
Dolphin	3300	3800	150	67	JAMSTEC	Japan
ISIS	6500	3000	190	30	Southampton Oceanography Centre	UK
Jason 2	6500	3000	150	30	WHOI	USA
Kaiko	11 000	5600	150	47	JAMSTEC	Japan
Kraken	1000	635	80	13	University of Connecticut	USA
Quest	4000	2450	200	100	University of Bremen	Germany
ROPOS	5000	2700	200	40	CSSF	Canada
Tiburon	4000	3357	499	30	MBARI	USA
Ventana	1500	2338	236	40	MBARI	USA
Victor	6000	4600	150	–	IFREMER	France

Fig. 3.3 Science ROVs. (a) Monterey Bay Aquarium Research Institute's ROV Tiburon (courtesy of MBARI), (b) IFREMER's ROV Victor, (c) Southampton Oceanographic Centre's ROV Isis (courtesy of SOC) and (d) University of Bremen, Quest ROV (courtesy of MARUM).

general pressure housing, but most cameras are externally mounted in their own pressure housings on pan and tilt motors. The pilot will have at least one camera with a wide angle lens, and additional cameras can be mounted to cover a variety of views including intensified tube and CCD cameras for viewing in far field or turbid waters, zoom lens cameras, or mini-CCD cameras which can be mounted in almost any place on the ROV (for example, the end of a manipulator). Larger scientific

ROVs have specialised tool skids and adaptations which may incorporate CTDs and other physico-chemical sensors, suction samplers, water bottles, core baskets, specialised cameras and other specialised sensors (Auster & Tusting, 1997).

Autonomous underwater vehicles

Autonomous underwater vehicles (AUV) are unmanned, untethered, battery-powered and have autonomous navigation capabilities (Fig. 3.4). There are some hybrid vehicles that are tethered by extremely thin umbilicals for data transfer (track control, status and sensing information) normally used for mine countermeasures. Some other vehicles may be partially controlled through a slow acoustic link. AUVs have been in a high state of development during the last five years with a small number of models being commercially available. The technology was originally driven from the military in the development of 'smart' torpedoes, but is currently being driven by exploration companies for large-area mapping. They can carry a small suite of scientific sensing instrumentation, for example, CTDs and sonar mapping systems. Surface vessels do not necessarily need to be specialised and the AUV can undertake a mission whilst the support vessel is doing other work. Vehicle navigation normally consists of an inertial navigation system using a mixture of internal compass, doppler speed logs, three-dimensional motion reference units and dead reckoning software. Some systems can surface periodically to take GPS location updates. Navigation can also be pre-programmed within a long baseline sonar navigation system. Camera systems can be mounted on the AUV, but the AUV must have the ability to provide a stable platform. Sensors, lighting and recording systems are extremely limited in terms of size and power availability.

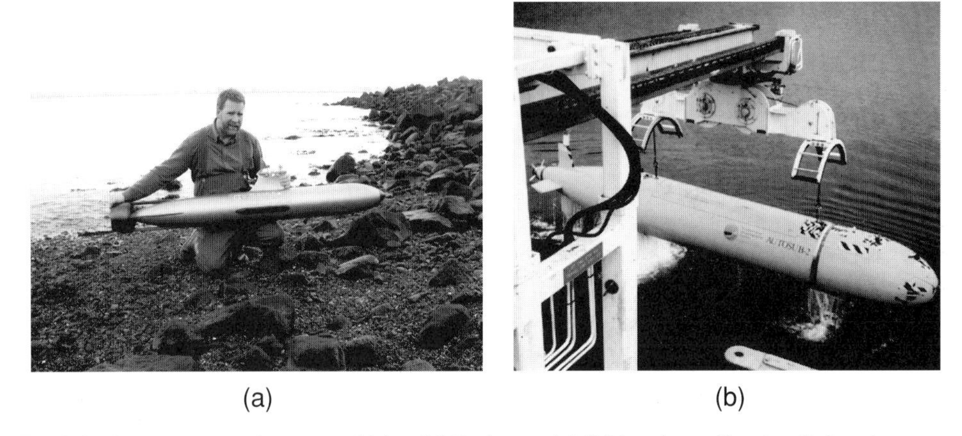

(a) (b)

Fig. 3.4 Autonomous underwater vehicles. (a) Gavia coastal AUV system with primarily imaging sensors (courtesy of Hafmynd) and (b) Autosub, large oceanographic AUV system with near bottom imaging capabilities (courtesy of SOC).

Manned submersibles

There are a number of manned submersibles available to the scientific community with a variety of depth ratings from 300 to 6000 m. A listing of the deeper-rated research submersibles is shown in Table 3.5. The submersible has certain advantages over an ROV of being untethered and being able to put a scientist onto the seabed to intervene through remote manipulation or to use the most sophisticated imaging system of all, human stereoscopic vision. Almost all submersible systems are fitted with video and still photographic systems, allowing a permanent record to be kept. Cameras are mounted externally, although some may be hand-operated through viewing ports. Deep submersibles have small viewing ports and external cameras may give a less restricted external view. Requirements are for small size and low power consumption. Hand-held camera systems can be run off their own batteries, but larger units (recorders and hull-mounted cameras) will be limited by having to be powered from the submarine battery system. Most research submersibles are equipped with a variety of external pelagic and benthic sampling and sensing equipment. The bottom time is generally limited to approximately 4 hours with a possibility of one to two dives per day.

Table 3.5 Major characteristics of deep-diving manned submersibles (abstracted from Rona, 2000).

Submersible name	Maximum depth (m)	Personnel (crew + observer)	Seabed time	Speed (km/h)	Country of operation
Alvin	4500	1 + 2	4–5 h	2.5	USA
Mir	6000	1 + 2	10–15 h	9.3	Russia
Nautile	6000	2 + 1	4–5 h	4.6	France
Cyanea	3000	2 + 1	5 h	3.7	France
Pisces V	2000	1 + 2	6 h	3.7	USA
Shinkai 2000	2000	2 + 1	4 h	5.6	Japan
Shinkai 6500	6500	2 + 1	4 h	4.6	Japan
SeaLink	914	2 + 2	4 h	2	USA
NR1	724	11 + 3	30 days	6.5	USA

Navigation and positioning of the carrier platform

It is at times essential to know the position of the carrier vehicle for tracking to a particular location. It should not be assumed that a towed carrier is positioned at an exact distance behind the towing vessel. Although expensive, acoustic beacon systems are the best solution for high accuracy positioning, where transponders are positioned on the seabed covering the work area. The carrier vehicle is equipped with a pinger and the support vessel with another transponder. By communicating with the transponders, the control system can locate the position of the pinger. Long baseline systems can work over a range of kilometres with metre or even sub-metre accuracy. With an ultrashort baseline system, the transponders are all located in one acoustic head deployed from the surface vessel. This is a less expensive but less accurate system. It can be used over a range of approximately a kilometre with accuracies in the range of 5–20 m. Accuracy and range for both systems are dependent on a number of factors including water layers, depth and bottom topography. Conditions will be less than optimal in shallow waters with an irregular bottom because of scattering and reflection of the acoustic signal. Acoustic beacon systems by themselves give a relative positioning in relation to the transponder head. For absolute positioning, the system can be linked to a GPS or DGPS satellite positioning system, giving the geographic coordinates of the carrier platform.

3.6 Special applications

Sediment profile imagery

Sediment profile imagery (SPI) or REMOTS (Remote Ecological Monitoring Of The Seafloor) utilises an imaging device in an inverted periscope (optical prism) that penetrates the sediment and facilitates the imaging of the sediment water interface and upper sedimentary layers (approximately 15 × 20 cm in area), allowing fine scale analysis of physical, chemical and biological features (Rhoads & Young, 1970; Rhoads & Cande, 1971; Rhoads & Germano, 1982, 1986). Figure 3.5 shows a typical SPI system. The camera system (photographic or digital) is mounted in a pressure housing whilst the distilled water filled prism is pressure-compensated. The flash head is normally mounted on the prism and great care must be taken to ensure overall even illumination over the faceplate. Direct measurements can be made from the images. Rhoads and Germano (1986) developed a computer-aided system for analysis of some parameters, such as grain size, surface boundary roughness, mean (apparent) redox depth, methane gas pockets, thickness of overlying (dredged/disposed) material, visible epifauna, tube density and type, faecal pellet layer, microbial aggregations, feeding voids, faunal dominants and

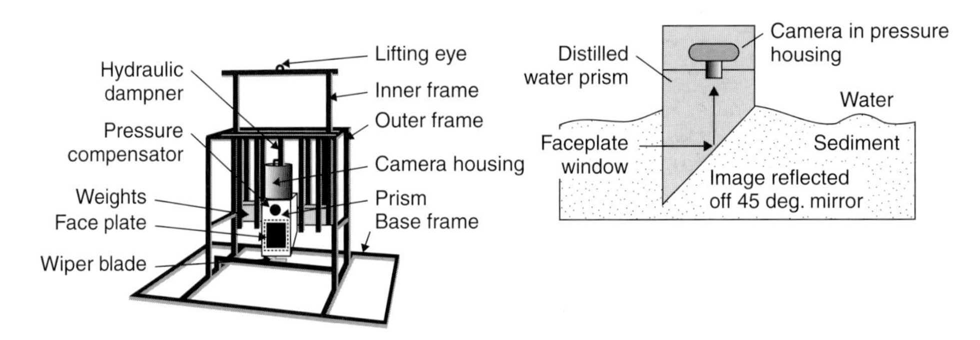

Fig. 3.5 Sediment profile imagery system (remote frame and detail of the head).

successional stages. This development made the retrieval of a variety of abiotic and biotic measures much easier and quicker compared with earlier methods, i.e. by cores. One of its primary uses is in large-area surveys where local hotspots need to be identified and sampled in more detail. Systems are frame-mounted and wire-deployed for remote use or can be diver-operated for directed sampling in shallow waters.

Laser technologies

Laser line scan systems (LLSS) work in ways similar to sidescan sonar with a swathe of seafloor being illuminated by a crossing laser beam and the reflected light gathered. As the towfish moves, individual lines build up a waterfall image of a seabed swathe. Unlike sonar, the system is more affected by water turbidity. The height of the towfish above the seabed determines the swathe width and desired resolution. Typically, at 5 m height a swathe of 7 m can be covered with a resolution of 1 cm. The system works at intermediate scales and resolution between sidescan and video. There are, however, very few systems available and the cost is very high. First generation LLSS produced monochrome two-dimensional images, whilst second generation systems include three-colour and three-dimensional systems (Carey et al., 2003).

Hobson et al. (2000) have developed the use of laser holography for *in situ* pelagic research (imaging and distribution). With limited use in benthic applications, the camera can be used for imaging of plankton and particulate matter and may have applications to the sediment–water interface. The system images a volume of water (up to 0.1 m^3) recording true, full-field three-dimensional optical images with resolution to 5 μm. Camera systems are still developmental as is the software needed for microscale spatial recording and analysis. New applications in the benthos may develop for this type of methodology in the future.

Application of medical technologies

X-ray imaging is mainly used in the laboratory to investigate the internal structure of sediment cores for both sedimentological and biogenic structures. In the same way, although with safer operation, ultrasound has been used in the laboratory for investigation of sediment structure in aquaria, and also ongoing activity in sediments and behavioural/physiological observation of organisms. In both cases, commercial equipment from medical applications has been used, although it is not necessarily suited to the marine environment. One further step has been in the use of computer axial tomography (CAT-scan), which allows a better estimation of sediment structures (step by step X-rays along a single axis allowing the build up of three-dimensional images). Its use in sediment stability studies has been shown by Michaud et al. (2003). This type of equipment is only available in hospitals and medical clinics, and therefore access to the technology is very limited.

3.7 Laboratory imaging

Video recordings are often used for teaching and demonstration purposes and most institutions have a stereoscope or microscope linked to a video recorder for this purpose. The use of video allows a number of observers to share the view of an object at the same time and it may be more economical than investing in a second viewing head for the stereo/microscope. It is also a useful technique for recording the behaviour of small organisms or recording living organisms for identification purposes where preservation techniques may make them unrecognisable (for example, turbellarians or platyhelminthes). The video camera will be a small sensor head without integral lens that can be attached through an optional C-mount connector on the microscope. Lighting is extremely important and the system will require adjustable high power spot lighting (100+ W fibre-optic light source). Positioning of the illumination source is very important with stereoscopy for either overall illumination, or highlighting with the use of shadow by side illumination. Circular fibre-optic sources or in-line light sources (light source shining down the optical path) are sometimes used, but they may be inadequate in spot power under high magnifications, and if the object is immersed in a fluid there may be unwanted reflections from the fluid surface. Biological samples are best observed under water, again to prevent unwanted reflections from wet surfaces.

Standard video cameras can also be used in the laboratory for observational and behavioural purposes, e.g. looking down over tanks or mesocosms. Care should be taken to ensure that there is adequate protection from electrical short-circuits and that the equipment should be protected from corrosion in humid environments. Recording equipment is best situated in a remote dry area.

Endoscopy and Boroscope relying on fibre-optic principles have been used in tank aquarium systems to view the inside of closed structures, for example, burrow systems. However, these systems are not widely used and suffer from corrosion. It is often better to arrange a view by using glass-walled tanks and external camera systems, where an animal may burrow against the glass wall allowing non-intrusive/non-disturbed recording from the other side of the wall.

Video cameras are increasingly used to illustrate research work for permanent record or public relations. It is always useful for the scientist to have a video camera on hand and he/she should have some familiarity with general video techniques and practice (including framing, lighting, stability, holding on shots and not to zoom in and out whilst recording unless absolutely necessary). For video production it is probably necessary to use professional services with detailed planning, rehearsal, taking and retaking of shots. Filming at this level cannot be done during a normal scientific working day and time should be put aside for this.

3.8 Image analysis

A large amount of data can be recovered from an image, ranging from anecdotal (description of a process) to semi-quantitative data (e.g. degree of coverage of an organism on a rock: a, none; b, low; c, medium; d, high; e, total) and finally, quantitative data. Quantitative data can be abstracted manually from a calibrated image by direct measurement and transformation. Computer software programs are readily available to undertake this automatically. Computer image analysis systems can be extensive, covering the entire operation from image collection to output of analysis. These systems tend to be expensive and it is more common for the scientist involved in image analysis to have his/her own image input and storage system and to utilise readily available software ranging from professional analysis software to simpler image processing software (e.g. Adobe PhotoShop) and shareware (e.g. NIH Image). Most software allow onscreen measurement or, in more detailed systems, the filtering of images, automatic abstraction of shapes or parts and their automatic measurement.

Consideration should be given to the intended application of the final images, whether they are for internal use, scientific publication or public domain publication. For high quality output, high quality original images as well as access to good printers are needed. A grabbed video image may be acceptable onscreen, but it will not print in sufficient detail for publication. Storage and archiving are also important as different storage formats have different shelf lives. Thermal print-outs from sidescan sonar records will be affected by temperature, digital media by external voltage changes or magnetic fields. CDs quoted as having an indefinite shelf life are in fact limited to some 25 years. Important images should be kept at least in duplicate in separate locations in fireproof conditions. The quality of the material should be periodically checked and recopied, if necessary, onto newer standardised archiving formats. A comprehensive review of principles on this subject can be found in Glasbey and Horgan (1995).

Afternote. Finally, we would like to add that with our increasing use of the digital world, almost every image is now manipulated in some way or other, whether it is a change in the storage resolution, enhance contrast or change the coloration. Ethically speaking, the scientist should always note when publishing that original images have been manipulated (if only to say that they have been enhanced).

References

AODC (1985) *Code of Practice for the Safe Use of Electricity Underwater*. International Marine Contractors Association, London (www.imca-int.com). AODC 035.

Auster, P.J. & Tusting, R.F. (Guest eds) (1997) Scientific sampling systems for submersible vehicles *Marine Technology Society Journal*, **31**(3), 83.

Blondel, P. & Murton, B.J. (1997) *Handbook of Seafloor Sonar Imagery*. Wiley Praxis. Series in Remote Sensing, 314 pp. John Wiley & Sons, Chichester.

Carey, D.A., Rhoads, D.C. & Hecker, B. (2003) Use of laser line scan for assessment of response of benthic habitats and demersal fish to seafloor disturbance. *Journal of Experimental Marine Biology and Ecology*, **285/286**, 435–452.

Foell, E.J. & Pawson, D.L. (1986) Photographs of invertebrate megafauna from abyssal depths of the north-eastern equatorial Pacific Ocean. *Ohio Journal of Science*, **86** (3), 61–68.

Glasbey, C.A. & Horgan G.W. (1995) *Image Analysis for the Biological Sciences.* John Wiley & Sons Ltd., Chichester, England.

Greenstreet, S.P.R., Tuck, I.D., Grewar, G.N., Armstrong, E., Reid, D.G. & Wright, P.J. (1997) An assessment of the acoustic survey technique, RoxAnn, as a means of mapping seabed habitat. *ICES Journal of Marine Science*, **54**, 939–959.

Hobson, P.R., Lampitt, R.S., Rogerson, A., Watson, J., Fang, X. & Krantz, E.P. (2000) Three-dimensional spatial coordinates of individual plankton determined using underwater hologrammetry. *Journal of Limnology and Oceanography*, **45**, 1167–1174.

Holme, N. (1984) Photography and television. In: *Methods for the Study of Marine Benthos* (Eds N. Holme & A.D. McIntyre), pp. 66–98. IBP Handbook 16, 2nd Edition. Blackwell Science, Oxford.

Kenny, A.J., Cato, I., Desprez, M., Fader, G., Schüttenhelm, R.T.E & Side, J. (2003). An overview of seabed-mapping technologies in the context of marine habitat classification. *ICES Journal of Marine Science*, **59**, 129–138.

Kidd, R.B. & Huggett, Q.J. (1980) Rock debris on abyssal plains in the northeast Atlantic – a comparison of epibenthic sledge hauls and photographic surveys. *Oceanologia Acta*, **4**, 99–104.

Michaud, E., Desrosiers, G., Long, B., de Moutety, L., Crémer, J.F., Pelletier, E., Locat, J., Gilbert, F. & Stora., G. (2003) Use of axial tomography to follow temporal changes of benthic communities in an unstable sedimentary environment (Baie des Ha! Ha!, Saguenay Fjord). *Journal of Experimental Marine Biology and Ecology*, **285/286**, 265–282.

Nilsson, H.C. & Rosenberg, R. (2000) Succession in marine benthic habitats and fauna in response to oxygen deficiency: analysed by sediment profile-imaging and by grab samples. *Marine Ecology Progress Series*, **197**, 139–149.

Parry, D.M., Kendall, M.A., Pilgrim, D.A. & Jones, M.B. (2003) Identification of patch structure within marine benthic landscapes using a remotely operated vehicle. *Journal of Experimental Marine Biology and Ecology*, **285/286**, 497–511.

Pilgrim, D.A., Parry, D.M., Jones, M.B. & Kendall, M.A. (2000) ROV image scaling with laser spot patterns. *Underwater Technology*, **24**, 93–103.

Rhoads, D.C. & Cande, S. (1971) Sediment profile camera for *in situ* study of organism-sediment relations. *Limnology. Oceanography*, **16**, 110–114.

Rhoads, D.C. & Germano, J.D. (1982) Characterisation of organism-sediment relations using sediment profile imaging: an efficient method of remote ecological monitoring of the seafloor (REMOT^{TM} system). *Marine Ecology Progress Series*, **8**, 115–128.

Rhoads, C. & Germano, D. (1986) Interpreting long-term changes in benthic community structure: a new protocol. *Hydrobiologia*, **142**, 291–308

Rhoads, D.C. & Young, D.K. (1970). The influence of deposit feeding benthos on bottom sediment stability and community trophic structure. *Journal of Marine Research*, **28**, 150–178.

Rona, P.A. (2000) Deep diving manned research submersibles. *Sea Technology*, **41**(12), 15–26.

Rumohr, H. (1995). Monitoring the marine environment with imaging methods. *Scientia Marina*, **59**, 129–138.

Shand, C.W. & Priestley, R. (1999) A towed sledge for benthic surveys. *Scottish Fisheries Information Pamphlet 22/1999*. 8 pp.

Solan, M., Germano, J.D., Rhoads, D.C., Smith, C.J., Michaud, E., Parry, D., Wenzhöfer, F., Kennedy, R., Henriques, C., Battle, E., Carey, D., Iocco, L., Valente, R., Watson, J. & Rosenberg, R. (2003) Towards a greater understanding of pattern, scale and process in marine benthic systems: a picture is worth a thousand worms. *Journal of Experimental Marine Biology and Ecology*, **285/286**, 313–338.

Smith, C.J. & Papadopoulou, K.N. (2003) Burrow density and stock size fluctuations of *Nephrops norvegicus* in a semi-enclosed bay. *ICES Journal of Marine Science*, **60**, 798–805.

Smith, C.J., Rumohr, H., Karakassis, I. & Papadopoulou, K.-N. (2003) Analysing the impact of bottom trawls on sedimentary seabeds with sediment profile imagery. *Journal of Experimental Marine Biology and Ecology*, **285/286**, 479–496.

Wernli, R. (2000) The present and future capabilities of Deep ROVs. *Marine Technology Society Journal*, **33**(4), 26–40.

Chapter 4
Diving Systems

C. Munro

4.1 Diving systems

All diving depends on a supply of breathing gas. Gas supply systems can be broadly classified into self-contained or SCUBA systems (properly known as self-contained breathing apparatus) and remotely supplied systems, where gas is piped to the diver through an umbilical tube.

SCUBA

The majority of scientific diving is undertaken using SCUBA systems. This system allows the diver greater freedom of movement than most remotely supplied systems. It is also the system taught by all recreational diving training agencies, the route by which most scientists initially learn to dive. SCUBA equipment is generally relatively lightweight, and low cost. Crucially for many scientific studies, the diver does not have to overcome the drag exerted by currents on a breathing gas umbilical (which can be considerable when the diver is operating a significant distance away from the umbilical tender). It is therefore much easier for the free-swimming diver to hover just above the seabed without being swept rapidly down-tide, or to maintain position on the seabed without being heavily weighted. These are crucial requirements for many scientific studies. SCUBA has the disadvantage that the dive profile is limited by the quantity of breathing gas a diver can carry.

Remotely supplied systems

Remotely supplied systems, where the diver's breathing gas is supplied via an umbilical either directly from a low-pressure compressor or from storage cylinders, is the most commonly used system for most civil engineering and offshore oil-industry divers. The remote supply is usually surface-supply, e.g. a compressor or air bank located on a boat or quayside, but it may also be a supply from a diving bell or a habitat to a tethered diver. Remote supply systems normally use positive pressure full-face band-masks or lightweight diving helmets fitted with hard-wire communication systems (see 'Voice communications' section). Such systems are also widely used by archaeological divers, who are often working within a small, well-defined area, frequently in very low visibility. They are, however, rarely used by biological divers, largely due to the considerable bulk and weight of the equipment and the limited freedom of movement the diver has (being tethered by the umbilical). Situations where surface-supply systems are used by biological divers include when working in very low visibility or polluted waters, such as sampling close to effluent outfalls, or where regulations require it (e.g. if conducting scientific diving around offshore oil or gas installations).

Hookah systems

Hookah systems (Fig. 4.1) are lightweight surface supply systems, whereby the diver is supplied with air from a small, portable compressor through a light air hose. These are mainly low-pressure systems, where the compressor output is around 6 bar (88 psi). Hookah systems are in common use in Australia, Asia and the Americas by recreational divers, with several systems aimed primarily at the recreational market being widely available (e.g. *Brownie Third Lung*, *Air-line* and *Air-Dive* systems). However, they are most frequently used by commercial marine harvesters, e.g. abalone, sea urchin and sea cucumber divers. Hookah systems offer the benefits of unlimited air supply without the excessive weight and considerable cost of conventional surface-supply equipment. Although used relatively rarely by biological divers, hookah systems are used at times to supply divers working out of the

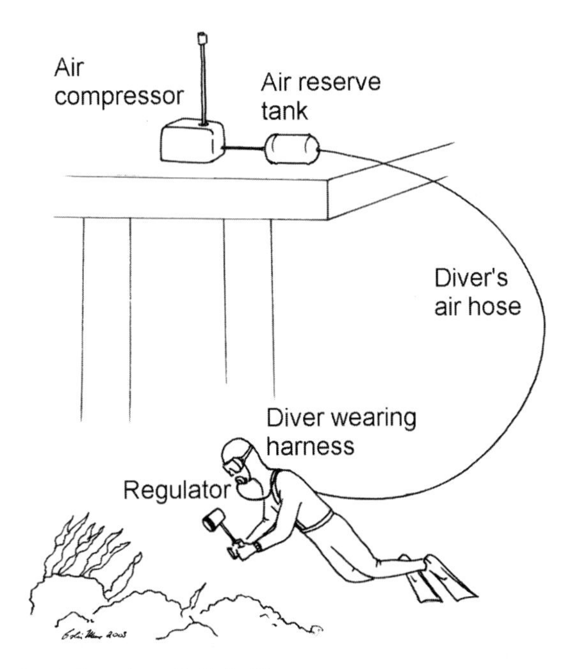

Fig. 4.1 Diver using hookah system. © Colin Munro

Aquarius underwater habitat (see the 'Saturation diving' and 'underwater habitats' section). As with conventional surface supply equipment, hookah systems are widely used by archaeological divers.

Breathing gas and supply systems

Air and nitrox

Most SCUBA systems use air as the breathing gas. However, the use of oxygen-enriched air (commonly known as nitrox) is growing in popularity for scientific studies (e.g. Dinsmore, 1989a, 1989b; Joiner, 2001; Holt, 2001; Munro, 2001; Stanton, 1991). At the time of writing, approximately half of the research dives undertaken at Dunstaffnage Marine Laboratory, Scotland, were conducted using nitrox mixes (Simon Thurston, personal communication, 2002). Nitrox contains higher-than-air percentages of oxygen (i.e. >21%). This reduces the partial pressure of nitrogen in the breathing gas, and so the amount of inert gas that will dissolve into the body at any given pressure. Consequently, the decompression penalty incurred is reduced. An additional potential benefit accrued relates to the narcotic effect of nitrogen when breathed at elevated partial pressures. It is well documented that the cognitive abilities of divers breathing compressed air at depths of 30 m or more are significantly impaired (e.g. Baddeley et al., 1968; Fowler et al., 1985; Rogers & Miller, 1989). Behnke et al. (1935) first identified increased partial pressures of nitrogen as the prime narcotic agent when breathed in compressed form. Thus, reduction of the partial pressure of N_2 breathed at depth may be expected to reduce the narcotic effect of that gas. Whilst rigorous scientific studies are lacking, many divers report experiencing reduced symptoms of narcosis when breathing nitrox mixes as opposed to air within the 25–40 m depth range (the range most suited for use of nitrox). A cost-benefit comparison of nitrox and air SCUBA systems in scientific diving operations is given in Holt (2001).

Heliox and trimix

Heliox, the combination of helium and oxygen was initially proposed by Professor Elihu Thompson, an American scientist better known for his work on high voltage generators and X-rays, as a means of reducing the narcosis divers experienced when breathing compressed air at depth. Unfortunately, helium was prohibitively expensive at that time. However, the discovery of vast reserves of helium in Texas natural gas wells in the early 1920s dramatically changed this. As at that time America virtually controlled all the world's helium reserves, the early work was conducted solely by the US Navy. Helium is approximately four times less narcotic than nitrogen, hence breathing heliox virtually eliminates the problem of inert gas narcosis for the diver operating at depth. Saturation and desaturation speeds of inert gases are inversely proportional to the square root of their atomic mass. Thus, body tissues will saturate or desaturate with helium some 2.7 times faster with nitrogen. Consequently, no-stop times will be shorter for heliox than air (one reason why heliox is an inappropriate gas for shallow dives). This is offset on deeper dives by shorter decompression stops, especially if nitrox is breathed during decompression. Helium conducts heat approximately six times faster than air. Thus, helium mixes should not be used for dry suit inflation. A common misconception is that breathing heliox cools the diver at a significantly greater rate than breathing air. Although it conducts heat faster, heliox has a lower heat capacity than air as helium is a less dense gas than nitrogen. Heliox is generally considered most appropriate for use as a breathing gas at depths of 60 m (200 ft) or greater.

Trimix, a mixture of oxygen, nitrogen and helium, is frequently used in place of heliox for deep diving. By replacing some of the helium in the inert gas component of the mix with nitrogen, the cost of breathing gas is significantly reduced. The trade-off is that a degree of nitrogen narcosis is introduced. High-pressure nervous syndrome (HPNS) is a condition causing tremors, dizziness, nausea and impairment of cognitive function that can occur at depths greater than 160 m (525 ft), believed to be due to hyper-excitation of nerve cells within the brain. There is some suggestion that the nitrogen in trimix mixtures may help ameliorate the effects of HPNS.

In 2001, America's National Oceanic and Atmospheric Administration (NOAA) Undersea Research Program ran a trimix and technical diving programme for diving scientists working on deep-water species, at the Caribbean Marine Research Centre in the Bahamas. Following on from this course, trimix was used for investigations of invertebrate and algal biodiversity, and chemical ecology, at working depths of 300–400 ft (91–122 m) on the Bahamian outer reef walls (Dr Marc Slattery, University of Mississippi, and Dr Michael Lesser, University of New Hampshire, personal communication, 2002). Pyle (2000) identified the deep regions (60–150 m) of coral reefs as a zone where fish biodiversity is still largely unrecorded due to the expense of deploying submersibles and their inherent limitations when operating in complex habitats. By conducting dives using trimix as the working depth breathing gas, Pyle and his collaborators have collected and identified 50 fish species new to science (e.g. Earle & Pyle, 1997; Gill et al., 1996; Pyle & Randall, 1992; Randall & Pyle, 2001a, 2001b; Pyle, 1996, 1998, 2000).

Open-circuit SCUBA

Since the invention of the aqualung in the early 1940s, most self-contained diving has been conducted using open-circuit diving systems, i.e. where breathing gas is delivered to the diver at close to ambient pressure, and exhaled gas is then vented into the surrounding water through a one-way valve. This is the principle of the early twin hose, single-stage demand valves, and is the way current single hose, twin-stage regulators still operate. For the diving scientist it has many advantages. The system is widely available, relatively cheap and robust. The operation of open-circuit SCUBA is also simple and requires little training (although competence in differing conditions, and with different gas mixes may require further training). Perhaps its greatest advantage for someone conducting work underwater that requires a high level of concentration is that it is a very reliable and almost totally automatic system, requiring only a modest level of monitoring during the dive. Given these advantages, it is likely that open-circuit SCUBA, containing air or nitrox gas mixes, will remain the system of choice for most scientific diving conducted in water less than 45 m depth, for the foreseeable future. However, open-circuit SCUBA systems have a number of disadvantages. Air bubbles vented into the water are noisy, and can influence the behaviour of many fish species and vagile invertebrates (e.g. Chapman et al., 1974; Sayer et al., 1996) making observation of natural behaviour difficult. Lobel (2001) notes that the near-field vibrations produced by exhaled bubbles are probably similar to the hydrodynamic disturbance caused by fast-moving predators. Open-circuit systems are also quite wasteful of breathing gas, as the exhaled oxygen and diluent gas (e.g. nitrogen in air) are lost from the system once exhaled by the diver. Thus, for work requiring extended dives or deep dives the diver must carry large amounts of compressed breathing gas. Additional gas cylinders worn by the diver not only make entry and exit from the water very physically demanding, but also restrict the diver's ability to manoeuvre freely over a study site without disturbing it, and make it difficult to carry equipment necessary for the study (e.g. cameras, sampling devices).

Rebreathers

The alternative to open-circuit diving systems is some form of rebreather. Rebreathers differ from open-circuit systems in that they are able to capture and re-use all or part of the diver's exhaled gas. The inspired and expired gas is maintained within a breathing loop, thus using the total volume of gas carried by the diver more efficiently. This consists of an inhalant and exhalent tube attached to the mouthpiece. A counterlung within the loop provides flexibility, allowing the loop volume to change as the diver inhales and exhales. Most systems incorporate two counterlungs, an inhalation and an exhalation bag, balancing breathing resistance. Exhaled gas passes through a CO_2 'scrubber' before passing onto the inhalant side of the loop. As a consequence, rebreathers greatly extend the potential dive duration of a given amount of breathing gas. They also dramatically reduce exhalent bubbles and consequently bubble noise. Rebreathers fall into three categories: oxygen rebreathers, semi-closed rebreathers and closed circuit rebreathers.

Oxygen rebreathers

In oxygen rebreathers, 100% oxygen is the only compressed gas carried by the diver. It is generally the simplest rebreather in construction, and its development occurred around the same time as open-circuit SCUBA. Whilst oxygen rebreathers are the most gas-efficient (in terms of the amount of gas carried that is available for metabolic use) they suffer from the severe limitation that they cannot be used deeper than around 6 m due to the risks of acute oxygen toxicity symptoms at partial pressures (pp) of O_2 greater than 1.6 bar (though the pressure at which such symptoms will occur is known to be highly variable, 1.4 bar is currently considered the maximum safe pp for working dives and 1.6 bar when the diver is at rest). Oxygen rebreathers have little use in current scientific diving procedures.

Semi-closed rebreathers

Semi-closed rebreathers (SCRs) consist of a breathing loop, to which breathing gas is supplied periodically from a compressed gas cylinder in order to compensate for oxygen consumed metabolically and gas lost from the system (e.g. through mouthpiece leakages). They allow a small amount of gas to vent periodically from the breathing loop, in order to prevent build-up of the inert component (normally nitrogen or helium) of the gas being breathed. Two principal types of SCR are manufactured: active addition and passive addition. Active addition is the simpler form, adding gas to the breathing loop at a constant, pre-set rate; passive addition SCRs link the rate of addition to the diver's breathing rate, a more complex but also more efficient design. Several SCR units have entered the recreational dive market in recent years. Shallow water systems primarily designed for nitrox diving are now comparable in price to conventional open-circuit SCUBA.

Closed-circuit rebreathers

Closed-circuit rebreathers (CCRs) differ from SCRs in having two separate gas cylinders; an oxygen cylinder and a diluent gas cylinder (generally air, nitrox, heliox or trimix). The oxygen partial pressure (ppO_2) is maintained at a relatively constant level, generally through the action of electronic sensors and servos. As oxygen is depleted from within the breathing loop, it is replaced from the oxygen cylinder. The diluent gas cylinder tops up the breathing loop as the gas in the breathing loop compresses with depth, and in some designs it also acts as an emergency open-circuit bail-out gas. Thus, in normal usage, gas is only lost from the breathing loop due to expansion as the diver ascends. Closed-circuit rebreathers are of more complex design than SCRs, and generally cost significantly more. The training requirement is also normally greater.

Use of rebreathers

Rebreathers have been used in a growing number of scientific studies. During studies of reef fish acoustic behaviour, Lobel (2001) found he was able to approach fish more closely, and that he observed greater numbers of fish and fish species, while using an SCR. Sayer and Thurston found that the numbers of fish observed during transect counts were significantly higher when using SCRs than when using

Fig. 4.2 Diver using closed circuit rebreather whilst surveying black coral off Hawaii. © Richard Pyle, University of Hawaii.

open-circuit systems, although the numbers of fish species were not (unpublished data, Martin Sayer, personal communication). Pyle (1999) used CCR systems to study and identify reef fish species living at depths between 60 and 90 m; Pence and Pyle (2002) also used CCR systems to conduct deep reef fish surveys around Fiji. Parrish and Pyle (2002) compared costs and efficiency of open-circuit SCUBA systems with those of CCR systems, whilst surveying commercial black coral beds off Hawaii (Fig. 4.2). They found that, for deep dives, the set-up time, costs of consumables and decompression penalties were significantly greater for open-circuit than for closed-circuit systems. However, these operational costs must be weighed against the higher initial purchase costs and training times.

4.2 Saturation diving and underwater habitats

Saturation diving works on the principle that once the diver's body tissues have become saturated with dissolved gas (for the pressure experienced at that depth), then extending the dive time creates no additional decompression penalty. By making excursions from a pressurised habitat, the diver's time on the seabed and the frequency of repeat excursions is limited only by cold and the capacity of his self-contained breathing equipment.

On September 14, 1962, Albert Falco and Claude Wesly began a week's immersion, living at a depth of 10 m (33 ft) off the coast of Marseilles in a large cylindrical habitat called *Diogenes*. This was Jacques Cousteau's *Conshelf I* project. Eight days earlier, and approximately 100 miles away, Robert Stenuit spent 24 hours living at 60 m (200 ft) depth, breathing oxy-helium in a habitat designed by American inventor Ed Link. These parallel trials were the first real attempts by scientists to live and work under pressure on the seabed for extended periods. *Conshelf I* and Link's '*Man-in-the-Sea*' programmes were followed by a series of more ambitious projects through the early and mid 1960s: *Conshelf II* and *III*; '*Man-in-the-Sea II*' and the US Navy's *Sealab* habitats extended both the depth and duration for working saturation divers. *Tektite*, a co-operative effort between the US Navy, NASA and the US department of the Interior, was the first subsea habitat from which significant marine biological studies were undertaken. This was followed by *Hydrolab*, which operated from 1966 to 1985 in the Bahamas and Caribbean. Over 600 researchers worked out of *Hydrolab* during this time.

Aquarius (Fig. 4.3), an undersea laboratory run by NOAAs National Undersea Research Program (NURP) and the University of North Carolina, was built in 1986. Originally deployed in the US Virgin Island, it was moved to the Florida Keys, off Key Largo, in 1993. Between 2000 and 2002, 17 biological research missions were undertaken, each lasting 8–9 days. *Aquarius* is stationed at approximately 16 m depth, 4.5 km offshore. A 10 m diameter Life Support Buoy moored directly above provides breathing gas, power and communication links to the habitat. The bottom times available to *Aquarius* divers are much greater than those of surface-orientated divers; NOAA has calculated that 60–70 days of surface-orientated diving would be required to provide the equivalent bottom time to a 10-day mission (Miller & Cooper, 2000). Diving excursions from the laboratory are conducted using protocols similar to those employed by cave divers, as the extended decompression required creates a physiological 'ceiling' preventing the divers from ascending. Researchers swim to and from their study sites, which may be up to 200 m distant from the laboratory, along radiating travel lines (Mark Hulsbeck, NURC research diver,

Fig. 4.3 The undersea laboratory *Aquarius*, a diving habitat off the coast of Florida. © University of North Carolina at Wilmington. Aquarius is owned by NOAA and is operated by the University of North Carolina at Wilmington.

personal communication, 2002). Saturation diving is particularly suited to studies requiring large amounts of in-water time over a period of days or weeks.

4.3 Data collection and recording

Since the ability of diving surveyors to recall observations made during a dive is known to be worse than the recall of similar observations made in the terrestrial situation (Godden & Baddeley, 1975), it is normally wise for surveyors to equip themselves with some means of directly recording *in situ* observations. A range of widely used methods is described below.

Written recording

Often the simplest methods are best. Plastic or laminate board slates are a cheap, low-tech option; they are easy to construct and to replace, when lost or damaged. Waterproof notepads are widely available from nautical and surveying supply companies. Though these will greatly increase the amount of data a diver can record during a dive, they are not as robust as slates, and pages may sometimes be lost underwater. In temperate and polar waters, cold quickly reduces manual dexterity, with the result that handwriting can be difficult to read when transcribing later. Where practical, a pre-prepared layout to the slate can greatly improve the dive efficiency (e.g. a species checklist with space for recording tally, abundance or density). A point to bear in mind is that pencils float, and so should be tied to the slate or the diver otherwise a pencil momentarily released is a pencil lost.

Audio recording

The surveyor's spoken observations underwater can be recorded either by direct recording, or by transmission to the surface. Use of communication equipment to record data enables surveyors to dive without the encumbrance of handheld slates or photography equipment. Audio data can be easily stored on tape, CD etc., but is very time-consuming to analyse. The current market in underwater communication equipment is expanding rapidly, so communication equipment is likely to be increasingly used as a tool for marine survey recording.

Direct recording usually consists of a waterproof housing strapped to the diver's cylinder, with a microphone mounted in a specialised mouthpiece. Gamble (1984) describes a bone conduction system in which the microphone is held against the diver' skull by the suit hood. Direct recording systems are relatively cheap and are ideal for use in surveys where divers are free-swimming on SCUBA.

Voice communication systems have long been standard requirements in commercial diving operations, but are still relatively infrequently used in scientific diving programmes. This is partly due to the fact that until quite recently, most voice

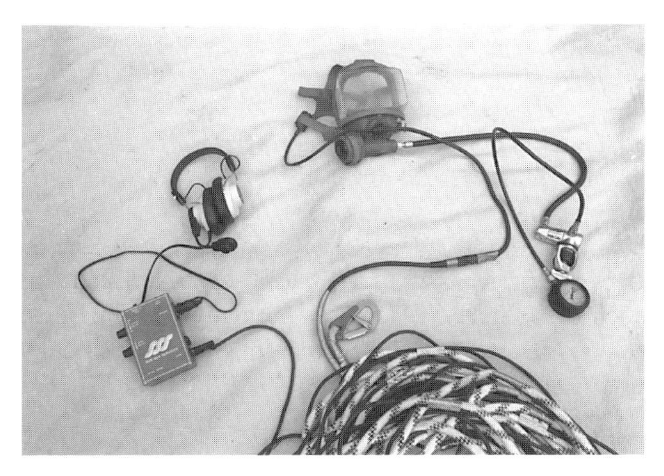

Fig. 4.4 Diver and surface units of a hard-wire voice communication system. © Colin Munro

communication systems were 'hard wire'; i.e. they required a physical connecting line from the diver to the surface transceiver (Fig. 4.4). The SCUBA diver then became a tethered diver, with all the limitations this involves. However, situations where tethered diving is required, or at least preferred, e.g. under ice or low visibility conditions, are also ones where the added safety of direct communication with the surface may be preferred. Hard-wire voice communications can be either two-wire (simplex) or four-wire (duplex or 'round robin') systems. Simplex systems allow the diver's speech and breathing to be continuously monitored by the surface team while there is no surface-to-diver communication. However, when the press-to-talk button on the surface set is depressed, the diver cannot be heard. Duplex systems are more expensive, but allow continuous communication both ways. At the time of writing, hard-wire communication systems still provide the clearest and most reliable diver-to-surface voice communication.

Through-water communication systems have been around for many years, but their reliability has not always been one hundred per cent. Most work by transmitting sound on a sideband carrier frequency. One problem is that the sound quality deteriorates rapidly over distance (sound intensity decreases with the square of distance due to spherical spreading) although many systems currently claim ranges of up to one kilometer in ideal conditions. Sound quality may also deteriorate through diffraction or reflection by seabed objects or thermoclines, or through interference from engine noise. That being said, through-water communication systems offer the great advantage of maintaining the freedom that SCUBA equipment provides.

Some through-water communication systems are available that can be used with normal SCUBA half-masks. These generally incorporate modifications to the diver's regulator mouthpiece in order to facilitate speech. Most through-water and hard wire systems are designed to work with positive pressure full-face masks, such as the *AGA Divator* or *Kirby Morgan EXO 26*. Most systems have the facility for all diver communication to be continuously recorded on audio tape.

Image recording

Still photography

As underwater photography is a vast subject on its own, this section will touch on only a few relevant points. It is assumed the reader is familiar with basic photographic principles, as there is no scope here to discuss systems and methodologies from first principles. For more detailed information on taking underwater photographs and the systems available, *The Manual of Underwater Photography*, by De Couet and Green (1989) is a good, if slightly dated text.

The camera is an extremely useful tool for the underwater biologist. It allows large amounts of data to be collected very rapidly, an important factor given that time underwater is usually quite limited. Images collected from the same area over time can be used to identify and track changes in habitat or species assemblage, species growth or behavioural patterns. Photographic images can be projected at high magnification to count or identify specimens which are difficult to see clearly underwater. They can also be used to record the appearance of species with which the researcher is not familiar. With the advent of image analysis software, the usefulness of photography to the diving scientist has increased considerably. Photographic images may be scanned, or digital images loaded directly on to a PC for analysis and comparison. This technique has particular applications for recording time-series data on growth or change.

Underwater photography has undergone considerable changes in the last 20 years, with significant improvement in technology and a wider range of systems now available. Yet, until very recently, only conventional film cameras provided sufficient image resolution, lens quality, choice of lens focal lengths, and full compatibility with suitable flash (strobe) attachments to be useful for producing detailed survey images. Digital cameras are currently improving rapidly, and can now be considered a serious choice. At the time of writing, 35 mm film still provides significantly better exposure latitude (i.e. the tonal range from white light to absence of light) than the equivalent dynamic range of images derived from digital cameras. However, the image resolution of better quality digital cameras, in the region of 5–6 megapixels, is broadly similar to that of 35 mm film (comparing effective pixels to resolved lines per inch of film). A further consideration is the ease of viewing and permanence of images. Film needs to be developed before the images can be viewed. For fieldwork in remote locations, this may mean waiting until the end of a period of fieldwork (although E6 slide film can be developed quite easily in the field). Thus, problems with the camera system or film may not be picked up until several days after everyone has gone home. Film may also need to be scanned and converted into an electronic image before detailed analysis can be conducted. Film can be easily damaged in dusty conditions, with particles embedding in the soft emulsion. However, film has the advantage that it is a stable medium with a very long life span, an important consideration for long-term monitoring programmes. Digital storage media are still a relatively young and rapidly evolving technology, and the stable life span of many formats has not yet been fully tested. A further consideration is that the area of the charge-coupled device (CCD) photodiode array in digital cameras is significantly smaller than the 24 × 36 mm frame of 35 mm film. One result of this is that, at any given camera-to-subject distance and angle of view, the depth of field will be significantly greater when using a digital camera. This difference in light-sensing area also has the effect of changing the angle of view obtained from lenses of a given focal length, when attached to digital or 35 mm film cameras. For example, on a housed 35 mm camera, a 20 mm lens will give an angle of view of around 90%. Most digital cameras will require a significantly shorter focal length to obtain the same angle of view. Precise figures cannot be given, as currently digital cameras, which differ in sensor areas between makes and models, lack the standardization of film cameras.

Given the above information, there are three fundamental choices available to the diving biologist: submersible 35 mm film cameras; housed 35 mm single lens reflex (SLR) film cameras and housed digital cameras. Waterproof compact cameras are not considered here as they are generally too limited to be of much use other than for producing snapshots of people working in shallow water. The medium-format housed cameras are also not of much use, although they can produce excellent images and have been used successfully for underwater monitoring (e.g. Lundälv, 1971, 1985). Medium-format cameras have major drawbacks in terms of size, weight, cost and limited number of frames per film roll. They also have limited close focussing ability and significantly reduced depth of field for equivalent lens angle of coverage. Given the improvement in the quality of 35 mm film stock that has occurred over the past two decades, the larger film area provided by medium-format systems no longer provides the earlier benefits. Digital camera systems are not discussed in detail here. The current lack of standardisation makes succinct description difficult, while the rapid advance in digital technology and the steady stream of new models and accessories becoming available means that a detailed review of digital camera systems would very quickly be outdated.

Submersible film cameras

The Nikonos camera has generally been the preferred choice for the diving scientist. Developed from the Calypsophot underwater camera designed by Belgian engineer Jean de Wouters for Jacques Cousteau, six Nikonos models have been made so far. Five of the six models are rangefinder design, i.e. the viewfinder does not look through the lens, and focussing is achieved by estimating the subject range. Nikonos cameras are widely used because of their compact size, robustness and the wide range of accessory lenses and strobes available. Nikonos lenses are also optically excellent. However, not being SLR, they suffer from parallax error between the viewfinder and lens, which can cause problems for the inexperienced user working at close range. Lack of through-the-lens focussing can also cause focussing problems for inexperienced users as the operator has to judge subject distance and also keep in mind the depth of field available at a particular setting.

Nikonos I and II models are no longer recommended for survey work as they were designed for use with bulb rather than electronic flash, which severely limits their usefulness. The Nikonos III, IVa and V models are still widely used. The Nikonos III is purely mechanical, with no in-built light metre. This lack of electronics makes the camera extremely robust and so suitable for harsh environments, especially when used in conjunction with a handheld light metre. It is however, a very old camera and servicing is becoming difficult. The Nikonos IVa has through-the-lens metering, but does not have TTL (through-the-lens) flash control. The Nikonos V has a built-in exposure metre, aperture priority automatic exposure and TTL flash control for suitable flashguns. The Nikonos V is still the most popular for survey work, although the electronic control is quite simple by today's standards (it was designed in the 1980s), and it is still a rangefinder camera. A further problem is that Nikon ceased production in 2001 with, as yet, no plans for a replacement model. The Nikonos RS was brought out in 1992 as the first underwater SLR. In addition to SLR capability, it possesses more sophisticated light and flash metering than the Nikonos V and a range of specifically designed lenses, including a wide-angle zoom and a macro lens. Unfortunately, the cost was significantly higher than similarly specified land SLRs and housings. Disappointing sales led to its being discontinued in 1996, although many are still in use at the time of writing. For the diving scientist the Nikonos RS possesses most of the advantages of a modern SLR, without the disadvantages of bulk associated with housed cameras.

Other 35 mm underwater cameras are available, but most are limited by the lack of accessories available. Exceptions are the Sea & Sea Motormarine and Seamaster cameras, for which a range of wide-angle and close up supplementary lenses and strobes are available.

Housed 35 mm SLR camera systems

Housed cameras offer significant advantages to the professional photographer. They allow precise framing of the subject, producing an aesthetically pleasing image. They also allow precise focussing on a particular part of the image (most relevant when using macro lenses with a narrow depth of field) either by manual through-the-lens focussing or using an autofocus facility. However, these are not necessarily the prime concerns of the diving scientist. One significant advantage is the facility to use a zoom lens, and thus to be able to photograph both small, close objects and take wide-angle views of habitats with the same set-up. For monitoring purposes, having a set focal length with which all replicate photographs are taken is very desirable (as discussed later, different focal lengths may dramatically change the appearance of a station, and cause differing degrees of distortion). Thus, a lens with a zoom facility being used by many workers involved in one monitoring programme may create data analysis problems that outweigh the benefits. Housings for cameras are becoming smaller, but they still tend to be larger than submersible cameras such as the Nikonos or Motormarine. Most are also less robust and more time-consuming to set up. These may be important considerations on a working survey vessel, where underwater photography is only a small aspect of the study and dry, clean space is at a premium. The bulk of camera housing (and the ease with which dome ports may be damaged) also tends to mean that they cannot simply be clipped on to the diver's buoyancy jacket as an ancillary tool while other parts of a study are being conducted. Camera housings are mostly constructed from aluminium alloy or clear plastic; very few are constructed using stainless steel. Although aluminium housings are undoubtedly more robust than those constructed from plastic, they are heavier (on the surface) and significantly more expensive. Plastic (polycarbonate, acrylic, ABS etc.) housings are much cheaper and are available for a far wider range of SLR camera models. As they are transparent, any leaks that occur can be spotted much more quickly than on metal housings. They tend to be generic in design, with more space around the housed camera, making them slightly bulkier than aluminium housings. The control shafts may not always move as smoothly at depth, due to slight distortions of the housing under pressure.

Photo-monitoring techniques

Photo-monitoring of individual colonies or species' assemblages using fixed location quadrats or transects is a widely employed technique for gathering data on temporal change within assemblages and linear growth of organisms (e.g. Green, 1980; Bullimore, 1983, 1986, 1987; Lundälv, 1985; Hiscock, 1984, 1986; Munro, 1998, 2000a, 2000b). For monitoring of epibenthic communities, the area covered by each photograph will be selected partly by the size and density of the species forming the community, and the degree of spatial heterogeneity observed. For photographing areas of around 20 cm × 15 cm or smaller, a Nikonos camera, supplementary close-up lens and framer used in conjunction with either a 28 mm or 35 mm prime lens, is generally the preferred choice. For larger areas, a wide-angle lens is normally used. The main choices are either a 15 mm or 20 mm lens for the Nikonos camera (giving an angle of view of c. 94° and 78°, respectively) or a 20 to 24 mm lens for a housed SLR (giving an angle of view between 94° and 84°).

Even with a wide-angle lens, the maximum area which can be covered will be constrained by water turbidity. English et al. (1997), in considering photo-monitoring on coral reefs, recommend using a Nikonos camera with 15 mm lens, located on a tetrapod where the ends of the legs form a one-metre square (camera to subject distance of 800 mm). In more turbid temperate waters, smaller frame sizes tend to be used, facilitating shorter camera-to-subject working distances. Photo-monitoring quadrat sizes of 666 × 500 mm have been used successfully for monitoring sessile epilithic fauna around Lundy Island, Bristol Channel, UK (Hiscock, 1986); while a frame size of 520 mm × 400 mm (camera-to-subject distance of 450 mm) was used for the more turbid, estuarine waters of Milforld Haven estuary, South Wales

Fig. 4.5 A photo-monitoring framer with camera and strobes attached, used to monitor epilithic faunal turfs. © Blaise Bullimore, Countryside Council for Wales.

(Munro, 1999). At Skomer Marine Reserve, also in South Wales, a 500 × 400 mm framer was used for photographing species within low hydroid/bryozoan dominated epilithic turfs (Fig. 4.5). Larger framer sizes (up to 1000 × 700 mm) are used for recording the presence of bigger colonies, e.g. massive bryozoans such as *Pentapora foliacea* (Blaise Bullimore, personal communication). All of the above systems used Nikonos cameras fitted with 15 mm lenses.

Accurate comparison of images from the same station taken at different times is highly dependent on the precision of camera positioning. Photo-transects can be quickly and simply constructed by stretching a tape or marked line between fixed points. However, this method rarely provides sufficient accuracy of positioning to ensure a good correlation of images between monitoring periods. As rock or coral surfaces are rarely perfectly flat, the transect will tend to pass above or below protrusions as it is being set up. Thus, on some occasions parts of the transect line will be higher than true horizontal, at other times lower. The longer the transect and more widely spaced the fixing points, the more this problem is exacerbated.

An additional consideration is that wide-angle lenses, in particular, suffer significantly from peripheral distortion and very steep perspective (i.e. subjects closer to the camera appear much larger than those more distant). Whilst this effect makes little difference to the aesthetic appeal of a photograph, it can cause significant problems when trying to make size comparisons. Incorrect placement of the camera, causing a shift in location of a colony within the image frame, can cause apparent growth or shrinkage where none has in fact occurred (Fig. 4.6). Steepened perspective also makes accurate growth measurements of three-dimensional colonies technically complex to achieve, even when the camera is precisely repositioned every time.

When photographing species growing on rock faces, a framer is frequently used to ensure the camera focal plane is parallel to the rock face, and camera-to-subject

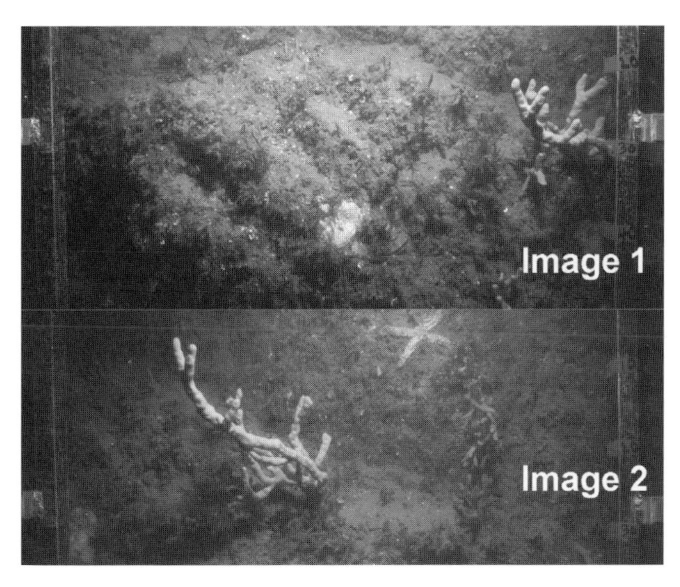

Fig. 4.6 Apparent growth of an erect sponge due to peripheral distortion and steepened perspective. © Colin Munro

distance is constant. The frame can also provide a scale bar and act as a moveable quadrat. Permanent markers (e.g. pins fixed into drilled holes or pitons hammered into crevices) help ensure precise positioning of the framer. These can be used to mark two corners of the photo-monitoring quadrat. For photographing sequential quadrats along a transect, a bar or rail firmly fixed to the rock, with locating marks along its length, works well. A 25 mm diameter pipe secured by pitons is one method employed at Skomer Marine Nature Reserve, South West Wales (B. Bullimore, Countryside Council for Wales, personal communication, 2002). The camera framer is hung on the rail by hooks, then aligned with the distance marks. This process is repeated along the rail. Methods for fixing attachment points or marks onto the seabed are discussed in more detail in the underwater site marking and relocation section.

When extracting measurements of species from photographs, either by direct measurement from prints or slides, or using image analysis software on digital images, a calibrating scale needs to be placed alongside the species of interest. Stainless steel or aluminium rules are convenient as they do not float or rust. However, polished metal is highly reflective and care needs to be taken when placing the rule to ensure that light from the flash does not bounce directly off the rule back into the camera lens, 'burning out' the calibrating scale on the photograph. Positioning of the scale as close to the subject as possible is critical, given the peripheral and perspective distortion that occur with wide-angle lenses. As an example, using a Nikonos and 15 mm lens system at a camera-to-subject distance of 300 mm, placement of the scale 40 mm in front of the subject will create an error in the region of 15%. For this reason, calibrating and extracting size measurements for three-dimensional colonies can be technically complex. Additionally, scale will

Fig. 4.7 Distortion inherent in a wide-angle lens causing changes across a photo-monitoring image. © Colin Munro

change from the centre to the periphery of the image (Fig. 4.7), thus measurements taken near the edge of the image should be treated only as approximations.

For photographing large, erect, planar species such as gorgonians and some branching sponges, a matt, opaque scaled background (Fig. 4.7) provides a calibration scale. Placing the scaled background directly behind the target colony isolates it from other colonies close behind. This helps prevent confusion as to whether a particular branch is part of the target colony or one growing directly behind it, a confusion that can easily occur when analysing photographs taken where colonies occur in dense aggregations.

The requirements for precise station marking in monitoring programmes and the analysis problems created by camera misalignment and methods to solve them are discussed in greater detail in Bullimore and Hiscock (2001), Munro (2000b) and Grist (2000).

Except in shallow and clear waters, artificial lighting is essential to expose images correctly, to ensure a good depth of field and to provide colour in the image. A flash with a beam angle that covers at least the entire angle of view of the lens used is needed (e.g. around 100° for a 15 mm lens). It should also have sufficient flash energy output (normally at least 80 Joules for coverage of a wide-angle lens) to allow a small aperture to be used, providing good depth of field. Dual flash set-ups help eliminate shadowing, which can often mask areas of the picture, especially where the rock surface has noticeable mounds and depressions.

Stereo-photography

Twin camera stereo-photographic systems have been used successfully in Scandinavia (e.g. Green, 1980; Lundälv, 1971, 1985) and as part of Skomer Marine Nature

Reserve's sublittoral monitoring programme (e.g. Bullimore, 1983, 1986). The system used at Skomer Marine Nature Reserve has a reference frame constructed from 25 mm box-section aluminium, filled with expanded polyurethane to prevent water ingress and add buoyancy. Twin Nikonos are mounted on the frame in such a way that they share the same focal plane and focal paths. The original system designed for the reserve's monitoring programme is fully described in Bullimore (1983) with additional modifications described in Bullimore (1986).

The distance between cameras, camera-to-subject distance and lens choice is determined to ensure a high degree of overlap between each pair of photographs (necessary for three-dimensional viewing) and that the reference frame lies completely within the field of view of each camera. Choice of camera-to-subject distance will be determined by the prevailing underwater visibility. Bullimore and Hiscock (2001) suggest that a minimum visibility of four times the camera-to-subject plane distance is required in order to gain useable images. Further details of system design and minimum monitoring areas for adequate community description are given in Bullimore and Hiscock (*op. cit.*). To view stereo photography paired images a stereo-comparator is used. A system developed for Skomer Marine Nature Reserve monitoring programme, using an adapted stereo microscope, is described by Bullimore (1987).

One advantage of stereo-photography over single camera photography is that three-dimensional images are generated, significantly increasing the number of species that can be identified within an image of faunal turf (B. Bullimore, personal communication, 2003). The twin viewpoints also help reduce the masking effect of canopy-forming species. Stereo-photography can also increase the potential for photogrammetric analysis of images. Digitisation and ortho-rectification of images are now widely used for broad-scale terrestrial and shallow water habitat mapping. To date it has been applied rarely, if ever, to photo-monitoring images taken by divers. However, this is likely to change given the growing availability and application of orthorectification and GIS software. As stereo-pair photographs enhance the information that can be extracted for orthorectified images, application of stereo systems may increase. However, stereo-photography systems are generally bulky and considerably more expensive than conventional mono systems. Synchronisation of camera shutter release mechanisms can also cause problems. Both camera shutters must be open simultaneously, and synchronised with flash firing. Trials conducted at Skomer Reserve found that mechanical means of simultaneously activating both camera shutters proved unreliable. This problem has been overcome at Skomer by setting camera 1 shutter on B, manually depressing the shutter and holding open while camera 2 shutter is fired. Camera 2 shutter is set to the flash synchronisation speed and is used to trigger the flashes. With practice, camera 1 shutter remains open around 0.5 s. (Bullimore & Hiscock, 2001). This is quite acceptable in turbid waters where ambient light levels are low enough to have little effect on film exposure in this time period; however, in clear shallow waters a better system would be required.

Video systems

The comments below reflect the situation at the time of writing. However, video technology and electronic data storage and analysis systems are evolving rapidly, and future improvements will inevitably address some of the current limitations of underwater handheld video.

Video is extremely useful for recording overall pictures of habitat appearance. By recording along a set path (e.g. a belt transect) a permanent record of an extensive site can be recorded. Video provides seamless images covering extensive areas, with greater similarity to the diver's viewpoint of such areas than the discrete snapshots provided by a photo-montage. Video systems will also normally resolve a better image in low ambient light conditions than film cameras will. On consumer camcorders the image format is smaller than 35 mm. Consequently, when used in wide-angle mode, the depth of field is very great and focussing underwater is normally unnecessary. However, video images are in general of significantly lower resolution than those of either film or digital still cameras.

The usefulness of video for survey and monitoring purposes will ultimately depend on image resolution. Analogue Hi8 and consumer digital (6.35 mm miniDV or 8 mm Digital8) camcorders are currently the main formats used for scientific handheld video recording. All of the above formats are capable of producing images acceptable for identification of conspicuous habitat features or larger and distinctive species. Digital formats have a higher resolution potential: Hi8 will resolve up to around 400–480 horizontal lines; DV and Digital8 will resolve around 540 lines (actually 720×480 pixels in NTSC format, 720×576 pixels in PAL format). However, these differences are largely academic as the final image viewed by the scientist will be influenced to a far greater extent by the clarity of the water, the quality of the lens (more or less directly proportional to the cost of the camcorder), the quality of the lighting used and the system used for copying and viewing the image. As an illustration of this, if the original Hi8 or digital tape is copied onto VHS tape for distribution and viewing, the VHS copy from either source will have a maximum resolution of less than 300 lines and will suffer significantly from colour bleed between adjacent regions of the picture. Similarly, if the digital image is further compressed (5:1 compression occurs within the camcorder) for storage on a PC hard drive or archiving on CD/DVD, then viewed images may again lose resolution. High quality video footage can occupy large amounts of disc storage space. For example, DV footage that is stored without further compression will occupy around 1GB for every for 4.5 minutes of footage. A number of CODECs (compression/decompression software or hardware) exist that maintain high quality whilst dramatically reducing storage space required. Currently the MPEG2 CODEC, which is compatible with DVD players, offers the best option for archiving video footage.

A potentially exciting development for the diving scientist is the new generation of high definition (HD) camcorders that is available since late 2003. Formerly only available to the top end of the broadcast market, these camcorders are now available at prices within the budget of most research scientists. Suitable housings are also available from a growing number of manufacturers. High Definition camcorders will record images of 720 lines progressive scan (720 × 1280 pixels) or 1080 interlaced (1080 × 1920 pixels), giving a significant improvement in image quality over DV camcorders.

Where video is used for quantifying numbers per unit area, or extracting size measurements, an additional factor to be considered is image distortion. As this problem is discussed in more detail in the photo-monitoring techniques section, it will suffice to say here that it is the author's experience that distortion tends to be a significantly greater problem with camcorder lenses than with 35 mm still camera lenses. This is likely to be due to a combination of the smaller image format, extreme wide-angle capability and cheaper plastic lenses used in consumer camcorders. Attempts to use video stills to measure gorgonian growth (with camcorder used in close-focus, wide angle-mode at a distance of approximately 25–30 cm) were abandoned due to the extreme peripheral distortion observed in the video image (Munro, 2000a).

Video belt transect surveys have been adopted by many statutory bodies and NGOs as part of their survey and monitoring protocol (e.g. the Australian Institute of Marine Science; the Joint Nature Conservation Committee of the UK; Hawaii's Coral Reef Assessment and Monitoring Programme; NOAA's National Undersea Research Programme). A widely adopted protocol for use on tropical waters with good visibility and high ambient light levels is described by Osborne and Oxley (1997). They suggest a discrete point sampling method. The surveyor swims at a constant speed along the transect, holding the camcorder at a set height (25 cm recommended) above the reef and the recording is done directly downwards. A 50 m long transect is recommended. The recorded images are analysed by playing back the tape and viewing on a suitable monitor with five points marked on the screen. The tape is stopped at set intervals and the species located underneath each of the five points is identified and logged. They suggest 40–80 pauses (giving 200–400 data points) along the transect to provide good estimates of total hard coral cover. For each lifeform/species category, percentage cover is estimated by dividing the number of points at which that group was recorded by the total number of data points, and multiplying by 100. A similar protocol developed by field biologists based at the Virgin Island Field Station is described in Rogers et al. (undated).

Video footage of permanent transects can be used for quantitative monitoring of lifeforms or larger, conspicuous species. It is also particularly useful for positively identifying subtle or qualitative changes in habitat that are difficult to confirm simply from the diver's descriptions, e.g. slight increases in sediment veneer overlying reefs. If using video transects for comparative monitoring of benthic species and habitats, it is imperative that the video path starts from the same end of the transect every time, as massive or branching species will look very different when viewed from different directions. A differing viewpoint may also change which low-lying colonies are visible and which are obscured. Use of autofocus is not recommended on current camcorder models, as this is an unnecessary drain on the camcorder's battery and may cause the lens to focus on drift weed or other objects floating past the camcorder. Wide-angle fixed focus, with focus set to around 0.3 m, appears to work well for habitat recording with most systems in most conditions. Depth of field is a function of lens angle of acceptance, aperture size and shutter speed. On most auto-exposure systems, it is recommended that the camcorder be pre-set to shutter priority (if available on the system) and to a relatively slow shutter speed, e.g. around 1/50th of a second. This should maintain optimal exposure and depth of field in most conditions. Recording should be started a few seconds before recording the subject of interest; this allows for the slight mechanical delay in tape transportation (pre-roll) and provides excess frames for any editing that may be required later. In low-light conditions, the use of lights will not only bring back colour to the image but also sharpen the image by (i) reducing the aperture size and thus increasing the depth of field and (ii) reducing 'noise' on the image from excessive signal amplification. However, by automatically reducing the aperture size, the area outside the lights' beam will become appreciably darker in the image, thus the total viewable area will be dramatically reduced. As video photo-sensors and videotape do not handle high contrast well, lights should ideally have a very wide beam and be fitted with diffusers to prevent 'hotspots' occurring in the picture. The use of two lights, with beams angled in slightly different directions, helps spread light more evenly and further reduce areas of image burnout. These points are expanded upon in a set of general guidelines for diver surveys of sublittoral epibiota using video are given in Munro (2001). Video belt transect surveys have also been widely adopted for fish surveys (see 'Fish survey techniques' and 'belt transects' section).

Video is extremely useful for helping to locate sites, by recording a swim-path to the study site (see 'underwater site marking and relocation' section). This is especially helpful for orientating workers who have not previously visited a site. It also has obvious applications in recording behaviour, either as part of a behavioural study (Fig. 4.8) or as an aid to identification (e.g. where the movement of a particular species is distinctive).

Fig. 4.8 Video system being used to record behaviour in the sea hare *Aplysia punctata*. © Colin Munro

4.4 Underwater survey and sampling techniques

Underwater site marking and relocation

General considerations

Accurate relocation of specific sites on the seabed is a requirement of many benthic studies, but is one of the biggest problems for the diving marine biologist. The problem is not so great in clear oceanic waters or when working on man-made structures such as piers or breakwaters, or very close to the shore, where a well-defined swim-path can be followed from the surface. However, in poor visibility, or offshore where the diver must descend from a boat, searching for a specific site frequently wastes valuable bottom time. Loss of data through failure to relocate a site is also something experienced by most diving scientists at some stage in their career.

When working offshore in any depth of water, it is always advisable to descend a shot or buoy line marking the site if the water is not sufficiently clear to allow the target to be seen from the surface. Modern position-fixing allows very accurate positioning of the support vessel above the selected site. However, a free-descending diver will drift horizontally some distance during the descent, even in quite slight currents. They are also likely to drift down-tide while on the surface performing final equipment checks before submerging. Thus, for example, in making a descent to around 20 m depth, a diver may easily reach the seabed off-station by 15 m or more. In poor visibility this can easily result in failure to locate the study site.

Standard Global Position System (GPS) receivers are now capable of ± 10 m accuracy for around 95% of the time. However, this does not necessarily always translate into vessel positioning and shot line deployment within 20 m of the study site. Human response time and vessel manoeuvrability will impose additional errors. Thus, where low visibility is expected, locating aids fixed to the seabed may be required to guide the diver to the study site. Snag-lines fixed close to the seabed, running either side across the tidal axis or (on a sloping seabed) along a depth contour, greatly increase the diver's chances of quickly locating the study site. Ideally these should indicate at which side of the site the diver is located (e.g. by using different colours or thicknesses of line, or tagged lines, either side). Snag lines may be augmented by high visibility sub-surface buoys attached along their length. Sub-surface buoys, fixed to rocks or heavy weights, rising 0.5–2 m above the seabed dramatically increase the diver's chances of spotting study sites in poor visibility. Rigid 6-inch diameter fishing floats used by commercial fishermen for trawl nets are ideal. The mid-water air space they form is also acoustically highly reflective. As a consequence they show up as a well on a vessels' echo sounder, providing useful confirmation of correct positioning. Metal structures may be constructed to help define and relocate the study area. Metal grids are commonly used in archaeological excavations, and the construction of a raised 4×4 m tubular galvanised steel grid to map seafan colonies is described in Munro (2000a). Large metal tripods have also been used to mark site and attach locator beacons at monitoring sites (e.g. Holt & Sanderson, 2001). Where steel structures are constructed to mark sites, it is recommended that a sacrificial zinc anode is attached to the structure in order to reduce corrosion.

Now that GPS is relatively inexpensive, surface sightings are less used. However, transits (a position line established by noting the alignment of two conspicuous landmarks) and bearings taken using a sighting compass provide good backups to electronic navigational aids. In areas where land is close and features conspicuous, sightings may be used as the main means of locating the vessel above the study site. It is recommended that photographs are taken of all landmarks forming transits or from which compass bearings have been taken.

In low visibility conditions, or where the seabed lacks distinctive features, it is advisable to employ as many complementary aids to site location as possible. Where underwater markers have been fixed, wide-angle photographs should be taken to show their appearance and location relative to seabed features. If visibility does not allow this, then detailed maps can be drawn illustrating distance and direction of the study site from other conspicuous features. Waterproof copies of diagrams and photographs may be made by encapsulating in plastic, allowing the divers to consult these underwater on subsequent monitoring visits. Video footage of the location and pathway between markers is also highly effective in aiding divers to form a mental map of the area and to recognise features along the pathway to the site.

Air drills and underwater fasteners

Diver-operated air drills may be used to create attachment points on rock or coral for lines, buoys or quadrats. Small, pistol-grip pneumatic drills constructed from stainless steel and powered by diving cylinders have been used to mark sublittoral monitoring sites around Lundy Marine Nature Reserve and Milford Haven estuary (Fig. 4.9). Suitable drills are manufactured by Ingersoll Rand, with a maximum

Fig. 4.9 A diver-operated air drill being used to mark a sublittoral monitoring station in Milford Haven. © Colin Munro/Sue Burton

chuck size of 13 mm. Using masonry drill bits, these can be used to drill holes several centimetres deep in limestone. They make little impression on solid rock, like granite but can be used to enlarge small crevices (albeit with considerable effort and numerous drill bits). Once holes have been drilled, these can be filled with a suitable underwater setting resin. Polyester resins formulated for bonding with stone and masonry work well, with those supplied in cartridges that can be applied using a mastic gun being easiest to use underwater, e.g. Fischer UN No 1866 or a similar type. Locating pins or eyebolts are then pushed into the resin and allowed to set. This provides a secure fitting that should last for several years provided it is not placed under significant load. Stainless steel, aluminium or brass fittings are recommended. The duration for which a full cylinder will power a drill obviously decreases with depth. At around 18 m, a 15 litre (120 ft^3) cylinder will empty in 8–10 minutes.

Acoustic pingers and receivers

Where visibility is very poor, the use of acoustic transponders may be necessary. Acoustic pingers that can be attached to the monitoring station are available from a number of manufacturers (e.g. Benthos Inc., Massachusetts; Sonardyne, Blackbushe, UK; Vemco Ltd., Nova Scotia; InterOcean Systems, San Diego). Acoustic pingers will emit pulses from a few days to several years. Many systems allow the signal power and frequency of pulses to be varied, hence the active lifespan of the pinger to be prolonged. Some models are activated by immersion, whilst others are user-activated. Delayed-activation pingers are also available, effectively extending their working life if initial relocation is not planned until some time after deployment. Relocation is achieved by a boat-mounted receiver (fixed beneath the water's surface), or a handheld receiver operated by the diver. The display on such receivers varies, typical systems having an LED display indicating direction to the acoustic beacon and signal strength. Many systems allow for transponders to have different 'address' codes, allowing the diver to locate and clearly identify a number of close-together stations. Typical stated locator ranges vary from 60 to 750 m, depending on system and conditions. Acoustic beacons have been successfully deployed on monitoring stations on Sarn Badrig Reef, North Wales (Holt & Sanderson, 2001) and Milford Haven Waterway.

By setting up a network of transponder beacons on the seabed, such systems can be used to map the location of features accurately across an extensive site. This is achieved by the diver moving to different seabed features, and recording the distance (displayed on the handheld receiver) to each beacon. Alternatively, beacons may be attached to the divers, allowing their location to be dynamically tracked as they travel across the site. Such systems have been developed primarily for the oil and gas industry, but have been adopted by underwater archaeologists for mapping wreck sites (e.g. the wreck of the *Coronation*, Cornwall, England, and the *Resurgam* submarine in Colwyn Bay, Wales). Full details of the system and methodology employed for these studies are given by Peter Holt, 3H Consulting Ltd., Plymouth, UK (website URL: www.threeh.demon.co.uk/TechniquesAcoustics.htm).

For open coast sites, it is generally recommended that surface buoys should not be used as a key method of relocation if this can be avoided. Unless attachment points and buoy lines are of very strong construction and secured to very heavy objects on the seabed, they are likely to disappear during storms. Additionally, constant movement through waves and tide will chafe through lines, and fouling growth may eventually cause the buoy to sink. Surface buoys are also attractive items for passing fishermen or others in small boats. If a surface buoy is used, it should not be attached directly to the study structure. Unless the structure weighs more than one tonne, strong wave action on the buoy can drag the structure off-station.

Sample collection

Corers

Core sampling is a cheap and simple means of collecting quantitative samples from soft sediment. In their simplest form, these are lengths of plastic or metal tubes of known diameter, which are pushed into the sediment by the diver. Where the sediment is coarse or consolidated, some workers suggest cores may be hammered into the sediment (e.g. Angel & Angel, 1967; Gage, 1975). However, extraction of an open-ended core will normally result in partial loss from the sample. To prevent this, it is usually necessary to dig around the core and seal the end (either with a bung or by hand) before attempting to withdraw it. Digging by hand in material that requires the cores to be hammered in can be difficult, especially if it has to be repeated several times during a dive.

It is normally recommended that the cores are fitted with rubber bungs or caps to retain the sample. In practice, these are often difficult to fit underwater, the cores being full of water, which is not compressible. An alternative is to slip plastic bags over either end, and retain with rubber bands. This is usually a two-person task (especially as the act of coring will often have severely reduced visibility), but can be accomplished quickly. Where cores can be capped very securely underwater, a mesh bag or similar item can be used to carry them. A more secure method is to place them within a crate. For a larger number of cores, or a large volume of cores, these can be tied to a surface buoy (Fig. 4.10). The divers can then ascend this line at the end of the dive, and the crate can be hauled up by the boat crew. Cores must, of course, be securely held in (e.g. by elasticated cord across the top of the crate) and the crate should be weighed to reduce the risk of it tipping while being hauled up.

Suction samplers

Diver-operated suction samplers are useful for excavating relatively large sediment samples. Airlifts have been used by underwater archaeologists since the 1950s

Fig. 4.10 A crate attached to a surface buoy being used to hold cores. © Colin Munro

(Du Plat Taylor, 1965) to clear overburden from wreck sites and to uncover artefacts. Airlifts work on the Venturi principle. Compressed air is fed into the lower end of a long rigid pipe which is normally guyed at an angle of 45° from horizontal. Air rises up the tube before escaping out of the far end. The entrained air causes a drop in pressure around the mouth of the pipe and so 'lifts' water and sediment up into the pipe (Fig. 4.11).

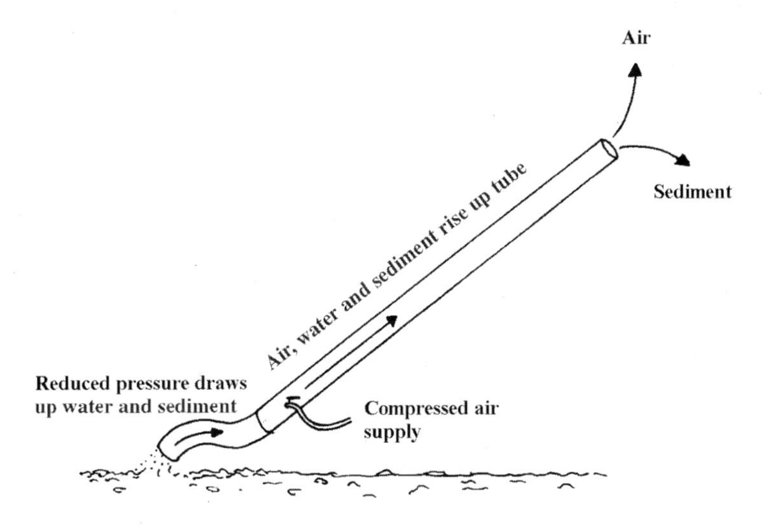

Fig. 4.11 Diagram illustrating the principle of an air lift or suction sampler. © Colin Munro

However, archaeologists' airlifts tend to be large, designed to move much larger amounts of sediment than biologists normally work with. They also need to be anchored in place to counteract their buoyancy whilst in operation. Consequently, lighter weight systems were designed, powered by diving air cylinders and easily moved into position and operated by two divers. Barnett and Hardy (1967) described a modified archaeological airlift, designed for sampling soft sediment infauna. This design incorporated a large diameter cylinder on the seabed end of the pipe, and a sieve and sampling bag on the far end. The sampling cylinder was pushed a few centimetres into the seabed, and the sediment was drawn up through the sieve, retained material then falling into the sampling bag. This system was further modified by Hiscock and Hoare (1973). They designed a smaller, portable system incorporating a flexible tube attached to the entry end of the sampling cylinder (Fig. 4.12). The Hiscock and Hoare sampler was designed primarily for sampling epilithic fauna, the sample chamber capacity being too small to be effective on most sediments. Quantitative sampling is achieved by sampling within a quadrat, using a paint scraper to dislodge attached species. More recent systems such as the bucket sampler (Figs 4.13 and 4.14) and the JW Miniature sampler (designed by John Woolford of Pembrokeshire, Wales) have been adopted by many of the Government Agencies within the UK and several marine survey consultancies. These designs have interchangeable sample retaining bags of differing mesh sizes, depending on the sediment type and sampling protocol. Bags can be changed underwater, allowing divers to take numerous samples. By the use of a reference cylinder attached to the end of the flexible pipe, or by sampling within a quadrat to a given depth, quantitative samples can be processed.

Suction samplers are particularly useful in mixed or coarse sediments, where cores or grabs may have difficulty penetrating (e.g. Robinson & Tully, 2000) or where large deep samples are required, such as collecting deep-burrowing

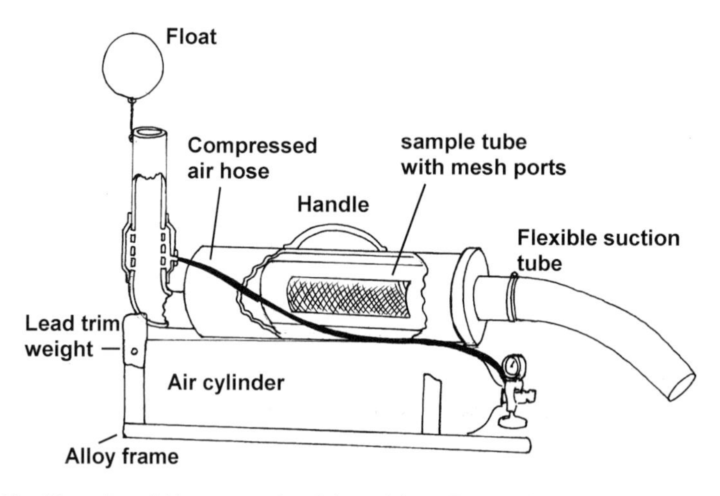

Fig. 4.12 The Hiscock and Hoare sampler. Adapted from Rostron (2001) after Hiscock and Hoare (1973).

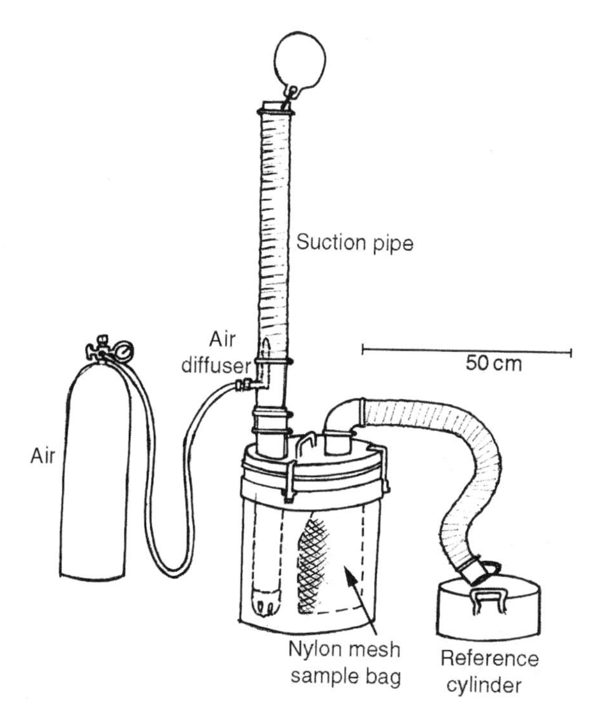

Fig. 4.13 A bucket suction sampler design. Adapted from Rostron (2001) after Hiscock (1987).

Fig. 4.14 Divers using a suction sampler. © Dale Rostron, Subsea Surveys

megafauna. The main disadvantages of suction samplers are that they can draw in animals from surrounding substrates thus inflating abundance estimates within the quadrat, and they may abrade animals as they are drawn in with the sediments (Simenstad et al., 1991). Much of the above information of sampler design is adapted from Rostron (2001), where additional details on system design and operation can be found.

Suckers (Yabby pumps and slurp guns)

Yabby pumps are long, narrow steel or plastic tubes containing a central plunger. They are used extensively in Australia to capture 'yabbies' (freshwater crayfish) by placing the end of the tube over the animal's burrow entrance and sharply raising the plunger. Yabby pumps have been adopted by numerous workers (e.g. Abed-Navandi & Dworschak, 1997; Abed-Navandi, 2000; Dworsschak & Jorg, 1993) to capture burrowing crustaceans underwater, especially thalassinid shrimps that excavate deep or complex burrows. Slurp guns operate on the same principle as yabby pumps, but are generally of larger diameter (Fig. 4.15). They are used for the collection of crevice fauna or fast-moving animals such as fish or shrimps (e.g. Lincoln & Sheals, 1979; Squires et al., 1999). Slurp guns have also been used for the collection of sea-ice algae. They have the advantage (compared to techniques working from the ice surface such as coring or cutting down through with an auger) of being able to sample several nearby sites rapidly. However, some material may be lost through leakage from the cylinder (Poulin, 2001).

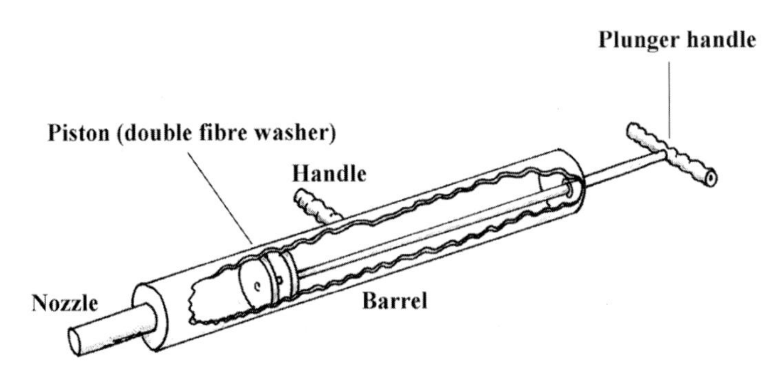

Fig. 4.15 Slurp gun design. © Colin Munro

Scrapers

Paint scrapers are useful tools for collecting low turf species such as hydroids or bryozoans, and encrusting sponges, colonial tunicates and barnacles. For very small animals, a pipette may be helpful for catching and transferring the detached specimens to plastic bags.

Resin casting

The structure of burrows in sediment can be investigated by resin casting. This technique was first developed by the marine geologist E.A. Schinn (Schinn, 1968), and has since been widely used by marine biologists studying burrowing megafauna (e.g. Atkinson & Nash, 1985, 1990; Dworschak & Jorg, 1993; Nickell & Atkinson, 1995; Nickell et al., 1995; Coelho et al., 2000). The method is fully described by Atkinson and Chapman (1984). Two main types of resin are used for this technique, polyester resins that use a small amount of catalyst and set relatively quickly (a few minutes to several hours) and epoxy resins that require larger amounts of catalyst and set more slowly (a few hours to several days). Polyester resins are used predominantly in temperate and polar regions. Epoxy resins are more durable, but considerably more expensive than polyester resins. Most epoxy resins will not set at low temperatures, and so are used chiefly in the tropics (R.J.A. Atkinson, personal communication).

The resin is carried in a watering can or similar container, with catalyst normally being added just prior to diving. The resin, being slightly denser than water, can be poured underwater into burrows (Fig. 4.16). A cylinder or box framing the burrow mouth helps direct the resin and also aids relocation. Inevitably, some small droplets of resin will be suspended in the water column during the pouring process and will adhere to the diver's equipment. For this reason cameras and similar equipment should be kept well clear during the pouring process. Once the resin has set (it is normal to leave overnight) the cast can be dug out of the sediment.

Fig. 4.16 Divers pouring resin into a burrow. © Colin Munro

Detailed survey methods

Manta tows

Manta boards are essentially large boards, generally constructed from wood or GRP, that act as hydrofoils. They are designed to be towed behind small boats, attached using a water-ski tow rope or similar. A diver or snorkeller hangs on to the board and is pulled through the water, either at the surface looking down or up to a couple of metres below the surface. Different designs will have handles or hand-holds for the diver, and some will allow the diver to angle the board in such a way that it rises or descends. An attached writing slate on which the diver can record data is incorporated in many designs.

Manta boards are used to survey larger areas than divers could cover unaided in a reasonable time. The level of detail recorded is limited, because the diver passes fairly rapidly over areas, and higher above the seabed than their normal free-swimming height. This is a low-cost, low-tech method useful for broadscale survey, mapping habitat type, assessing the extent of widespread, highly visible impacts or identifying areas for more detailed survey. A protocol developed by the Australian Institute of Marine Science (AIMS) for surveying coral reef features is described by English et al. (1997).

Tows should be conducted at a constant speed. This should be slow enough that the diver does not have to put undue effort into holding on rather than recording (normally 1.5 knots or less). Speed should also take into account the horizontal visibility. Whilst good design will help minimise unbalanced pressure on upper and lower surfaces, greater speeds may also cause the manta board to tend to lift towards the surface, or worse, to dive towards the seabed. Resisting this tendency can be extremely tiring and distracting for the diver. When the diver is being towed below the surface, a means of communication between the diver and the surface team on the boat must be developed. Care should be taken to ensure that octopus valves are tucked out of the way, as these can free-flow due to the rate of water flowing past them rapidly depleting the diver's air supply.

Manta tows have been used extensively to estimate coral cover and assess the numbers of and damage (feeding scars) caused by crown-of-thorns starfish, *Acanthaster planci* (e.g. Kenchington & Morton, 1976; Fernandes et al., 1993; CRC Reef Research Centre, 2001). Recording sheets with pre-printed species, habitat or condition categories are recommended by English et al. (1997) as this greatly speeds up recording. Manta board tows have also been used successfully in good visibility in temperate waters, e.g. for surveying habitats within Port Erin Bay (Riley & Holford, 1965) and for identifying the location of maerl beds within the Firth of Clyde (personal observation). More complex towed vehicles have been developed for specific purposes, such as the semi-enclosed two-man vehicle used by Main and Sangster (1978) for observing mobile fishing gear.

Transect surveys

By quantifying the survey area, transect surveys greatly facilitate gathering of data on species percentage cover and density. By setting more or less precise (depending on protocol used) limits on the area considered, they also help make comparisons between study areas more valid. The more commonly used types of quadrat and transect survey are described below.

Line intercept transect

The line intercept transect (LIT) is a widely used method for assessing percentage cover within the benthic community of coral reefs (e.g. Marsh et al., 1984, Bradbury et al., 1986). The majority of workers using this technique have used a 20 m tape, marked in distance increments, simply stretched out along the reef following a depth contour. Survey divers then record the distance along the transect at which each species or lifeform transition occurs (Fig. 4.17). This process is usually repeated at different depths, with five replicate transects at each depth. Percentage cover of each species or lifeform is calculated by dividing the cumulative lengths of the

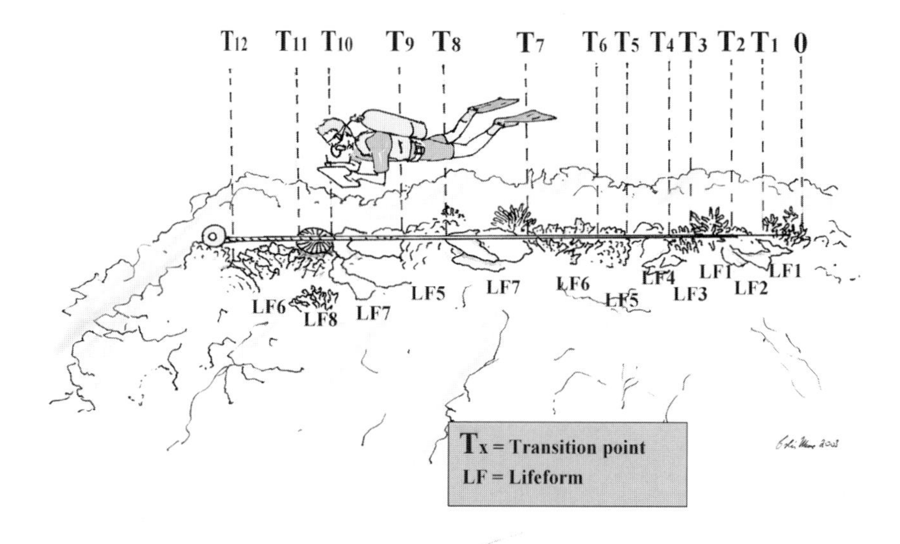

Fig. 4.17 Line Intercept Transect survey. © Colin Munro

target species or lifeform by the length of the transect and multiplying by 100. A formal protocol for use of this method for recording coral reef lifeforms data has been developed by the Australian Institute of Marine Science, and is described in English et al. (1997). This is a simple and inexpensive method, suitable for use by non-specialists when classifying species at lifeform level. However, some lifeforms are difficult to standardise and the method is considered inappropriate for monitoring small changes in community structure (English et al., 1997).

Line point transect

The Line Point Transect is similar in design to the LIT, but with marked points along the length of the transect line. All benthic species lying directly underneath points are recorded. Unlike the LIT method, size of colonies is not recorded. Liddel and Ohlhorst (1987) used transects marked at 200 mm intervals. They recommended a minimum of five 20 m long transects per site. They considered the results generated by this method to be comparable with those of the LIT, but requiring less survey time. However, because of the limited number of discrete survey points along the transect, uncommon small species are likely to be missed. Because of its simplicity, the line point transect survey method is frequently used by volunteer diver programmes (e.g. *Reef Check*, Reef Check Survey Instruction Manual).

Chain transect

The chain transect is used to gather data on species presence and abundance, and to provide a measure of seabed relief. A line transect, marked in one-metre increments, is tensioned between two stakes set so that the transect runs parallel the reef crest or shore. Starting at one end of the transect, a lightweight chain (1.5–2 m long) is then

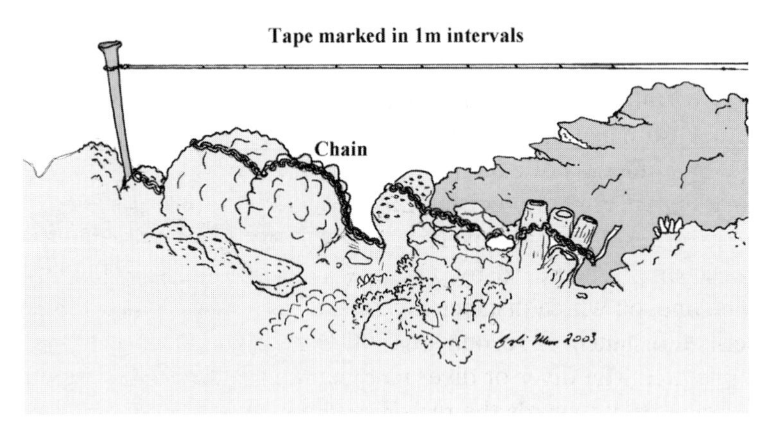

Fig. 4.18 Chain transect. © Colin Munro

draped along the reef, below the line, following the contours of reef (Fig. 4.18). The length of chain covering each species or lifeform is then measured by counting links. The number of links under each metre of line transect is also recorded. Where the reef is multi-layered, e.g. where tabular coral forms a tier directly over encrusting forms, the total length of all surfaces under the chain should be measured. By calculating the ratio of total chain length to transect length, an indication of reef rugosity is obtained.

This method provides a very detailed picture of reef composition and, unlike the LIT, takes the three-dimensional nature of reef colonies into account. Against this, chain transects are slow to deploy and to survey. They are most suited for surveying areas dominated by massive or encrusting lifeforms, and are inappropriate for use on areas supporting large numbers of highly branched species of otherwise structurally complex species (e.g. *Acropora* spp., gorgonians or branching sponges). They may also be inappropriate in areas supporting many fragile species. Chain surveys have been widely used for surveying and monitoring coral reefs in the Caribbean and Florida Keys (e.g. Meier & Porter, 1992; Rogers et al., 1997; Rogers & Miller, undated). A detailed protocol for use of the chain survey method on coral reefs is described in Rogers et al. (1994).

Belt transect

Belt transects are essentially long quadrats laid, or visualised, on the seabed. The length and width of the transect used will depend on the size and density of the species being recorded, the underwater visibility and to a certain extent the depth (limiting time available for deployment and recording). Belt transects are useful for recording sessile species that occur in low densities or are patchily distributed. They are also extensively used for counting coral reef fish (see 'Fish survey techniques' section).

In its simplest form, a belt transect consists of a diver or dive pair swimming along a compass bearing for a set number of fin-beats (or set time), recording target species within a set width either side of the diver's course (e.g. Bunker, 1985). This method allows a large number of transect surveys to be conducted quickly, with very little set-up time required. However, such transects are poorly defined: the diver's course may be influenced by currents and individual divers may cover differing distances and estimate transect width differently. It is therefore most useful where collecting large amounts of data in a short period of time is more important to the study than a high degree of precision in defining the study area.

A more precise area can be covered by laying a measured line along the seabed. This is most simply achieved by attaching the line on a diver's reel to a fixed weight. The line, on which distance increments have been marked, is then run out at a prescribed distance. A second small diving weight attached to the reel helps keep the line taut. The diver or diver pair can then return to the beginning of the line and survey back towards the reel. The width surveyed either side of the line can be better controlled by the divers carrying a metre rule to verify the boundaries. By dropping a shot line within the survey area, then randomly selecting a number of compass bearings along which the transect should be laid, several unbiased transects can be surveyed in rapid succession. Where the seabed is uneven, a taut transect line can be raised some distance off the bottom in places. Additionally, where the transect is long, it may not be practical to try and get the line taut. In these situations a weighted line is more appropriate. This can be made by crimping small lead weights at intervals along the line. For very long transects that are to remain in place for some time, leaded rope (with lead wire woven in) can be bought from commercial fishing outlets.

Belt transects with fully defined boundaries are more time-consuming to lay but are useful where very accurate counts of colonies or individuals per unit area are required. A simple, effective method uses two lengths of tubular steel (scaffold pole sections being ideal) as boundaries at either end of the corridor. Polypropylene line, with weights crimped on at intervals, forms the sides of the corridor. The steel ends provide rigidity and weight to keep the side lines taut once the transect is laid (Fig. 4.19). Repeatedly folding the lines back upon themselves, and securing with

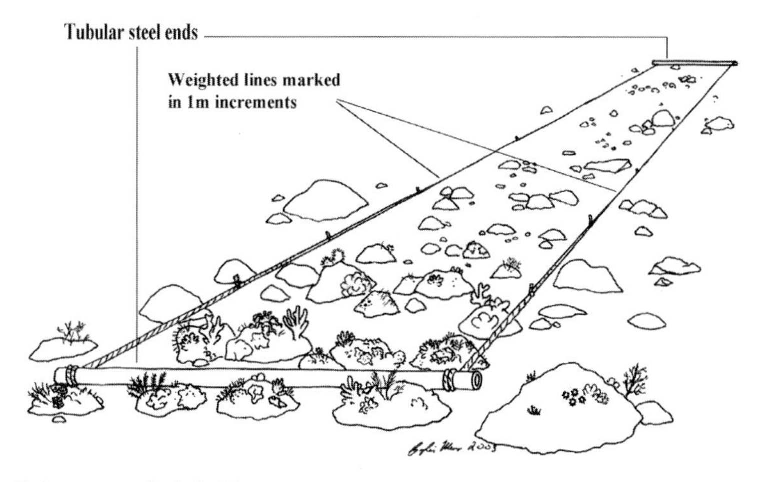

Fig. 4.19 Belt transect. © Colin Munro

rubber bands, makes the transect more manageable for carrying from the surface to the survey site and quick to lay once in place. On completion of the survey the divers attach a buoy line to one end, and the transect may be hauled from the surface. In relatively low visibility, a one-metre wide belt transect has proved most efficient (e.g. Munro, 1992, 1996). A two-metre wide belt, with a central line dividing it into two one-metre corridors allows larger areas to be surveyed, while still requiring little deployment time. By replacing the steel ends with weighted plastic tubes, the transect can be hung from pitons or bolts fixed on a vertical face. This can act as a very large quadrat for surveying or monitoring low-density species on vertical rock habitats. Fully defined belt transects are particularly suitable for video surveys, if the lines marking the lateral edges to the transect are marked in distance increments and kept within the camcorder's field of view at all times.

Quadrats

Quadrats may be random or fixed. Randomly placed quadrats allow more powerful statistical analysis to be applied to data collected, and are often most appropriate for surveying habitats that are fairly uniform, such as seagrass beds, maerl beds or mud habitats (e.g. counting burrow densities). Reef habitats are generally too heterogeneous for random quadrats to describe their species composition adequately without an enormous investment of time. On a purely practical level, the placement of quadrats is also usually partly determined by the nature of the rock or coral surface. Areas containing large crevices or high ridges will be unsuitable as (i) it will not be possible to place the quadrat flush against the rock surface and (ii) the crevice or ridge may significantly increase the surface area beneath the quadrat. Thus, placement of fixed quadrats on reefs is generally a compromise between unbiased representation of the reef community and practical constraints.

For surveying or monitoring reef-building corals, 1 m^2 quadrats are recommended by most workers (e.g. English et al., 1997; Rogers et al., 1994), with smaller or larger quadrats used depending on size and spatial distribution of target organisms. On temperate reef systems, 0.25 m^2 (50×50 cm) are preferred: (i) because generally lower visibility makes surveying or photography of larger areas difficult; (ii) because most temperate rocky reef species are quite small and often cryptic, thus systematic counting within a larger area would be extremely time-consuming and prone to error; and (iii) because larger size quadrats would prove difficult to position flush against most rocky reefs.

Moore et al. (1999) concluded that for sublittoral rock habitats, less than ten 0.1 m^2 (32×32 cm) quadrats were required to provide a good, overall description of the epilithic community, and to give reasonable estimates of mean abundance for a few dominant species (i.e. in terms of numbers per unit area or percentage cover). They also found, however, that a much larger number of quadrats would be required to give reasonable estimates of mean abundance for most epilithic species present.

Bohnsack (1979), studying marine canal wall biota in the Florida Keys, considered that ten 0.6 m^2 quadrats were sufficient to describe the community present.

Differing growth forms (e.g. foliose algae, encrusting sponges, octocorals with interconnected stolons) and dramatic differences in size and density between juveniles and adults (e.g. barnacles or mussels) can present considerable difficulties for the surveyor attempting to record accurate and statistically comparable data for community monitoring. Strung quadrats, whereby the area within the quadrat is subdivided into a grid of squares, can be used to provide better estimates of percentage cover or for frequency counts. An advantage of frequency counts is that they provide a single measure of abundance that is applicable to different growth forms. Moore et al. (1999) found that it took up to twice as long to survey a set unit area using frequency counts as it did recording percentage cover.

Quadrats have the advantage in that they are particularly suitable for photo-monitoring. By projecting a slide of the quadrat on to a gridded surface, or scanning the transparency and similarly overlaying a grid using photo-editing or GIS software, much of the analysis can be conducted post-dive. This has the advantage that difficult species can be checked against relevant texts or with specialists in that group. Conversely, the ability to check the surface texture, three-dimensional shape and to move aside shading species is lost (although stereo-photography allows a degree of three-dimensional viewing). For monitoring programs, a combination of *in situ* counts or percentage cover estimates complimented by photo-monitoring is recommended (e.g. English et al., 1997; Munro, 1999). Comparisons between 0.1 m^2 quadrats and belt transect records suggest that quadrats are much more time-consuming to survey, but are better at recording common but inconspicuous species. Conversely, they tend to miss or under-record larger, low density species (Moore et al., 1999).

Plotless and rapid survey techniques

Several rapid methodologies, and methodologies where no defined survey area is required, have been developed for surveying or assessing benthic habitats. Many of these protocols are country or region-specific, having been developed specifically for local habitats or conditions, or the requirements of particular organisations. Some of the more widely used methods are described below.

Rapid assessment protocol techniques

The rapid assessment protocol (RAP) is a methodology developed specifically for assessment of degradation of reefs in the Western Atlantic and in the Gulf of Mexico. The protocol focuses on the principal reef-building corals, algae (macroalgae and encrusting corallines) and *Diadema* urchins (considered the key grazers on reefs in this region). The core method is to estimate the percentage of living, dead, bleached and diseased coral on the reef, by recording along a 10 m transect. Algal cover is estimated by recording percentage cover within a 25 × 25 cm quadrat at 2 m intervals along the transect. Canopy height of the algae is also recorded, and used to estimate algal biomass. *Diadema* numbers are estimated by using the laid 10 m line as a one-metre wide belt transect (using a metre rule for guidance) and counting all specimens within that belt. Transects are placed randomly, with a minimum of six per site considered necessary for proper assessment. Fish are recorded by using either a 30 m long, 2 m wide, belt transect or a random swim whereby all fish species are recorded and relative abundance estimated. The above is merely an outline of the protocol. Full details of the methodology are given in Steneck et al. (2000) and AGGRA methodology, version 3.1 (Atlantic and Gulf Rapid Reef Assessment, 2000).

MNCR methodology

In the UK, the Joint Nature Conservation Committee's Marine Nature Review team (MNCR) has developed a standard, semi-quantative, methodology for surveying both intertidal and subtidal benthic flora and fauna. This is essentially a roving diver method, designed to be used by experienced biologists only. Divers, working in pairs, initially record the substrate, depth and relief of the habitat. They then swim randomly across the seabed, recording species observed and biogenic features (e.g. burrows or mounds in sediment) as they go. Abundances of species and physical features are recorded separately for each habitat the divers pass across. The methodology is based around the SACFOR scale. Using this, species are recorded, either in terms of percentage cover or density (depending on growth form) in six logarithmic steps (designated Superabundant; Abundant; Common; Frequent; Occasional; Rare). The actual density or percentage ranges cover assigned to these codes depends on the size range of the species being recorded. For example, large solitary ascidians would most likely fall into the 3–15 cm high category. For such a species a density of 1–9 per 100 m^2 would be classed as Occasional, whilst species over 15 cm high (such as a gorgonian) occurring at this density would be classed as Frequent. The MNCR SACFOR scale is reproduced in Table 4.1.

This system is not sufficiently quantitative to be used for monitoring purposes. Lack of temporal or spatial boundaries tends to make the assessment of abundances too subjective. This is exacerbated by limitations of the SACFOR scale, inevitable in any compromise between simplicity and encompassing all the species. For example, the lowest abundance category 'Rare' equates to 1–5% cover; for species <1 cm tall; however, on most temperate reef systems many small epilithic species that appear relatively common to the diver will actually cover far less than 1 percent of the rock surface (e.g. 400 *Dendrodoa grossularia* tunicates per m^2 would occupy approximately 1% of the surface area). However, this methodology does offer considerable advantages. It is suitable for use in all benthic habitats. It requires no pre-survey set-up time and provides detailed species lists and descriptive data that

Table 4.1 The MNCR SACFOR scale. © Joint Nature Conservation Committee, UK.

	Growth form		Size of individuals/colonies					
% cover	Crust/ meadow	Massive/ turf	<1 cm	1–3 cm	3–15 cm	>15 cm	Density	
>80%	S		S				>1/0.001 m^2 (1 × 1 cm)	>10 000/m^2
40–79%	A	S	A	S			1–9/0.001 m^2	1000–9999 m^2
20–39%	C	A	C	A	S		1–9/0.01 m^2 (10 × 10 cm)	100–999 m^2
10–19%	F	C	F	C	A	S	1–9/0.1 m^2	10–99 m^2
5–9%	O	F	O	F	C	A	1–9 m^2	
1–5% or density	R	O	R	O	F	C	1–9/10 m^2 (3.16 × 3.16 m)	
<1% or density		R		R	O	F	1–9/100 m^2 (10 × 10 m)	
					R	O	1–9/1000 m^2 (31.6 × 31.6 m)	
						R	<1/1000 m^2	

Use of the MNCR SACFOR abundance scales

The MNCR cover/density scales adopted from 1990 provide a unified system for recording the abundance of marine benthic flora and fauna in biological surveys. The following notes should be read before their use.

1. Whenever an attached species covers the substratum and percentage cover can be estimated, that scale should be used in preference to the density scale.
2. Use the massive/turf percentage cover scale for all species, excepting those given under crust/meadow.
3. Where two or more layers exist, for instance foliose algae overgrowing crustose algae, total percentage cover can be over 100% and abundance grade will reflect this.
4. Percentage cover of littoral species, particularly the fucoid algae, must be estimated when the tide is out.
5. Use quadrats as reference frames for counting, particularly when density is borderline between two of the scale.
6. Some extrapolation of the scales may be necessary to estimate abundance for restricted habitats such as rockpools.
7. The species (as listed above) take precedence over their actual size in deciding which scale to use.
8. When species (such as those associated with algae, hydroid and bryozoan turf or on rocks and shells) are incidentally collected (i.e. collected with other species that were superficially collected for identification) and no meaningful abundance can be assigned to them, they should be noted as present (P).

can be used to categorise and compare sites. Survey and recording methods are described in Hiscock (1996).

Fish survey techniques

Although fish survey techniques include many of those described earlier, there are also a number of methods used specifically for recording fish. The particular method selected will, in part, depend on the nature of the fish species being recorded (e.g. schooling, cryptic). The majority of fish survey techniques using divers have been developed for use on coral reefs. This is largely because the generally good visibility prevailing there allows divers to identify fish from a reasonable distance, and, in addition, the complexity of the habitat complexity is better suited for divers than most remote survey techniques.

The most widely used methods fall into three categories: belt transects, stationary counts and random swims. Some of the more commonly used variations of these methods are described below.

Belt transect

As with sessile benthic organisms, belt transect surveys are suited to recording species occurring in low densities or those that are patchily distributed. The method was first described by Brock (1954) and remains a commonly used technique (e.g. Craik, 1981; Russ, 1984a, 1984b). For recording fish, the belt transect is visualised as a box-section corridor, the width, and 'ceiling' height determined by target species and visibility. Rogers et al. (1994) recommend a 2 m wide transect for smaller and cryptic species, and up to 5 m for larger species such as groupers and snappers, with a transect length of between 50 and 100 m. They emphasise that swimming speeds should be constant and standardised in order to ensure good comparison between surveys. A similar protocol for conducting coral reef fish census using belt transects has been developed by the Great Barrier Reef Marine Park Authority (GBRMPA, 1978, 1979). They recommend a 5 m wide, 5 m high and 50 m long corridor within which all target species are recorded, with a 1 m wide transect being used to record more sedentary juvenile fish. Both methods are fully described in English et al. (1997).

In temperate waters, fish species diversity tends to be much lower than on coral reefs, but underwater visibility also tends to be much poorer. In shallow, rocky areas the presence of kelp stands adds to the difficulty of conducting visual censuses. On complex rocky habitats off West Scotland, Sayer and Thurston (unpublished data) used 10 m long by 2 m wide transects to survey active fish. A standardised speed along the transect of 2 m min^{-1} was maintained by the divers. Comparing results with other techniques, and with activity profiles for fish over 24-hour periods gained from underwater television, they concluded that diver surveys do not yield absolute numbers for fish present. However, relative comparisons between years and seasons can be made if a standardised technique is adopted.

Video belt transects are increasingly being used to replace or complement diver observation belt transects (e.g. Parker et al., 1994; Sinclair Knight Merz, 1999; Thrush et al., 2002; Booth & Beretta, 2002). Video offers the obvious benefits that tapes can be replayed and played at slow speed. This means that identification does not need to be made instantaneously, and estimates of numbers can be re-checked.

Random swim survey protocols

The basis of these methods is that a diver, or divers, swim(s) over the survey site in randomly chosen directions, either for a set period of time or within a given area. A number of variations on this principle have been developed by different workers. Widely used in the Western Atlantic and Caribbean, the Random Swim Technique (also known as the Rapid Visual technique) uses survey duration as the limiting parameter. The method does not record information on species density, only the number of species recorded during the dive. Once a species is recorded, it is then ignored and the surveyor actively looks for unrecorded species. The survey duration is subdivided into discrete units (Rogers et al., 1994, recommend a 50-minute dive divided into five 10-minute intervals). A measure of relative abundance is gained by scoring species according to the time interval in which they are first recorded (e.g. species recorded between 0–10 minutes will score 5, species recorded between 41–50 minutes will score 1). This method has the advantage that it can be applied over reefs where the relief would make deployment of a transect difficult. A disadvantage is that it provides few quantitative data. It is considered to provide better data on total species richness than either stationary or belt transect techniques. Variations of this technique are described in Jones and Thompson (1978), Kimmel (1985) and Rogers et al. (1994).

The Roving Diver Technique is a protocol widely used for collecting data on coral reef fish using volunteer divers (e.g. Jeffrey et al., 2001). The method requires the divers to swim around within 100 m radius of the dive start point. All fish species observed are logged and abundance assessed on a four-category logarithmic scale (1; 2–10; 11–100; >100). Comparing roving diver and transect survey methods, Schmitt et al. (2002) found that a greater number of rare species was recorded with the roving diver method, but that both methods provided a more complete overall species assessment than either method was able to provide in isolation. Detailed methodology for the Roving Diver Technique, and a comparison with the stationary survey method, is given in Bohnsack (1996).

Stationary technique

This method, developed by Bohnsack and Bannerot (1986), was designed to provide standardised quantitative data on relative abundance and frequency of occurrence of all fish species recorded at a site. A brief description of the method is given below, with full details provided in Rogers et al. (1994). Sites are randomly selected and pertinent features such as depth, structural complexity, substratum and underwater visibility recorded. A 15-m line (ideally at tape measure) is then laid taut on the seabed. This forms the diameter of a visualised cylinder, extending from the seabed to the surface. The surveyor moves to the mid-point of the line and begins to scan the water within the imaginary cylinder. During the initial five minutes of the survey, only species presence is recorded. It is assumed that by the end of this period, most of the species that will be recorded during the survey will have already been logged. The remaining survey time is spent working through the species list generated, recording numbers of each species and estimating their length. The total survey time should not exceed 15 minutes. A variation of this method proposed by Kimmel (1993) records all species during the entire 15-minute period (as opposed to only those noted in the first five minutes). The argument for this is that some cryptic and sedentary species will adapt to the diver's presence over time, and thus may only be observed in the later stages of the survey. The stationary technique relies on divers being able to identify species quickly. While it reduces bias due to disturbance from moving divers, cryptic species will tend to be under-recorded. This technique also has limited use because of poor visibility.

References

Abed-Navandi, D. (2000) Thalassinideans new to the fauna of Bermuda and the Cape Verde Islands. *Annalen des Naturhistorischen Museums in Wien*, **102**B, 291–299.

Abed-Navandi, D. & Dworschak, P.C. (1997) First record of the thalassinid *Callianassa truncata*, Giard & Bonnier, 1890 in the Adriatic Sea (Crustacea: Decapoda: Callianassidae). *Annalen des Naturhistorischen Museums in Wien*, **99**B, 565–570.

Angel, H.H. & Angel, M.V. (1967) Distribution pattern analysis of a benthic community. *Helgoländer wissenshaftliche Meeresuntersuchungen*, **29**, 439–459.

Atkinson, R.J.A. & Chapman, C.J. (1984) Resin casting: a technique for investigating burrows in sublittoral sediments. *Progress in Underwater Science,* **9**, 15–25.

Atkinson, R.J.A. & Nash, R.D.M. (1985) Burrows and their inhabitants. *Progress in Underwater Science,* **10**, 109–115.

Atkinson, R.J.A. & Nash, R.D.M. (1990) Some preliminary observations on the burrows of *Callianassa subterranea* (Montagu) (Decapoda: Thalassinidea) from the west coast of Scotland. *Journal of Natural History,* **24**, 403–413.

Atlantic and Gulf Rapid Reef Assessment (2000) AGGRA methodology, version 3.1. Atlantic and Gulf Rapid reef Assessment, MGG-RSMAS, University of Miami, Florida. Website: http://coral.aoml.noaa.gov/agra/method/method2.doc

Baddeley, A.D., Defigueredo, J.W., Curtis, J.W.M. & Williams, A.N. (1968) Nitrogen narcosis and performance underwater. *Ergonomics*, **11**, 157–164.

Barnett, P.R.O. & Hardy, B.L.S. (1967) A diver operated quantitative sampler for sand macrofaunas. *Helgolander wissenschaftliche Meeresuntersuchungen*, **15**, 390–398.

Behnke, A.R., Thomson, R.M. & Motley, E.P. (1935) The psychological effects from breathing air at 4 atmospheres pressure. *American Journal of Physiology*, **112**, 554–558.

Bohnsack, J.A. (1979) Photographic quantitative sampling of hard-bottom benthic communities. *Bulletin of Marine Science*, **29**(2), 242–252.

Bohnsack, J.A. (ed.) (1996) Two visually based methods for monitoring coral reef fishes. In: *A Coral Reef Symposium on Practical, Reliable, Low Cost Monitoring Methods for Assessing the Biota and Habitat Conditions of Coral Reefs* (Eds M.P. Crosby, G.R. Gibson, & K.W. Potts). Office of Ocean and Coastal Resource Management, NOAA, Silver Spring, MD.

Bohnsack, J.A. & Bannerot, S.P. (1986) A stationary visual census technique for quantitatively assessing community structure of coral reef fishes. Department of Commerce, NOAA Technical Report NMFS 41.

Booth, D.J. & Beretta, G.A. (2002) Changes in a fish assemblage after a coral bleaching event. *Marine Ecology in Progress Series*, **245**, 205–212.

Bradbury, R.H., Loya, Y., Reichelt, R.E & Williams, W.T. (1986) Patterns in the structural topology of benthic communities on two corals reefs of the central Great Barrier Reef. *Coral Reefs*, **4**, 161–167.

Brock, V.E. (1954) A preliminary report on a method of estimating reef fish populations. *Journal of Wildlife Management*, **18**, 297–308.

Bullimore, B. (1983) Skomer Marine Nature Reserve subtidal monitoring project: development and implementation of low-budget stereo-photographic monitoring of rocky subtidal epibenthic communities. Unpublished report to the Nature Conservancy Council, Peterborough, UK.

Bullimore, B. (1986) Skomer Marine Nature Reserve subtidal monitoring project: photographic monitoring of subtidal epibenthic communities, August 1984–November 1985. Unpublished report to the Nature Conservancy Council, Peterborough, UK.

Bullimore, B. (1987) Skomer Marine Nature Reserve subtidal monitoring project: Photographic monitoring of subtidal epibenthic communities 1986. Unpublished report to the Nature Conservancy Council, Peterborough, UK.

Bullimore, B. & Hiscock, K. (2001) Procedural guideline No. 3–12. Quantitative surveillance of sublittoral rock biotopes and species using photographs. In: *Marine Monitoring Handbook* (Eds J. Davies, (senior ed.), J. Baxter, M. Bradley, D. Connor, J. Khan, E. Murray, W. Sanderson, C. Turnbull, & M. Vincent) Joint Nature Conservation Committee, Peterborough, UK. [Online] Available from: <http://www.jncc.gov.uk/marine/mmh/Pg%203-12.pdf>

Bunker, F. (1985) A brief review of the ecology of the sea fan *Eunicella verrucosa* and a survey of the Skomer population in 1985. Unpublished report to the Nature Conservancy Council, Peterborough, UK.

Chapman, C., Johnstone, A., Dunn, J. & Creasey, D. (1974) Reactions of fish to sound generated by diver's open-circuit underwater breathing apparatus. *Marine Biology*, **27**, 357–366.

Coelho, V.R., Cooper R.A. & de Almeida, Rodrigues S. (2000) Burrow morphology and behaviour of the mud shrimp *Upogebia omissa* (Decapoda: Thalassinidea: Upogebiidae). *Marine Ecology Progress Series*, **200**, 229–240.

Craik, G.J.S. (1981) Underwater survey of coral trout *Plectropomus leopardus* (Serranidae) populations in the Capricornia Section of the Great Barrier Reef Park. *Proceedings of the Fourth International Coral Reef Symposium*, **1**, 53–58.

C.R.C Reef Research Center. Crown of Thorns starfish on the Great Barrier Reef. Current state of knowledge, April 2001. Cooperative Research Center of the Great Barrier Reef World Heritage Area, James Cook University, Townsville, Australia.

De Couet, H. & Green, A. (1989) *The Manual of Underwater Photography*. Verlag Christa Hemmen, Weisbaden.

Dinsmore, D.A. (1989a) The operational advantages of enriched air nitrox versus air for research diving. *Marine Technology Society Journal*, **23**, No. 4, 42–46.

Dinsmore, D.A. (1989b) NURC/UNCW nitrox diving program, pp. 75–84. In: *Workshop on Enriched Air Nitrox Diving* (Eds R.W. Hamilton, D. Crosson & A.W. Hulbert) National Undersea Research Program Research Report 89(1). [NURC/UNCW R-89-20]
Du Plat Taylor J. (Ed.) (1965) *Marine Archaeology*, Hutchinson, London, UK.
Dworschak, P.C. & Jorg, A.O. (1993) Decapod burrows in mangrove-channel and back-reef environments at the Atlantic Barrier reef, Belize. *Ichnos*, **2**, 277–290.
Earle, J.L. & Pyle, R.L. (1997) *Hoplolatilus pohlei*, a new species of sand tilefish (Perciformes: Malacanthidae) from the deep reefs of the D'Entrecasteaux Islands, Papua New Guinea. *Copeia*, **2**, 383–387.
English, S., Wilkinson, C. & Baker, V. (1997) *Survey Manual for Tropical Marine Resources*. Second Edition. Australian Institute of Marine Science. Townsville, Australia.
Fernandes, L., Moran, P.J. & Marsh, H. (1993) A system for classifying outbreaks of crown-of-thorns starfish as a basis for management. Abstract. *Proceedings of the Seventh International Coral Reef Symposium, Guam*, **2**, 803–804.
Fowler, B., Ackles, K.N. & Porlier, G. (1985) Effects of inert gas narcosis on behavior – a critical review. *Undersea Biomedical Research*, **12**, 269–402.
Gage, J.D. (1975) A comparison of the deep sea epibenthic sledge and anchor-box dredge samplers with the van Veen grab and hand coring by divers. *Deep-Sea Research*, **22**, 693–702.
Gamble, J.C. (1984) Diving. In: *Methods for the Study of Marine Benthos* (Eds N.A. Holme, & A.D. McIntyre), Second Edition, pp. 99–139. Blackwell Scientific Publications, Oxford, UK.
GBRMPA (1978) Workshop on reef fish assessment and monitoring. GBRMPA Workshop Series No. 2, Great Barrier Reef Marine Park Authority, Townsville, Australia.
GBRMPA (1979) Workshop on coral trout assessment techniques. GBRMPA Workshop Series No. 3, Great Barrier Reef Marine Park Authority, Townsville, Australia.
Gill, A.C., Pyle, R.L. & Earle, J.L. (1996) *Pseudochromis ephippiatus*, new species of dottyback from southeastern Paupa New Guinea (Teleostei: Perciformes: Pseudochromidae). *Revue Francais Aquariologie*, **23**, No. 4, 97–100.
Godden, D. & Baddeley, A.D. (1975) Context dependent memory in two natural environments on land and underwater. *British Journal of Psychology*, **66**, 325–331.
Green, N. (1980) *Underwater stereophotography applied in ecological monitoring. Report 1. Methods and preliminary evaluation*. Norwegian Institute for Water Research, Oslo.
Grist, N. (2000) A review of aspects of the use of electronic methods for the analysis of monitoring photographs. Unpublished report by Unicomarine Ltd. to English Nature (Devon, Cornwall & Isles of Scilly), Truro, UK.
Hiscock, K. (1984) Sublittoral monitoring at Lundy: July 28th to August 4th, 1984. Unpublished report from the Field Studies Council Oil Pollution Research Unit to the Nature Conservancy Council, Peterborough, UK.
Hiscock, K. (1986) Marine biological monitoring at Lundy: March 26th to 29th, July 26th to August 2nd 1986. Unpublished report from the Field Studies Council Oil Pollution Research Unit to the Nature Conservancy Council, Peterborough, UK.
Hiscock, K. (1987). Subtidal rock and shallow sediments using diving. In: *Biological surveys of Estuaries and Coasts* (eds. J.M. Baker & W.J. Wolff) pp. 198–237.
Hiscock, K. (Ed.) (1996). Marine Nature Conservation review: rationale and methods. Unpublished report by the Joint Nature Conservation Committee, Peterborough, UK.

Hiscock, K. & Hoare, R. (1973) The ecology of sublittoral communities at Abereiddy Quarry, Pembrokeshire. *Journal of the Marine Biological Association of the UK*, **55**, 833–864.

Holt, R.H.F. (2001) Monitoring and technological developments in scientific SCUBA diving. In: *The Establishment of a Programme of Surveillance and Monitoring for Judging the Condition of the Features of Pen Llyn a'r Sarnau cSAC: 1. Progress to March 2001* (Eds W.G. Sanderson, R.H.F. Holt & L. Kay). Countryside Council for Wales contract science report No. 380.

Holt, R.H.F. & Sanderson, B. (2001) Procedural Guideline No. 6-2. Relocation of intertidal and subtidal sites. In: *The establishment of a programme of surveillance and monitoring for judging the condition of the features of Pen Llyn a'r Sarnau cSAC: 1. Progress to March 2001* (Eds W.G. Sanderson, R.H.F. Holt & L. Kay). Countryside Council for Wales contract science report No. 380.

Jeffrey, C.F.G., Patten-Semmens, C., Gittings, S. & Monaco, M.E. (2001) Distribution and sighting frequency of reef fishes in the Florida Keys National Marine Sanctuary. Marine Sanctuaries Conservation Series MSD-01-1. U.S. Department of Commerce, National Oceanic and Atmospheric Administration, Marine Sanctuaries Division, Silver Spring, MD.

Joiner, J.T. (Ed.) (2001) *NOAA Diving Manual: Diving for Science and Technology.* Fourth edition. Best Publishing Company, AZ, USA.

Jones, R.S. & Thompson, M.J. (1978) Comparison of Florida reef fish assesmblages using a rapid visual technique. *Bulletin of Marine Science*, **28**, 159–172.

Kenchington, R.A. & Morton, B. (1976) Two Surveys of the Crown of Thorns Starfish Over a Section of the Great Barrier Reef. Australian Government. Public Service, Canberra.

Kimmel, J.J. (1985) A new species-time method for visual assessment of fishes and its comparison with established methods. *Environmental Biology of Fishes*, **12**, 23–32.

Kimmel, J.J. (1993) Suggested modifications to diver-oriented point counts for fishes. *Proceedings of the 73rd Annual Meeting, American Society of Ichthyologists and Herpetologists*, Austin, TX, USA.

Liddell, W.D. & Ohlhorst, S.L. (1987) Patterns of reef community structure, North Jamaica. *Bulletin of Marine Science*, **40**, 311–329.

Lincoln, R.J. & Sheals, J.G. (1979) *Invertebrate Animals – Collection and Preservation.* British Museum (Natural History) and Cambridge University Press.

Lobel P.S. (2001) Fish bioacoustics and behavior: passive acoustic detection and the application of a closed-circuit rebreather for field study. *Marine Technology Society Journal*, **35**, 19–28

Lundälv, T. (1971) Quantitative studies on rocky bottom biocoenoses by underwater photogrammetry. A methodical study. *Thalassia Jugoslavica*, **7**, No. 1, 201–208.

Lundälv, T. (1985) Detection of long-term trends in rocky sublittoral communities: Representativeness of fixed sites. In: *The Ecology of Rocky Coasts* (Ed. P.G. Moore), pp. 329–345. Hodder and Stoughton, London.

Main, J. & Sangster, G.I. (1978) A new method for observing fishing gear using a towed wet submersible. *Progress in Underwater Science*, **3**, 259–267.

Marsh, L.M., Bradbury, R.H. & Reichelt, R.E. (1984) Determination of the physical parameters of coral distributions using line transect intersect data. *Coral Reefs*, **2**, 175–180.

Meier, O.W. & Porter, J.W. (1992) Methods comparison for monitoring Floridian reef coral populations. *American Zoologist*, **32**, 625–640.

Miller, S.L. & Cooper, C. (2000) The *Aquarius* Underwater Laboratory: America's 'Inner Space' Station. *Marine Technology Society Journal*, **34**, No. 4, 69–74.

Moore, J., James, B. & Gilliland, P. (1999) Development of a monitoring programme and methods in Plymouth Sound cSAC: application of diver and ROV techniques. Unpublished report by Cordah Limited to English Nature, Peterborough, UK.

Munro, C.D. (1992) An investigation into the effects of scallop dredging in Lyme Bay. Unpublished report for the Devon Wildlife Trust, Exeter, UK.

Munro, C.D. (1996) Lyme Bay potting impacts study. In: *A Study on the Effects of Fish (Crustacea/Mollusc) Traps on Benthic Habitats and Species* (Eds N.C. Eno & D. MacDonald). Final report to the European Commission Directorate General XIV Fisheries. Study contract No. 94/076.

Munro, C.D. (1998) Lundy Marine Nature Reserve subtidal monitoring 3rd–6th May 1998. Unpublished report by Marine Biological Surveys consultancy to English Nature. ENRR No. 294. Peterborough, UK.

Munro, C.D. (1999) Monitoring of the rocky sub-littoral of Milford Haven. May-July 1998. Unpublished report by Marine Biological Surveys consultancy to Milford Haven Waterway Environmental Monitoring Steering Group, Milford Haven, UK.

Munro, C.D. (2000a) East Tennants Reef Seafan Monitoring Project: annual report March 2000. Unpublished report by *Reef Research*, East Tennants Reef seafan monitoring study. Crediton, UK. [Online] Available from: <http://www.reef-research.org/pdf_rpts/Rpt_ETR_01_Mar_2000.pdf>

Munro, C.D. (2000b) Lundy Marine Nature Reserve subtidal monitoring 25th–27th September 1999. Unpublished report by Marine Biological Surveys consultancy to English Nature, Peterborough, UK.

Munro, C.D. (2001) Procedural guideline No. 3–13. *In situ* surveys of sublittoral epibiota using hand held video. In: *Marine Monitoring Handbook* (Eds. J. Davies, (senior ed.), J. Baxter, M. Bradley, D. Connor, J. Khan, E. Murray, W. Sanderson, C. Turnbull, & M. Vincent) Joint Nature Conservation Committee, Peterborough, UK. [Online] Available from: <http://www.jncc.gov.uk/marine/mmh/Pg%203-13.pdf>

Munro, L. (2001) East Tennants Reef seafan study – Report to Project AWARE (UK), August 2001. Reef Research Report 8/2001 ETR 03. Reef Research, Crediton, UK. [Online] Available from: <http://www.reef-research.org/pdf_rpts/RR_Rpt_8_2001_ETR_03.pdf>

Nickell, L.A. & Atkinson, R.J.A. (1995) Functional morphology of burrows and trophic modes of three thalassinidean shrimp species, and a new approach to the classification of thalassinidean burrow morphology. *Marine Ecology Progress Series*, **128**, 181–197.

Nickell, L.A., Atkinson, R.J.A., Hughes, D.J., Ansell, A.D. & Smith, C.J. (1995) Burrow morphology of the echiuran worm *Maxmuelleria lankesteri* (Echiura: Bonelliidae), and a brief review of burrow structure and related ecology of the Echiura. *Journal of Natural History*, **29**, 871–885.

Osborne, K. & Oxley, W.G. (1997) Sampling benthic communities using video transects, in English, S., Wilkinson, C. & Baker, V. 1997. *Survey Manual for Tropical Marine Resources*. Second Edition. Australian Institute of Marine Science. Townsville, Australia.

Parker, Jr., R.O., Chester, A.J. & Nelson, R.S. (1994) A video transect method for estimating reef fish abundance, composition and habitat utilization at Gray's Reef National Marine Sanctuary, Georgia. *Fishery Bulletin,* **92**, 787–799.

Parrish, F.A. & Pyle, R.L. (2002) Field comparison of open-circuit SCUBA to closed-circuit rebreathers mixed-gas diving operations. *Marine Technology Society Journal,* **36**, No. 2, 82–91.

Pattengill-Semmens, C.V. & Semmens, B.X. (in prep.) Status of coral reefs of Little Cayman and Grand Cayman, British West Indies, in 1999 (Part 2: fishes). *Atoll Research Bulletin*, 2002.

Pence, D. and Pyle, R.L. (2002) University of Hawaii dive team completes Fiji deep reef fish surveys using mixed-gas rebreathers. *The Slate*. A news publication of the American Academy of Underwater Sciences.

Poulin, M. (2001) Marine Biodiversity Monitoring – protocol for monitoring sea-ice algae. A report by the Marine Biodiversity Monitoring Committee (Atlantic Maritime Ecological Science Cooperative, Huntsman Marine Science Centre) to the Ecological Monitoring and Assessment Network of Environment, Canada. [Online] Available from: <http://www.eman-rese.ca/eman/ecotools/protocols/marine/seaicealgae/intro.html>

Pyle, R.L. (1992) The peppermint angelfish *Centropyge boylei*, n.sp. Pyle and Randall. *Freshwater and Maine Aquarium*, **15** No. 7, 16–18.

Pyle, R.L. (1996) How much coral reef biodiversity are we missing? *Global Biodiversity*, **6**, No. 1, 3–7

Pyle, R.L. (1998) Use of advanced mixed-gas diving technology to explore the coral reef 'Twilight Zone'. pp. 71–88. In: *Ocean Pulse: A Critical Diagnosis* (Eds. J.T. Tanacredi, & J. Loret). Plenum Press, New York.

Pyle, R.L. (1999) Mixed gas closed circuit rebreather use for identification of new reef fish species from 200–400 fsw. In: Proceedings of the AAUS Technical Diving Forum: Assessment and feasibility of technical diving operations for scientific exploration. American Academy of Underwater Sciences.

Pyle, R.L. (2000) Assessing undiscovered fish biodiversity on deep coral reefs using advanced self-contained diving technology. *Marine Technology Society Journal* **34**, No. 4, 82–91.

Pyle, R.L. & Randall, J.E. (1992) A new species of *Centropyge* (Perciformes: Pomacanthidae) from the Cook Islands, with a redescription of *C. boylei. Revue Francais Aquariologie*, **19**, No. 4, 115–124.

Randall, J.E. & Pyle, R.L. (2001a) Three new species of labrid fishes of the genus *Cirrhilabrus* from islands of the tropical Pacific. *Aqua*, **4**, No. 3, 89–98.

Randall, J.E. & Pyle, R.L. (2001b) Four new serranid fishes of the anthiine genus *Pseudoanthias* from the South Pacific. *Raffles Bulletin of Zoology*, **49**, No. 1, 19–34.

Reef Check, undated. Reef Check survey instruction manual version 5.1. [Online] Available from: <http://www.reefcheck.org/manual.pdf>

Riley, J.D. & Holford, B.H. (1965) A sublittoral survey of Port Erin Bay, particularly as an environment for young plaice. *Report of the Marine Biological Station of Port Erin (1964)*, **77**, 49–53.

Robinson M. & Tully O. (2000) Seasonal variation in community structure and recruitment of benthic decapods in a sub-tidal cobble habitat. *Marine Ecology Progress Series*, **206**, 181–191.

Rogers, C.S., Garrison, G., Grober, R., Hillis, Z-M. & Franke, M.A. (1994) *Coral reef monitoring manual for the Caribbean and Western Atlantic.* National Park Service, Virgin Islands National Park. St. Thomas, Virgin Islands. [Online] Available from: <http://www.fcsc.usgs.gov/Monitoring_Manual.pdf>

Rogers, C.S., Garrison, R. & Grober-Dunsmore, R. (1997) A fishy story about hurricanes and herbivory: seven years of research on a reef in St. John, US Virgin Island. *Proceedings of the 8th International Coral Reef Symposium*, **1**, 555–560.

Rogers, C.S. & Miller, J. (undated) Coral bleaching, Hurricane damage, and benthic cover on coral reefs in St. John, US Virgin Island: a comparison of surveys with the chain transect method and videography. Proceedings of the National Coral Reef Institute Conference, April 1999. [Online] Available from: <http://www.cpacc.org/download/vidchain.pdf>

Rogers, C.S., Miller, J. & Waara, R.J. (undated) Tracking changes on a reef in the US Virgin Islands with videography and SONAR: a new approach. *Proceedings of the 9th International Coral Reef Symposium.* [Online] Available from: <http://www.FCSC.usgs.gov>

Rogers, W.H. & Moeller, G. (1989) Effects of brief, repeated hyperbaric exposures on susceptibility to nitrogen narcosis. *Undersea Biomedical Research*, **16**, 227–232.

Rostron, D.M. (2001) Procedural guideline No. 3–10: sampling marine benthos using suction samplers. In: *Marine Monitoring Handbook* (Eds. J. Davies, (senior ed.), J. Baxter, M. Bradley, D. Connor, J. Khan, E. Murray, W. Sanderson, C. Turnbull, & M. Vincent) Joint Nature Conservation Committee, Peterborough, UK [Online] Available from: <http://www.jncc.gov.uk/marine/mmh/Pg%203-10.pdf>

Russ, G.R. (1984a) Distribution and abundance of herbivorous grazing fishes in the Central Great Barrier Reef. I: Levels of variability across the entire continental shelf. *Marine Ecology Progress Series*, **20**, 23–34.

Russ, G.R. (1984b) Distribution and abundance of herbivorous grazing fishes in the Central Great Barrier Reef. II: Patterns of zonation of mid-shelf and outer shelf reefs. *Marine Ecology Progress Series*, **20**, 35–44.

Sayer, M.D.J., Thurston, S.R., Nickell, L.A., Duncan, G.D. & Gale, A.V. (1996) Use of rebreathers in fish research. In: *Underwater Research and Discovery* (Proceedings of the 3rd Underwater Science Symposium), pp. 35–38. Aberdeen: Society for Underwater Technology.

Schinn, E.A. (1968) Burrowing in recent lime sediments of Florida and the Bahamas. *Journal of Palaeontology,* **42**, 879–894.

Schmitt, E.F., Sluka, R.D. & Sullivan-Sealey, K.M. (2002) Evaluating the use of roving diver and transect surveys to asses the coral reef fish assemblage off southeastern Hispanola. *Coral Reefs*, **21**, 216–223.

Simenstad, C.A., Tanner, C.D., Thom, R.M. & Conquest, L.L. (1991) Estuarine Habitat Assessment Protocol. Puget Sound Estuary Program. Environmental Protection Agency Report 910/9-91-037.

Sinclair Knight Merz (1999) Video Transect Analysis of Subtidal Habitats in the Dampier Archipelago. Unpublished report to the Western Australian Museum of Natural Science. [Online] Available from: <http://www.museum.wa.gov.au/discovery/naturewatch/dampier/reports/reports.html>

Squires, H.J., Ennis, G.P. & Dawe, G. (1999) On biology of the shrimp *Eualu pusiolus* (Kroyer, 1841) (Crustacea, Decapoda) at St. Chad's, Newfoundland. *NAFO Scientific Council Studies*, **33**, 1–10.

Stanton, G. (1991) The future of special gas mixtures in scientific diving at Florida State University. In: *Progress in Underwater Science*, **16**, 123–137.

Steneck, R.S., Ginsburg, R.N., Kramer, P., Lang, J. & Sale, P. (2000) Atlantic and Gulf Rapid Reef Assessment (AGRRA): A species and spatially explicit reef assessment protocol. 9th International Coral Reef Symposium, Bali, Indonesia.

Thrush, S.F., Schultz, D., Hewitt, J.E. & Talley, D. (2002) Habitat structure in soft-sediment environments and abundance of juvenile snapper *Parus auratus. Marine Ecology Progress Series*, **245**, 273–280.

Chapter 5
Macrofauna Techniques

A. Eleftheriou and D.C. Moore

5.1 Littoral observation and collection

The study of the intertidal fauna and flora is in some ways easier than that of subtidal areas, but since the habitat is subjected to both aerial and aquatic climates, environmental factors influencing distribution are more complex.

When visiting the shore it should be remembered that low tide is a quiescent period for many animals which are active only when covered by the sea. Also, certain predators invade the shore at different periods of the tidal cycle: fish and crustaceans at high tide, and humans, birds, insects, and occasionally rats or cattle, at low tide.

On sandy or muddy shores estimates of the standing crop of macrofauna and flora are made by driving a square sheet metal frame of appropriate area (e.g. 0.1 m^2 or 0.25 m^2) into the substratum (Bakus, 1990), the sediment within the frame being excavated to the desired depth. Plastic or metal tubes, which remove an undisturbed core of sediment (Desprez & Ducrotoy, 1987; Sylvand, 1995), are also widely used: a large version (0.1 m^2) which is slid into the sediment to remove a sample to a depth of 10 cm has been successfully employed on some beaches (Grange & Anderson, 1976).

Preliminary excavations should be made to find a suitable sampling depth. On some shores the majority of species and individuals occur in the top 15 cm of depth, in others it may be necessary to excavate to >30 cm, or even deeper for certain crustaceans and bivalves.

A plentiful supply of water is required for sieving; sometimes a nearby stream may be utilised, but it should be noted that fresh water may damage the more delicate organisms. If there is no stream in the vicinity and the low watermark is quite far away, it may be possible to arrange sampling at a time when the water's edge is fairly close to the sampling position. Subsequent sorting is made easier if all traces of fine material can be washed out of the sample when sieving.

Types of sieve mesh are discussed on pp. 204–206. On muddy or fine sandy shores it may be possible to use an aperture as small as 0.5 mm, but on coarse sand or gravel shores a coarser mesh of 1.0 or 1.4 mm may be necessary. The selected mesh size (0.5 mm) is indicative for the majority of benthic organisms. However,

Fig. 5.1 The Riley push-net for shallow water sampling.

when sampling for benthic larvae or juveniles it might be necessary to use a 0.25 or even a 0.125 mm mesh. Since most small individuals are found in the top 5–10 cm it may be possible to use a fine sieve for the surface layers, passing the deeper layers through a coarser mesh (Pamatmat, 1968). In this way much unnecessary labour may be avoided. Organisms which pass through the smallest mesh employed can be collected by taking smaller volumes of deposit with a Perspex or glass coring tube, and adopting the methods described for meiofauna in Chapter 6 where finer mesh sizes are used. The sieved material is preserved and further processed in the laboratory.

At high tide, fish and crustaceans invade the intertidal zone. These may be sampled qualitatively by shrimping net, beam trawl (Edwards & Steele, 1968), or by a small sledge such as that described by Pullen et al. (1968). The Riley push-net (Fig. 5.1) may be used in shallow water, either just below low tidemark, or on the flooded beach. It consists of a light metal frame construction seating on two skids with a net of 5.0–10.00 mm knot to knot mesh leading to a codend of finer mesh. The foot rope is weighted down by lead weights and in front of the mouth a number of tickler chains are attached to the frame skids to help stir up the organisms in the sediment. If operated at a standard speed and for a fixed time it can produce comparative data on small active animals such as shrimps or juvenile flatfish. Some small sand-burrowing crustaceans swim freely in the overlying water at high tide, returning to their zones in the sand as the tide retreats (Watkin, 1941). These animals can be sampled by nets such as those described by Colman and Segrove (1955), Macer (1967) and Eleftheriou (1979).

It is usual to survey an intertidal flat by means of traverses running from high to low water mark, with sampling stations sited at regular intervals. Where possible, two or more samples should be taken at each station, as a measure of the variability of the populations. For intensive and repetitive investigations, the shore can be divided by intersecting transects to provide samples at regularly spaced intervals. Because the surface of an intertidal flat is often more or less uniform in appearance, the question of possible bias in selecting the position of the traverses and of individual stations does not usually arise, although the presence of irregularities on the beach should be carefully observed, since these are often associated with turbulent conditions at particular tidal levels. On smaller areas, such as pocket beaches, the profile may be much more obviously irregular, and the various features – ridges and runnels, streams, pools of fresh or salt water – should be taken into account when sampling. Examples of surveys are given by McIntyre and Eleftheriou (1968), Eleftheriou and McIntyre (1976) and Eleftheriou and Robertson (1988).

In atidal beaches or on shores with a very small tidal range such as the Baltic or the Mediterranean, the littoral zone under investigation for the fauna and flora is greatly compressed to a narrow zone along the edge of the water. Sampling techniques and overall methodology remain the same as described above, with stations taken above, at and below the swash zone extending a certain distance in the infralittoral but not exceeding 1–2 m depth. The very shallow water (1–2 m) in some shores can be investigated by handheld devices such as core tubes (Maitland, 1969; Baker et al., 1977; Kanneworff & Nicolaisen, 1983).

On rocky shores, sampling is carried out by means of a square frame of heavy gauge wire laid on the substratum, the animals within the frame being counted directly, or estimated in terms of percentage cover/abundance scale of the surface (for size of sampling unit, see Chapter 1). Sometimes estimates may be made *in situ*, otherwise growths must be scraped off for subsequent examination in the laboratory. For larger organisms, a frame of 1 or 0.25 m^2 is suitable, but for smaller organisms or where the rock surface is irregular frames of 0.1 m^2 (316 × 316 mm) should be used. A flexible frame might be appropriate on some rock surfaces. For the sampling of filamentous algae and their associated fauna in shallow water, a square-quadrat box-sampler with a sampling area of 1109 cm^2 is described by the Finnish IBP-PM Group (1969) and other methods used by divers, as described in Chapter 4, may sometimes be appropriate.

For counting small organisms such as barnacles, squares of 0.01 m^2 (100 × 100 mm) are suitable. For such counts a piece of thick (6 mm) Perspex exactly 100 mm square and etched with a grid of 10 mm squares is convenient, allowing smaller areas to be counted and also minimising the possibility of missing individuals or counting them twice. Besides organisms living on the rock surface, estimates should be made of the crevice-living and boring species, and of those sheltering among weeds. The importance of photography (Chapter 4) for non-destructive recording of seashore species and habitats must be emphasised.

However, the size of the sample (quadrat) should be determined by the population under investigation and, in particular, its spatial distribution. There is considerable debate concerning the choice of the optimal size of the sample which could range from an arbitrary choice to a statistically valid replication. In *Ecoscope* (2000) it is pointed out that 'there is no simple rule for calculating the optimal size (quadrat)'. Andrew and Mapstone (1987) present a useful discussion on the subject while Pringle (1984), having reviewed several papers, concluded that a 0.25 m^2 area should be appropriate for sampling marine organisms.

Stations should be spaced out at regular distances (or height intervals) along traverses from high to low water mark. The lack of uniformity of most rocky shore habitats will often make it necessary to make a number of estimates in different types of habitat (e.g. rock pool, rocks exposed/sheltered from waves/sun, crevices, under stones) at each station. Since such habitats cannot normally be selected by predetermined measurements, the question of bias will arise. Bias cannot be entirely eliminated when making such estimates, and indeed on a rocky shore which can have an almost infinite variety of microhabitats, it is questionable whether it is possible to take a sample which is in any degree representative of a wider area. Space does not permit detailed consideration of the literature on intertidal methods, but a useful discussion of methods for both sediment and rocky shores is given by Gonor and Kemp (1978), and in the separate contributions by Jones (1980) and Jones et al. (1980) (in Price et al., 1980). Methods for coral reefs are described in Stoddart and Johannes (1978), Marsh et al. (1984) and Bradbury et al. (1986) (see also Chapter 4).

A semi-quantitative method for surveying both intertidal and subtidal biota and organisms has been developed by the Joint Nature Conservation Committee and published in a series of reviews (Marine Nature Conservation Reviews – MNCR). Details of this methodology can be found in Chapter 4 as well as in Hiscock (1996). On the other hand, a large-scale study of the narrow inshore zone of the world's oceans at depths of less than 20 m has been initiated by the Census of Marine Life (www.coml.org) through the Natural Geography Inshore Areas project (NAGISA). Building on-site selection criteria and sampling protocols, now available online, this project's aims are to achieve coverage with standardised techniques that will provide a biodiversity baseline for future comparisons.

Position fixing and levelling on the shore

The position of stations or transects on the shore may be fixed by standard surveying techniques as outlined, for example, by Southward (1965), and by Jones (1980). A cheap and simple level which may be used by one person is described by Kain (1958), and other levelling methods are described by Emery (1961), Stephen (1977) and Nelson-Smith (1979). Fuller treatment of levelling techniques is given by the Admiralty (1948), Ingham (1975) and Pugh (1975). Other relevant references are Kissam (1956), Morgans (1965) and Zinn (1969).

For repeat sampling, positions on a rocky shore may be marked with paint, marks chiselled into the rock, expanding bolts inserted into crevices, holes drilled in the rock, or by small concrete blocks cast *in situ*. Boalch et al. (1974) used an underwater-setting resin compound for fixing metal bolts into holes drilled into rocks on the shore.

On sediment shores, positions may be marked by posts driven deeply into the sediment. However, such posts may cause some alteration to the environment and may also attract the attention of beach users resulting in parts being removed or damaged as well as the sediment around the posts receiving more trampling than elsewhere. In such instances it is advisable to position reference marks on rocks or other permanent structures, from which locations are determined by tape measure, and/or transect lines.

Since the second edition of this handbook (Holme & McIntyre, 1984), positioning technology has developed to the point where small, handheld satellite based positioning systems are readily available. The Global Positioning System (GPS), and its more accurate sibling, Differential GPS (DGPS) which uses a fixed terrestrial reference position, can give accuracies up to less than 5 m over the surface of the planet. The removal of 'Selective Availability' from the satellite system and the introduction of the European Geostationary Navigation Overlay System (EGNOS), a European version of the American Wide Area Augmentation System (WAAS) means that even small handheld units can now achieve the same accuracy.

Systems such as those produced by Garmin International Inc. (www.garmin.com) can log positions on demand and can therefore be used to plot the geography of beaches and bays accurately, by the simple expedient of the researcher walking along the boundaries they wish to map. Position files can then be downloaded to a suitable computer for plotting (see pp. 210–212), creating accurate and reproducible maps of the survey location upon which details of faunal or floral abundance can be superimposed. In addition, many of the units now carry altitude sensors and electronic compasses which, once correctly calibrated, may be of use in various survey activities.

For intertidal organisms, the duration of exposure at each low tide is important. This may be roughly determined from the zonation of plants and animals on the shore, and by observations of the length of time for which the selected sites are exposed over a number of low tides. Where tidal data are available the positions may be levelled to a benchmark or other fixed mark of known height; it will then be possible to calculate the mid-tide level of the beach as well as the exposure from data given in the tide tables. On sedimentary beaches the ensuing beach profiles can provide important information on the erosional and depositional cycles and therefore can be used to determine both the short and long-term stability of the shore. Consideration should be given to the use of high precision theodolites and laser based levelling surveyors (such as those manufactured by Leitz/Wild) employed by civil engineers and cartographers. These instruments are very expensive but it may be possible to hire them for survey use.

In estuaries or in parts of the world where accurate tidal data are not available, a series of observations on a graduated tide pole should be carried out; positions on the shore can then be levelled in relation to this or to a fixed point above the shore. Allowance should be made for spring and neap tides, the effects of wind and barometric pressure, and for river outflow in estuaries. It should not be assumed that the tide follows a symmetrical harmonic curve, nor that tidal heights are necessarily the same on the two tides of one day (see official *Admiralty Manual of Tides* or various online tidal predictors, such as provided by Admiralty EasyTide (http://easytide.ukho.gov.uk)). On wave-exposed coasts levels are elevated through spray and swash, so that plants and animals will tend to occur much higher than on sheltered shores.

5.2 Remote collection

There are a number of reviews in which the equipment and techniques used for sampling the benthos are described (e.g. Gunter, 1957; Thorson, 1957; Holme, 1964; Hopkins, 1964; Longhurst, 1964; Reys, 1964; Bouma, 1969, Kajak, 1971; Lamotte & Bourliere, 1971; Eagle et al., 1978; Gray et al., 1991; Rumohr, 1999) and a comprehensive bibliography of benthic samplers given by Elliott et al. (1993).

While the choice of equipment depends largely on local conditions (size of ship, power and capabilities of lifting gear, whether sampling in exposed or sheltered conditions, depth of water, bottom deposit, and type of sample required), the multiplicity of samplers which have been described is evidence not only of such factors, but also of a widespread dissatisfaction with existing methods of collection. Because the many samplers which are available have been described and discussed in reviews such as those listed above, this chapter will not attempt to cover the whole field nor to give a historical review, but rather guide readers towards the most suitable instruments for their own particular purposes: to this end a summary of the attributes of selected samplers is given in Table 5.2 (p. 200). There are some grounds (notably those of rocks and boulders) which cannot be adequately sampled by any of the instruments listed here. At best, they can be sampled only by dredge, which may prove inadequate even as a qualitative sampler. Such habitats are often better investigated by diver observation from a submersible (Chapter 7), or by photography and television (Chapter 4).

A combination of different techniques and the application of different samplers is therefore essential for obtaining a meaningful synthesis of the different biota and animal populations. In the search for more information, sampling is quite frequently not confined to the limits of the continental shelf (200 m) but well beyond it in the deeper areas of the continental slope (>200 m). Consequently, some of the sampling gear used has a universal application, as it is used as much in 'shallower' water as in deeper areas (Agassiz and otter trawls, some sledges and sleds, multicorers and box corers). The methods and samplers used only in the deep sea can be found in Chapter 7.

Trawls

Beam, Agassiz, otter and shrimp trawls may be used for qualitative sampling of the epifauna. These nets are designed to skim over the surface of the bottom, and because of the large area covered, are useful for collecting scarcer members of the epifauna, species of fish, cephalopods and crustaceans associated with the sea bottom. The efficiency of such gear, in terms of numbers of animals captured in relation to those in the area swept by the net, is generally low, and is selective for particular species. Attempts at quantifying results from trawl catches are considered on pp. 173–174.

The *beam trawl* is still used commercially for fishing of shrimps, prawns and flatfish. The mouth of the net is held open by a wooden or metal beam of 2–10 m length, with metal runners at either end. The net is a fairly long bag, of mesh about 12.5 mm knot-to-knot, the lower leading edge of which is attached to a weighted chain, forming a ground rope which curves back behind the top of the net attached to the beam so that fish and invertebrates disturbed by the ground rope cannot escape upwards. In commercial beam trawls several rows of tickler chain are fixed in front of the ground rope in order to increase sediment disturbance and the emergence of fish and invertebrates (Creutzberg et al., 1987; Kaiser et al., 1994).

The *Agassiz* or *Blake* trawl (Fig. 5.2) is virtually a double-sided beam trawl with a double weighted chain designed to collect samples in places where it is not possible to control which way up the trawl lands on the bottom (Agassiz, 1888). Compared to the beam trawl it suffers from the disadvantage that the ground and head ropes, being interchangeable in function, are necessarily of the same length. Consequently, few fish are caught by the Agassiz trawl, which is not, therefore, used commercially.

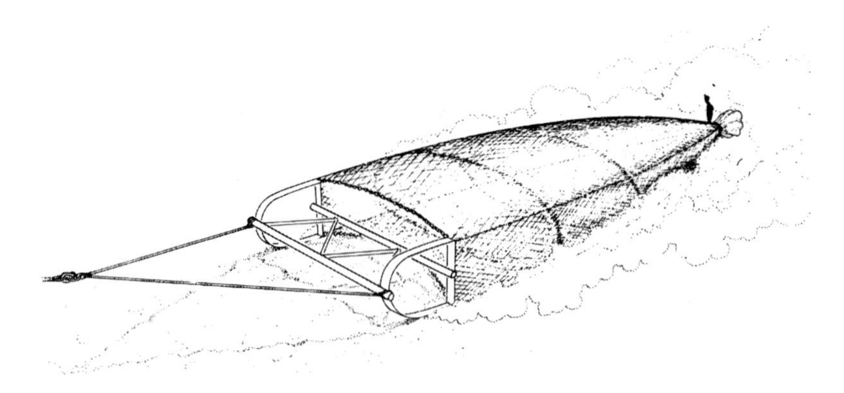

Fig. 5.2 The Agassiz double-sided beam trawl.

Both beam and Agassiz trawls can be towed on a pair of bridles attached to a single tow-rope or wire. Unfortunately, the cross-bars or beams are liable to be damaged if they meet an obstruction, and a weak link should be used, especially when working on unexplored grounds. Carey and Heyamoto (1972) describe a beam trawl with a flexible connection between beam and runners, which allows momentary collapse of the net when an obstruction is encountered and subsequent easy retrieval from a single bridle.

Otter trawls (see Fig. 7.5c, Chapter 7) used for commercial fishing also capture members of the invertebrate epifauna, but because of the rather large mesh size only the larger animals are retained. Besides those animals retained in the cod-end of the trawl, many small organisms may be found enmeshed in the body of the trawl. These must be removed. The rigging, shooting and working of otter trawls is a specialised subject; instructions for making up a small trawl are given by Steven (1952), while scaled-down otter trawls with finer meshes have effectively been used for the collection of small fish and epifaunal invertebrates (Eleftheriou, 1979). Details of various types of fishing gear designs have been published in the FAO (1972) catalogue, while further references are given by von Brandt (1978). Galbraith et al. (2004) give a beautifully illustrated introduction to commercial fishing gears and methods used in Scotland. One important development in estimating the capture efficiency of commercial trawls was the application of underwater photography or TV. Dyer et al. (1982) and Cranmer et al. (1984) successfully applied UW photography to their studies on the benthic invertebrates of the North Sea.

Bottom sledges

Many types of sledge have been designed for sampling the epifauna and the hyperbenthos as well as larger members of the plankton immediately above the bottom of the sea. Some are little more than plankton nets on runners (e.g. Myers, 1942), but Ockelmann's detritus-sledge has a tickler-chain to stir up newly settled benthic invertebrates (Ockelmann, 1964). Others have opening and closing mechanisms, and sometimes a flow metre to measure the quantity of water filtered (Bossanyi, 1951; Beyer, 1958 (illustrated in Holme, 1964); Frolander & Pratt, 1962; Macer, 1967; Bieri & Tokioka, 1968; Yocum & Tessar, 1979; Sorbe, 1983). A sledge net for hand towing in the intertidal zone is described by Colman and Segrove (1955).

An improved version of Macer's sled-mounted 'supra-benthic' sampler (Macer-GIROQ) has been developed by Brunel et al. (1978), who used it successfully to depths of 200 m (Fig. 5.3). It was used to provide quantitative estimates of megabenthos in the Georges Bank with a sheet steel collection box sampling at a depth of 25 cm (Thouzeau & Vine, 1991). Further changes were made by Dauvin et al. (1995) while Rowell et al. (1997) made further modifications by the addition of stabilising wings, wider runs, a video system and a double odometer.

A modified version of the Rothlisberg and Pearcy (1977) epibenthic sledge was constructed by Fossa et al. (1988) for sampling the epifauna at moderate depths (40–120 m). The sledge, which is of standard design, is equipped with a pneumatic opening and closing device which ensures sampling only when the sledge is in contact with the sea floor. Other modifications of the Rothlisberg and Pearcy (1977) epibenthic sledge were provided by Buhl-Jensen (1986) and Brattegard and Fossa

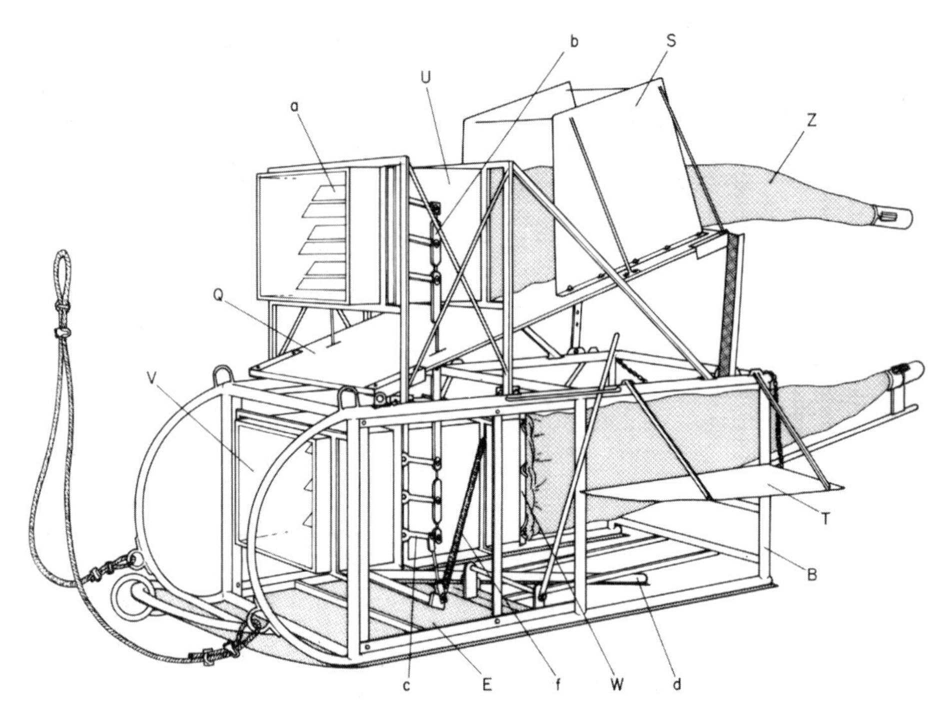

Fig. 5.3 The Macer-GIROQ epibenthic sledge as developed by Brunel et al. (1978). B, tubular chassis; E, sheet-metal gliding plate, turned upward at front; Q, adjustable wooden depressor; S, vertical fin; T, horizontal fin; U, wooden box at front of upper net; V, wooden box at front of lower net; W, metal strip for attachment of net; Z, zooplankton net; a, shutter closing mechanism; b, adjustable control link; c, crank lever; d, lever operating closing mechanism; f, closing spring.

(1991) for sampling the peracarid epifauna at greater depths (130–380 m). The use of weights, an opening and closing mechanism and a protective mat ensured a more efficient sampling and easier operational handling.

A number of workers used samplers with superimposed nets (Hesthagen, 1978; Sorbe, 1983; Dauvin & Lorgeré, 1989; Dauvin et al., 1995) or with additional nets mounted on either side of the original nets (Oug, 1977), in order to study the distribution of hyperbenthos in the first metre above the sea floor. Koulouri et al. (2003), using the basic frame of the Shand and Priestley (1999) photosled (towed trawl simulator sledge, TTSS2), developed a benthic sledge with superimposed nets, opening and closing mechanism, odometer and underwater television system for the study of the small invertebrate communities living above the sediment-water interface which is severely disturbed by fishing gear activities (Fig. 5.4). In a similar way, Sirenko et al. (1996) studied the suprabenthos by a benthopelagic sampler net frame attached to the upper part of a 3 m Agassiz trawl. It should be pointed out that all the epibenthic sledges sample the bottom layers at various distances from the ocean floor. Inevitably, sampling too near the bottom sediment results in contamination of the samples with both sediment and organisms living near the sediment surface. Many authors overcome this problem by combined or additional sampling which allows them to apportion the organisms to the appropriate biota.

Fig. 5.4 The towed trawl simulator sledge (TTSS2) (Koulouri et al., 2003). 1, groundrope; 2, opening-closing mechanism; 3, electronic controller; 4, TV camera; 5, collectors; 6, odometer; 7, super-imposed nets.

Epibenthic sledges have a heavy frame enclosing the net, and are particularly useful for deep sea sampling (see Chapter 7). There are also several instruments designed for deep sea sampling which are an intermediate between the Agassiz trawl and the dredge in that they consist of a net attached to a rectangular steel frame fitted with runners (Menzies, 1962) (see Fig. 7.10e, Chapter 7).

Dredges

Dredges have a heavy metal frame and are designed for breaking off pieces of rock, scraping organisms off hard surfaces, or for limited penetration and collection of sediments and fauna. A small dredge with runners used at Plymouth (Fig. 5.5) has been extensively employed for sampling the epifauna on the continental shelf.

The *naturalists' or rectangular dredge* (Fig. 5.6) is a useful instrument for exploratory purposes as it can obtain samples on a variety of grounds. One of the dredge arms is attached directly to the tow-rope, the other being joined to it by a few turns of twine, which act as a weak link to release the dredge should it become fast on the seabed. It is important not to use too many turns of twine, since, particularly with synthetic twine, it may then be too strong to part should the need arise. More sophisticated links, involving a metal shear pin (Fig. 5.7) are available

Fig. 5.5 Dredge with runners as used by the Marine Laboratory at Plymouth. The width is 1 m and the net is protected by canvas strips on either side of the net.

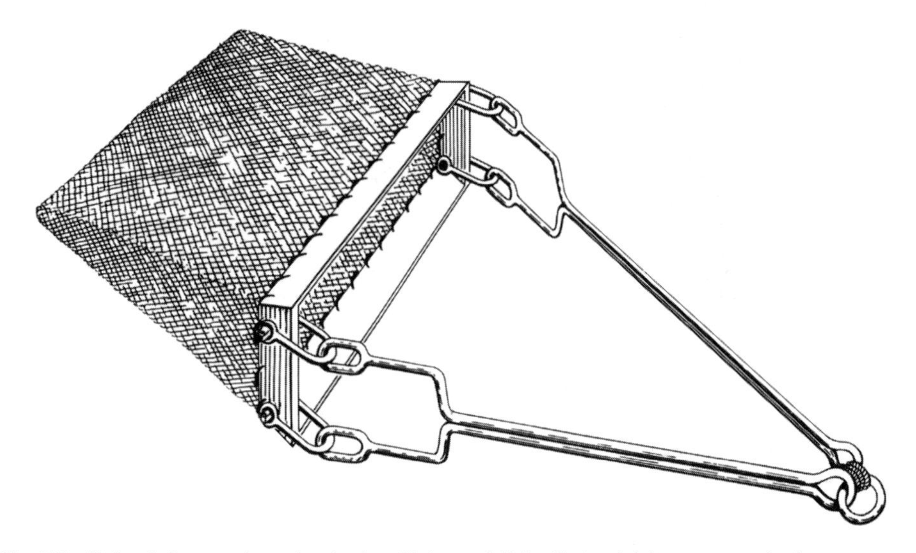

Fig. 5.6 Naturalist's or rectangular dredge. Note weak link of twine joining one arm to ring.

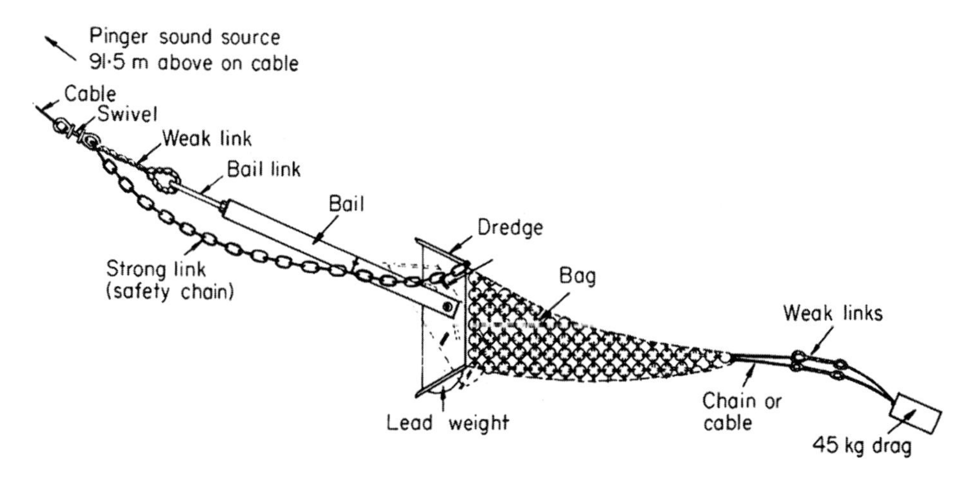

Fig. 5.7 Rock dredge with bag of interlaced metal rings. Note arrangement of weak link, safety chain and swivel. (Redrawn from Nalwalk et al., 1962.)

commercially. An alternative arrangement is to attach a safety line consisting of a piece of chain or wire from the towing point to some point towards the back of the dredge, for retrieval when the weak link parts. A swivel should always be inserted between towing warp and the dredge (or any other sampling gear), and this must be of a ball- or roller-bearing type, turning freely under load, when the towing warp is of wire.

The dredge net is usually about half as deep as it is wide, the mesh varying according to circumstances. Machine-made netting of mesh 10–12 mm knot-to-knot is generally suitable, but material used for shrimp trawls is usually too light in construction. For collection of sediment, the bag can be lined with an inner bag of sacking, stramin or burlap. Impervious material such as canvas should not be used, as water must drain away when the dredge comes on board. Where a net bag is required (e.g. to avoid washing out of the sample when hauling), the net should be an open-ended sleeve with a cod-end tied at the bottom with a rope which is untied to release the sample. Another simple but effective dredge widely used in the Mediterranean for the collection of organisms and bottom material is the Charcot dredge (Picard, 1965)

On some types of sea bottom a net may raise an unnecessarily large sample and burst when a heavy load is brought up. Chafing of the net while being towed along the bottom can be minimised by fitting sheets of rubber, hide or canvas (Fig. 5.5) on either side of the net bag.

When operating from a small boat where the dredge is hand-hauled, a rectangular dredge frame of length 300–380 mm is large enough. For ease of hand-hauling this should be towed on a rope of at least 10 mm, and preferably 12 mm, diameter. For larger boats with a power hoist a dredge frame of 450–600 mm is suitable, and 750–1300 mm is used for trawler-sized vessels. On ships equipped with winches, wire ropes will be used for working the dredge, the diameter of the wire being related to the size of the ship and the depth of water, bearing in mind that, in spite of weak links, very severe strains can be exerted on the towing warp from time to time. On the continental shelf it is usual to tow with warp equal to 2.5–3 times the depth of water, but in the deep sea a factor <1.5 times may be possible by using heavy weights or diving plates on the towing line in front of the gear and where an acoustic pinger is used to show when the dredge is on the bottom (see Chapter 7). When dredging is being carried out, the ship should drift or steam slowly (l–2 knots), a check being made on the performance of the dredge by the angle of the warp and any jerks or vibrations, strain being recorded, if possible, on a strain gauge. The dredge should normally be towed on the bottom for 5–10 minutes, but on some grounds it may fill up at once and can be hauled up almost immediately. Special techniques for deep-sea dredging are given in Chapter 7.

Many types of dredge have been produced for different purposes. For geological sampling there are sturdy rock dredges, often with teeth (Nalwalk et al., 1962 – see Fig. 5.7; Boillot, 1964) and Clarke (1972) has described a heavy dredge for sampling in mixed boulder and mud substrata. Rock dredges typically have bags of metal rings or wire grommets, similar to those used on oyster and scallop dredges. These may be lined with a finer mesh of synthetic netting, if desired. Where digging into the sediment is required, dredges which are bowed, oval or circular in shape, are more effective than those with a straight edge, but the penetrating powers of most are limited, except in soft mud, so that they typically sample only the shallower-burrowing members of the infauna. Dredges with teeth are more effective for certain purposes (Baird, 1955, 1959), and the use of inclined steel diving plates to make the dredge sink more quickly and to help maintain contact with the bottom is worth considering.

A specialised dredge mounted on a sledge adapted for the study of juvenile molluscs in the superficial sediment layer, designed by Thouzeau and Hily (1986) and subsequently improved by Thouzeau and Lehay (1988) was used for the study of young scallops (Thouzeau & Lehay, 1988; Thouzeau & Vine, 1991). The dredge, which is equipped with video camera, odometer and interchangeable nets had a tested efficiency of $>80\%$ according to the size of the target organisms.

Although seldom providing satisfactory quantitative samples, trawls and dredges are indispensable sampling instruments and should always be carried on board ship. Dredges, in particular, are invaluable for preliminary surveys to discover the nature of the bottom and its fauna and may be the only instruments which can be used under adverse sea conditions; being mechanically unsophisticated they are a standby when more complex equipment has broken down, and on some grounds they may be the only means of obtaining a sample and would then be the prime means of investigation.

Notwithstanding the misgivings felt about the semi-quantitative aspect of dredges, quantitative collection of the larger-sized epifauna and infauna of low abundance on soft sediments has been satisfactorily achieved by Bergman and van Santbrink (1994). By using the Deep Digging Dredge (Triple D) rigged with an interchangeable blade to collect a slice of the benthic sediments, accurate and reliable results on the benthic assemblies of the soft sediments were obtained. The sediments were sieved and retained in a net attached to the frame of the metal box connected to the dredge frame provided with a pair of runners and a measuring wheel.

Semi-quantitative estimates with trawl and dredge

When trawling it may be possible to standardise the conditions and duration of tow, so as to obtain estimates of population density which are of value for comparative purposes, and such hauls are commonly used for estimating demersal fish populations. However, it is not easy to estimate the exact time during which the gear is on the bottom: both trawls and dredges will continue to drag along the bottom for some time after hauling has commenced. Carey and Heyamoto (1972) used a time-depth recorder to monitor the behaviour of a trawl on the bottom, and methods involving acoustic signals are described by Laubier et al. (1971) (Figs 7.3 and 7.4, Chapter 7) and also by Rice et al. (1982). In order to determine the actual bottom contact of otter trawls, a digital timing device (BENSEM – Benthic Sampler Effectiveness Measurer) has been developed and mounted on the otter trawl door, improving trawl reliability and consistency of sampling (Diener & Rimer, 1993). It is well known that trawls and dredges sample only a fraction of the fauna lying on the surface of the seabed (e.g. Mason et al., 1979) and very few of the burrowing animals, so that results merely represent minimum densities on the ground. Several trawls and dredges (Rothlisberg & Pearcy, 1977; Sorbe, 1983; Fossa et al., 1988) have a mechanically activated device for opening and closing the sampler.

A number of workers have fitted measuring wheels on trawl or dredge frames in order to measure distance covered (e.g. Belyaev & Sokolova, 1960; Riedl, 1961; Wolff, 1961; Gilat, 1964; Richards & Riley, 1967; Bieri & Tokioka, 1968; Carey & Heyamoto, 1972; Pearcy, 1972; Carney & Carey, 1980; Rice et al., 1982). The performance and success of such wheels seem to vary, and their success is complicated by the fact that some sampling instruments tend to progress by leaps and bounds (Baird, 1955; Menzies, 1972). In addition, the wheel may jam or malfunction for all or part of the tow, giving a false reading. The fitting of an odometer wheel on each side of the frame might seem to overcome this problem, but in spite of this, Carney and Carey (1980) found considerable variation between readings from wheels fixed on either side of a beam trawl. They considered that slippage might result in under-reading of distance by as much as 40%. If a continuous record of rotation can be obtained, so that it is known that the wheel is rotating throughout the tow, a reasonably satisfactory measure of distance should be obtained (Holme & Barrett, 1977; Thouzeau & Hily, 1986; Thouzeau & Lehay, 1988), although Rice et al. (1982) have some doubts over the accuracy of a continuously recording wheel attached to their epibenthic sledge. In some cases, the sampling gear equipped with an underwater TV or photography system (Dyer et al., 1982; Cranmer et al.,

1984; Holme & Barrett, 1977; Thouzeau & Hily, 1986; Thouzeau & Lehay, 1988; Thouzeau & Vine, 1991) provides direct assessment of the ocean floor morphology and of the abundance and distribution of the larger epibenthic organisms. These can be compared against the samples taken by the sampling gear.

Where the dredge bag is lined with closely woven material to retain the sediment, semi-quantitative data in terms of numbers of animals per unit volume of sediment can be obtained. The results are naturally dependent on the depth of penetration, but under difficult conditions where grab samples cannot be taken, this may be the best that can be achieved.

Anchor dredges

Forster's anchor dredge (Forster, 1953) is an invaluable instrument for semi-quantitative sampling of sands and other firmly packed deposits. This dredge has an inclined plate intended to dig in deeply at one place, and it is not towed, as are other dredges. It is shot by allowing the ship to drift while warp equal to five times the depth is gradually paid out; the warp is then made fast, and the strain exerted as the ship is brought to a standstill drives the dredge into the sand to a depth of up to 25 cm. This dredge is best used from a small launch fitted with a power hoist, since when used from a larger ship there is a tendency for it to be jerked out of the sediment. This also occurs if insufficient length of warp is paid out. The sample theoretically taken by Forster's anchor-dredge is wedge-shaped in section and approximately the same area as the digging plate, so that it may be used for semi-quantitative studies. A double-sided anchor dredge (Holme, 1961) capable of working either way up is shown in Fig. 5.8.

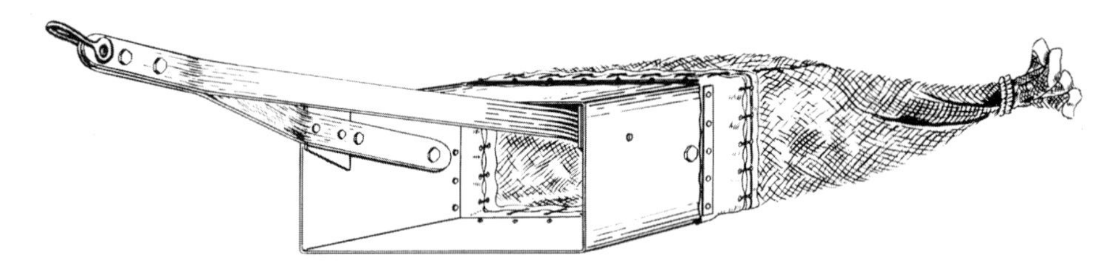

Fig. 5.8 Double-sided anchor-dredge with wishbone towing arms free to swivel or can be locked to one side if required.

Thomas (1960) describes a modified anchor dredge with an adjustable digging plate and a self-sifting mesh net. Because this dredge is intended to be towed through the sediment the results obtained are non-quantitative, but it appears to be a useful collecting instrument for deeper-burrowing animals.

Sanders et al. (1956) describe a rather different type of anchor dredge (see Fig. 7.10c, Chapter 7) for use on the shelf and the deep sea. This is double-sided with two angled digging plates between which is a wide horizontal plate that limits penetration to 11 cm. Since the dredge is very heavy it is assumed to sink consistently to this depth in the sediment, and, therefore, to sample to a constant depth. Once the bag is full, further material is rejected from the mouth of the dredge. Because of the observed consistency of operation, at least on silt-clay grounds, this dredge has been employed for quantitative studies. A lighter model is described in Sanders (1956).

A modified version of the Sanders' anchor-dredge (the *Anchor-Box Dredge*) was developed by Carey and Hancock (1965). This samples an area of 1.3 m^2, to a depth of 10 cm, and has been successfully used to depths of 2800 m, where it was worked with a warp to depth ratio of 1.39:1. A smaller version of this dredge with a total capacity of 40 litres was used by Probert (1984).

Grabs

Quantitative samples of animals inhabiting sediments are usually taken by grab. The grab, which is lowered vertically from a stationary ship, captures slow-moving and sedentary members of the epifauna and infauna to the depth excavated.

There has been much discussion on the depths to which animals burrow into the sea floor (MacGinitie, 1935, 1939; Thorson, 1957; Holme, 1964). The majority inhabit the top 5–10 cm, but some burrow more deeply (Barnett & Hardy, 1967; Kaplan et al., 1974; Thayer et al., 1975). Exceptionally, some crustaceans have been found to burrow to depths $\geq$3 m (Pemberton et al., 1976; Myers, 1979). Few grabs are designed to dig deeper than 15 cm, and, in practice, many dig to less than 10 cm in firmly packed deposits (Table 5.2, p. 199). Thus, a grab may be an adequate sampling instrument for some grounds (Ankar, 1977b), while on others it may leave significant elements of the fauna unsampled below the depth of bite.

If a grab is used as the prime means of investigation it would be advisable, at the pilot survey stage, to make comparative hauls with a deeper-digging instrument such as the Forster anchor dredge, a box corer or a suction sampler (details in this Chapter) in order to check whether there are deeper-burrowing individuals out of the reach of the grab. If this should prove to be the case, one of these methods should be used from time to time to supplement the information obtained by the grab. The sampling efficiency of grabs is further discussed in the efficiency of capture section (p. 194).

For sampling the macrofauna, grabs covering a surface area of 0.1 or 0.2 m^2 are commonly employed, several samples being taken to aggregate to 0.5 or 1.0 m^2 per station. Samples of this total size are usually considered adequate for quantitative determinations of the commoner species, measurements of population abundance and biomass, but do not adequately sample scarcer animals, which are often members of the epifauna. Moreover, some fast-moving species escape the grab altogether. It is, therefore, advisable to supplement grab estimates of the epifauna by hauls with an epifaunal bottom sledge, an Agassiz or beam trawl, or by underwater photography, television or diving.

The *Petersen grab* used by C.G.J. Petersen for investigations in the Danish fjords at the beginning of the century is the prototype from which many modifications and

Fig. 5.9 Petersen grab in sampling position on the seafloor. After the release hook has actuated, an upward pull exerted on the central chain closes the two buckets of the grab. (After Hardy, 1959.)

improvements have been made (Petersen & Boysen Jensen, 1911). It consists of two buckets (Fig. 5.9) hinged together, which are held in an open position during lowering. The top of each bucket has a gauze-covered window to allow water to escape while the grab is closing. However, this offers some resistance to the rapid lowering of the gear, which is desirable when sampling in deeper water. When on the bottom, the lowering rope slackens, allowing a release hook to operate so that on hauling the two buckets (Fig. 5.9) close together before the grab leaves the bottom. The disadvantages of the Petersen grab for sampling in other than soft muds and sheltered waters have often been discussed (e.g. Davis, 1925; Thorson, 1957; Holme, 1964). These relate to premature operation of the release during descent due to momentary slackening of the rope as the ship rolls, failure to penetrate sufficiently deeply into the sediment, losses due to the jaws not closing completely and inadequate sampling due to an oblique upward pull when closing due to the drift of the ship on station. For these reasons the Petersen grab has not been much in use in recent years, many workers choosing alternative sampling gear. Nevertheless, it has been used, mostly in comparative studies (Pearson et al., 1985; Bagge et al., 1994) with satisfactory results.

The *Campbell grab* (Hartman, 1955) is similar to the Petersen grab, but its greater efficiency is due to its larger size (0.55 m^2 sampling area, contrasted with 0.1 or 0.2m^2 for the Petersen Grab), and greatly increased weight (410 kg).

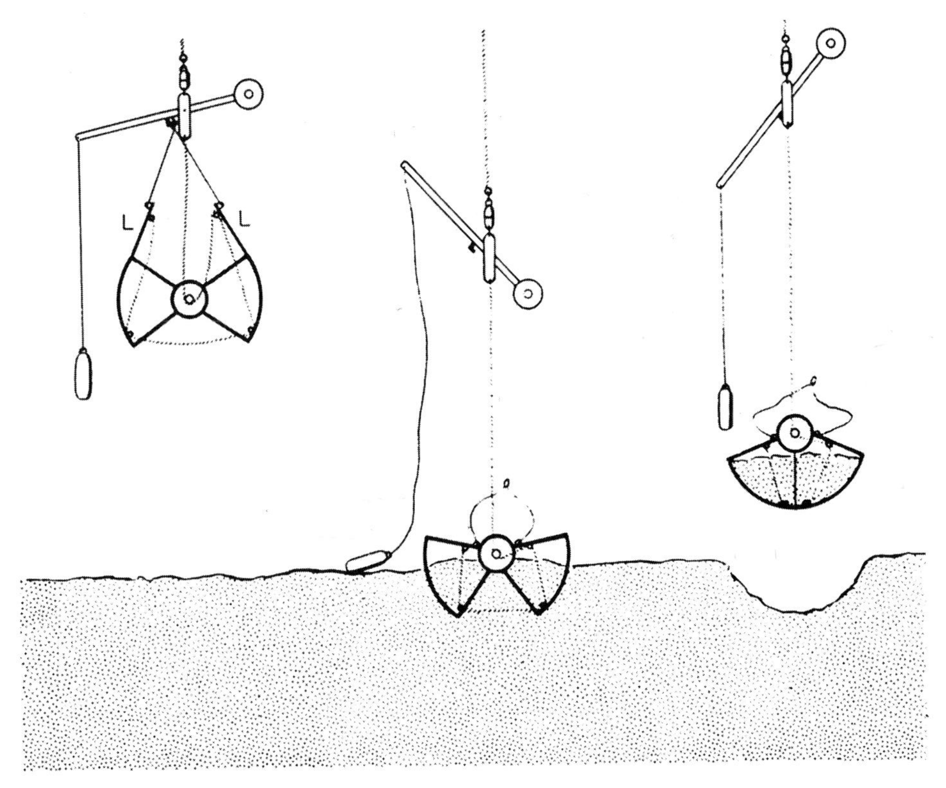

Fig. 5.10 Okean grab. Sequence in sampling operation. Note the counterweight release and the lids (L) of the two buckets, which are open during descent. (Redrawn from Lisitsin and Udintsev, 1955.)

The *Okean grab* (Lisitsin & Udintsev, 1955) is basically a Petersen grab but with the tops of the buckets having hinged doors which are held open during the descent (Fig. 5.10) and which close when the grab reaches bottom. Very rapid rates of lowering are possible with the Okean grab, which has, in addition, a counterweight mechanism to prevent tripping in mid-water. Nevertheless, it does suffer from the same problems as the Petersen grab.

The *van Veen grab* (van Veen, 1933) improves on the Petersen grab in having long arms attached to each bucket, thus giving better leverage for closing (Fig. 5.11). The arms also tend to prevent the grab from being jerked off the bottom, should the ship roll as the grab is closing. On the other hand, the arms may pull the grab to one side if, through drift of the ship the upward pull for closing is oblique. The sampling efficiency of this grab in different sediments has been tested by Christie (1975), Ankar (1977a), Ankar et al. (1979) and Kingston (1988). The van Veen grab has been adopted as the standard sampler for benthic investigations in the Baltic Sea (Dybern et al., 1976) but some improvements to the mechanism have been proposed by Sjolund and Purasjoki (1979). Under open-sea conditions, the counterweight release mechanism described by Lassig (1965) can be employed.

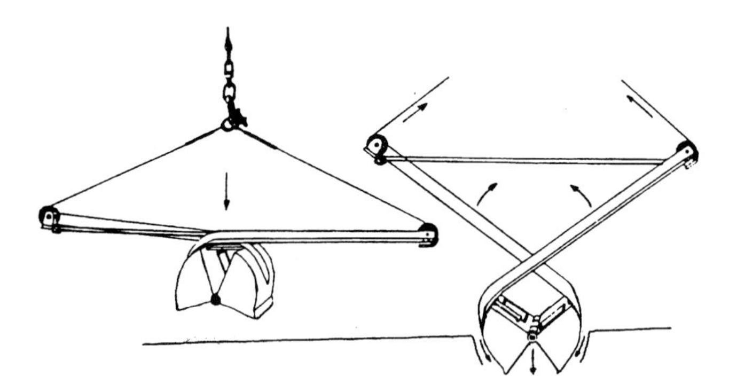

Fig. 5.11 Long arm, warp rigged van Veen grab in sampling operation. (Redrawn from Rumohr 1999.)

The *Ponar grab* (Powers & Robertson, 1967), basically a sediment sampler used in limnological work, is closed by a scissor action of the arms attached to the buckets. It appears to be an easy-to-handle efficient sampler under sheltered conditions, but the release mechanism is not suited for open-sea use. It has been increasingly but erroneously used for macrofaunal sampling.

The *Hunter grab* (Hunter & Simpson, 1976) is a robust and compact grab with jaws which are extended to form levers, with the upper surfaces providing lids which allow a free flow of water through the sampler during the descent. There is no evidence that it is used by the main body of researchers.

The *Smith–McIntyre grab* (Smith & McIntyre, 1954) was designed for sampling under the difficult conditions often encountered when working from a small boat in the open sea. This grab has hinged buckets mounted within a stabilising framework (Fig. 5.12) and powerful springs to assist penetration of the sediment. Trigger plates on either side of the frame ensure that the grab is resting flat on the bottom before the springs are released. Closing of the grab is completed as hauling commences by cables linked to the arms attached to each bucket. The Smith–McIntyre grab covers an area of 0.1 m^2, and on firm sands penetrates to about the same depth as the 0.1 m^2 van Veen grab, but the greater reliability of its release makes it preferable for open-sea use.

A number of workers have adopted the Smith–McIntyre grab as their standard sampling instrument, but some consider it complicated and, because of the spring-loaded mechanism, occasionally dangerous to use. The *Day* grab (Day, 1978) is a popular alternative (Fig. 5.13) as it represents an attempt to simplify the design of this type of instrument. It incorporates a frame to keep the grab level on the seabed, and two trigger plates for actuating the release, but there are no springs to force the hinged buckets into the bottom. Penetration is assisted by the greater weight of the sampler. The Day grab seems to sample as efficiently as the Smith–McIntyre grab (Tyler & Shackley, 1978), and is preferred by some workers because of its greater simplicity.

A large hydraulically powered grab, with a sampling area of 0.5 m^2, mounted on a frame and equipped with a video imaging system consisting of two video cameras

Fig. 5.12 Smith–McIntyre grab. Above in open position ready for lowering; below in closed position. Note the trigger plates on either side, both of which must be in touch with the bottom before the release is actuated. The threaded studs with butterfly nuts are for attachment of lead weights. (Photograph by A.D. McIntyre.)

(VideoGrab) and a high-resolution acoustic imaging system (DRUMS – Dynamically Responding Underwater Matrix Scanner) is designed to take undisturbed samples of the upper 10–25 cm of the ocean floor and has the distinct advantage for the operator to have visual contact with the ocean floor and therefore control the operation and take reliable samples (Rowell et al., 1997).

The *orange peel grab* described by Reish (1959b) and modified by Briba and Reys (1966) is a large sampler which Thorson (1957) did not consider as a satisfactory quantitative sampler. As there are no reports of its use in the scientific literature, it may be considered as obsolete.

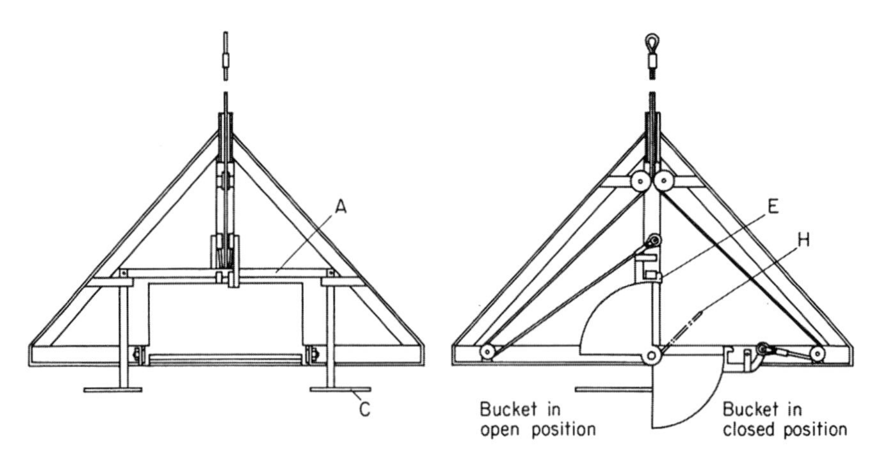

Fig. 5.13 The Day Grab. Left, end view, open for lowering; right, side view, one bucket open, the other closed. On reaching the seabed the two pressure plates (c) are pushed upward, releasing the transverse beam (A) so that the hooks (E) holding the buckets open are released. The buckets are closed by tension on the two cables, the hinged flap (H) allowing water to escape during the descent but acting as a cover during hauling. (From Day, 1978.)

The *Baird grab* (Baird, 1958) was designed for sampling the epifauna of oyster beds, but has also been used in the open sea. It has two inclined digging plates which are pulled together by springs and levers (illustrated in Holme, 1964). It covers an area of 0.5 m^2 and has been found to dig into sediments quite well, having applications for sampling the infauna where a sample from a large area is required. Since the surface of the sample is not covered, there may be some washing out during hauling. The Baird grab had a very limited use and there are no recent references to the use of this sampler.

The *Hamon grab* (Fig. 5.14) samples an area of about 0.29 m^2 by means of a rectangular scoop rotating through 90°. It appears to be particularly effective in coarse, loose sediments, but, because of its mode of action, may not sample entirely quantitatively. Dauvin (1979) reports that in spite of its considerable weight (350 kg) and size (height 2 m) it is easy to handle on board ship, and is capable of sampling the deeper infauna of sediments, below 10 cm.

The *Holme grab* (Holme, 1949) samples by means of a single semi-circular scoop rotating through 180°. This design minimises loss of material while hauling, and a later model (Holme, 1953) with two independently operating scoops reduces any tendency towards sideways movement during digging. However, there is no evidence that this sampler has been extensively used.

The *Grizzle–Stegner grab* (Grizzle & Stegner, 1985) is another quantitative grab for shallow water sampling which combines features of several grabs such as the design of a van Veen grab mounted on a Smith–McIntyre frame. The grab has many attractive characteristics but despite the authors' assurances about its overall performance and reliability, it could be said that it has remained untried by the rest of the scientific community. Similarly, the Kingston Hydrostatic Grab designed by

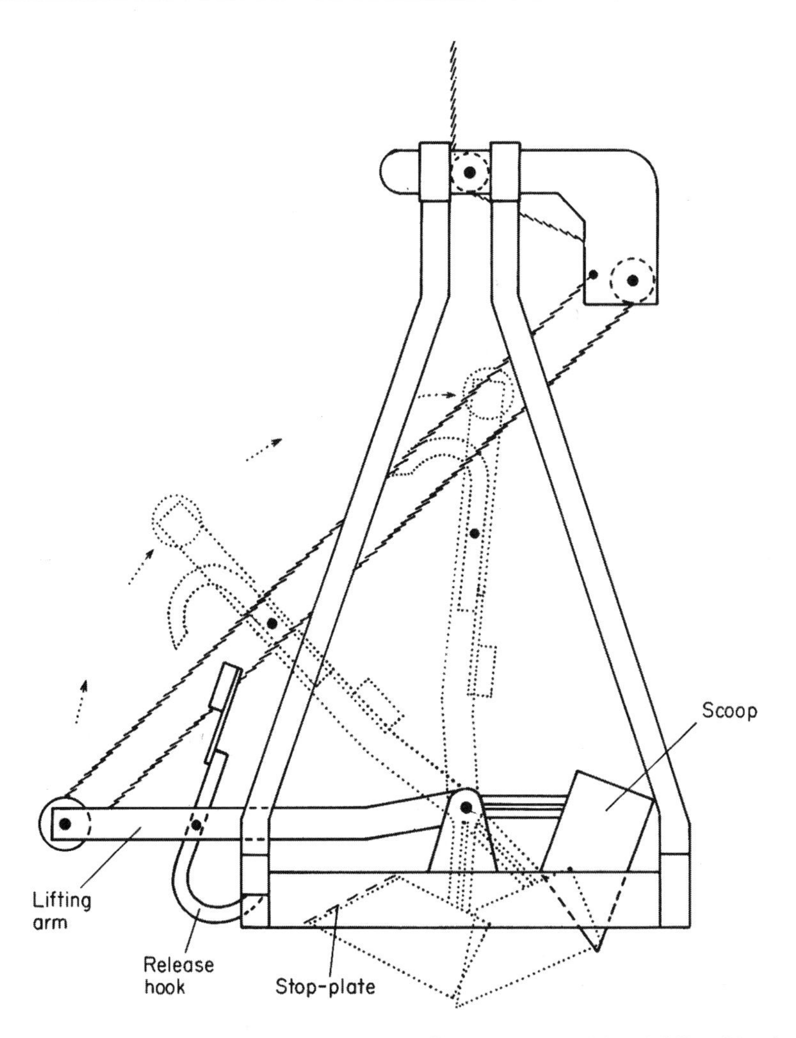

Fig. 5.14 Hamon grab, showing mode of action. The lifting arm rotates through 90° to drive the scoop through the sediment, closing against the stop plate. (After Dauvin, 1979.)

the Heriot Watt Institute of Offshore Engineering (Remote Technology Ltd.), UK, utilising hydrostastic pressure to open and close specially improved buckets has also remained relatively untried.

The *Shipek sediment sampler* (Hydro Products Spec.Bull.3 pp) has a single semi-circular scoop activated by powerful springs. Covering an area of only 0.04 m^2, this instrument is rather small for macrofauna investigations, and its sampling action tends to destroy any layering present in the sediment. Its main use is by geologists to obtain a small sample of bottom sediment. A similar spring-activated grab was described by Franklin and Anderson (1961) which, despite the large sample volume (2.5 l) and the powerful closing mechanism, nevertheless results in substantial loss of the superficial sediments.

Many small grabs such as the *Ekman* and *Birge-Ekman grab* (Blomqvist, 1990; see bibliography in Elliott et al., 1993) are designed for use from a small boat on soft sediment, but since they cover an area of only about 0.04 m^2, are not very suitable for macrofauna sampling. However, a modified Birge-Ekman grab has been designed by Rowe and Clifford (1973) for use by SCUBA divers or from deep submergence vessels, while Hakanson (1986) improved the Ekman grab by incorporating an automatic closing mechanism.

Box samplers and corers

Because of their reliability, box corers and samplers have been extensively used in the deep sea (see Chapter 7). However, their use in shallower water studies is equally relevant, hence a short outline of these samplers is given here.

The *Reineck box sampler* (Reineck 1958, 1963) consists of a rectangular corer supported in a pipe frame with a hinged cutting arm which is pulled down to close the bottom of the tube (Fig. 5.15). This instrument samples an area 20 × 30 cm to a depth of 45 cm and weighs 750 kg. A similar instrument is described by Bouma and Marshall (1964). There have been a number of other modified versions of the original box samplers: Hessler and Jumars (1974) describe a 'spade corer' covering an area of 0.25 m^2, further modified by Jumars (1975) in having the sample

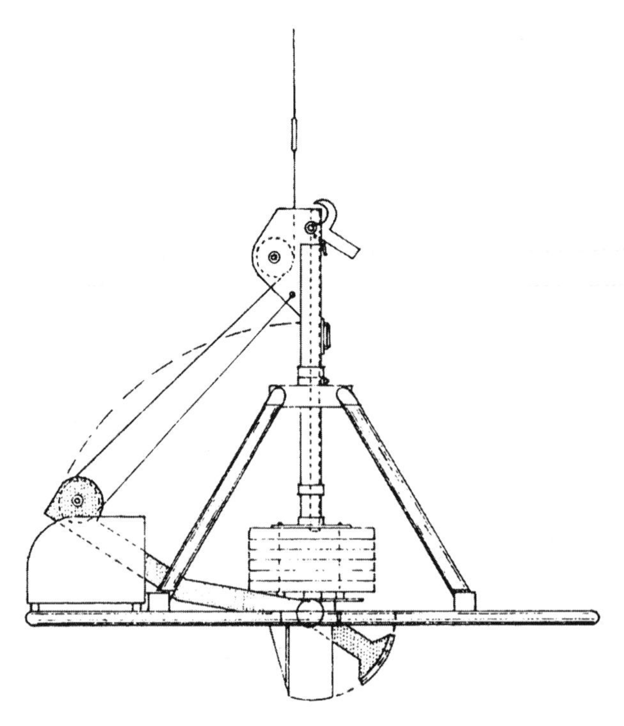

Fig. 5.15 Reineck box sampler. The rectangular coring tube is closed by a spade actuated by pulling up on the lever on the left. (Redrawn from Reineck, 1963.)

subdivided into 25 contiguous subcores. Farris and Crezée (1976) improved the sealing of the corer so that it gave better retention of coarse sand cores, together with the overlying water. Another type of box corer, sampling 30 × 30 cm, is described by Jonasson and Olausson (1966). Another version (GOMEX) of this box corer was designed by Boland and Rowe (1991) with considerable improvements with its tripping and closing mechanisms. Box corer samples compared very favourably with other quantitative samplers. The *IOS box corer* described by Peters et al. (1980) has direct lowering control from the ship for penetration of the sediment, allowing elimination of the main framework and of a piston assembly previously needed to control penetration.

Box corers provide a means of obtaining deep and relatively undisturbed samples suitable for evaluation of macrofauna from a variety of sediments, and are also suitable for deep-sea use. The surface of the bottom, and supernatant water, appear to be taken without undue disturbance, as evidenced by the persistence of animal tracks and burrows and sessile forms in their living positions in the samples (Hessler & Jumars, 1974).

Owing to their large size and great weight box corers are difficult to work, requiring a large vessel and calm conditions for safe deployment. The above authors state that launch and recovery are always challenging and potentially dangerous. Finally, box corers are very expensive, although they are rarely lost and seldom damaged.

Another much lighter box corer has been designed by the Institute of Oceanographic Sciences Deacon Laboratory (IOSDL) of the United Kingdom. The closing mechanism is activated upon contact with the bottom by two closely fitting shovels mounted on pivot arms. The extent of usage of the apparatus is not known.

Despite the opinions of several authors that they consider the deployment of box corers is difficult and potentially dangerous, nevertheless, because of the large sampling area, the undisturbed sample and their deep penetration, box corers have become standard oceanographic samplers widely used in many benthic surveys.

In order to overcome the problem of effective and meaningful sample replication, essential for providing a better basis for describing benthic assemblages, a new generation of multiple corers initially inspired by the Craib corer prototype (Craib, 1965) has been designed. Barnett et al. (1984) pioneered the development of multiple corers by producing a sampler in a range of sizes and weights and capable of taking 4–12 core tubes. The sampler is lowered into the sediment by a hydraulic damper mounted on a supporting frame enabling sampling to be carried out with minimum disturbance. The very positive characteristics of the Craib corer such as reliability in operation, penetration depth and minimum disturbance are maintained. However, having recognised that the Craib corer takes rather too small samples, mainly for meiofaunal and/or geochemical work, it was subsequently modified by Barnett (1998–1999) in order to take cores of different diameters from 65 to 114 mm. Despite some of the drawbacks encountered, the corer has been successfully used, and there is the prospect of still further improvements.

A much smaller and lighter sampler, the minicorer, nearly identical to the Barnett multiple corer (Barnett et al., 1984) has been designed and used by the Alfred-Wegener Institute of Bremerhaven. However, the triggering mechanism is different and it can simultaneously take up to four deep and undisturbed samples. It has been successfully used in conjunction with a CTD system and an acoustic pinger.

A nine-core multibox corer, attached to a circular frame, sampling an area of 0.22 m^2 over 2–3 m^2 of seafloor is described by Gerdes (1990) (see Chapter 7). The sampling boxes, activated by wire cables, penetrate into the sediment supported by the weight of the frame and they are sealed by a shear device upon lifting. It has been tried successfully in Arctic waters but it does not appear to have received wide acceptance.

There are a number of other corers which can be used in certain situations. These include the *Haps corer*, a frame-supported bottom corer with a round box and a flat spade (Kanneworff & Nicolaisen, 1973) and a number of narrower diameter corers more appropriate for meiofauna or geological studies that are referred to in Chapter 6. However, a Haps corer with the same characteristics was successfully used for the study of macrofauna by Bagge et al. (1994).

Large diameter (>10 cm) gravity, piston or vibrocorers designed for geological work (McManus, 1965; Reineck, 1967; Burke, 1968; Pearson, 1978; see also Blomqvist, 1991; Elliott et al., 1993) are, on the whole, untried for biological work. However, their deep penetration, relatively large sampling area and effective core catchers (Kermabon et al., 1966; Burke et al., 1983) are features which could make them suitable for macrofauna sampling.

Suction samplers

A number of samplers employ suction, either to force a coring tube down into the substratum or to draw the sediment and its fauna up into a tube leading to some form of self-sieving collector.

The *Knudsen sampler* (Knudsen, 1927) was the first sampler to use suction to take a core of a size suitable for sampling the macrofauna. A pump attached to the top of a wide coring tube (36 cm diameter × 30 cm long) is actuated by unwinding cable from a drum when the sampler is on the seabed. On lifting, the coring tube is inverted as it comes out of the bottom, thus retaining the sample. The Knudsen sampler meets many of the specifications for the perfect sampler but as it falls over on the ocean floor it can be used only under calm conditions. Despite the improvement made by Barnett (1969) there is no evidence that this sampler has been used in recent years.

The same principle of suction sampling was adopted by Kaplan et al. (1974) who describe a manual and an automatic sampler for use to water depths of 3.5 m, and by Thayer et al. (1975), who describe a more complex device with a similar depth capability. Brown et al. (1987) also designed a vacuum sampler for the sampling of diverse types of substrata in shallow streams. A significant advantage of this

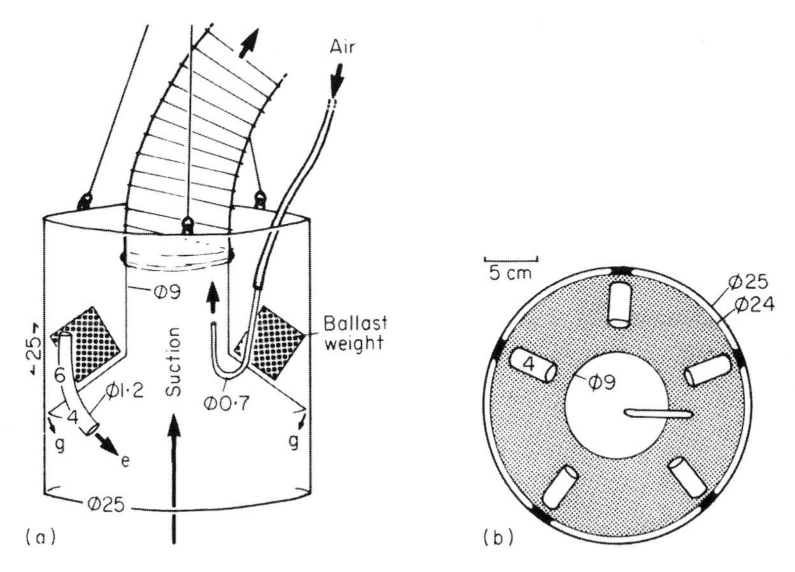

Fig. 5.16 Suction sampler of Emig (1977). (a) in profile; (b) from below. Air introduced through the small tube produces suction in the central tube, through which sediment and fauna is drawn up. The five compensating tubes (e) and the gap between cylinder and cone (g) are provided not only to enable the tube to dig into the substratum but also to help bring the sediment into suspension so that it is more easily collected. Diameters of the various tubes, in centimetres, are shown.

sampler over other suction samplers lies in its successful recovery of undamaged organisms during pumping.

A number of samplers use pumped water to suck up samples of sediment and fauna. For example, the *Benthic Suction Sampler* of True et al. (1968) employs a jet of water acting through a venturi to suck a coring tube of 0.1 m^2 area into the bottom, the sediment and fauna being drawn up into a wire-mesh collecting basket. The instrument is powered by a submerged electric motor or by pressure hose from the ship, and has even been successfully operated in deep water from a submersible. Other remotely controlled suction samplers are described by Emig and Lienhart (1971) and Emig (1977) (Fig. 5.16), while diver-controlled suction samplers are described in Chapter 4. Van Arkel and Mulder (1975) describe a handheld corer, working on a counterflush system, which can be used from a small boat in shallow water (Fig. 5.17). A modification of this system, which enables the corer to suck itself into the sediment, is described by Mulder and van Arkel (1980). An instrument for deep sampling to 65–70 cm in intertidal sands is described by Grussendorf (1981), and another shallow water sampling system is described by Larsen (1974). Remote sampling by suction devices on difficult substrata has also been designed and successfully applied. The Drake and Elliott (1982) sampler has been designed to sample stony bottoms in freshwater biota, while a portable lightweight dredge for quantitative sampling was successfully used in a range of substratum material from very fine to very coarse gravel and coralline rubble (Brook, 1979). It should be noted that a large number of suction samplers are designed for sampling in the shallow waters of inshore areas in estuaries, lagoons and in freshwater habitats.

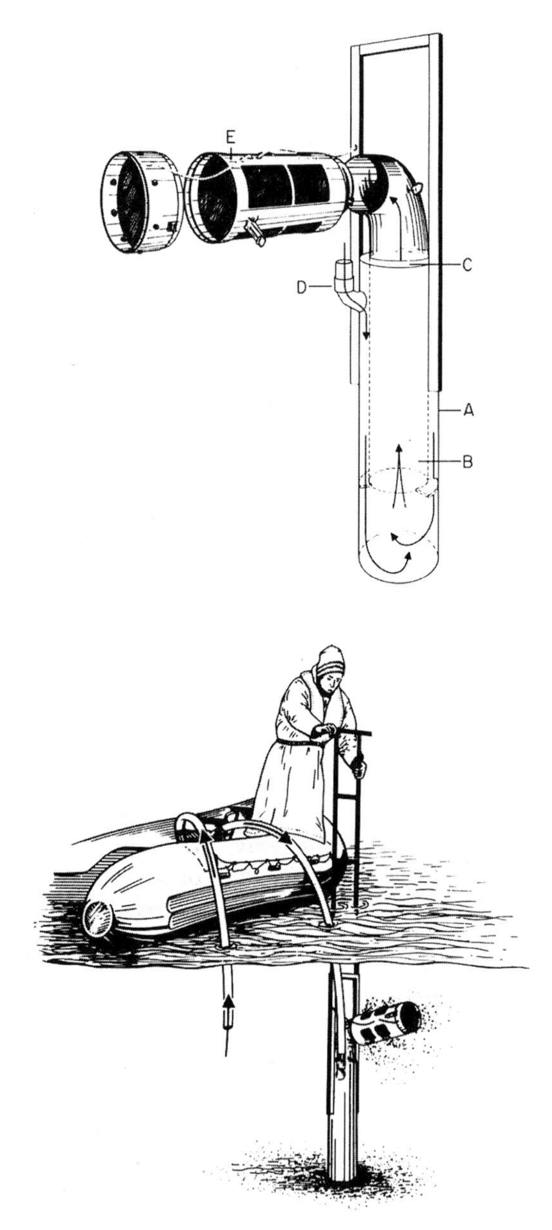

Fig. 5.17 Suction sampler of van Arkel and Mulder (1975). The sampler consists of two concentric pipes (A and B) united at the top (C). Water is injected through D. During operation the sampler is pushed into the sediment, a mixture of water, sediment and organisms passing up pipe B to the cylindrical sieve E.

Some suction samplers are towed slowly along the bottom, sediment and fauna being sucked into a collecting bag. These include the vacuum sled of Allen and Hudson (1970), used for making quantitative estimates of young pink shrimps buried in the sand. In addition, there are hydraulic dredgers used for commercial harvesting of shellfish (e.g. Pickett, 1973). There are a number of diver-operated suction samplers (see Chapter 4).

Other methods of sampling

Some species are not readily taken by conventional sampling gear such as dredges, trawls or grabs, either because they occur too sparsely to be represented in samples covering a limited area or because they live in habitats inadequately sampled by the instrument employed. Alternative methods, which do not necessarily sample other members of the fauna, are available for such species. Techniques for underwater photography and television, of special value in estimating scarce members of the epifauna, are described in Chapter 4.

After storms, burrowing animals are often washed in from shallow water on to the beach, and this may be the best means of obtaining some deep-burrowing species not readily taken by ship-borne samplers. Empty mollusc shells and other recent remains cast up on the beach are usually some guide to the nature of the shallow-water offshore fauna. Occurrence of burrowing species on the sediment surface following dinoflagellate blooms has been noted by Dyer et al. (1983) and others.

Fish stomachs often contain deep-burrowing or active members of the benthos seldom taken by sampling instruments, but as the fish are likely to have been feeding selectively, little idea of the abundance of the prey species can be obtained by this means. Similarly, the stomachs of birds feeding in estuaries and salt marshes may show the presence of otherwise unreported species.

Species of fish and crustaceans which hide away in rock crevices may be taken by using one of the chemical fish collectors which drive them out of their hiding places. Some such as Rotenone are toxic, but the Quinaldine compounds (Gibson, 1967) have an anaesthetic action and usually cause no permanent damage. The range of chemical techniques available, together with other methods such as use of explosives, are reviewed by Lagler (1978), and by Russell et al. (1978) for coral reefs. An electrofishing technique, claimed to be less selective for sampling rock lobsters in that 'soft' individuals which have moulted are also taken, is described by Phillips and Scolaro (1980).

Free-living invertebrates may be captured in baited traps or with light lures. Different types of trap are described in the commercial fisheries literature, including Davis (1958), von Brandt (1972), FAO (1972), Kawamura and Bagarinao (1980) and Motoh (1980). A net trap for sampling epibenthic crustaceans of coral reef systems operated by a diver was designed by Carleton and Hamner (1989). The trapping effectiveness of this device was found to be several orders of magnitude greater when compared with towed sleds and plankton nets. Traps for invertebrates and small fish, some with light, are described by Zismann (1969), Beamish (1972), Espinosa and Clark (1972); Ervin and Haines (1972), Thomas and Jelley (1972) and Haahtela (1969, 1974).

Traps for the deep sea are described in Chapter 7.

There are many active animals, both fish and invertebrates, which are poorly sampled because they actively avoid trawls and other towed gear. There is still a need for the development of methods for their assessment, although photographic methods give good results for some species (e.g. Kanneworff, 1979). An attempt to sample such populations by a cage lowered on to the bottom is described by van Cleve et al. (1966) and, on a smaller scale, the use of throw traps for sampling small fish in shallow marshes is described by Kushlan (1981). References to small drop and pull-up traps and dropnets are given by Aneer and Nellbring (1977). Pots and creels of different design are baited traps set down on the seabed to catch primarily commercially important species such as crabs, lobsters, etc. However, they also inadvertently catch many other invertebrates such as gastropods, starfish and other predators. Collectors of benthic animals settling on hard substrata have been used experimentally, the results of which can be found in a large bibliography. Substrate settlement units allow the assessment of recruitment responses of epifaunal and infaunal invertebrates (Martel & Chia, 1991).

5.3 Working sampling gear at sea

Continental shelf

In the earlier part of this chapter, comments were made on the working of dredges and other towed gear. Grabs and other instruments operated vertically require a different technique. When using a grab of moderate weight (up to, say, 100–150 kg) it is important not to use too thick a wire for lowering: many grabs are not hydrodynamically shaped and sink rather slowly, so that if too heavy a wire is used it may form a loop below the grab, causing kinking or entanglement. For the same reason the grab should be lowered at steady speed with gentle braking (or reverse torque) on the winch. This also helps to prevent the wire slackening as the ship rolls, which can trip the release prematurely. A light grab can be worked on a 6 mm wire, but some workers prefer to use a neutrally buoyant rope, which overcomes some of the above difficulties.

As soon as bottom is reached, paying out should be stopped, and hauling should commence. Any delay will increase wire angle if the ship is drifting, causing the instrument to be pulled out obliquely, so that it samples less effectively. It is, however, important to haul in slowly and steadily until the sampler has left the bottom because, with most grabs, closing is completed as hauling commences, so that if the warp is suddenly pulled up the grab will tend to be jerked out of the bottom while it is still closing. In addition, great strains are produced in releasing a sampler which has dug deeply into the sediment, and these may cause the gear to buckle or the warp to part.

On the other hand, many grabs are not stable on landing and direct observation by photography or video have shown sideways tipping and somersaulting of the grab, which can put serious doubts on the quality of sampling (Kingston, 1988).

The main problems with towed gear relate to the rate of paying out of the wire, the total length paid out, and the ship's course and speed throughout the operation. Failure to take a sample may be due either to paying out too much warp, or at too high a speed, so that the wire becomes entangled with itself or with the gear, or too little may be paid out so that bottom is never reached. For dredges and Agassiz trawls the gear can either be lowered from a stationary ship, or the ship can go ahead slowly while shooting is underway, which is essential for working otter trawls (Laubier et al., 1971). A heavy weight attached to the wire some distance ahead of the dredge aids descent, and gives a more horizontal pull on the gear, so reducing the length of wire to be paid out. Little and Mullins (1964) have shown that diving plates increase the speed of descent of a beam trawl, and reduce the length of wire required.

Even when the trawl or dredge has reached the bottom it may not function satisfactorily. Menzies (1972) and Menzies et al. (1973) describe results from mounting a forward-facing time-lapse camera in the mouth of a 1.52 m trawl. It was found that in a 90-minute tow the net was fishing normally for only about 4 minutes, and that most of the time was spent twisting, flipping and flopping over the seabed, spilling its contents back on the sea floor, or being buried mouth first in the mud. The use of pingers and direct video observation would seem to overcome many of the uncertainties attached to the achievement of satisfactory results with such gear.

Exploration of canyons on the continental slope requires special care. Where the bottom is steeply sloping it may not be possible to get a reliable depth sounding, because of echoes from the sidewalls of the canyon, and it may be difficult to place the gear in the required location on the seabed. Unless the canyon is particularly well charted, and accurate position fixes can be obtained, working any type of gear may be hazardous. Because of the risk of loss it is inadvisable to use expensive instruments such as pingers, and indeed pinger signals may give a false impression by failing to give a return echo from vertical cliff-faces towards which the gear may be drifting.

For general purpose collecting on the slope and in canyons a sturdy rock-collecting dredge (Fig. 5.7), with a safety link can be used. This should be lowered vertically from the ship, which then drifts or steams slowly towards the canyon wall. Manoeuvering of the ship at slow speeds is more easily accomplished if the ship is steaming against the surface current. Exploration of canyons can only be carried out in good weather, but even then frequent losses of gear are to be expected.

Diving submersibles and ROVs have been used for exploration of the sea floor and have been instrumental in important discoveries of sea life and conditions on the sea floor (see Chapter 7 concerning their importance in the exploration and investigation of hydrothermal vents).

Recovery of lost gear

With the increasing sophistication of underwater gear and the development of 'instrument packages' of electronic equipment, often purpose-made, the action to be taken should such valuable gear be lost or become fouled on the seabed ought to be anticipated. The advent of widely available and easily used multi-beam sonar systems means that it is now possible (and advisable) to survey with great accuracy sampling locations prior to starting a sampling programme. Such surveying can identify areas possibly unsuited to the deployment of such instrument packages and, possibly more importantly, with their sub-metre resolution should be able to locate 'lost' equipment, if necessary.

Loss of gear can be minimised by ensuring that all wires, shackles, swivels, etc. are in good condition and of appropriate size, and that weak links will fail when required to do so. Very often loss occurs through parting of the wire or associated components close to the underwater towing point either: (a) at the sea surface, through increasing strains as the gear and sample are brought out of the water, or the gear is accidentally hoisted right up into the towing block through failure to stop the winch; (b) on the seabed through entanglement with underwater obstructions, the side walls of canyons, etc. or (c) when recovering gear from on or in the sea floor, being hoisted by an oblique upward pull, thus increasing the strain which may result in the wire parting.

Apart from precautions to minimise loss, the following guidelines are suggested to aid recovery:

(1) Name, address, fax and telephone numbers and e-mail should have been conspicuously marked on the gear.
(2) On towed instruments a light line (length equal to 2–3 times depth) with marker float is attached to the gear. This will greatly aid underwater location by a diver, particularly near the limit of diving depth.
(3) A similar pop-up marker, automatically released after a time delay, with light and radio beacon, may be used.
(4) The possibility of designing the equipment so that an instrument package can be released independently of the main framework should be considered. This might be brought up on a light line, or on a pop-up buoy system with pre-set time-delay.
(5) On vertically lowered gear it may be difficult to attach a marker line because of the likelihood of it twisting around the lowering wire. Under such circumstances, and whenever valuable equipment is involved, an acoustic transponder (pinger) which can be interrogated by sonar from the ship will allow location after an interval of weeks or even months. Below diving depths the possibility of recovery will depend on the value of the lost gear, and whether a submersible can be deployed for salvage.

It is suggested that by common practice, if not by law, gear on the seabed which has been clearly marked as indicated above is not 'lost', and could not be the subject of a salvage claim. It is assumed that at the time of loss the ship's master would log the position, taking account of any transit lines to the shore. In the event of none of the suggested procedures being carried out, an anchored marker float (which should have been made ready in advance) should be dropped over the side without delay.

Grapnels and trawls have, on occasion, been used with success for recovery of lost gear. For more expensive instruments and samplers, insurance cover, if available, should be considered although premiums may prove to be uneconomical for many projects.

5.4 Efficiency of benthos sampling gear

The efficiency with which a sampler operates is related both to its design and to its mode of operation. Sampling efficiency is a useful concept only when referring to quantitative or semi-quantitative gear.

Dredges and trawls

Most dredges, defined as collecting instruments which are towed along the bottom, are at best semi-quantitative. When used to collect fauna living on or just above the bottom, the efficiency of a dredge, judged by its ability to capture all the animals within its sweep, is usually low. The performance will vary with the configuration and nature of the bottom and since several types of ground may be encountered on any given tow, it is difficult to allow for such variations. Other complicating factors are the behaviour of the ship, the length of warp, and the speed of towing – increased speed above a low level usually reduces catches. Further, the type of warp used can affect performance. Wire, with a weight in water greater than the drag, can increase the effective weight of the dredge on the bottom, while rope, with the drag greater than the weight and a consequent backward and upward catenary, can produce a lift. The fitting of depressors or diving plates, tickler chains and the proper use of teeth on the leading edge can increase efficiency (Baird, 1959). Attempts have been made to employ odometers to measure the exact distance covered by the dredge on the sea bottom (see p. 173), which should help to quantify results.

The behaviour of the animals themselves is also of significance and if sampling is related to only one species, with consequent reduction in the range of habitat and behaviour encountered, it should be possible to select the appropriate sampler or to make appropriate gear modifications to increase efficiency. Yet Dickie (1955) has shown that dredges specifically designed to capture scallops had an efficiency of only 5% on uneven inshore grounds and just over 12% on smoother offshore areas. However, somewhat higher efficiencies were reported by Chapman et al. (1977). Again, juvenile stages of a flatfish have been the subject of population studies in Britain, and gear has been developed for their capture. Since they occur on relatively flat sandy grounds in shallow water where their behaviour can be observed by divers, it may be expected that high gear efficiencies should be possible. Riley and Corlett (1965) used a 4 m beam trawl and found it worked best when towed at a speed of 35 m per minute with three tickler chains attached. Creutzberg et al. (1987) studied the effect of different numbers of tickler chains on beam trawl catches.

Efficiency ranged from 33 to 57%, and even for fish in their first year of life it varied considerably at different times of the year, depending on the size and age of the fish. In other experiments, Edwards and Steele (1968) found a catching efficiency of a 2 m beam trawl for plaice to range between 23 and 37%. Efficiencies of this order, however, seem to be exceptional for dredges, and when the total fauna is considered, a value nearer 10% is probably more realistic (Richards & Riley, 1967). One important development in estimating the capture efficiency of commercial trawls lay in the application of diving techniques (Hemmings, 1973; High et al., 1973), originally by divers clinging on the headline of the trawl, and subsequently by the use of a two-man submersible. This permits easy and safe observation as well as the use of underwater television and photography (Chapter 4). Caddy (1971) and Chapman et al. (1979) made field observations concerning the swimming behaviour of pectinid bivalves in response to dredging, while Eleftheriou and Robertson (1992) assessed experimentally the efficiency of the scallop dredges.

Grabs

The concept of efficiency is more meaningful for grabs, which are lowered on (ideally) a vertical warp from a stationary ship with dynamic positioning, to take a deposit sample of a given surface area. In this context, the term 'efficiency' tends to be used loosely to cover various purely functional aspects of an instrument's general performance and digging characteristics, as well as its ability to produce an acceptable picture of animal density and distribution. Although all these uses of the term are related, they should not be confused.

Performance

Considering first the functional aspect, this refers primarily to the ability of a grab to perform consistently and correctly, according to its design, in all conditions of deposit, depth, and weather. These features are perhaps better covered by the word 'reliability' rather than 'efficiency' and are best judged by the volume of deposit collected, so that an instrument which is filled to capacity on every haul would be regarded as completely reliable within the limits of its design. The first requirement for reliability is that the grab must land on the bottom in a condition in which it will operate properly. Any grab which is activated by the slackening of the warp when the gear strikes bottom will tend to be set off in mid-water by the roll of the ship, and so may be difficult to use in bad weather or deep water. An instrument which is not stable when in an upright position on the bottom, or which must remain at rest for a time to collect the sample may be upset by strong currents or by an oblique upward pull on the warp. Once correctly on the seabed, a grab-type instrument covers a known surface area, and, assuming that it can be raised by a reasonably vertical warp and not pulled laterally off the bottom (the skill and experience of the operator is often important here, and comments on the use of grabs at sea are given on pp. 188–189), then the extent to which it attains its maximum depth and, therefore greatest volume of sample, depends largely on the weight of the instrument and the nature of the substratum. On soft mud most grabs will fill completely, but on the most difficult grounds of hard-packed sand conventional grabs will merely scrape the surface unless they are adequately weighted, and even those instruments which achieve some initial penetration by spring-loading (Smith & McIntyre, 1954) will be raised off the bottom if they are not heavy enough. However, Riddle (1989) showed that the ratio of initial penetration to weight indicated that weight is not the only factor that influences the grab performance. If all of these factors are satisfactory the volume of deposit will serve as an index of the depth of penetration, but will not give an absolute measure unless the exact profile of the bite is known.

Given that a particular grab is fully reliable, as discussed above, one would further wish to know, still dealing with the functional aspect, how well its design allows it to collect the deposit below the surface area which it covers. This is the sense in which Birkett (1958) used the term 'efficiency', which he defined as the ratio between the volume of sediment collected and the theoretical volume, which is calculated by multiplying the area covered by the deepest penetration depth. This could perhaps be called the index of digging performance.

In recent years a generation of suction samplers has been introduced, working on the principle that the deposit can be raised by jets of air or water (Drake & Elliott, 1983). Such samplers tend to have a high digging performance (especially if operated by divers, see Chapter 4) since they can lift all of the deposit to a given penetration depth from within their area of operation. In contrast, it was considered until recently that the biting profile of grabs with horizontally placed spindles was more or less semi-circular and that the digging performance of such grabs was, therefore, low, since the deepest part of the bite sampled only a fraction of the surface area spanned by the open jaws. Observations by Gallardo (1965), Lie and Pamatmat (1965) and Ankar (1977a), however, indicate that the profile is more nearly rectangular, and that divergences from a true rectangle can be explained in terms of the closing mechanisms of the various grabs. Thus the Petersen and the chain-rigged van Veen, which have an upward leverage as the jaws close, tend to leave a hump of deposit in the middle of the sampling area, while the Smith–McIntyre, with a downward pull on the arms as the jaws close, digs deeper in the middle of the sampling area and thus takes a rather larger volume than the chain-rigged van Veen, for the same degree of penetration. These grabs thus appear to have higher digging performance than had been supposed, and on hard-packed sands this can be increased by the addition of extra weights. In contrast, Riddle (1989) (Fig. 5.18) studying the bite profiles of six grabs (Petersen, chain-rigged van Veen, short arm warp-rigged van Veen, long arm warp-rigged van Veen, Day, Smith–McIntyre) concluded that only the two designs of warp rigged van Veen grabs took deep and parallel-side bites. The long-arm warp-rigged van Veen performed best by taking deep samples over the whole of the sampling area. Riddle (1989) also concluded that the bucket design is of primary importance in the performance of grabs.

Re-designing the bucket shape in order to optimise penetration and upward movement on closure is suggested.

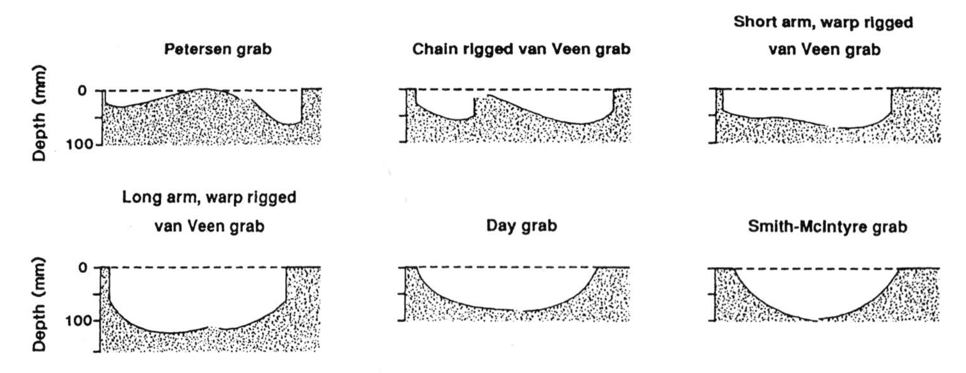

Fig. 5.18 The bite profiles of six grab samplers tested on medium to fine sands (only one bite profile is shown out of three for each type of grab). (Redrawn from Riddle, 1989.)

Efficiency of capture

The second aspect of efficiency is more complex and relates to the ability of the gear to collect the fauna so as to give a reasonably accurate picture of its density and distribution. This may be called the efficiency of capture. Few studies have been made of how efficient a particular grab is in capturing all the fauna below a given surface area. This was attempted by Lie and Pamatmat (1965), who compared collections taken with a 0.1 m^2 van Veen grab at high tide with hand-dug samples of the same unit area at low tide. They showed that for the most abundant species there were significant differences between the two sets of samples in only 8 out of 37 cases.

Efficiency of capture may be defined as the ratio of the number of animals in the volume of deposit collected by the grab to the number present in the same volume *in situ*. By using as the denominator of this ratio the same volume as was collected rather than the theoretical volume based on a rectangular bite, the ratio excludes the functional efficiency of the grab and deals only with its ability to capture the fauna available to it. Long and Wang (1994), comparing the capture efficiency of benthic sampling devices, recommended that differences in abundance (minimum detectable difference) and based on the mean to the standard deviation ratio rather than on absolute differences.

One possible cause of low efficiency of capture is the downrush of water caused by the grab's descent, which may disturb the surface of the deposit and result in the loss of superficial fauna. Smith and McIntyre (1954) considered that the use of gauze windows on the upper surface of their grab reduced this, producing higher catches of small crustaceans on sandy grounds, and this is supported by experimental work by Wigley and Emery (1967), who studied the behaviour of grabs by motion pictures.

On muddy grounds, disturbance of the surface layers may present a serious problem, as has been shown by Andersin and Sandler (1981): comparison of the efficiency of two types of van Veen grab showed that large windows at the upper surface of the buckets minimised the shock wave, resulting in increased efficiency by an average of 50% for small amphipods and by 80% for small polychaetes, as compared to a grab with small windows. It should be noted that many grabs and corers in use have been provided with gauze windows on the upper surface of the instrument. It is nevertheless still probable that most sampling instruments, even when landed carefully on the bottom, will not sample the fine surface layer adequately. Apart from this aspect, even a sampler which fulfils the criteria of reliability as described above, i.e. which consistently takes its maximum volume, may not be the most suitable for every task. For example, one with a low or medium maximum penetration will not sample adequately the deep burrowing animals, and an instrument which covers only a small surface area, no matter how efficient as a machine, may be quite unsuitable for sampling widely dispersed species.

In conclusion, selection of the most 'efficient' grab involves consideration of reliability, digging performance and capture efficiency which are, in turn, influenced by such factors as the size of the sample required, type of deposit, depth of water, prevailing weather conditions, handling facilities available, and the experience of the operator. The influence of such factors described above was exemplified in Heip's work (Heip et al., 1985) where no statistical difference between the macrofauna of the silty sediments of the North Sea, sampled by van Veen and box corer, was found.

Corers

From the purely functional aspect, coring devices have a higher digging performance and most tend to be relatively reliable in that a particular instrument of a given weight will usually provide consistently similar lengths of core, depending on the type of sediment sampled. The main difficulty may be loss of the sample during ascent, and on sandy grounds a core retainer is usually required.

As a means of providing quantitative data on the fauna, the main criterion of a corer's efficiency must be the accuracy with which the collected core represents the sediment column as it was *in situ*. Only the larger versions of the different types of corer used in meiofaunal and geochemical studies are suitable for the efficient sampling of macrofauna. With wide cores (>10 cm diameter), core shortening due to wall friction, observed with the narrow diameter corers, is not noticeable and undisturbed cores can be obtained. Design of the more recent corers minimises the downwash effect which reduces the risk of losing the superficial layer of sediment.

In conclusion, the most efficient corer will be the one which takes relatively undisturbed samples, penetrates deeply and shows the smallest loss of surface layers. Box corers, conforming to most of these characteristics have a distinct advantage over many other core samplers (see also Blomqvist, 1991).

Comparative efficiency

Many workers have tested sampling devices and provided information as to their comparative efficiency. The criteria used in these comparisons are:

(1) The digging characteristics of the sampler (depth of penetration, volume of sediment, degree of disturbance).
(2) The efficiency of capture in order to give a representative picture of the density and distribution of the fauna.
(3) Technical characteristics of the samplers (ease of manipulation, weight, ease of access to the sample, safety aspects, mechanical reliability, etc.).

A selection of works dealing with the comparative efficiency of different samplers and sampling methods is given in Table 5.1 and a more complete bibliography of works on the comparative efficiency of samplers can be found in Elliott et al. (1993).

5.5 Choice of a sampler

The choice of a sampler must necessarily be a compromise, based on requirements of the survey, working conditions and the availability of suitable gear (see Blomqvist, 1991). Table 5.2 represents an attempt to show the main characteristics of the samplers described in this chapter. Choice may often be restricted by weight, and the heavier equipment commonly used for deeper work (large grabs, epibenthic sledges, box corers) can only be worked from large research vessels which have the necessary lifting equipment (see Chapter 7). Many of the lighter samplers, on the other hand, have serious limitations imposed by their being less efficient, having more limited penetration into firm sediments, or sampling too small an area.

Sampling of firmly packed sediments presents particular problems, since many instruments fail to penetrate sufficiently deeply to sample all the fauna. If the objective is to sample to ≥20 cm, the choice is limited to a few samplers, all of which have restrictions on their suitability. The Forster anchor dredge is a useful instrument for this purpose, but it samples most effectively when used in shallow water and from a small launch, and its samples must be considered as only semi-quantitative. Box corers and multicorers are only suitable for use from large ships, and some grabs and corers take samples that are of too small an area (0.015 m^2) for general macrofauna studies. Others take samples of adequate depth and surface area, but they are unsuitable for open-sea conditions and for deeper water. Some suction samplers, whether remote or diver-operated, are mostly restricted to shallow water, and generally sample less well in cohesive muds.

Where quantitative samples are required, using an instrument of only moderate weight, it must be accepted that penetration of firm sediments is unlikely to be adequate to capture the deeper burrowers. Choice is limited to a few grabs, notably the warp-rigged van Veen, Smith–McIntyre, the Day grab, some corers and the

Table 5.1 Guide to literature in which efficiency of different sampling gear and methods is compared. Cross-referencing is obtained by reading vertically and horizontally from the name of each type of gear. The categories are broad, and where different sizes or variants of the same sampler have been tested against each other, the references are shown in brackets after the name. However, comparative hauls between different types of otter trawl are not included. Descriptions and references to the different types of sampling gear are in Chapter 5.

	Otter trawl	Agassiz trawl	Beam trawl	Dredge	Anchor dredge	Epi-, Suprabenthic sledges	Petersen grab	Campbell grab	Okean grab	Van Veen grab [1, 3, 7]	Ponar grab	Hunter grab	Smith–McIntyre grab	Day grab	Orange-peel grab
Agassiz trawl															
Beam trawl															
Dredge			18												
Anchor dredge															
Epi-, Suprabenthic sledges					16,53										
Petersen grab			18	18											
Campbell grab						31									
Okean grab							24								
Van Veen grab [1, 3, 7]		27			16	16	7,13, 17,26, 42,44,55		2						
Ponar grab							13,34, 39			13,51, 62					
Hunter grab															
Smith–McIntyre grab					11	65	17,34, 41,50, 55			6,17, 41,47, 51	34,51	23			
Day grab													43		
Orange-peel grab										35,51	22,51		8,29,51		
Baird grab													19		

Table 5.1 (*Continued*)

Hamon grab													54													
Holme grab					53	53	20,21			35					35											
Shipek grab							39			51	15,39,51		51		51											
Ekman grab [4, 15]							4,13,39			13,38	13,15 22,39 63				22				15,39							
Box/Spade corers					57					5,51,61	51		40,51		51				51							
LUBS sampler						31		31																		
Knudsen sampler							25			45																
Corers					16	16	39,56,58			16,38 52,64	15,39								15,39	15,33 38,39						
Suction samplers [28]							21			9	59		29		28,29,36			21						60		
Photography	12,66	27		49		37		14,30 32,48		27																
Submersibles	46																									46

The references are numbered as follows: 1. Andersin and Sandler (1981); 2. Ankar et al. (1978); 3. Ankar et al. (1979); 4. Bakanov (1977); 5. Beukema (1974); 6. Bhaud and Duchêne (1977); 7. Birkett (1958); 8. Bourcier et al. (1975); 9. Christie and Allen (1972); 10. Dickie (1955); 11. Dickinson and Carey (1975); 12. Dyer et al. (1982); 13. Elliott and Drake (1981); 14. Emery et al. (1965); 15. Flannagan (1970); 16. Gage (1975); 17. Gallardo (1965); 18. Gilat (1968); 19. Higgins (1972); 20. Holme (1949); 21. Holme (1953); 22. Hudson (1970); 23. Hunter and Simpson (1976); 24. Ivanov (1965); 25. Johansen (1927); 26. Kutty and Desai (1968); 27. McIntyre (1956); 28. Massé (1967); 29. Massé et al. (1977); 30. Menzies et al. (1963); 31. Menzies and Rowe (1968); 32. Owen et al. (1967); 33. Paterson and Fernando (1971); 34. Powers and Robertson (1967); 35. Reys (1964); 36. Reys and Salvat (1971); 37. Rice et al. (1982); 38. Rowe and Clifford (1973); 39. Sly (1969); 40. Smith and Howard (1972); 41. Smith and McIntyre (1954); 42. Thamdrup (1938); 43. Tyler and Shackley (1978); 44. Ursin (1954); 45. Ursin (1956); 46. Uzmann et al. (1977); 47. Wigley and Emery (1967); 48. Wigley and Emery (1967); 49. Wigley and Theroux (1970); 50. Wildish (1978); 51. Word (1976); 52. Jensen (1981); 53. Huberdeau and Brunel (1982); 54. Dauvin (1979); 55. Riddle (1989); 56. Bagge et al. (1994); 57. Probert (1984); 58. Baker et al. (1977); 59. Long and Wang (1994); 60. Stoner et al. (1983); 61. Heip et al. (1985); 62. Ellis and Jones (1980); 63. Howmiller (1971); 64. Thayer et al. (1975); 65. Koulouri et al. (2003); 66. Cranmer et al. (1984).

Table 5.2 Suitability of different sampling gear for various applications.

Gear	Weight	Sample		Quantitative?	Depth of sample (firm sand)	Deposit			Sea depth			Difficult sea conditions	Ship size
		Width (m)	Area (m^2)			Rock/ stones	Firm sand	Mud	Shallow	Shelf	Deep sea		
Beam trawl	L-M	2–10		SQ†	0	O	+	+	+	+		+	SML
Agassiz trawl	L-M	2–4			0	O	+	+	+	+	+	+	SML
Otter trawl	H	>10			0	O	+	+	+	+		+	SML
Macer-GIROQ sampler	H	0.5		Q	0	O	+	+	+	+			SML
Epibenthic sled (Hessler and Sanders)	H	0.8			0	O	+	+			+		ML
Epibenthic sledge (TTSS2)	M	0.58		Q	0	O	+	+	+	+	+	O	SML
Triple-D dredge	H	1.5		Q	1	O	+	+	+	+	+	O	ML
Rectangular dredge	L	0.3–1.3			0	+	+	+	+	+	+	+	SML
Small Biology Trawl (Menzies)	L	1.0			0	O	+	+	+	+	+		ML
Anchor dredge (Forster)	L	0.5		SQ	3	O	+	+	+	+	O		SM
Anchor dredge (Thomas)	L	0.6			2	O	+	+	+	+			SM
Small anchor dredge (Sanders)	L	0.29		SQ	1	O	+	+	+	+	O		SM
Anchor dredge (Sanders et al.)	H	0.57		SQ	2	O	+	+		+	+		ML
Anchor-box dredge	H	0.5	1.33	SQ	1	O	+	+		+	+		ML
Petersen grab	L		0.1*	Q	1	O	O	+	+	+		O	SM
Campbell grab	H		0.55	Q	2	O	+	+		+	+		ML
Okean grab	L		0.08*	Q	1	O	+	+	+	+	+		SML
van Veen grab	L		0.1*	Q	1	O	+	+	+	+		O	SM
Ponar grab	L		0.055	Q	1	O	+	+	+	+	O		SM
Hunter grab	L		0.1	Q	1	O	+	+	+	+	O		SM
Smith–McIntyre grab	L		0.1	Q	1	O	+	+	+	+			SM
Day grab	L		0.1	Q	1	O	+	+	+	+			SM
Orange-peel grab	M		Various	Q	1	O	+	+	+	+	+		ML
Baird grab	L		0.5	Q	2	O	+	+	+		O	O	SM
Hamon grab	H		0.29	SQ	2	O	+	+	+	+			ML
Holme grab	M		2 × 0.05	Q	1	O	+	+	+	+			M

Table 5.2 *(Continued)*

Gear	Weight	Sample		Quantitative?	Depth of sample (firm sand)	Deposit			Sea depth			Difficult sea conditions	Ship size
		Width (m)	Area (m^2)			Rock/ stones	Firm sand	Mud	Shallow	Shelf	Deep sea		
Shipek grab	L		0.04	Q	1	O	+	+	+	+	O		SM
Birge–Ekman grab	L		0.04	Q	M1	O		+	+	+			SM
Reineck box sampler	H		0.06*	Q	3	O	+	+	+	+	+		ML
LUBS sampler	M		0.06–0.25	Q	M2	O		+		+	+		ML
Haps corer	L		0.015	Q	M3	O		+	+	+			SM
Knudsen sampler	M		0.1	Q	3	O	+		+	+	O	O	SM
Suction sampler (True et al.)	L		0.1	Q	3	O	+		+	+	+‡	O	SM
Suction sampler (Kaplan et al.)	L		0.1	Q	3	O	+	+	+	O	O	O	S
Suction sampler (Thayer et al.)	L		0.07	Q	3	O	+	+	+	O	O	O	S
Flushing sampler (van Arkel)	L		0.02	Q	3	O	+	+	+	O	O	O	S
Photography	L		<2	Q	0	+	+	+	+	+	+	O	ML
Television	L	1–2		Q	0	+	+	+	+	+		O	ML
Submersible observation	H	2–10		SQ†	0	+	+	+	+	+	+		
Traps	L	X	x		0	+	+	+	+	+	+		SML

General applications: +, suitable; blank, possible application; O, unsuitable.
Weight (total with any additional weights included): L, <100 kg, M, 100–200 kg; H, >200 kg.
Sample width and area: * Other sizes available. x, traps sample an indefinite area.
Quantitative? Q, quantitative; SQ, semi-quantitative; SQ† , semi-quantitative if odometer wheel fitted.
Depth of sample (penetration of sampler into firm sand): 0, surface sample only; 1, 1–10 cm penetration; 2, 10–20 cm penetration; 3, >20 cm penetration; M, above penetration depths but in soft mud only.
Sea depth: shallow, diving depth (i.e. <30m); shelf, 30–200 m; deep sea, >200 m (i.e. slope and abyss); ‡ from submersible.
Difficult sea conditions (most sampling gear cannot be used under severe conditions of swell, waves or currents): +, instruments likely to obtain a sample under such conditions; O, these instruments can only be used under calm conditions and/or the absence of strong currents.
Ship size, S, launch with power hoist; M, trawler; L, large research vessel.

smaller version of box corers. Most other grabs sample too small an area, do not dig sufficiently deeply, or have other drawbacks not necessarily indicated in Table 5.2. The Baltic Marine Biologists have adopted the 0.1 m^2 van Veen grab as the standard sampling instrument for macrofauna (Dybern et al., 1976), but this is not the best choice under more open-sea conditions, for which purposes some workers have adopted the Smith–McIntyre grab, being substituted by its modified version of the Day grab. The latter is of simpler construction than the former, and is without springs, making it safer in use. It appears to dig about as deeply as the Smith–McIntyre grab.

5.6 Treatment and sorting of samples

A benthos sample usually consists of a volume of sediment from which the animals must be extracted. Macrofauna samples may vary in size from a few to many litres, and the extraction process is often divided into two stages, the first being carried out in the field with a view to reducing the bulk of material to be taken back to the laboratory, where the second stage of separation takes place.

Initial treatment

At sea, the standard procedure is to receive the bottom sampler on deck on a wooden or metal-lined sieving table or hopper, which should be designed in such a way that the sampler can be emptied and the contents washed through a sieve without loss of material. A measure of the total volume of deposit collected is often required, since this helps in assessment of the performance of the gear. The volume may be measured either by using a dip stick before emptying the sampler, or by arranging that the contents pass into a graduated container before being sieved. Methods of dealing with the sample have been reviewed by Holme (1964), and vary from the simple arrangement of McNeely and Pereyra (1961) which consists of washing the sample through a nest of sieves with a hose, to the elaborate set-up described by Durham (in Hartman, 1955) in which a mechanical shaker agitates a graded series of screens under a set of sprinkler heads. Similarly, Rumohr (1999) provided a useful overview of sieves and sieving procedures for the sediments. A small hopper (Fig. 5.19) for general use has been described by Holme (1959) but there is probably no single set-up which would be satisfactory for the full range of sampling instruments and working conditions, and it may be desirable to modify an existing pattern appropriately to suit particular needs, or to construct a new unit such as the combined grab cradle and wash trough (Fig. 5.20) designed by Carey and Paul (1968) for the Smith–McIntyre grab. A more recent design of a sieving table with hand-controlled water sprinkler is given by Rumohr (1999) on a design provided by G. Fallesen.

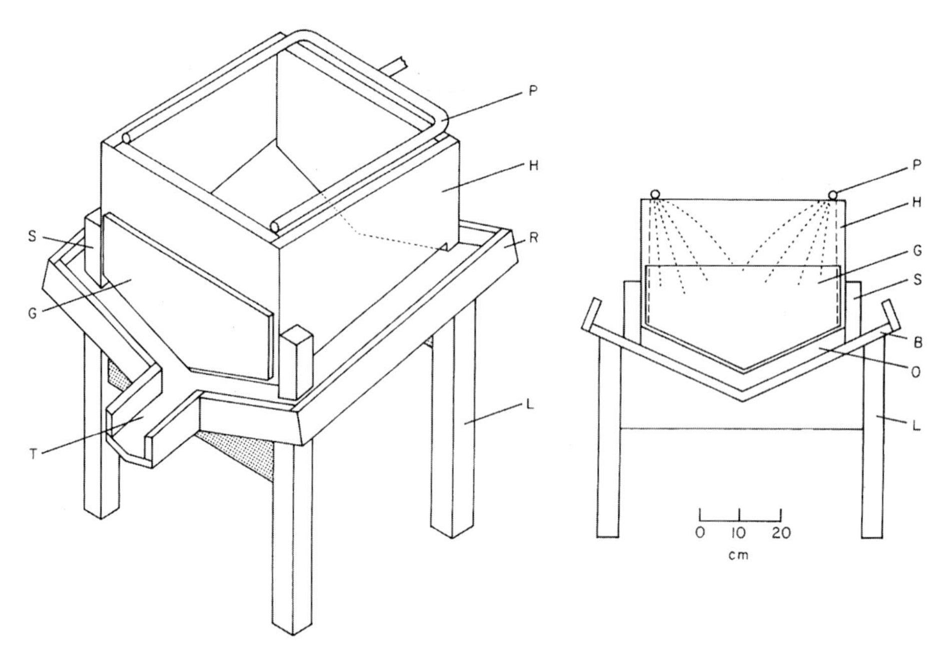

Fig. 5.19 Small hopper for general treatment of sediment samples (Holme 1959). Left, general view; right, cross-section. P, pipes supplying jets along the top of hopper; H, side-wall of hopper; R, retaining wall at side of base (B); T, spout; G, rising gate; S, short legs supporting hopper off base; L, legs; O, gap between hopper and base.

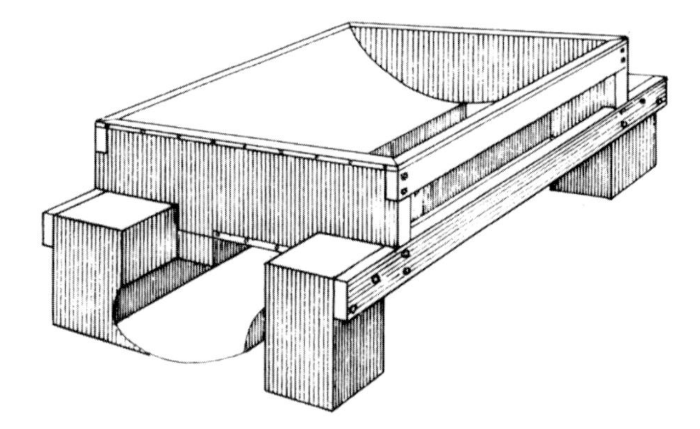

Fig. 5.20 Combined grab cradle and wash trough. (Redrawn from Carey and Paul, 1968.)

Other mechanical means for washing large numbers of samples are given by Pedrick (1974), who describes a non-metallic device suitable for pollution studies. For the processing of large samples (over 20–30 l) an elaborate system has been used by M.H. Thurston and R.G. Aldred (Institute of Oceanographic Sciences) (Fig. 5.21). The sample is placed in a large tub and, after preliminary hand picking of the larger animals, it is passed through a series of sieve baskets (16, 2 and 0.5 mm mesh) where it is separated by washing and agitation into three separate size classes. Each size class is kept separately and is transferred to containers for fixation. The

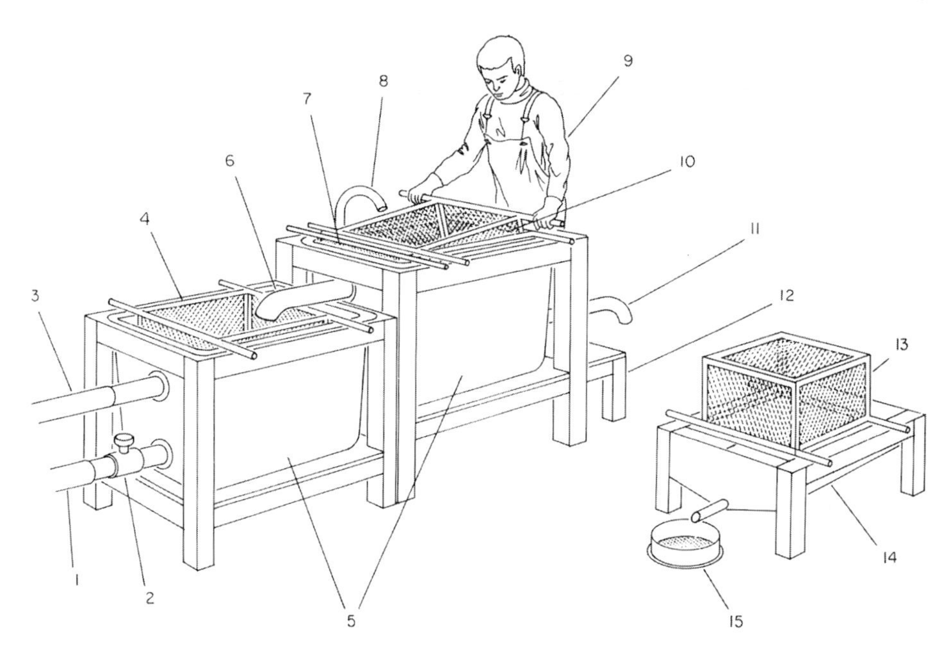

Fig. 5.21 Equipment for washing and sieving sediment samples (Institute of Oceanographic Sciences). 1, hose for draining lower tank; 2, valve; 3, outflow hose for lower tank; 4, 500 μm mesh basket; 5, fiberglass tanks; 6, outflow for upper tank; 7, 2 mm mesh basket; 8, water supply; 9, operator agitating sieve basket; 10, 16 mm mesh basket; 11, outlet and valve for draining upper tank; 12, retractable step; 13, inverted sieve basket; 14, basket washing trough with V-section, sloping bottom; 15, sieve.

system, which is operated by one or two persons, enables large samples (200–300 l or more) to be processed in a reasonably short time.

The Wilson Autosiever (Fig. 5.22), described in Proudfoot et al. (2000), employs a water sprinkling system which gently washes the underside of the sieve. This is claimed to lead to more efficient operation in terms of time and the condition of the retained specimens, and maintaining the apertures clear can also minimise retention of meiofaunal organisms. The autosiever introduces, for the first time, a level of reproducible automation which can also reduce the potential safety risk of back problems for operators. Proudfoot et al. did not find significantly increased faunal statistics, but they describe a case study from IRTU in Northern Ireland which found increased numbers of species being returned from samples processed through the autosiever.

The surface area of the sieve may depend to some extent on the design of the sieving table or hopper, but, unless the sample is to be washed through slowly in small sections, the sieve must be of a certain minimum size to allow an adequate sieving area and to prevent clogging by the sediment. For samples of more than a few litres, a sieve of at least 30 × 30 cm is desirable. Washing can be done by single or multiple jets from a hose, but unless this is very gentle and, therefore, time-consuming, it can damage the animals. A gentler method, described by Sanders et al. (1965) utilises a large vessel with a spout about one-third of the way up. A continuous stream of water passes through the vessel, carrying the animals over into

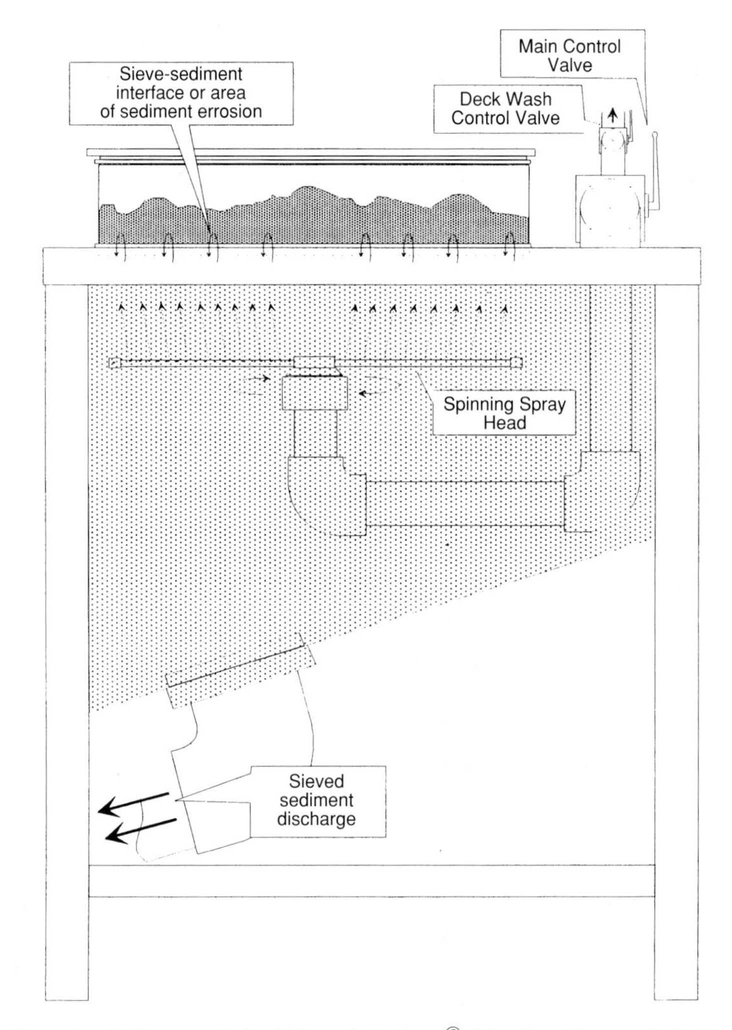

Fig. 5.22 Operational diagram of the Wilson Autosiever© (Gardline Surveys Ltd) (Redrawn from Proudfoot et al., 2000.)

a sieve. If the animals are required in especially good condition sieving should be done by hand, the sieve being gently agitated in water so that flow takes place from below as well as from above. Rumohr (1999) describes a small sieve holder which allows the transfer of the sieve residue to the sample container while retaining the quality of the sampled specimens.

The mesh size of the sieve is of critical importance and should be determined at an early stage of planning. Sieves with either round or square holes may be used, but a square mesh is to be preferred since it has a higher percentage open area, and this type is in general use for soil analysis so that a wide range of sizes to standard specifications is available. In practice, mesh sizes have varied from 2 to 0.5 mm, and even finer apertures have occasionally been used to capture juvenile stages or to make maximum use of deep-sea samples. The effects of different meshes on results have been referred to by McIntyre (1961) and Driscoll (1964). Jonasson (1955) showed that for one particular species a small decrease in mesh size, from 0.62 to 0.51 mm, resulted in a 47% increase in numbers, and stressed that the use of too large a mesh could produce an erroneous picture of seasonal peaks in animal numbers. Lewis and Stoner (1981), testing the retention of 0.5 and 1.0 mm meshes, found that only 55–77% of the total macrofauna was retained by the 1.0 mm mesh. Nalepa and Robertson (1981) found that use of a 0.595 mm mesh resulted in serious underestimates of the abundance of freshwater macrobenthos, although 97% of the biomass was retained. They concluded that the optimum mesh size chosen should be small enough (e.g. 0.10 mm) to retain all or most of the individuals of the taxa studied. In a more detailed study, Reish (1959a) passed five grab samples from a shallow water muddy ground through a series of 11 sieves with apertures ranging from 4.7 to 0.15 mm. His data have been recalculated in Table 5.3 to show cumulative percentages for the main species and groups. If only molluscs were required, a screen of 0.85 mm, which separated about 95% of the individuals in his samples, would have been suitable (see also Schlacher & Wooldridge, 1996; Dunkerschein et al., 1996), but a very much finer mesh was needed for nematodes and crustaceans. Analyses of polychaetes into species show the variation which can occur within a single taxonomic group: about 95% of the *Lumbrineris* were found on the 1.0 mm mesh, but to attain this level of separation for *Cossura candida* required a mesh of 0.27 mm. When the overall assemblage is considered it appears that a 0.27 mm mesh was needed to collect about 95% of all individuals. On the other hand, if only biomass is required, Reish (1959a) found that >90% was retained on the 1.4 mm sieve. While these results clearly apply only to the particular ground studied, they emphasise the importance of correct selection of sieve mesh, according to the purpose of the survey.

The Baltic Marine Biologists have standardised the mesh used in their studies at 1.00 mm (Dybern et al., 1976; Ankar et al., 1979), with the recommendation that a 0.50 mm mesh should be used in addition, whenever possible. Rumohr (1999) recommends the use of a 1.00 mm mesh for descriptive surveys and the use of a 0.5 mm mesh for special purposes such as the determination of production of bottom-living organisms.

In general, it is suggested that a 0.5 mm sieve should be used for macrofauna separation, but since this may retain too large a volume of material on coarse grounds, a compromise may have to be made, the final mesh selected being related to the grade of deposit and the size of the organisms to be separated. The use of a mesh corresponding to one of those in International Standard (ISO) is recommended, and a choice should be made from one of the following apertures: 2.00: 1.40; 1.00; 0.71; or 0.50 mm.

In areas where coarse deposits necessitate the use of wide-meshed sieves, it is recommended that additional small samples be sieved through a finer mesh to assess the losses occurring through the coarser sieve.

Table 5.3 Number of specimens retained on graded screens, and cumulative percentages. (Calculated from Reish, 1959a.)

	Mesh sizes (mm)											
	4.7	2.8	1.4	1.0	0.85	0.7	0.59	0.5	0.35	0.27	0.15	Total
Nematoda						1	2	6	26	456	90	581
%						0.2	0.5	1.5	6.0	72.0	100.0	
Nemertea	2	7		6	3	5	2		1			
%	7.7	34.6	57.7	69.2	88.5	96.2	100.0					
Polychaeta:												
Lumbrineris spp.	8	10	33	9	3							63
%	12.7	28.6	81.0	95.2	100.0							
Dorvillea articulata			22	52	30	5	4	2	2	2		119
%			18.5	62.2	87.4	91.6	95.0	96.6	98.3	100.0		
Prionospio cirrifera	1	6	23	100	115	30	18	10		1		304
%	0.3	2.3	10.0	42.8	80.6	90.1	96.4	99.8	99.8	100.0		
Capitita ambiseta	3	6	29	104	109	23	21	13	2			310
%	1.0	2.9	12.3	45.8	81.0	88.4	95.2	99.4	100.0			
Cossura candida			1	11	129	100	157	265	88	105	10	866
%			0.1	1.4	16.3	27.8	46.0	76.6	86.7	98.8	100.0	
Other Polychaeta	10	23	27	38	31	10	11	7	4	5	2	168
%	6.0	19.6	35.7	58.3	76.8	82.7	89.3	93.4	95.8	98.8	100.0	
Crustacea	3				2		1	3	5	3		17
%	17.6	17.6	17.6	17.6	29.4	29.4	35.3	52.9	82.4	100.0		
Mollusca	5	4	7	5	2			1				24
%	20.5	37.5	66.7	87.5	95.8	95.8	95.8	100.0				
Pisces	1											
%	100.0											
Total	33	56	148	322	426	171	214	308	127	572	102	2479
%	1.3	3.5	9.7	22.5	39.7	46.6	55.3	67.7	72.8	95.9	100.0	

Preservation

Having reduced the sample to a manageable size by initial sieving, the sorting of animals from the residue can proceed. If the final extraction is carried out soon after collection, and near the sampling site, sorting of living material may be possible, with the advantage that movement of the animals helps in their detection, especially if they are small. But it is frequently necessary, after initial sieving in the field, to preserve the collections for later sorting. In such circumstances it is important to ensure adequate labelling. It may be convenient to number the tops or sides of jars with a waterproof marker, but even if this is done, a properly annotated paper, or synthetic paper label, strong enough to withstand water and preservatives should be placed inside the jar. Alternatively, plastic sheet with a matt surface which can be marked with pencil can be used, and this is particularly suitable for sediment samples.

Formalin is normally used for the initial preservation, and this can conveniently be diluted with sea water. While a 10% solution of commercial formalin (equivalent to 4% formaldehyde) is suitable for histological purposes, the strength may be reduced to between 2.5 and 5% formalin for general storage, provided that the volume of preserving fluid is considerably greater than that of the specimens (Rumohr, 1999). In very large samples, perhaps containing much gravel, care should be taken to see that not only is there sufficient preservative, but that it is adequately mixed through the sample. Since formalin tends to become acid with storage and so cause damage to the specimens, a buffer such as borax (sodium tetraborate) or hexamine (hexamethylene tetramine = Urotropine) is often added to the formalin. Unbuffered formalin is known to erode molluscan shells and affect biomass as a result of dissolving lipids and fatty tissues. These substances have been criticised for causing disintegration of labels, or for producing a precipitate, and for plankton samples sodium acetate is sometimes used. The addition of marble chips to the formalin is often used as minimum precaution to prevent the development of acidic conditions. It should be noted that formaldehyde is a toxic compound and should be handled with caution. Material exposed to formaldehyde should be thoroughly washed in tap water and ventilation should be provided in confined spaces in order to eliminate any formalin vapours.

Although bulk treatment is satisfactory for general samples, particular animals required in good condition should be extracted and dealt with individually. It is an advantage to narcotise highly contractile animals before fixation, allowing subsequent preservation in an extended condition. An account of anaesthetic agents is given by Steedman (1976) and by Lincoln and Sheals (1979). Alcohol is often used for later storage of samples, but it is less satisfactory for initial field preservation because of its volatility. Mixing alcohol with sea water causes a precipitate, and it may cause the separation of lamellibranchs from their shells. A mixture of 70% ethanol and 5% glycerine is often used for permanent storage.

The use of 'Dowcil 100' in 10% solution or Kohrsolin releases formaldehyde only in the presence of proteins and has many advantages over formalin because it does not give off irritant fumes. It is also preferable to alcohol because it is neither flammable nor volatile. Its rather high cost has prevented its wider usage so far. Moreover, the use of an alternative preservative other than the traditional formaldehyde has been resisted by most researchers because of the possibility of it producing different effects, thus potentially compromising valid comparisons between sample series.

Detailed information on the use of fixatives, preservatives and buffering agents is provided by the SCOR working group (Steedman, 1976) and by Lincoln and Sheals (1979). Recovery of the organisms from the sediment sample is followed by further processing involving identification, enumeration, weighing, etc., for general or specific purposes. Some of the taxonomic work is greatly assisted by the application of different chemical agents. Berlese's fluid, lactophenol and glycerol have all been used by histologists to clear tissues of soft-bodied macrofauna. In common with most fixatives and preservatives there are safety concerns which must be addressed in the use of these materials. For all practices the use of a ventilation system in close proximity to the workstation is recommended. Safety practices vary from country to country but the manufacturers recommendations, contained within the materials safety data sheets should be followed explicitly. After the final processing has been completed the organisms should be transferred to a preservation fluid, such as 70–80% alcohol, or a solution of propylene phenoxetol for storage.

Subsequent sorting

If the study is restricted to major species or to large individuals, hand-sorting may be straightforward. This is best done in glass trays below which black or white material can be inserted to provide varying backgrounds suitable for distinguishing different types of animals. If every individual must be extracted this can be a time-consuming task, which may severely restrict the extent of sampling. It is often possible to divide a sample into fractions by agitating the light material into suspension and pouring it through a fine sieve. This separates small animals (such as crustaceans and polychaetes) together with fine debris, leaving large or heavy animals behind in the main sample.

Bulk staining of samples with vital stains (Rose Bengal, rhodamine B, eosin, etc.) to facilitate sorting has sometimes been used, and a counter-staining technique for samples containing large quantities of detritus is described by Williams and Williams (1974), where the primary stain, Rose Bengal or Lugol's iodine, is counterstained with chlorazol black E to provide a high colour contrast between the animals and the detritus in the samples. Hamilton (1969) used fluorescence for faster sorting of freshwater organisms from sediment and detritus: organisms stained with a dye (rhodamine B) fluoresce when examined under long-wave ultra violet light. Of all the stains, Rose Bengal (4 gl^{-1} of 36% formaldehyde) is the most widely used, although there is some opposition to its widespread application as it may obscure diagnostic features used in species identification. The use of methylene blue during the identification process allows analysts to determine the fine differences in some integuments in soft bodied macrofauna, particularly polychaetes, and has the advantage of being a temporary stain with only a short-lasting effect. See also Methods for meiofauna in Chapter 6.

Two methods sometimes used to ease the work of sorting macrofauna are flotation and elutriation. Flotation is based on differences in specific weight between the organisms and the sediment – application of a medium of a suitably high density causes the animals to float free of heavier debris. Liquids such as carbon tetrachloride (Birkett, 1957; Dillon, 1964; Whitehouse & Lewis, 1966) sugar solutions (Anderson, 1959; Kajak et al., 1968; Fast, 1970; Lackey & May, 1971), ZnCl, (Sellmer, 1956; Mattheisen, 1960) and many others have been applied with varying degrees of success. Unfortunately, organic detritus also floats, making these methods unsuitable for muds and silts without adjustment of the specific gravity of the flotation medium. Other disadvantages are that most of these techniques are messy, and in the case of carbon tetrachloride, dangerous if inhalation of vapour occurs (Dillon, 1964). Thus, the liquid must always be covered with water and the operation must be carried out with adequate ventilation. De Jonge and Bouwman (1977) used a colloidal silica polymer (Ludox™) to separate nematodes from sediment and detritus (see Chapter 6); this technique has also been used for the successful separation of macrofauna species.

Elutriation involves passing a continuous upward stream of water through the sample in a container with an overflow to a fine collecting screen. The flow rate is adjusted so that the water agitates the sample and carries up the small animals but not the sediment. Several apparatuses suitable for macrofauna (Lauff et al., 1961; Pauly, 1973; Worswick & Barbour, 1974) utilise both water and air jets. Barnett's fluidised sand bath (Fig. 5.23) uses an upward current of water to separate animals from large quantities (up to 20 1) of sediment. Sand uniformly fluidised by the passage of water provides a dense flotation medium into which the sample is tipped. Organisms float at the sand/water interface and are collected with a special sieve while the lightest ones are retained on the fine screen in the overflow. The average separation time is approximately 10 minutes and the claimed efficiency, 98–100%. Nevertheless, this apparatus does not seem to have been used by the majority of benthos researchers.

5.7 Data recording

Analysis of benthic infaunal surveys often results in large volumes of data in the form of species by sample (or station) matrices and the supporting environmental physical and chemical data. Rees et al. (1990) suggest that initial recording of such data should be in the form of the open book method which will prevent

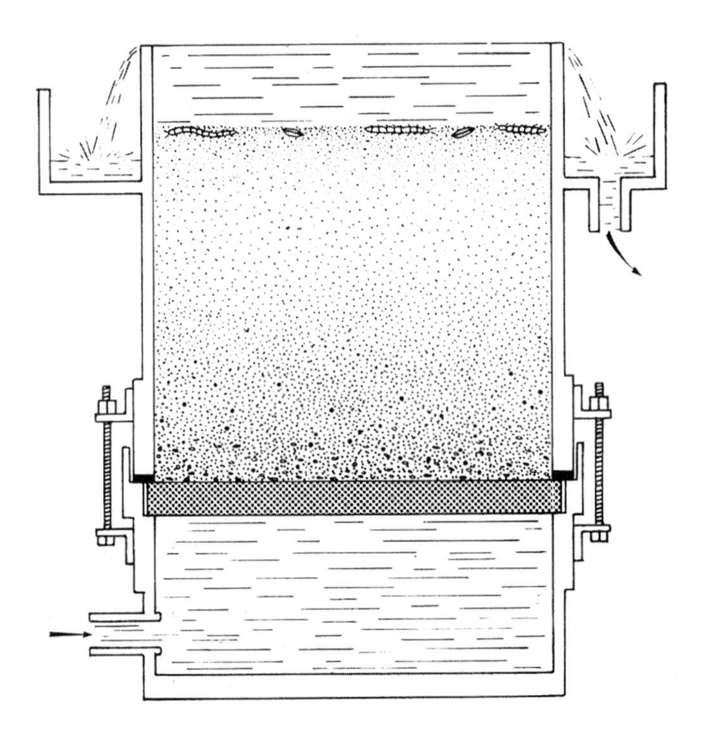

Fig. 5.23 General diagram of Barnett's fluidised bath. Water entering at the bottom passes upward via a ceramic sheet through the sandbed.

more inexperienced workers from forcing identification data into previously defined sheets.

Where large amounts of information on species occurrence, with supporting environmental data, are collected, there is likely to be a need for use of a computerised system for information storage, sorting and retrieval. Modern desktop computers effectively offer infinite storage solutions for benthic surveys, with 40–80 GB hard drives now being common, offering more storage space than could be filled in any lifetime of analysis. (Raw data derived from *all* extant oil related benthic surveys in the North Sea (from 1974 to 2000), and their associated chemicals parameters was recently produced on a single CD-ROM (640 MB) with space to spare (UKOOA, 2001).)

The use of databases and spreadsheets has become a daily occurrence since the earlier edition of this book and this technology continues to develop faster than any other aspect of benthic methodology, perhaps, apart from developments in acoustic mapping.

Databases provide a robust and reliable method of storing data and, when properly used, can ensure such details as taxonomic nomenclature are entered both efficiently and correctly. This can be achieved in a number of ways including the use of code lists for species names. When entered into the database, these codes are translated into the full, correct taxonomic binomial. Many laboratories find the use of a multi-level system as the most efficient method, with an in-house system relating to those species actually encountered in their analyses being based on three- or four-letter couplets. (In general, it is easier for the human brain to recall a letter-based code than a purely number-based system.) This in-house code is then related to a comprehensive taxonomic listing, usually based on a numerical code, for correct hierarchical taxonomic ordering for tabulation and presentation purposes.

Taxonomic listings such as those produced by the Marine Conservation Society for the UK can provide excellent reference material for localised regions but for a fully hierarchical list, the NODC Taxonomic list from NOAA is of most use. Although apparently tortuous, the majority of this manipulation is conducted in the background and the analyst is unaware of the complex operations being performed, but some initial effort in setting up the system is required.

Standardisation of records requires agreement on the scientific names for genus and species. Reference can be made to names used in a standard text, or a list, with author references, must be specially prepared. Quality assurance systems such as the Nation Marine Biology Analytical Quality Control scheme, conducted under the auspices of the MPMMG has sought to standardise species identification within active laboratories within the UK, with some degree of success.

Earlier editions of this book recognised that there was a requirement for textual data to be available in relation to many of the records maintained in most surveys. These data are often required for the expert interpretation of survey data and also to provide a method for historical comparison of possible anomalies between years. Modern database applications will allow the storage of such information and the addition of fields after defining the original database is now much more simple. Search and selection facilities and relating benthic faunal data with environmental factors also held in the same database application are also easily done.

Database applications such as UNICORN from Unicomarine are widely used for data input and storage and can also be used as an aid to identification by holding taxonomic details of specific characters, along with a photographic record of such characters. Databases are not designed to perform complex mathematical tasks, but they can be easily programmed to export selected data for analysis either in a standard spreadsheet or in a specialised analytical application.

Spreadsheets are invaluable for the data manipulation required to produce basic faunal indices, but more complex analyses are best undertaken within dedicated applications such as those produced by Plymouth Marine Laboratory (PML) UK and Cornell University. The increase in storage power of the desktop computer has also been accompanied by an increase in their computing power. This means that powerful multivariate analytical tools are now available to most benthic ecologists. It is not the purpose of this chapter to describe in detail such applications but we would refer the reader to PRIMER (Plymouth Routines In Multivariate Ecological Research) from PML (which includes Multi-Dimensional Scaling (MDS) and Canonical Analysis) and TWINSPAN and DECORANA from Cornell.

A more recent innovation is the use of Geographic Information Systems (GIS) in the presentation of benthic faunal data. This allows spatial display of data at a variety of resolutions from global to local. Applications such as ArcView/ArcInfo and MapInfo can access data directly from database and spreadsheet applications and display them in a variety of spatial resolutions and geographical projections. Some even allow data to be presented against the backdrop of Admiralty Charts where these are available. Certain relationships within the data can also be displayed within these applications.

There are many applications which can undertake, in an objective manner, the previously subjective technique of contouring data over survey areas, allowing environmental factors to be directly compared with biological data, and to elicit possible relationships which have previously been latent.

Many of the above techniques are very robust and will frequently continue an analysis even in the absence of some elements of the data. It is therefore imperative that an element of expert interpretation must continue to be involved in data analysis.

Acknowledgements

We are indebted to many people who have helped quite considerably with the finalisation of this chapter. First of all reviewing and rechecking all the relevant literature on the subject would not have been possible without the help of the Marine Laboratory Library Staff in Aberdeen. C. Dounas and Y. Koulouri processed the Figures and the Tables and Margaret Eleftheriou using her long experience advised on the final presentation of the chapter. V. Mara provided support with the duplication of this chapter as well as the whole manuscript. To all of them we express our gratitude. We are also indebted to the following for permission to reproduce copyright material: Blackwell Science (Figs. 5.1–5.3, 5.5–5.10, 5.12–5.17, 5.19–5.21, 5.23 and Tables 1–3), Elsevier Publications (Fig. 5.18), I. Wilson (Fig. 5.22), M. Kouratoras for redrawing Fig. 5.11 and C. Dounas for kindly providing some original material for Fig. 5.4. This chapter includes some material written for the previous editions (1971, 1984) by N.A. Holme, A.D. McIntyre and A. Eleftheriou.

References

Admiralty (1948) *Admiralty Manual of Hydrographic Surveying*. Second Edition., Hydrographic Department, Admiralty, London, 572 pp.

Agassiz, A. (1888) Three cruises of the United States Coast and Geodetic Survey Steamer 'Blake'. Vol. 1. *Bulletin of the Museum of Comparative Zoology at Harvard College*, **14**, 314 pp.

Allen, D.M. & Hudson, J.H. (1970) A sled-mounted suction sampler for benthic organisms. *United States Fish and Wildlife Service Special Scientific Report-Fisheries*, **614**, 5 pp.

Andersin, A.B. & Sandler, H. (1981) Comparison of the sampling efficiency of two van Veen grabs. *Finnish Marine Research*, **248**, 137–142.

Anderson, R.O. (1959) A modified flotation technique for sorting bottom fauna samples. *Limnology and Oceanography*, **4**, 223–225.

Andrew, N.L. & Mapstone, B.D. (1987) Sampling and the description of spatial pattern in marine ecology. *Oceanography and Marine Biology Annual Review*, **25**, 39–90.

Aneer, A. & Nellbring, S. (1977) A drop-trap investigation of the abundance of fish in very shallow water in the Asko area, northern Baltic proper. In: *Biology of Benthic Organisms. Eleventh European Symposium on Marine Biology, Galway, October 1976.* (Eds B.F. Keegan, P. O'Ceidigh & P.J.S. Boaden), pp. 21–30. Pergamon Press, Oxford.

Ankar, S. (1977a) Digging profile and penetration of the van Veen grab in different sediment types, *Contributions from the Asko Laboratory, University of Stockholm, Sweden*, **16**, 22.

Ankar, S. (1977b) The soft bottom ecosystem of the Northern Baltic proper with special reference to the macrofauna. *Contributions from the Asko Laboratory, University of Stockholm, Sweden*, **19**, 62 pp.

Ankar, S., Andersin, A.B. Lassig, J., Norling, L. & Sandler, H. (1979) Methods for studying benthic macrofauna. An intercalibration between two laboratories in the Baltic Sea. *Finnish Marine Research*, **246**, 147–160.

Ankar, S., Cederwall, H., Lagzdins, G. & Norling, L. (1978) Comparison between Soviet and Swedish methods of sampling and treating soft bottom macrofauna. Final report from the Soviet-Swedish expert meeting on intercalibration of. biological methods and analyses. Asko July *5–12*, 1975. *Contributions from the Asko Laboratory, University of Stockholm, Sweden*, **23**, 38 pp.

Bagge, O. Nielsen E. & Mellergard, S. (1994) Comparative bottom sampling using Haaps and Petersen samplers. ICES, k:42, 20 pp.

Baird, R.H. (1955) A preliminary report on a new type of commercial escallop dredge. *Journal du Conseil Permanent international pour l'Exploration de la Mer*, **20**, 290–294.

Baird, R.H. (1958) A preliminary account of a new half square metre bottom sampler. *International Council for the Exploration of the Sea, Shellfish Committee, C.M.* 1958/70, 4.

Baird, R.H. (1959) Factors affecting the efficiency of dredges. In: *Modern Fishing Gear of the World* (Ed. H. Kristjonnson), pp. 222–224. Fishing News (Books), London.

Bakanov, A.I. (1977) Comparative evaluation of the effectiveness of different dredges. *Gidrobiologicheskii Zhurnal, Kiev*, **13**(2), 97–103.

Baker, J.H., Kimball, T. & Bedinger, C.A. (1977) Comparison of benthic sampling procedures: Petersen grab vs. Mackin corer. *Water Research*, **11**, 597–601.

Bakus, G.J. (1990) *Quantitative Ecology and Marine Biology*. Balkema, Rotterdam.

Bandy, O.L. (1965) The pinger as a deep-water grab control. *Undersea Technology*, **6**(3), 36.

Barnett, P.R.O. (1969) A stabilising framework for the Knudsen bottom sampler. *Limnology and Oceanography*, **14**, 648–649.

Barnett, P.R.O. (1998–1999) *The core issue?* University Marine Biological Station, Millport, 30th Annual Report, 1999–2000, 14.

Barnett, P.R.O. & Hardy, B.L.S. (1967) A diver-operated quantitative bottom sampler for sand macrofaunas. *Helgolander wissenschaftliche Meeresuntersuchungen*, **15**, 396–398.

Barnett, P.R.O., Watson J. & Connelly, D. (1984) A multiple corer for taking virtually undisturbed samples from shelf, bathyal and abyssal sediments. *Oceanologica Acta*, **7**(4), 399–408.

Beamish, R.J. (1972) Design of a trapnet for sampling shallow-water habitats. *Technical Report, Fisheries Research Board of Canada*, **305**, 14 pp.

Belyaev, G.M. & Sokolova, M.N. (1960) On methods of quantitative investigation of deep-water benthos. *Trudy Instituta Okeanologii, Akademiya Nauk SSSR*, **39**, 96–100 (in Russian).

Bergmann, M., & van Santbrink, J.W. (1994) A new benthos dredge (Triple D) for quantitative sampling infauna species of low abundance. *Netherlands Journal of Sea Research*, **33**(1), 129–133.

Beukema, J.J. (1974) The efficiency of the van Veen grab compared with the Reineck box sampler. *Journal du Conseil Permanent international pour l'Exploration de la Mer*, **35**, 319–327.

Beyer, F. (1958) A new bottom living Trachymedusa from the Oslo Fjord. Description of the species, and a general discussion of the life conditions and fauna of the fjord deeps. *Nytt magasin for zoologi*, **6**, 121–143.

Bhaud, M. & Duchene, J-C. (1977) Observations sur l'efficacité compareé de deux bennes. *Vie et Milieu, Ser. A*, **27**, 35–54.

Bieri, R. & Tokioka, T. (1968) Dragonet II, an opening-closing quantitative trawl for the study of microvertical distribution of zooplankton and the meio-epibenthos. *Publications of the Seto Marine Biological Laboratory*, **15**, 373–390.

Birkett, L. (1957) Flotation technique for sorting grab samples. *Journal du Conseil Permanent international pour l'Exploration de la Mer*, **22**, 289–292.

Birkett, L. (1958) A basis for comparing grabs. *Journal du Conseil Permanent international pour l'Exploration de la Mer*, **23**, 202–207.

Blomqvist, S. (1990) Sampling performance of Ekman grabs – in situ observations and design improvements. *Hydrobiologia*, **206**, 245–254.

Blomqvist, S. (1991) Quantitative sampling of soft bottom sediments: Problems and solutions. *Marine Ecology Progress Series*, **72**, 295–304.

Boalch, G.T., Holme, N.A., Jephson, N.A. & Sidwell, J.M.C. (1974) A resurvey of Colman's intertidal traverses at Wembury, South Devon. *Journal of the Marine Biological Association of the United Kingdom*, **54**, 551–553.

Boillot, G. (1964) Géologie de la Manche occidentale. Fonds rocheux, dépots quaternaires, sédiments actuels. *Annales de l'Institut Oceanographique*, **42**, 1–219.

Boland, G.S. & Rowe, G.T. (1991) Deep-sea benthic sampling with the GOMEX box corer. *Limnology and Oceanography*, **36**(5), 1015–1020.

Bossanyi, J. (1951) An apparatus for the collection of plankton in the immediate neighbourhood of the sea-bottom. *Journal of the Marine Biological Association of the United Kingdom*, **30**, 265–270.

Bouma, A.H. (1969) *Methods for the Study of Sedimentary Structures*. John Wiley & Sons, New York.

Bouma, A.H. & Marshall, N.F. (1964) A method for obtaining and analyzing undisturbed oceanic bed samples. *Marine Geology*, **2**, 81–99.

Bourcier, M., Masse, H., Plante, R., Reys, J.P. & Tahvildari, B. (1975) Note préliminaire sur l'étude comparative des bennes Smith–McIntyre et Briba-Reys. *Rapports et proces-verbaux des Réunions. Commission Internationale pour l'Exploration scientifique de la Mer Méditerranée*, **23**(2), 155–156.

Bradbury, R.H., Loya, Y., Reichert, R.E. & Williams, W.T. (1986) Patterns in the structure topology of benthic communities on two coral reefs of the central Great Barrier Reef. *Journal of Wildlife Management*, **4**, 161–167.

Brattegard, T. & Fossa, J.H. (1991) Replicability of an epibenthic sampler. *Journal of the Marine Biological Association, UK*. **71**, 153–166.

Briba, C. & Reys, J.P. (1966) Modification d'une benne 'orange peel' pour les prélèvements quantitatifs du benthos de substrats meubles. *Recueil des Travaux de la Station Marine d'Endoume*, **41**(57), 117–121.

Brook, I.M. (1979) A portable suction dredge for quantitative sampling in difficult substrates. *Estuaries*, **2**(1), 54–58.

Brown, A.V., Schram M.D. & Brussock, P.P. (1987) A vacuum benthos sampler suitable for diverse habitats. *Hydrobiologia*, **153**(3), 241–247.

Brunel, P., Besner, M., Messier, D., Poirier, L., Granger, D. & Weinstein, M. (1978) Le traîneau suprabenthique Macer-GIROQ; appareil amélioré pour l'échantillonage quantitatif etagé de la petite faune nageuse au voisinage du fond. *Internationale Revue der Gesamten Hsydrobiologie*, **63**, 815–829.

Buhl-Jensen, L. (1986) The benthic amphipod fauna of the west Norwegian continental shelf compared with the fauna of five adjacent fjords. *Sarsia*, **71**, 193–208.

Burke, J.C. (1968) A sediment coring device of 21-cm diameter with sphincter core retainer. *Limnology and Oceanography*, **13**, 714–718. (Collected Reprints, Woods Hole Oceanographic Institution, 2149.)

Burke, J.C., Casso, S.A. & Hamblin, R.E. (1983) *Tripod modification of sphincter corer: Construction, operation, core extrusion and sampling efficiency*. Technical Report. Woods Hole Oceanographic Institute.

Caddy, J.F. (1971) Efficiency and selectivity of the Canadian offshore scallop dredge. *ICES Shellfish and Benthos Committee* **K25**, 8.

Carey, A.G. & Hancock, D.R. (1965) An anchor-box dredge for deep-sea sampling. *Deep-Sea Research*, **12**, 983–984.

Carey, A.G. & Heyamoto, H. (1972) Techniques and equipment for sampling benthic organisms. In: *The Columbia River Estuary and Adjacent Ocean Waters: Bioenvironmental Studies* (Eds A.T. Pruter & D.L. Alverson), pp. 378–408. University of Washington Press.

Carey, A.G. & Paul, R.R. (1968) A modification of the Smith–McIntyre grab for simultaneous collection of sediment and bottom water. *Limnology and Oceanography*, **13**, 545–549.

Carleton, J.H. & Hamner, W.M. (1989) Resident mysids: Community structure, abundance and small-scale distributions in a coral reef lagoon. *Marine Biology*, **102**(4), 461–472.

Carney, R.S. & Carey, A.G. Jr. (1980) Effectiveness of metering wheels for measurement of area sampled by beam trawls. *Fishery Bulletin, National Marine Fisheries Service, NOAA, Seattle, USA*, **18**, 791–796.

Chapman, C.J., Main, J., Howell, T. & Sangster, G.I. (1979) The swimming speed and endurance of the queen scallop *Chlamys opercularis* in relation to trawling. *Progress in Underwater Science*, **4**, 57–72.

Chapman, C.J., Mason, J. & Kinnear, J.A.M. (1977) Diving observations on the efficiency of dredges used in the Scottish fishery for the scallop, *Pecten maximus* (L.). *Scottish Fisheries Research Report*, **10**, 15 pp.

Christie, N.D. (1975) Relationship between sediment texture, species richness and volume of sediment sampled by a grab. *Marine Biology*, **30**, 89–96.

Christie, N.D. & Allen, J.C. (1972) A self-contained diver-operated quantitative sampler for investigating the macrofauna of soft substrates. *Transactions of the Royal Society of South Africa*, **40**, 299–307.

Clarke, A.H. (1972) The Arctic dredge, a benthic biological sampler for mixed boulder and mud substrates. *Journal of the Fisheries Research Board of Canada*, **29**, 1503–1505.

Colman, J.S. & Segrove, F. (1955) The tidal plankton over Stoupe Beck Sands, Robin Hood's Bay (Yorkshire, North Riding). *Journal of Animal Ecology*, **24**, 445–462.

Craib, J.S. (1965) A sampler for taking short undisturbed marine cores. *Journal du Conseil Permanent International pour l'Exploration de la Mer*, **30**, 34–39.

Cranmer, G.J., Dyer, M.F. & Fry, P.D. (1984) Further results from headline camera surveys in the North Sea. *Journal of the Marine Biological Association, UK*. **64**(2), 335–342.

Creutzberg, F., Duineveld, G.C.A. & van Noort, G.J. (1987) The effect of different numbers of tickler chains on beam-trawl catches. *Journal du Conseil International permanent pour l'Exploration de la Mer*, **43**, 159–168.

Dauvin. J.C. (1979) *Recherches Quantitatives sur le Peuplement des Sables Fins de la Pierre Noire, Baie de Morlaix, et sur la Perturbation par les Hydrocarbures del'AMOCO-CADIZ'*. Thèse du Diplome de Docteur de 3ᵉCycle, Université Pierre et Marie Curie, Paris, 251 pp.

Dauvin, J.C. & Lorgeré, J.C. (1989) Modifications du traîneau Macer-GIROQ pour l'amélioration de l'échantillonage quantitative étage de la faune suprabenthique. *Journal de la Recherche Oceanographique*, **14**, 65–67.

Dauvin, J.C., Sorbe, J.C. & Lorgèré, J.C. (1995) Benthic boundary layer macrofauna from the upper continental slope and the Cap Ferret Canyon (Bay of Biscay). *Oceanologica Acta*, **18**, 113–122.

Davis. F.M. (1925) *Quantitative studies on the fauna of the sea bottom. No. 2. Results of investigations in the southern North Sea, 1921–24*. Fishery, Investigations Ministry of Agriculture Fisheries and Food, Series 2, **8**(4), 50 pp.

Davis, F.M. (1958) *An Account of the Fishing Gear of England and Wales* (Fourth Edition). Fisheries Investigations, London, Series 2, S (4), 50 pp.

Day, G.F. (1978) The Day grab – a simple sea-bed sampler. *Report of the Institute of Oceanographic Sciences* No. 52.

De Jonge, V.N. & Bouwman, L.A. (1977) A simple density separation technique for quantitative isolation of meiobenthos using the colloidal silica Ludox-™. *Marine Biology*, **42**, 143–148.

Decker, C.J. & O'Dor, R. (2002). A Census of Marine Life: Unknowable or Just Unknown? *Oceanologica Acta*, Vol. 25, No. 5, 179–186.

Desprez, M. & Ducrotoy, P.P. (1987) Nouvelle méthode de prélèvement des coques (*Cerastoderma edule*) et de la faune associée. *Les Nouvelles du GEMEL, Saint-Valery-sur-Somme*, **4**, 5 pp.

Dickie, L.M. (1955) Fluctuations in abundance of the giant scallop *Placopecten magellanicus* (Gmelin) in the Digby area of the Bay of Fundy. *Journal of the Fisheries Research Board of Canada*, **12**, 797–857.

Dickinson, J.J. & Carey, A.G. Jr, (1975) A comparison of two benthic infaunal samplers. *Limnology and Oceanography*, **20**, 900–902.
Diener, D. & Rimer, J.P. (1993) New way benthic sampling. *Proceedings of the Marine Technology Society Conference, 1993*, Washington, pp. 70–75.
Dillon, W.P. (1964) Flotation technique for separating faecal pellets and small marine organisms from sand. *Limnology and Oceanography*, **9**, 601–602.
Drake, C.M. & Elliott, J.M. (1982) A comparative study of three air-lift samplers used for sampling benthic macro-invertebrates in rivers. *Freshwater Biology*, **12**, 511–533.
Drake, C.M. & Elliott, J.M. (1983) A new quantitative air-lift sampler for collecting macroinvertebrates on stony bottoms in deep rivers. *Freshwater Biology* **13**, 545 –559.
Driscoll, A.L. (1964) Relationship of mesh opening to fauna1 counts in a quantitative benthic study of Hadley Harbor. *Biological Bulletin, Marine Biological Laboratory Woods Hole*, **127**, 368.
Dukerschein, J., Gent, R. & Sauer, J. (1996) Recovery of macro-invertebrates by screening in the field: a comparison between coarse (1.18 mm) and fine (0.60 mm) mesh sieves. *Journal of Freshwater Ecology*, **11**, 61–65.
Dybern, B.I., Ackefors, H. & Elmgren, R. (1976) Recommendations on methods for marine biological studies in the Baltic Sea. *The Baltic Marine Biologists*, **1**, 98 pp.
Dyer, M.F., Fry, W.G., Fry, P.D. & Cranmer, G.J. (1982) A series of North Sea benthos surveys with trawl and headline camera. *Journal of the Marine Biological Association of the United Kingdom*, **62**, 297–313.
Dyer, M.F., Pope. J.G., Fry, P.D., Law. R.J. & Portmann, J.E. (1983) Changes in fish and benthos catches off the Danish coast in September 1981. *Journal of the Marine Biological Association of the United Kingdom*, **63**, 767–775.
Eagle, R.A., Norton, M.G., Nunny, R.S. & Rolfe, M.S. (1978) The field assessment of effects of dumping wastes at sea: 2. Methods. *Fisheries Research Technical Report*, Lowestoft, **47**, 24 pp.
Ecoscope (2000) *A species and habitats monitoring handbook*, Volume 2: Habitat monitoring. Research, Survey and Monitoring Review. Scottish Natural Heritage, Edinburgh.
Edwards, R. & Steele, J.H. (1968) The ecology of O-group plaice and common dabs at Loch Ewe. I. Population and food. *Journal of Experimental Marine Biology and Ecology*, **2**, 215–238.
Eleftheriou, A. & McIntyre, A.D. (1976) The intertidal fauna of sandy beaches – a survey of the Scottish coast. *Scottish Fisheries Research Report*, **6**, 61 pp.
Eleftheriou, A. (1979) *The ecology and biology of the shallow hyperbenthos in a sandy bay*. Ph.D. thesis, Aberdeen University, UK.
Eleftheriou, A. & Robertson, M.R. (1988) The intertidal fauna of sandy beaches – A Survey of the East Scottish Coast. *Scottish Fisheries Research Report* **38**, 52 pp.
Eleftheriou, A. & Robertson, M.R. (1992) The effects of experimental scallop dredging on the fauna and physical environment of a shallow sandy community. *Netherlands Journal of Sea Research (Proceedings, 26th European Marine Biology Symposium. Biological Effects of Disturbances on Estuarine and Coastal Marine Environments)*, **30**, 289–299.
Elliott, J.M., Tullett, P.A. & Elliott, J.A. (1993) A new bibliography of samplers for freshwater benthic invertebrates. *Freshwater Biological Association*, Occasional Publication, **30**, 92.
Elliott, J.M. & Drake, C.M. (1981) A comparative study of seven grabs used for sampling benthic macroinvertebrates in rivers. *Freshwater Biology*, **11**, 99–I 20.

Ellis, D.V. & Jones, A.A. (1980) The Ponar grab as a marine pollution monitoring sampler. *Canadian Research*, June/July, 23–25.

Emery, K.O. (1961) A simple method of measuring beach profiles. *Limnology and Oceanography*, **6**, 90–93.

Emery, K.O., Merrill, A.S. & Trumbull, V.A. (1965) Geology and biology of the sea floor as deduced from simultaneous photographs and samples. *Limnology and Oceanography*, **10**, 1–21 *(Collected Reprints, Woods Hole Oceanographic Institute, 1965(1), No.* 1508.)

Emig, C.C. (1977) Un nouvel aspirateur sous-marin, à air comprimé. *Marine Biology*, **43**, 379–380.

Emig, C.C. & Lienhart, R. (1971) Principe de l'aspirateur sous-marin automatique pour sédiments meubles. *Vie et Milieu* (Suppl.), **22**, 573–578.

Ervin, J.L. & Haines, T.A. (1972) Using light to collect and separate zooplankton. *Progressive Fish Culturist*, **34**, 171–1 74.

Espinosa, L.R. & Clark, W.E. (1972) A polypropylene light trap for aquatic invertebrates. *California Fish and Game*, **58**, 149–152.

FAO (1972) *Catalogue of Fishing Gear Designs*, Fishing News (Books), London.

Farris, R.A. & Crezeé, M. (1976) An improved Reineck box for sampling coarse sand. *Internationale Revue der Gesamten Hydrobiologie*, **61**, 703–705.

Fast, A.W. (1970) An evaluation of the efficiency of zoobenthos separation by sugar flotation. *Progressive Fish Culturist*, **32,** 212–216.

Finnish IBP-PM Group (1969) Quantitative sampling equipment for the littoral benthos. *Internationale Revue der Gesamten Hydrobiologie*, **54**, 185–193.

Flannagan, J.F. (1970) Efficiencies of various grabs and corers in sampling freshwater benthos. *Journal of the Fisheries Research Board of Canada*, **27**, 1691–1700.

Forster, G.R. (1953) A new dredge for collecting burrowing animals. *Journal of the Marine Biological Association of the United Kingdom*, **32**, 193–198.

Fossa, J.H., Larsson, J. & Buhl-Jensen, L. (1988) A pneumatic, bottom-activated, opening and closing device for epibenthic sledges. *Sarsia*, **73**(4), 299–302.

Franklin, W.R. & Anderson, D.V. (1961) A Bottom Sediment Sampler. *Limnology and Oceanography*, **6**(2), 233–234.

Frolander, H.F. & Pratt, I. (1962) A bottom skimmer. *Limnology and Oceanography*, **7**, 104–106.

Gage, J.D. (1975) A comparison of the deep sea epibenthic sledge and anchor-box dredge samplers with the van Veen grab and hand coring by divers. *Deep-Sea Research*, **22**, 693–702.

Galbraith, R.D., Rice, A. & Strange, E.S. (2000) An introduction to commercial fishing gear and methods used in Scotland. *Scottish Fisheries Information Pamphlet*, **25**, 43.

Gallardo, V.A. (1965) Observations on the biting profiles of three 0.1 m^2 bottom-samplers. *Ophelia*, **2**, 319–322.

Gerdes, D. (1990) Antarctic trials of the multi-box corer, a new device for benthos sampling. *Polar Record*, **26**(156), 35–38.

Gibson, R.N. (1967) The use of the anaesthetic quinaldine in fish ecology. *Journal of Animal Ecology*, **36**, 295–301,

Gilat, E. (1964) The macrobenthonic invertebrate communities on the Mediterranean continental shelf of Israel. *Bulletin de l'Institut Oceanographique, Monaco*, **62**(1290) 46 pp.

Gilat, E. (1968) Methods of study in marine benthonic ecology. *Colloque Comité du Benthos, Comité international pour l'Exploration de la Mer Méditerranée, Marseille, Norember 1963*, 7–13.

Gonor, J.J. & Kemp, P.F. (1978) Procedures for quantitative ecological assessments in intertidal environments. *U.S. Environmental Protection Agency, Office of Research and Development, Ecological Research Series*, EPA-600/3-78-087. Corvallis, Oregon, 103 pp.

Grange, K.R. & Anderson, P.W. (1976) A soft-sediment sampler for the collection of biological specimens. *Records New Zealand Oceanographic Institute*, **3**, 9–13.

Gray, J.S., McIntyre, A.D. & Stirn, J. (1991) Manual of methods in aquatic environment research Part II. Biological Assessment of marine pollution with particular reference to benthos. *FAO Fisheries Technical Paper*, **324**, 49 pp.

Grizzle, R.E. & Stegner, W.E. (1985) A new quantitative grab for sampling benthos. *Hydrobiologia*, **126**(1), 91–95.

Grussendorf, M.J. (1981) A Bushing-coring device for collecting deep-burrowing infaunal bivalves in intertidal sand. *Fishery Bulletin, National Marine Fisheries Service, NOAA, Seattle, USA*, **79**, 383–385.

Gunter, G. (1957) Dredges and trawls. *Memoirs of the Geological Society of America*, **67**(1), 73–78.

Haahtela. I. (1969) Methods for sampling scavenging benthic Crustacea especially the isopod *Mesidotea entomon(L.)* in the Baltic. *Annales Zoologici Fennici*, **15**, 182–185.

Haahtela, I. (1974) The marine element in the fauna of the Bothnian Bay. *Hydrobiological Bulletin*, **8**, 232–241.

Håkanson, L. (1986) Modifications of the Ekman sampler. *Internationale Revue der Gesampten. Hydrobio*logie, **71**, 719–721.

Hamilton, A.C. (1969) A method of separating invertebrates from sediments using longwave ultraviolet light and fluorescent dyes. *Journal of the Fisheries Research Board of Canada*, **26**, 1667–1672.

Hardy, A.C. (1959) *The Open Sea: Its Natural History. Part II. Fish and Fisheries. New Naturalist Series*, **37**, 322. Collins, London.

Hartman, O. (1955) Quantitative survey of the benthos of San Pedro Basin, Southern California. Part I. Preliminary results. *Allan Hancock Pacific Expeditions*, **19**(1), 185 pp.

Heip, C., Brey, T., Creutzberg F., Dittmer J., Dorjes J., Duineveld G., Kingston P., Mair H., Rachor E., Rumohr H, Thielemanns L. & Vanosmael C. (1985) Report on the intercalibration exercise on sampling for macrobenthos, *ICES*.

Hemmings, C.C. (1973) Direct observations on the behaviour of fish in relation to fishing gear. *Helgolander Wissenschaftliche Meeresuntersuchungen*, **24**, 348–360.

Hessler, R.R. & Jumars, P.A. (1974) Abyssal community analysis from replicate box cores in the central North Pacific. *Deep-Sea Research*, **21**, 185–209.

Hesthagen, I.H. (1978) The replicability of sampling the hyperbenthic region by means of Beyer's 50 cm epibenthic closing net. *Meeresforschungen*, **26**(1–2), 1–10.

Higgins, R.C. (1972) Comparative efficiencies of the Smith–McIntyre and Baird grabs in collecting *Echinocardium cordatum* (Pennant) (Echinoidea: Spatangoida) from a muddy substrate. *Records New Zealand Oceanographic Institute*, **1**, 135–140.

High, W.L., Ellis, I.E. Schroeder, W.W. & Loverich, G. (1973) Evaluation of the undersea habitats – Tektite II. Hydro-Lab, and Edalhad – for scientific saturation diving programs. *Helgolander Wissenschaftliche Meeresuntersuchungen*. **24**, 16–44.

Hiscock, K. (1996) *Marine Conservation Review; Rationale and Methods*. Unpublished Report by the Joint Nature Conservation Committee, Peterborough, UK.

Holme, N.A. (1949) A new bottom-sampler. *Journal of the Marine Biological Association of the United Kingdom*, **28**, 323–332.

Holme, N.A. (1953) The biomass of the bottom fauna in the English Channel off Plymouth. *Journal of the Marine Biological Association of the United Kingdom*, **32**, 1–49.

Holme, N.A. (1959) A hopper for use when sieving bottom samples at sea. *Journal of the Marine Biological Association of the United Kingdom*, **38**, 525–529.

Holme, N.A. (1961) The bottom fauna of the English Channel. *Journal of the Marine Biological Association of the United Kingdom*, **41**, 397–461.

Holme, N.A. (1964) Methods of sampling the benthos. *Advances in Marine Biology*, **2**, 171–260.

Holme, N.A. & Barrett, R.L. (1977) A sledge with television and photographic cameras for quantitative investigation of the epifauna on the continental shelf. *Journal of the Marine Biological Association of the United Kingdom*, **57**, 391–403.

Holme, N.A. & McIntyre, A.D. (Eds) (1984) *Methods for the study of marine benthos*. IBP Handbook, 16. Second Edition, Blackwell Scientific Publications, Oxford.

Hopkins, T.L. (1964) A survey of marine bottom samplers. *Progress in Oceanography*, **2**, 215–256.

Howmiller, R.P. (1971) A comparison of the effectiveness of Ekman and Ponar grabs. *Transactions of the American Fisheries Society*, **100**(3), 560–564.

Huberdeau, L. & Brunel P. (1982) Efficacité et selectivité faunistique comparée de quatre appareils de prélèvements endo-, epi- et suprabenthiques sur deux types de fonds. *Marine Biology*, **69**, 331–343.

Hudson, P.L. (1970) Quantitative sampling with three benthic dredges. *Transactions of the American Fisheries Society*, **99**, 603–607.

Hunter, W. & Simpson, A.E. (1976) A benthic grab designed for easy operation and durability. *Journal of the Marine Biological Association of the United Kingdom*, **56**, 951–957.

Ingham, A.E. (Ed.) (1975) *Sea Surveying*. John Wiley & Sons, London.

Ivanov, A.I. (1965) Underwater observations of the functioning of sampling equipment for benthos collections (Petersen, Okean dredges). *Okeanologiia*, **5**(5), 917–23 (in Russian); *Oceanology*, **5**, 119–24 (English translation).

Jensen, K. (1981) Comparison of two bottom samplers for benthos monitoring. *Environmental Technology Letters*, **2**, 81–84.

Johansen, A.C. (1927) Preliminary experiments with Knudsen's bottom sampler for hard bottom. *Meddelelser fra Kommissionen for Havundersøgelser, Ser. Fisk*, **8**(4), 6.

Jonasson, A. & Olausson, E. (1966) New devices for sediment sampling. *Marine Geology*, **4**, 365–372.

Jonasson, P.M. (1955) The efficiency of sieving techniques for sampling freshwater bottom fauna. *Oikos*, **6**, 183–207.

Jones, W.E. (1980) Field teaching methods in shore ecology. In: *The Shore Environment. Volume 1: Methods*. (Eds J.H. Price, D.E.G. Irvine & W.F. Farnham), pp. 19–44. Systematics Association Special Volume 17(a). Academic Press, London, 321 pp.

Jones, W.E., Bennel, S., Beveridge, C., McConnell, B., Mack-Smith, S., Mitchell, J. & Fletcher, A. (1980) Methods of data collection and processing in rocky intertidal monitoring. In: *The Shore Environment. Volume 1: Methods* (Eds J.H. Price, D.E.G. Irvine &

W. F. Farnham), pp. 137–170. Systematics Association Special Volume 17(a). Academic Press, London, 321 pp.

Jumars, P.A. (1975) Methods for measurement of community structure in deep-sea macrobenthos. *Marine Biology*, **30**, 245–252.

Kain, J.M. (1958) Observations on the littoral algae of the Isle of Wight. *Journal of the Marine Biological Association of the United Kingdom*, **31**, 769–780.

Kaiser, M.J., Rogers, S.I. & McCandless, D.T. (1994) Improving quantitative surveys of epibenthic communities using a modified 2-m beam trawl. *Marine Ecology Progress Series*, **106**(1–2), 131–138.

Kajak, Z. (1971) Benthos of standing water. In: *A Manual on Methods for the Assessment of secondary Productivity in Fresh Waters* (Eds W.T. Edmondson & G.G. Winberg), pp. 25–65. IPB Handbook No. 17. Blackwell Scientific Publications, Oxford, 358 pp.

Kajak, Z., Dusoge, K. & Prejs, A. (1968) Application of the flotation technique to assessment of absolute numbers of benthos. *Ekologia Polska, Series A*, **29**, 607–619.

Kanneworff, P. (1979) Density of shrimp *(Pandalus borealis)* in Greenland waters observed by means of photography. *Rapports et Procès-verbaux des Réunions, Conseil International pour l'Exploration de la Mer*, **175**, 134–138.

Kanneworff. E. & Nicolaisen. W. (1973) 'The Haps'. A frame-supported bottom corer. *Ophelia*, **10**, 119–128.

Kanneworff. E. & Nicolaisen. W. (1983) A simple, hand-operated quantitative, bottom sampler. *Ophelia*, **22**(2), 253–255.

Kaplan, E.H., Welker, J.R. & Krause, M.G. (1974) A shallow-water system for sampling macrobenthic infauna. *Limnology and Oceanography*, **19**, 346–350.

Kawamura, G. & Bagarinao, T. (1980) Fishing methods and gears in Panay Island, Philippines. *Memoirs of Faculty of Fisheries, Kagoshima University*, **29**, 81–121.

Kermabon, A., Blavier, P., Cortis, V. & Delauze, H. (1966) The 'Sphincter' corer: a wide-diameter corer with water-tight core-catcher. *Marine Geology*, **4**, 149–162.

Kingston, P. (1988) Limitations on offshore environmental monitoring imposed by sea-bed sampler design. *Advances in Underwater Technology*, **16/31**, 273–281.

Kissam, P. (1956) *Surveying*, Second Edition. McGraw-Hill Book Company, New York, 482 pp.

Knudsen, M. (1927) A bottom sampler for hard bottom. *Meddelelser fra Komissionen for Havundersøgelser, ser. Fisk*, **8**(3), 4 pp.

Koulouri, P.T., Dounas, C.G. & Eleftheriou, A. (2003) A new apparatus for the direct measurement of otter trawling effects on the epibenthic and hyperbenthic macrofauna. *Journal of the Marine Biological Association of the United Kingdom*, **83** 1363–1368.

Kushlan, J.A. (1981) Sampling characteristics of enclosure fish traps. *Transactions of the American Fisheries Society*, **110**, 557–562.

Kutty, M.K. & Desai, B.N. (1968) A comparison of the efficiency of the bottom samplers used in benthic studies off Cochin. *Marine Biology*, **1**, 168–171.

Lackey, R.T. & May, B.E. (1971) Use of sugar flotation and dye to sort benthic samples. *Transactions of the American Fisheries Society*, **100**, 794–797.

Lagler, K.F. (1978) Capture, sampling and examination of fishes. In: *Methods for Assessment of Fish Production in Fresh Waters* (Ed. T. Bagenal), pp. 6–47 IBP Handbook No. 3 (Third edition). Blackwell Scientific Publications, Oxford, 365 pp.

Lamotte, M. & Bourlière, F. (1971) *Problèmes d'Ecologie: l'échantillonnage des peuplements animaux des milieux aquatiques.* Comité francais du Programme Biologique International. Masson et Cie, Paris, 294 pp.

Larsen, P.F. (1974) A remotely operated shallow water benthic suction sampler. *Chesapeake Science*, **15**, 176–178.

Lassig, J. (1965) An improvement to the van Veen bottom grab. *Journal du Conseil Permanent International pour l'Exploration de la Mer*, **29**, 352–353.

Laubier, L., Martinais, J. & Reyss, D. (1971) Opérations de dragages en mer profonde. Optimisation du trait et determination des trajectoires grace aux techniques ultrasonores. *Rapport Scientifiques et Techniques, Centre National Pour l'Exploitation des Océans*, **3**, 23 pp. (French and English texts).

Lauff, G.M., Cummins, K.W., Eriksen, C.H. & Parker, M. (1961) A method for sorting bottom fauna samples by elutriation. *Limnology and Oceanography*, **6**, 462–466.

Lewis, F.G. III & Stoner, A.W. (1981) An examination of methods for sampling macrobenthos in seagrass meadows. *Bulletin of Marine Science*, **31**, 116–124.

Lie. U. & Pamatmat, M.M. (1965) Digging characteristics and sampling efficiency of the 0.1 m^2 van Veen grab. *Limnology and Oceanography*, **10**, 379–384.

Lincoln, R.J. & Sheals, J.G. (1979) *Invertebrate animals. Collection and Preservation.* British Museum (Natural History), 150 pp.

Lisitsin, A.I. & Udintsev, G.B. (1955) A new type of grab. *Trudy Vsesoyuznogo gidrobiologicheskogo obshchestva*, **6**, 217–222 (in Russian).

Little, F.J. & Mullins, B. (1964) Diving plate modification of Blake (beam) trawl for deep-sea sampling. *Limnology and Oceanography*, **9**, 148–150.

Long, B.G. & Wang, Y.G. (1994) Method for comparing the capture efficiency of benthic sampling devices. *Marine Biology*, **121**, 397–399.

Longhurst, A.R. (1964) A review of the present situation in benthic synecology. *Bulletin de l'Institut Océanographique de Monaco*. **63**(1317), 54 pp.

Macer, T.C (1967) A new bottom-plankton sampler. *Journal du Conseil Permanent International pour l'Exploration de la Mer*, **31**, 158–163.

MacGinitie. G.E. (1935) Ecological aspects of a California marine estuary. *American Midland Naturalist*, **16**, 629–765.

MacGinitie, G.E. (1939) Littoral marine communities. *American Midland Naturalist*, **21**, 28–55.

Maitland, P.S. (1969) A simple corer for sampling sand and finer sediments in shallow water. *Limnology and Oceanography*, **14**, 151–156.

Marsh, L.M., Bradbury, R.H. & Reichert, R.E. (1984) Determination of the physical parameters of coral distributions using line transects intersect data. *Coral Reefs*, **2**(4), 175–180.

Martel, A. & Chia, F.-S. (1991) Drifting and dispersal of small bivalves and gastropods with direct development. *Journal of Experimental Marine Biology and Ecology* **150**(1), 131–147.

Mason, J., Chapman, C.J. & Kinnear, J.A.M. (1979) Population abundance and dredge efficiency studies on the scallop, *Pecten maximus* (L.). *Rapports et procès-verbaux des réunions. Conseil International pour l'Exploration de la Mer*, **175**, 91–96.

Masse, H. (1967) Emploi d'une suceuse hydraulique transformée pour les prélèvements dans les substrats meubles infralittoraux. *Helgolander Wissenschaftliche Meeresuntersuchungen*, **15**, 500–505.

Masse, H., Plante, R. & Reys, J.P. (1977) Etude comparative de l'efficacité de deux bennes et d'une suceuse en fonction de la nature du fond. In: *Biology of Benthic Organisms*. (Eds B.F. Keegan, P. O'Ceidigh & P.J.S. Boaden), pp. 433–41 *Eleventh European Symposium on Marine Biology. Galway, October* 1976. Pergamon Press, Oxford, 630 pp.

Mattheisen, G.C. (1960) Intertidal zonation in populations of *Mya arenaria*. *Limnology and Oceanography*, **5**, 381–388.

McIntyre, A.D. (1956) The use of trawl, grab and camera in estimating marine benthos. *Journal of the Marine Biological Association of the United Kingdom*, **35**, 419–429.

McIntyre, A.D. (1961) Quantitative differences in the fauna of boreal mud associations. *Journal of the Marine Biological Association of the United Kingdom*, **41**, 599–616.

McIntyre, A.D. & Eleftheriou, A. (1968) The bottom fauna of a flatfish nursery ground. *Journal of the Marine Biological Association of the United Kingdom*, **48**, 113–142.

McManus, D.A. (1965) A large-diameter coring device. *Deep-Sea Research*, **12**, 227–232.

McNeeley, R.L. & Pereyra, W.T. (1961) A simple screening device for the separation of benthic samples at sea. *Journal du Conseil Permanent International pour l'Exploration de la Mer*, **26**, 259–262.

Menzies, R.J. (1962) The isopods of abyssal depths in the Atlantic Ocean. In: *Abyssal Crustacea*. (Eds J.L. Barnard, R.J. Menzies & M.C. Bacescu), pp 79–206, Vema Research Series, 1. Columbia University Press, New York, 223 pp.

Menzies, R.J. (1972) Current deep benthic sampling techniques from surface vessels. In: *Barobiology and the Experimental Biology of the Deep Sea* (Ed. R.W. Brauer), pp. 164–169. University of North Carolina, 428 pp.

Menzies, R.J., George, R.Y. & Rowe, G.T. (1973) *Abyssal Environment and Ecology of the World Oceans*. John Wiley & Sons, New York, 488 pp.

Menzies, R.J. & Rowe, G.T. (1968) The LUBS, a large undisturbed bottom sampler. *Limnology and Oceanography*, **13**, 708–714 *(Contribution from Florida State University, 224)*.

Menzies, R.J., Smith, L. & Emery, K.O. (1963) A combined underwater camera and bottom grab: a new tool for investigation of deep-sea benthos. *Internationale Revue der Gesamten Hydrobiologie*, **48**, 529–545.

Morgans, J.F.C. (1965) A simple method for determining levels along seashore transects. *Tuakara*, **13**, 3.

Motoh, H. (1980) Fishing gear for prawn and shrimp used in the Philippines today. *Technical Report Aquaculture Department, Southeast Asian Fisheries Development Centre*, **5**, 43 pp.

Mulder, M. & Arkel, M.A., Van (1980) An improved system for quantitative sampling of benthos in shallow water using the flushing technique. *Netherlands Journal of Sea Research*, **14**, 119–122.

Myers, AC. (1979) Summer and winter burrows of a mantis shrimp, *Squilla empusa*, in Narragansett Bay, Rhode Island (USA). *Estuarine and Coastal Marine Science*, **8**, 87–98.

Myers, E.H. (1942) Rate at which Foraminifera are contributed to marine sediments. *Journal of Sedimentary Petrology*, **12**, 92–95 (*Collected Reprints, Woods Hole Oceanographic Institution, 314*).

Nalepa, T.F. & Robertson, A. (1981) Screen mesh size affects estimates of macro- and meio-benthos abundance and biomass in the Great Lakes. *Canadian Journal of Fisheries and Aquatic Sciences*, **38**, 1027–1034.

Nalwalk, A.J., Hersey, J.B., Reitzel, J.S. & Edgerton, H.E. (1962) Improved techniques of deep-sea rock dredging. *Deep-Sea Research*, **8**, 301–302 *(Collected Reprints, Woods Hole Oceanographic Institution, 1216).*

Nelson-Smith, A. (1979) Monitoring the effect of oil pollution on rocky sea shores. In: *Monitoring the Marine Environment.* (Ed. D. Nichols), pp. 25–53. Institute of Biology Symposia No. 24. Institute of Biology, London, 205 pp.

Ockelmann, K.W. (1964) An improved detritus-sledge for collecting meiobenthos. *Ophelia*, **1**, 217–222.

Oug, E. (1977) Faunal distribution close to the sediment of a shallow marine environment. Sarsia, **63**(2), 115–121.

Owen, D.M., Sanders, H.L. & Hessler, R.R. (1967) Bottom photography as a tool for estimating benthic populations. In: *Deep-Sea Photography* (Ed. J.B. Hersey), pp. 229–234. The Johns Hopkins Oceanographic Studies, **3**, 310.

Pamatmat, M.M. (1968) Ecology and metabolism of a benthic community on an intertidal sandflat. *Internationale Revue der Gesamten Hydrohiologie*. **53**, 211–298. (Department of Oceanography, University of Washington, Contribution 427).

Paterson, C.G. & Fernando, C.H. (1971) A comparison of a simple corer and an Ekman grab for sampling shallow water benthos. *Journal of the Fisheries Board of Canada*, **28**, 365–368.

Pauly, D. (1973) Über ein Gerät zur Vorsortierung von Benthosproben. *Berichte der Deutschen Wissenschaftlichen Kommission für Meeresforschung*, **22**, 458–460.

Pearcy, W.C. (1972) Distribution and die1 changes in the behaviour of pink shrimp, *Pandalus jordani*, off Oregon. *Proceedings of the National Shellfisheries Association*, **62**, 15–20.

Pearson, M. (1978) A lightweight sediment sampler for remote field use. *Journal of Sedimentary Petrology*, **48**(2), 653–656.

Pearson, T.H., Josefson, A.B. & Rosenberg, R. (1985) Petersen's benthic stations revisited. Is the Kattegat becoming eutrophic? *Journal of Experimental Marine Biology and Ecology*, **92**(2–3), 157–206.

Pedrick, R.A. (1974) Nonmetallic elutriation and sieving device for benthic macrofauna. *Limnology and Oceanography*, **19**, 535–538.

Pemberton, G.S., Risk, M.J. & Buckley, D.E. (1976) Supershrimp: deep bioturbation in the Strait of Canso, Nova Scotia. *Science, New York*, **192**, 790–791.

Peters, R.D., Timmins, N.T., Calvert, SE. & Morris, R.J. (1980) The IOS box corer: its design, development, operation and sampling. *Institute of Oceanographic Sciences*, **106**, 7 pp. (unpublished manuscript).

Petersen, C.G.J. & Boysen Jensen, P. (1911) Valuation of the sea. I. Animal life of the sea bottom, its food and quantity. *Report from the Danish Biological Station*, **20**, 81 pp.

Phillips, B.F. & Scolaro, A.B. (1980) An electrofishing apparatus for sampling sublittoral benthic marine habitats. *Journal of Experimental Marine Biology and Ecology*, **47**, 69–75.

Picard, J. (1965) Recherches qualitatives sur les biosciences marines des substrats dragables de la région marseillaise. *Recueil des Travaux de la Station Marine d'Endoume*. Bull **36** (Fasc 52) 1–160.

Pickett, G. (1973) The impact of mechanical harvesting on the Thames estuary cockle fishery. *Laboratory Leaflet, MA FF*, **29**, 21.

Powers, C.F. & Robertson, A. (1967) Design and evaluation of an all-purpose benthos sampler. *Great Lakes Research Division, Special Report*, **30**, 126–131.

Price, J.H., Irvine, D.E.G. & Farnham, W.F. (Eds) (1980) *The Shore Environment. Volume 1. Methods*, Systematics Association Special Volume 17(a). Academic Press, London, 362 pp.

Pringle, J.D. (1984) Efficiency estimates for various quadrat sizes used in benthic sampling. *Canadian Journal of Fisheries and Aquatic Sciences*, **41**(10), 1485–1489.

Probert, P.K. (1984) A comparison of macrofaunal samples taken by box-corer and anchor-box dredge. *New Zealand Oceanographic Institute, Records* **4**(13), 150–156.

Proudfoot, P.K., Elliott, M., Dyer, M.F., Barnett, B.F., Allen, J.H., Proctor, N.L., Cutts, N., Nikitik, C., Turner, G., Breen, J., Hemmingway, K.L. & Mackie, T. (2000) Proceedings of the Humber Benthic Field Methods Workshop, Hull University 1997. *Collection and Processing of Macrobenthic Samples from Soft Sediments: A Best Practice View*, 47 pp.

Pugh, J.C. (1975) *Surveying for Field Scientists*. Methuen, London, 230 pp.

Pullen, E.J., Mock, C.R. & Ringo, R.D. (1968) A net for sampling the intertidal zone of an estuary. *Limnology and Oceanography*, **13**, 200–202.

Rees, H.L., Moore, D.C., Pearson, T.H., Elliott., Service, M., Pomfret, J. & Johnson, D. (1990) Procedures for the monitoring of marine benthic communities at UK sewage sludge disposal sites. *Scottish Fisheries Information Pamphlet*. Aberdeen, **18**, 79 pp.

Reineck, H.E. (1958) Kastengreifer und Lotrohre ‘Schnepfe’, Gerate zur Entnahme ungestörter, orientierter Meeresgrundproben. *Senckenbergiana lethaea*, **39**, 42–48; 54–56.

Reineck, H.E. (1963) Der Kastengreifer. *Natur und Museum*, **93**, 102–108.

Reineck, H.E. (1967) Ein Kolbenlot mit Plastik-Rohren. *Senckenbergiana lethaea*, **48**, 285–289.

Reish, D.J. (1959a) A discussion of the importance of screen size in washing quantitative marine bottom samples. *Ecology*, **40**, 307–309.

Reish, D.J. (1959b) Modification of the Hayward orange peel bucket for bottom sampling. *Ecology*, **40**, 502–503.

Reys. J.-P. (1964) Les prélèvements quantitatifs du benthos de substrats meubles. *Terre et la Vie*, 1, 94–105. *(Recueil des Travaux de la Station Marine d’ Endoume*, **46** (18)).

Reys, J.-P. & Salvat, B. (1971) L’échantillonage de la macrofaune des sédiments meubles marins. In: *Problèmes* d’Ecologie: *l’échantillonage des peuplements animaux des milieux aquatiques* (Eds M. Lamotte & F. Bourlière), pp. 185–242 Comité francais du Programme Biologique International. Masson et Cie, Paris.

Ricc, A.L., Aldrcd, R.G., Darlington, E. & Wild, R.A. (1982) A quantitative estimation of the deep–sea megabenthos; a new approach to an old problem. *Oceanologica Acta*, **5**, 63–72.

Richards, S.W. & Riley, G.A. (1967) The benthic epifauna of Long Island Sound. *Bulletin of the Bingham Oceanographic Collection*, **19**, 89–135.

Riddle, M.J. (1989) Bite profiles of some benthic grab samplers. *Estuarine, Coastal and Shelf Science*, **29**(3), 285–292.

Riedl, R. (1961) Etudes des fonds vaseux de l’Adriatique. Méthodes et résultats. *Recueil des Travaux de la Station Marine d’Endoume*, **23**, 161–169.

Riley, J.D. & Corlett, J. (1965) The numbers of O-group plaice in Port Erin Bay. *Report of the Marine Biological Station, Port Erin*, **78**, 51–56.

Rothlisberg, P.C. & Pearcy, W.G. (1977) An epibenthic sampler used to study the ontogeny of *Pandalus jordani* (Decapoda, Caridea). *Fishery Bulletin*, **74**, 990–997.

Rowe, G.T. & Clifford, CH. (1973) Modification of the Birge-Ekman box corer for use with SCUBA or deep submergence research vessels. *Limnology and Oceanography*, **18**, 172–175.

Rowell, T.W., Schwinghamer, P., Chin-Yee, M., Gilkinson, K., Gordon Jr., D.C., Hartgers, E., Hawryluk, M., McKeown, D.L., Prena J., Reimer, D.P.,Sonnichsen, G., Steeres, G., Vass, W.P., Vine, R & Woo, P. (1997) Grand Banks otter trawling experiment: III. Sampling Equipment, Experimental Design and Methodology. *Canadian Technical Report of Fisheries and Aquatic Sciences*, **2190**, 36 pp.

Rumohr, H. (1999) Soft bottom macrofauna: Collection, treatment and quality assurance of samples. *ICES Techniques in Marine Environmental Sciences*, **27**, 19 pp.

Russell, B.C., Talbot, F.H., Anderson, G.R.V. & Goldman, B. (1978) Collection and sampling of reef fishes. In: *Coral Reefs: Research Methods*. (Eds D.R. Stoddart & R.E. Johannes), pp. 329–345. *UNESCO Monographics on Oceanographic Methodology*, **5**, 581 pp.

Sanders, H.L. (1956) Oceanography of Long Island Sound, 1952–1954. X. The biology of marine bottom communities. *Bulletin of the Bingham Oceanographic Collection*, **15**, 345–414.

Sanders, H.L., Hessler, R.R. & Hampson, G.R. (1965) An introduction to the study of deep-sea benthic faunal assemblages along the Gay Head-Bermuda transect. *Deep-Sea Research*, **12**, 845–867. *(Collected Reprints, Woods Hole Oceanographic Institution*, 1666.)

Schlacher, T.A. & Wooldridge, T.H. (1996) How sieve mesh size affects sample estimates of estuarine benthic macrofauna. *Journal of Experimental Marine Biology and Ecology*, **201**(1–2), 159–171.

Sellmer, G.P. (1956) A method for the separation of small bivalve molluscs from sediments. *Ecology*, **37**, 206.

Shand, C. & Priestley, R. (1999) A towed sledge for benthic surveys. *Fisheries Research Services, Scottish Fisheries Information Pamphlet*, **22**, 8 pp.

Sirenko, B.I., Markhaseva, E.L., Buzhinskaya, G.N., Golikov, A.A., Menshutkina, T.V., Petryashov, V.V., Semenova, T.N., Stepanjants, S.D. & Vassilenko, S.V. (1996) Preliminary data on suprabenthic invertebrates collected during the RV Polarstern cruise in the Laptev Sea. *Polar Biology*, **16**(5), 345–352.

Sjolund, T. & Purasjoki, K.J. (1979) A new mechanism for the van Veen grab. *Finnish Marine Research*, **246**, 143–146.

Sly, P.G. (1969) Bottom sediment sampling. *Proceedings of the 12th Conference on Great Lakes Research*, 883–898 *(Collected Reprints. Canada Center for Inland Waters, 2–23)*.

Smith, K.L. & Howard, J.D. (1972) Comparison of a grab sampler and large volume corer. *Limnology and Oceanography*, **17**, 142–145.

Smith, W. & McIntyre, A.D. (1954) A spring-loaded bottom sampler. *Journal of the Marine Biological Association of the United Kingdom*, **33**, 257–264.

Sorbe, J.C. (1983) Description d'un traîneau destiné a l'échantillonage quantitatif étage de la faune suprabenthique néritique. *Annales de l'Institut Oceanographique de Paris* (NS), **59**(2), 117–126.

Southward, A.J. (1965) *Life on the Sea-Shore*. Heinemann, London, 153 pp.

Steedman, H.F. (Ed.) (1976) Zooplankton fixation and preservation. *Monographs on Oceanographic Methodology, UNESCO, Paris*, **4**, 350 pp.

Stephen, W.J. (1977) A one-man profiling method for beach studies. *Journal of Sedimentary Petrology*, **47**, 860–863.

Steven, G.A. (1952) *Nets. How to Make, Mend and Preserve Them.* Routledge and Kegan Paul, London, 128 pp.

Stoddart, D.R. & Johannes, R.E. (Eds) (1978) *Coral Reefs: Research Methods. Monographs on Oceanographic Methodology, UNESCO, Paris*, **5**, 581 pp.

Stoner, A.W., Greening, H.S., Ryan, J.D. & Livingston, R.J. (1983) Comparison of macrobenthos collected with cores and suction sampler in vegetated and unvegetated marine habitats. *Estuaries*, **6**(1), 76–82.

Sylvand, B. (1995) *La Baie des Veys Littoral Occidental de la Baie de Seine, (Manche) 1972–1993: structure et évolution a long-terme d'un écosystème benthique intertidal de substrat meuble sous influence estuarienne.* Thèse de Doctorat d'Etat, Universite de Caen, 397 pp.

Thamdrup, H.M. (1938) Der van Veen-Bodengreifer. Vergleichsversuche uber die Leistungsfahigkeit des van Veen- und des Petersen-Bodengreifers. *Journal du Conseil Permanent International pour l'Exploration de la Mer*, **13**, 206–212.

Thayer, G.W., Williams, R.B., Price, T.J. & Colby, D.R. (1975) A large corer for quantitatively sampling benthos in shallow water. *Limnology and Oceanography*, **20**, 474–480.

Thomas, M.L.H. (1960) A modified anchor dredge for collecting burrowing animals. *Journal of the Fisheries Research Board of Canada*, **17**, 591–594.

Thomas, M.L.H. & Jelley, E. (1972) Benthos trapped leaving the bottom in Bideford River, Prince Edward Island. *Journal of the Fisheries Research Board of Canada*, **29**, 1234–1237.

Thorson, G.E. (1957) Sampling the benthos. *Memoirs of the Geological Society of America*, **67**(1), 61–73.

Thouzeau, G. & Hily, L. (1986) A.QUA.R.E.V.E.: une technique nouvelle d'échantillonage quantitatif de la macrofaune epibenthique des fonds meubles. *Oceanologica Acta*, **9**(4), 507–513.

Thouzeau, G. & Lehay, D. (1988) Variabilité spatio-temporelle de la distribution de la croissance et de la survie des juvéniles de *Pecten maximus* (issus des pontes 1985, en baie de Saint-Brieux). *Oceanologica Acta*, **11**(3), 267–283.

Thouzeau, G. & Vine, R. (1991) L'échantillonae du megabenthos: Technique dévelopée sur le Georges Bank (Atlantique nord-ouest). *Comptes Rendus de l'Académie des Sciences de Paris, Ser. III*, **312**(12), 607–613.

True, M.A., Reys, J.-P. & Delauze, H. (1968) Progress in sampling the benthos: the benthic suction sampler. *Deep-Sea Research*, **15**, 239–242.

Tyler, P. & Shackley, S.E. (1978) Comparative efficiency of the Day and Smith–McIntyre grabs. *Estuarine and Coastal Marine Science*, **6**, 439–445.

UKOOA (2001) Marine Environmental Surveys Database on the UKCS – UK Benthos, Aberdeen (CD Rom or download)

Ursin, E. (1954) Efficiency of marine bottom samplers of the van Veen and Petersen types. *Meddeleleser Fra Danmarks Fiskeri-og Havundersogelser*, N.S. **1**(7), 8 pp.

Ursin, E. (1956) Efficiency of marine bottom samplers with special reference to the Knudsen sampler. *Meddeleleser Fra Danmarks Fiskeri-og Havundersogelser*, N.S. **1**(14), 6 pp.

Uzmann, J.R., Cooper, R.A., Theroux, R.B. & Wigley, R.L. (1977) Synoptic comparison of three sampling techniques for estimating abundance and distribution of selected megafauna: submersible vs camera sled vs otter trawl. *Marine Fisheries Review*, **39**(12), 11–19.

van Arkel, M.A. & Mulder, M. (1975) A device for quantitative sampling of benthic organisms in shallow water by means of a flushing technique. *Netherlands Journal of Sea Research*, **9**, 365–370.

van Cleve, R., Ting, R.Y. & Kent, J.C. (1966) A new device for sampling marine demersal animals for ecological study. *Limnology and Oceanography*. **11**, 438–443.

van Veen, J. (1933) Onderzoek naar het zandtransport von rivieren. *De Ingenieur*, **48**, 151–159.

von Brandt, A. (1972) *Fish Catching Methods of the World*. Fishing News (Books), West Byfleet, Surrey, 240 pp.

von Brandt, A. (1978) Bibliography for fishermen's training. *FA0 Fisheries Technical Paper*, **184**, 176 pp.

Watkin, E.E. (1941) Observations on the night tidal migrant Crustacea of Kames Bay. *Journal of the Marine Biological Association of the United Kingdom*, **26**(8), 1–96.

Whitehouse, J.W. & Lewis, B.G. (1966) The separation of benthos from stream samples by flotation with carbon tetrachloride. *Limnology and Oceanography*, **11**, 124–126.

Wigley, R.L. & Emery, K.O. (1967) Benthic animals, particularly *Hyalinoecia* (Annelida) and *Ophiomusium* (Echinodermata), in sea-bottom photographs from the continental slope. In: *Deep-Sea Photography* (Ed. J.B. Hersey), pp. 235–349. The Johns Hopkins Oceanographic Studies, **3**, 310 pp.

Wigley, R.L. & Theroux, R.B. (1970) Sea-bottom photographs and macrobenthos collections from the continental shelf off Massachusetts. *United States Fish and Wildlife Service, Special Scientific Report-Fisheries*, **613**, 12 pp.

Wildish, D.J. (1978) Sublittoral macro-infaunal grab sampling reproducibility and cost. *Technical Report, Fisheries and Marine Service, Canada*, **770**, 4 pp.

Williams, D.D. & Williams, N.E. (1974) A counterstaining technique for use in sorting benthic samples. *Limnology and Oceanography*, **19**, 152–154.

Wolff, T. (1961) Animal life from a single abyssal trawling. *Galathea Reports*, **5**, 129–162.

Word, J.Q. (1976) Biological comparison of grab sampling devices. *Southern California Coastal Water Research Project. Annual report for the year ending 30 June 1976*, 189–194.

Worswick, J.M. & Barbour, M.T. (1974) An elutriation apparatus for macro-invertebrates. *Limnology and Oceanography*, **19**, 538–540.

Yocum, W.L. & Tessar, E.J. (1979) Sled for sampling benthic fish larvae. *Progressive Fish-Culturist*, **42**(2), 118–119.

Zinn, D.J. (1969) An inclinometer for measuring beach slopes. *Marine Biology*, **2**, 132–134.

Zismann, L. (1969) A light–trap for sampling aquatic organisms. *Israel Journal of Zoology*, **18**, 343–348.

Chapter 6
Meiofauna Techniques

P. J. Somerfield, R. M. Warwick and T. Moens

6.1 Introduction

Meiofaunal organisms are mobile metazoans which are smaller than macrofauna and larger than microfauna. The size boundaries of meiofauna are based on the standardised mesh apertures of sieves with 500 μm (or 1000 μm) as upper, and 63 μm (or 42 μm) as lower, limits. All fauna passing through the coarser mesh and retained on the finer one may be considered to be meiofauna. Note that this definition does not include protozoa (which are not metazoans). For the purposes of this chapter we will refer to a mesh aperture of 63 μm, although if a smaller mesh is used to extract the fauna the same mesh or a smaller one should be used for all subsequent procedures. Such a division of the fauna on the basis of size could be considered to be somewhat arbitrary, but studies on the size spectra of benthic animals (Schwinghamer, 1981b; Warwick, 1984, 1989) show that the division is real, as species in benthic communities have a genuine bimodal size distribution. Furthermore, these differently sized components of the fauna are also ecologically distinct, in that the meiofauna tend to be separated from macrofauna in terms of their reproduction (all meiofauna are *in situ* breeders), dispersal (no meiofaunal organisms have a specific dispersal phase), and life histories (most meiofaunal juveniles resemble the adults). Despite this, meiofaunal research as a discipline began relatively late, the term meiofauna first being used as recently as 1942 (Mare, 1942). For decades meiofaunal research was seen as something separate from general 'benthic ecology', with its emphasis on macrofauna. This was, in part, a result of the perceived difficulty in conducting meiofaunal research, which apparently required specialised equipment for sample collection and taxonomic expertise beyond the scope of most laboratories. Happily, this situation is changing. Great efforts have been made to simplify, and to some extent standardise, meiofaunal research techniques for routine benthic analyses. An increased research effort worldwide has seen the adoption of some equipment and techniques as acceptable, and the rejection of others. Simplified pictorial keys have been developed, which allow even non-specialists to attempt the identification of many major taxa to the level of putative species. As a result, meiofaunal research has become what it always should have been, namely an integral part of benthic research, rather than a rather esoteric pastime for a small number of specialists.

One sign of the growth and maturation of meiofaunal research has been the appearance of books dedicated to the subject. Two books of particular note are those by Higgins and Thiel (1988) and Giere (1993). Both books present comprehensive treatments of methods used in meiofaunal research. Our intention is not to repeat all that is included therein. Rather, we aim to present methods which are currently commonly used for meiofaunal research, although where specialised or more up-to-date techniques are available we aim to mention these in passing as well.

6.2 Sample collection

Any instrument or method suitable for sampling macrofauna will, in principle, also be suitable for the smaller organisms. The main difference between sampling for macrofauna and meiofauna is that, because of the much higher numerical density of the latter, smaller samples are usually adequate. These can be obtained by sub-sampling from a larger volume but this may introduce errors or inaccuracies, so there are advantages in collecting a sample that can be examined entire.

Intertidal sediments

Quantitative sampling

In sediments, coring is the best quantitative sampling technique. Corers have a known cross-sectional area and may be made from tubes or pipes of any rigid material. Transparent (e.g. Perspex) or translucent materials have the advantage that the sediment sample can be viewed in the tube. In some deposits, where stones or shells are present, or when very deep cores are required in single lengths, the tube may need to be hammered into the bottom and a metal corer is then probably required. The choices of core diameter and sampling depth, and, therefore, the sample volume, depend on the requirements of a particular study. In many sandy intertidal habitats a diameter of 2–4 cm has been found to give samples which can be sorted entire, and cores of about 1 cm diameter are suitable for estuarine muds where densities are usually high. Short cores are usually adequate on muddy grounds, since most of the animals live in the upper layers, and there is little life below 6–8 cm (Rees, 1940; McIntyre, 1968). On sand, however, with interstitial species extending to great depth, cores of 20–30 cm may be required to collect the bulk of the fauna. Animals are known to exist in considerable numbers below 50 cm (Renaud-Debyser, 1963), and even below 1 m in exposed beaches (McLachlan et al., 1979), where the species are likely to be different from those in the surface layers. Corers should have smooth internal and external surfaces to reduce friction, and the lower edge should be beveled to aid penetration of the sediment. To enable removal of intact sediment cores, an air- or watertight closure such as a rubber stopper is required at the upper end. In well-consolidated sediments such as fine mud, a bottom closure is not generally required, providing that the diameter of the core is not excessive. In less well-consolidated sediments such as coarse sand, it may be necessary to close the bottom of the core-tube prior to withdrawing it from the sediment. This is generally achieved by inserting the core, closing it at the top, and then digging away the sediment next to the core tube until a stopper can be put in place and the core tube removed from the sediment.

Care should be taken to avoid compaction of sediments. As the core is forced into the sediment friction between the sediment and the walls of the corer may compress the sediment within the tube, altering the volume of sediment sampled and potentially biasing the reconstruction of vertical profiles. Another potential problem is that animals may be drawn down from the upper to lower levels as the corer is inserted into the sediment. Compaction and drawing-down of material may be minimised by using a core with a smooth surface, or by using a corer with a larger internal diameter. Applying suction to the top of the corer as it is inserted may also reduce compaction. For long core tubes in intertidal sand, for example, a gentle suction may be applied by mouth as the corer is inserted. More conveniently, and more commonly, a corer with a piston may be employed. A small but simple handheld piston-corer may be made by cutting the needle end from a plastic disposable syringe (e.g. Chandler & Fleeger, 1983) and beveling the cut end with a file. To sample with such a device, the piston should initially be near the lower end of the tube (Fig. 6.1). The tube of the corer should then be pushed slowly into the sediment to the required depth, or a little further. It is helpful if markings indicating the required depth are made on the outside of the tube. The piston provides suction, which prevents core compression during sampling and retains the sample within the tube as it is withdrawn. If the corer is inserted beyond the required sample depth, the piston can be used to extrude the excess sediment, which may then be removed with a knife or similar implement. The piston may then be used to extrude the sample into a suitable container for fixation.

In removing the sample from a standard core tube, it is best to allow the core to slide down into the collecting vessel, since this causes less disturbance than pushing it out with a piston. Fenchel (1967) describes a method by which the top stopper of the core tube is replaced by a bored cork fitted with a short glass tube attached to a length of rubber tubing closed by a clamp. Opening and closing the clamp allows the core to slide down inside the tube step-wise. Alternatively, more control may be obtained by compressing and releasing the top rubber bung with the fingers. If the deposit is of dry sand and the column tends to break rather than slide evenly, a small quantity of filtered sea water carefully added to the top helps to keep the column intact. In samples from muddy ground, the lower part of the core may form a plug of clay, and the sample must then be pushed out from below with a piston. This method is less satisfactory, since it compresses the core and may cause water and particles to mix from one layer to another, although, this may be ameliorated by using potter's clay to seal cores from below (Cleven, 1999).

The vertical distribution of meiofauna in the sediment may be of interest. In sediments such as coarse sand, where core-compression is unlikely to be a problem, cores may be extruded from below using an appropriately sized piston and sliced off

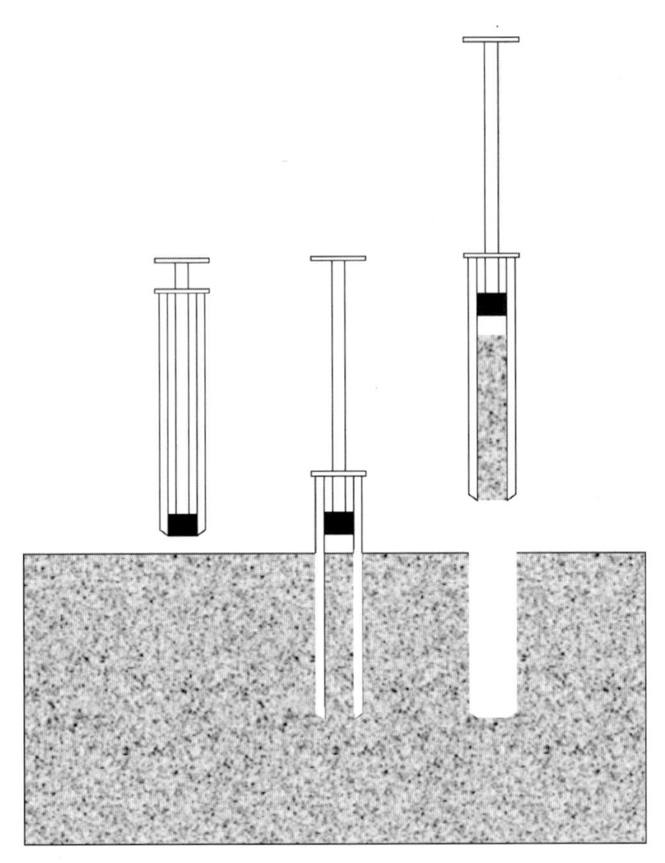

Fig. 6.1 Technique for collecting a sediment core using a piston-corer such as a sawn-off disposable syringe. The piston is placed near to the sediment surface and held steady as the core tube is pushed into the sediment. Suction from the piston holds the sediment in place as the core is withdrawn.

in appropriate layers. In sediments where the surface layers are poorly consolidated, such as soft muds, it may not be possible to extrude cores without loss. Cores may be allowed to slip downwards out of the core tube in controlled steps by carefully manipulating the top closure, allowing them to be cut into appropriate intervals without causing additional compaction. Markings at appropriate intervals on the piston or on the core tube help to ensure that cores are cut in the right places. As meiofauna are known to migrate in standing sediments, cores should be processed as soon after collection as possible. For very deep vertical profiles, specialised corers (e.g. Renaud-Debyser, 1957) may be necessary. Alternatively, a pit can be dug and core samples taken from the vertical face (Pollock, 1970). The majority of studies of the vertical distribution of meiofauna have examined variation in community structure between slices 1 and 2 cm thick. Very small-scale profiles may be more appropriate if interactions between species, such as competition, are considered. Cores may be sectioned in slices as thin as 1 mm using a modified core-tube attached to a micrometer (Fig. 6.2). Such a device was used by Joint et al. (1982) to

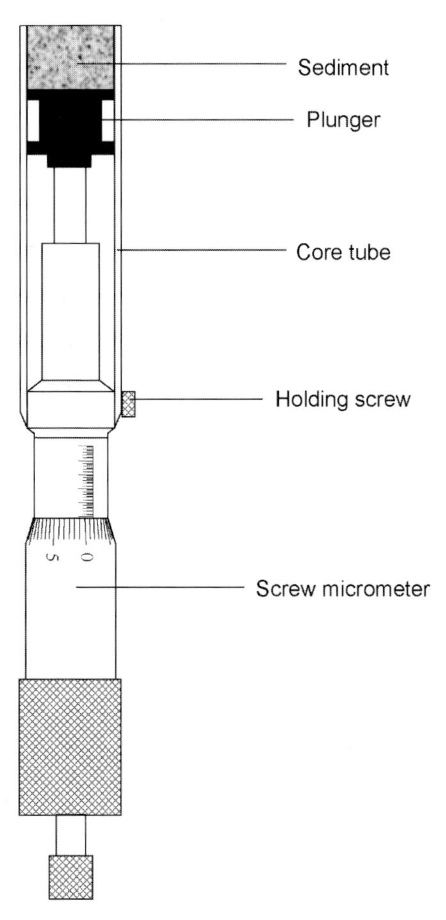

Fig. 6.2 Sampler constructed from a sawn-off disposable syringe mounted on a micrometer screw. Turning the screw raises the plunger, and the distance by which it is raised can be read off the vernier scale. Sediment sections are sliced off the top of the core.

examine interactions between diatoms, meiofauna and bacteria in the surface layers of an intertidal sand flat, by Warwick and Gee (1984) to compare fine-scale vertical distributions in sands and muds, and by Fleeger and Gee (1986) in an experimental evaluation of interspecific competition in harpacticoid copepods.

Qualitative sampling

If specific animals are required in large numbers, or if only particular groups are to be studied, it may be useful to deal with quite large volumes of deposit. For qualitative sampling on intertidal sediments, a general impression of the meiofauna can be obtained by simply digging a trench and allowing water to accumulate in it. Actively swimming animals are then scooped from the water with a fine plankton net, and the more sedentary or adhesive forms are sampled by collecting sand grains from the trench. A refined version of this procedure, the 'methode des sondages' is often used on tideless beaches to concentrate animals from the level of the water table (Delamare-Deboutteville, 1960). Alternatively, sand can be added to filtered sea water (preferably containing anaesthetic such as 6%, i.e. 73.2 $g \cdot l^{-1}$, $MgCl_2$) in a bucket, stirred, and the supernatant poured through a fine sieve.

Subtidal sediments

Quantitative sampling

When sampling sublittorally, a basic choice must be made between getting the investigator to the seabed, or bringing some of the seabed to the surface, prior to sampling. A variety of methods is available for the collection of meiofauna from shallow sublittoral environments, including direct collection using SCUBA, and remote sampling using a range of samplers such as grabs and coring devices (Blomqvist, 1991; Fleeger et al., 1988; Kramer et al., 1994; Somerfield & Clarke, 1997).

In shallow water, where SCUBA divers can work, some of the intertidal techniques can be highly satisfactory. Divers usually obtain superior samples because they are able to position the samplers with care and to insert the core slowly (McIntyre, 1971). Also, by being able to observe the sampling activity the investigator may gain insights into the nature of the site and the details of the sampling. In areas beyond the practical reach of SCUBA, cores may be taken using submersibles or remotely controlled vehicles (Thiel & Hessler, 1974; Thistle, 1978). Various remote samplers have been constructed for sampling in shallow water by attaching a core tube to the end of a pole. Associated with the core tube is a closure mechanism, generally only at the top of the tube, which is either operated automatically as the core tube is pushed into the sediment (e.g. Frithsen et al., 1973) or is operated remotely from the surface. These corers are light, simple to operate, but limited to water depths of 4 m or so.

Although a range of samplers suitable for use from small boats have been designed, few of them have been adopted for general use. Good design features for a remote meiofaunal sampler are a controlled arrival at the sediment surface and open flow-through tubes to reduce water disturbance and bow wave effects, slow penetration of the sediment and tubes of sufficient size to reduce core-compression, and trip and closure mechanisms that do not interfere with water flow through the tube or disturb the sediment before or during penetration. A bow wave is formed when water flows around rather than through the sampler as it is lowered. As the device reaches the sediment surface, this wave causes a flow which washes surface material and meiofauna out of the area which is to be sampled, introducing bias. Sediments with an easily resuspended surface or a flocculent layer are the most difficult to sample.

Deliberate corers (Fig. 6.3) are wire-lowered devices which consist of a supporting frame and a moveable sampling head which carries one or more core tubes. The frame is lowered until it settles on the seabed with the core tubes a short distance (30 cm or so) above the sediment surface. The sampling head is automatically

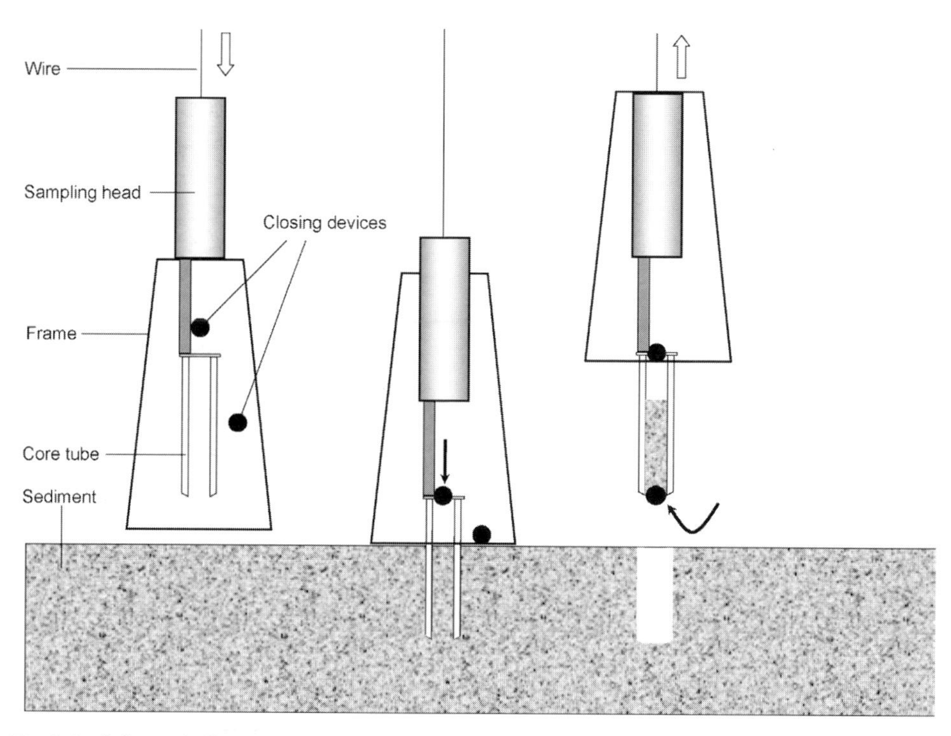

Fig. 6.3 Schematic illustrating the operation of a deliberate corer. The frame is lowered until it settles on the seabed with the open core tubes a short distance above the sediment surface. The moveable sampling head is automatically released and slowly lowers the core tubes into the sediment, at a speed controlled by a damping device. Once the core tubes have penetrated the sediment, an automatic trip device triggers the closure mechanisms at the top of each core. As the wire is hauled up the cores are withdrawn from the sediment and a further device closes the bottom of each core.

released and lowers the core tubes into the sediment, at a slow speed controlled by some sort of damping device such as a piston. Once the core tubes have penetrated the sediment, an automatic trip device triggers the closure mechanisms at the top of each core (generally, a spring-loaded lid with a rubber seal). As the wire is hauled the cores are withdrawn from the sediment. In sediments through which water can flow, such as coarse sand, the suction induced by the top closure alone will not be sufficient to retain the cores. A further device is therefore required to close the bottom of each core tube. Various types have been designed, but those that are external to the core tubes are preferable.

One of the most commonly used deliberate corer designs is that of Craib (1965), which carries a single core tube, approximately 5 cm in diameter, which is closed at its lower end by a rubber ball on a lever that swings the ball into the lower end of the tube as it is withdrawn from the sediment. Samples of 15 cm length are produced, and good samples have been reported from sediments ranging from hard sand to soft mud. The apparatus weighs 44 kg and can be used from a small boat. Its main disadvantage is that, since it must stand upright for a short period while the tube is penetrating, there may be difficulties in bad weather, on uneven bottoms, or in deep water. A much larger device, the Barnett–Watson multiple corer (Barnett et al., 1984), takes up to 12 cores simultaneously. This sampler requires a large vessel for its operation, but is suitable for sampling in very deep water. Each core tube is closed by a lower lid mounted on a long lever that falls onto the sediment after triggering and swings into the closing position as the tubes rise above the sediment surface. Intermediate-sized multiple corers are now widely available. These are broadly based on the design of Barnett et al. (1984) but are generally smaller and take fewer (generally four) cores, often closed by a spring-loaded sliding plate. Modern designs are flexible, in that tubes of a range of different sizes may be mounted. Additional instrumentation may also be carried, along with video cameras and lights. Such devices do not require large vessels for their operation, but may still take samples in deep water. Smaller deliberate corers may struggle to penetrate sandy sediments, in which case additional weights may need to be added to the frame.

Box-corers are, as their name suggests, large deliberate corers that take a square core. The closure mechanism is a large spade which is levered under the lower end of the core as it is lifted, there being no seal applied to the top of the core (although the surface of the core is protected from water movement by sealed doors). Although widely used for the retrieval of 'undisturbed' sediment samples, there is evidence for bow-wave induced bias in the operation of such samplers (Blomqvist, 1991; Bett et al., 1994), and it is probable that they do not routinely take samples with a quality equal to smaller deliberate corers such as the Craib corer (Fleeger et al., 1988).

Although grabs are widely used for quantitative sampling (see Chapter 5), many workers have severe reservations about the quality of such samples (e.g. Elmgren, 1973; Heip et al., 1977; Blomqvist, 1991). Grabs are usually constructed with two jaws which are forced together to enclose a sample of the sediment. The grab must be such that it can be opened from the top, allowing vertical cores to be taken by hand. Cores should be taken only when it has been possible to keep the grab upright with little disturbance to the surface sediment of the sample. Jaw penetration and closure disturb and compress the sediment, while the relatively solid construction of the majority of types may induce a bow wave which disturbs the sediment surface prior to sampling (see Chapter 5). Even in grab types with open space inside them, such as the van Veen grab, the disturbance may be strong.

The normal method of collecting a sub-sample of sediment from a box-core or a grab for meiofaunal work is to take cores using a handheld corer such as described for the collection of intertidal sediments. Box-cores and grabs are designed for sampling macrofauna, and therefore collect sediment from an area that is large in relation to the small-scale patchiness of meiofaunal organisms. For this reason, it is advisable to collect at least three such subsamples and to pool them into a single sample, the contents of which are then more representative of the contents of the sampler as a whole, rather than one small part of them.

Much has been written about the relative merits of different sampling methods for meiofauna. The general conclusion is that different samplers provide samples of differing quality. Cores collected by hand are generally considered to be the best, both intertidally and sublittorally. If samples cannot be collected directly, then a sampling method that collects a contained core directly from the seabed is the next best. Least preferred are sampling methods that attempt to retrieve a quantity of undisturbed sediment from which subsamples may be collected using a hand corer on the surface, such as subsampling from boxcorers or grabs. Of the latter, boxcorers are generally considered to be preferable (Fleeger et al., 1988), and it has been said that grab sampling should be avoided, if possible, for quantitative meiofaunal sampling (Fleeger et al., 1988). It should be noted, however, that dogma should not be allowed to dictate the sampling methods used for a particular piece of work. There may be cases where subsampling from a grab is a sensible option, depending on the prevailing conditions, the type of sediment to be sampled, the ancillary equipment available on the ship, and most importantly, the particular scientific question that the investigator is addressing (Somerfield et al., 1995). Differences between samplers are often attributed to bow wave effects (e.g. Blomqvist, 1991; Bett et al., 1994). Bow wave effects are undoubtedly real but it should be remembered that a host of other factors can also influence the quality of samples during and after retrieval from the seabed (Somerfield & Clarke, 1997).

Qualitative sampling

The epibenthic meiofauna, and species living in the surface centimetre of flocculent bottom material, can be collected non-quantitatively in large numbers for taxonomic and other purposes with epibenthic sledges. This type of apparatus can also be used in a semi-quantitative way to investigate the seasonal appearance of the so-called 'temporary meiofauna' – the juvenile stages of the macrofauna, which are often patchily distributed and usually restricted to the superficial deposits. These sledges have wide runners designed to disturb and skim off the sediment surface, which is then collected in a fine net (Mortensen, 1925; Ockelmann, 1964). For collecting kinorhynchs, Higgins (1964) fitted a rake-like apparatus at the leading edge behind which was a plane-blade that, like a carpenter's plane, removed the upper layer of substratum.

Secondary substrata

Quantitative sampling

Substrata other than sediments have meiofauna living on them. Examples include plants such as algae or seagrasses; sessile animals such as hydroids, bryozoans, sponges, barnacles and corals; hard substrata such as rock, and even motile animals such as crustaceans, echinoderms and molluscs. It is almost impossible to sample such substrata quantitatively. It may be possible to detach a portion of the substratum from a known area, or to collect a standard volume of the substratum, but in either case the degree to which samples may be regarded as quantitative is questionable. Such problems may be exacerbated by the fact that many meiofaunal organisms have quite specific microhabitat requirements, as such they may not be distributed evenly throughout the substratum (Somerfield & Jeal, 1996). To overcome these problems, one approach is to use artificial substratum units (ASUs). These may be manufactured to a standardised design and deployed according to a standardised protocol in such a way that they are directly comparable. Generally, they are constructed to mimic particular naturally occurring secondary substrata, such as seagrass blades (Bell & Hicks, 1991), algae (Edgar, 1991), pneumatophores (Gwyther & Fairweather, 2002), or secondary substrata with a network of interstices resembling sponges, holdfasts or algal tufts such as are ubiquitous in shallow-water areas with hard substrata (Gee & Warwick, 1996). The latter may usefully be mimicked using nylon-mesh pan-scourers (Schoener, 1974) and, after an appropriate period of deployment, support communities of meiofauna which are comparable to those in the adjacent substrata (Cummings & Ruber, 1987; Gee & Warwick, 1996). When secondary substrata (natural or artificial) are retrieved underwater it is necessary to contain the animals within the structure as it is collected, for example, by surrounding the substratum with a plastic bag prior to detaching it. Relatively little quantitative work has been undertaken on the meiofauna of aufwuchs or hard substrata. Suction samplers (e.g. Tanner et al., 1977) may be useful, although most samples are collected by scraping a known area.

Qualitative sampling

The majority of methods for sampling secondary substrata are qualitative in nature. Any sampling method that collects secondary substrata may be used, although those that minimise the extent to which the substrata are subjected to strong water flows are to be preferred. Examples of collecting methods include collection by hand, collection by epibenthic or rock dredges, sledges and trawls, or collection by dragging a plankton net through an area of secondary substrata such as a seagrass bed.

6.3 Fixation and preservation

Preservation and extraction techniques depend on the degree to which taxa are to be identified. Certain groups, sometimes referred to as 'hard' meiofauna, such as nematodes, copepods, ostracods and kinorhynchs remain identifiable after rough preservation in the sediment using 4% formaldehyde. 'Soft' meiofauna (gastrotrichs, turbellarians, etc.) are, however, difficult to recognize after this treatment, and live extraction and examination are essential for these (Table 6.1).

Fixation of meiofauna is normally done by bulk fixation of samples. Although alternatives such as microwave fixation or gluteraldehyde (Pfannkuche & Thiel, 1988) have been suggested, bulk fixation usually involves the use of formalin. The fixation method used must, however, be appropriate both for the taxonomic group

Table 6.1 Special methods of collection and extraction for particular meiofaunal taxa. For further details see Higgins and Thiel (1988).

Taxon	Method
Cnidaria	Extract alive. Usual decantation procedures
Turbellaria	Extract alive. Handle with fine paintbrush or pipette. Decantation often works without anaesthetic. If sediment sample is allowed to stagnate, may concentrate on sediment surface. Seawater ice technique effective.
Gnathostomulida	As for Turbellaria
Nemertea	Extract alive. Usual decantation procedures.
Nematoda	Standard methods of collection and extraction (alive or preserved).
Gastrotricha	Extract alive using relaxation-decantation or seawater ice. Handle with fine micropipette rapidly (they are frustratingly adhesive to surfaces).
Rotifera	Extract alive using relaxation-decantation or seawater ice.
Loricifera	Standard methods of collection and extraction (alive or preserved).
Priapulida	Standard methods of collection and extraction (alive or preserved).
Kinorhyncha	Standard methods of collection and extraction (alive or preserved) are appropriate, but an effective way of extracting large quantities from offshore muds is to pass a stream of fine air bubbles (from an aquarium airstone) through a suspension of the sediment, which brings them to the surface film. This does not work for intertidal species (their cuticle is not hydrophobic).
Polychaeta	Standard methods of collection and extraction (alive or preserved), but alive preferable (many species fragile).
Oligochaeta	Standard methods of collection and extraction (alive or preserved).
Tardigrada	Standard methods of collection and extraction (alive or preserved). Narcotisation of live material essential prior to decantation (cling strongly to sand grains): brief freshwater shock is effective.
Ostracoda	Standard methods of collection and extraction (alive or preserved).
Mystacocarida	Standard methods of collection and extraction (alive or preserved). Narcotisation not necessary for live extraction (do not adhere to sand grains).
Copepoda and other micro-crustaceans	Standard methods of collection and extraction (alive or preserved).
Halacaridae	Usual methods of collection. Best extracted by decantation of preserved material (resistant to anaesthetics or freshwater shock).
Bryozoa, Gastropoda, Holothuroidea, Tunicata	These taxa have a few meiofaunal representatives present in low abundance. Best extracted in large non-quantitative samples by live elutriation after narcotisation.

of interest (Table 6.2) and for the purpose of the study. For standard analyses of 'hard' taxa, samples should be fixed with 10% formalin as soon as possible after collection, and stored in 4% formalin. Formalin to be used as a fixative should be buffered at a minimum pH of 8.2 and diluted to a 10% solution using filtered seawater. For large bulk samples including sediments it may be preferable to add 40% stock solution directly to the samples, and then to make up the volume of the sample with filtered seawater to achieve the required concentration. Invert the sealed container several times to ensure that the fixative solution and the sample are adequately mixed. Formalin fixation of samples should last a week or more, but samples may be preserved for longer in the fixative, if required. Despite its widespread use, formalin is dangerous and care should be taken to ensure that

Table 6.2 Special methods of preparation and examination of material for particular meiofaunal taxa. For further details see Higgins and Thiel (1988).

Taxon	Method
Cnidaria	Examine alive. Contract with normal preservation: narcotise before fixation, then shock fix in hot strong formalin.
Turbellaria	Examine alive. Anaesthetise, restrict to small drop of liquid, then fix with fast acting fixative (Bouin's, gluteraldehyde). Serial anatomical reconstruction necessary for descriptions: embed in wax or resin.
Gnathostomulida	As for Turbellaria
Nemertea	Examine alive. Taxonomic descriptions require histological work, as for Turbellaria.
Nematoda	Fix in 10% formalin (4% formaldehyde). Transfer by slow evaporation to anhydrous glycerol for whole mounts (note pigmentation, e.g. 'eyespots', first, as this may disappear). Mount on normal glass slides or Cobb slides (see main text).
Gastrotricha	Examine alive. Fix after relaxation in $MgCl_2$ with Bouin's fluid. Embedding and serial sectioning necessary for some taxonomic work.
Rotifera	Fix in 5–10% formalin. Slow evaporation into pure glycerol for mounting as for nematodes.
Loricifera	Fix in 2% gluteraldehyde of 6–8% formalin. Transfer by slow evaporation to anhydrous glycerol for whole mounts. Mount on normal glass slides or Cobb slides (see main text).
Priapulida	Examine alive or fixed. Transfer by slow evaporation to anhydrous glycerol for whole mounts.
Kinorhyncha	Fix in 6–10% formalin. Mounting in a modified Hoyer's solution is recommended, but slow evaporation into glycerol is also adequate. Mount on normal glass slides or Cobb slides (see main text).
Polychaeta	Examination of live specimens advantageous. Fix in 10% formalin. After at least 24 hours, store in 70–80% alcohol. Dissection of single segments of fixed specimens with a pair of parapodia often necessary for identification.
Oligochaeta	Fix in 10% formalin, then transfer to 70–80% alcohol. Whole mounts in glycerol usually sufficient for identification
Tardigrada	Fix and preserve in 3–7% formalin. Mount by slow evaporation into anhydrous glycerol.
Ostracoda	Fix in 10% formalin. For identification, specimens dissected with microneedles or sharpened tungsten wire. Shell removed and kept dry in a covered slide. Appendages dissected out and mounted in polyvinyl lactophenol.
Mystacocarida	No special fixation technique. Store in ethanol after fixation.
Copepoda and other micro-crustceans	Fix in 4–5% formalin. Transfer to 70% alcohol for long term storage. Specimens can sometimes be identified from whole mounts in glycerol sufficiently thick that the specimen can be rolled by pushing the coverslip. However, dissection and mounting of limbs is often essential: this requires skill and patience, and is best learned from a practising expert rather than from books.
Halacaridae	Fix and store in 70% alcohol. For critical examination must be cleared: pierce body with tungsten needle and squeeze out body contents in a drop of lactic acid (gentle warming helps the clearing process). Permanent mounts can be made in glycerine on Cobb slides.
Bryozoa, Gastropoda, Holothuroidea, Tunicata	Examine live. Details of specialised treatment of preserved material can be found in Higgins and Thiel (1988).

adequate health and safety precautions are taken to minimise the risk of inhalation of fumes or contact with the solution. Sediment samples are usually brought back to the laboratory for extraction.

6.4 Sample processing

Extraction

In order to examine and count meiofauna, the animals must be extracted from the sediment. Extraction methods vary according to the type of sediment and depend on whether extraction is to be qualitative – to obtain representative specimens – or quantitative, i.e. to extract every organism possible for detailed counts. Extraction can be done on either fresh or preserved samples. If the vertical distribution of the fauna is to be studied, it is essential that the sample should be divided into appropriate sections immediately on collection, since changes within the sample (e.g. in packing, water content, temperature) can produce rapid alterations in the vertical distribution of the fauna.

For qualitative extraction, the sample can be allowed to stand in seawater in the laboratory and the deposit examined at intervals when organisms which come to the surface (often aggregated away from the light) can be pipetted off. Stirring the sample, or bubbling air through it, brings certain types of animals (such as most kinorhynchs and some small crustaceans which have hydrophobic cuticles) to the surface film of the water, from where they can be scooped off using a wire loop or blotting paper. A bicycle pump attached to an aeration block by flexible tubing can be used in the field to bubble samples. For deep-living fauna, the sample may be shaken and decanted, preferably using an anaesthetic as described below.

Techniques for quantitative extraction fall into two broad categories. Those like decantation, elutriation and flotation, which rely on the different rates of sedimentation of organisms and sediment particles, are suitable for both living and preserved material. Techniques which employ an environmental gradient drive living animals out of the sediment.

Sediment

Preserved material

Fixed sediment samples contain a mixture of formalin and sediment components such as silt, clay, sand grains and organic detritus, in addition to the fauna. Although the following processes are often referred to as 'extraction of the fauna', in reality what we are endeavouring to do is to extract various sediment components, so as to end up with the fauna from the original sample and as little else as possible. Although these techniques are generally adequate for preserved samples, if possible occasional live samples should be examined to help with the interpretation of the preserved material, and to allow identification of the more delicate forms. Some taxa with dense shells (bivalve larvae, ostracods) are not adequately extracted by these techniques which rely on the differential settlement of animals and sediment components in water or other fluids.

Prior to processing samples, it is important to check that the laboratory freshwater supply does not contain meiofauna. To do this, run tapwater through a 63 μm sieve for 5–10 minutes and check the contents of the sieve (if any) under a binocular microscope. If meiofauna are present it is necessary to make some arrangement to have it removed. Many of the following techniques involve washing and concentrating meiofauna on sieves with freshwater. Attaching a flexible tube to the freshwater tap is highly recommended, as it greatly increases one's control over the direction and strength of the flow. Meiofauna are small, so do not use a strong jet of water, and splashing must be avoided.

A problem with extracting meiofauna from some fine sediments is the presence of cohesive lumps of clay and faecal pellets, which are difficult to sieve out and which may conceal animals. With preserved samples, pre-treatment to improve the sieving efficiency is possible. Thiel et al. (1975) used ultrasonic treatment, by means of either an ultrasonic bath or a probe, which effectively broke up sediment aggregations and improved sieving efficiency without undue damage to the meiofauna. Barnett (1980) found that freezing in a domestic freezer for 24 hours, followed by thawing, broke down resistant sediments into small particles, which could then be further fragmented with the water softening agent 'Calgon'. A solution made by adding approximately 100 g of Calgon to 500 ml of the sample in formalin was effective, the mixture being stored for 24 hours with intermittent shaking.

An initial washing of the sample with freshwater on a 63 μm sieve removes the finer sediment components, silt and clay, and much of the formalin. Take care not to overload the sieve; larger muddier samples may need to be sieved in smaller amounts. Puddling may help if the sieve becomes clogged. Continue washing until the water passing through the sieve is relatively clear. For some fine sediments, especially if the initial samples are small, this is sufficient to reduce the sample to a quantity that can be extracted with Ludox.

Some types of sample may contain significant quantities of larger material such as leaf or paper fragments. It may be helpful in such circumstances to remove this material by washing the sample through a 1mm sieve nested on top of the 63 μm sieve. Extreme care is necessary not to clog the smaller mesh as it is not in direct view. The contents of the 1mm sieve should be checked under a binocular microscope to ensure that any meiofauna have been washed out of it.

If, after initial washing, the sample contains appreciable quantities of sand, this can be removed by a decantation extraction. Wash the remaining sample into a 1 litre widemouth stoppered measuring cylinder (no more than 150 ml of sediment at a time) and fill the cylinder to above the 1 litre mark with fresh water. Put in the stopper and invert the cylinder 5–10 times to distribute the sediment evenly throughout the volume. Allow to stand briefly, so that dense particles (mainly sand) drop out, then carefully pour the supernatant onto a 63 μm sieve. Repeat 3–6 times. If the next process is to be a flotation extraction with Ludox (see below) do not be too concerned if some fine sand appears on the 63 μm sieve. It is good practice to check an aliquot of the sediment remaining in the cylinder for meiofauna before discarding it, particularly if the sample is one of the first to be processed from a new survey, to assess the efficiency of the extraction.

The key to flotation extraction (or density separation, as it is sometimes called) is to suspend the fauna and sediment in a dense fluid that has a specific gravity very close to that of the animals. The animals are therefore neutrally buoyant, and remain in suspension, but sediment components are negatively buoyant and slowly sink. Various media have been used – sugar solutions (Anderson, 1959; Heip et al., 1974; Higgins, 1977), magnesium sulphate (Lydell, 1936), sodium chloride (Lyman, 1943), zinc chloride (Sellmer, 1956) and carbon tetrachloride (Dillon, 1964). The disadvantage of most of them, such as NaCl or sucrose, is that they have a very high osmotic potential and can, therefore, damage some of the fauna. These media have now been largely replaced by the use of the colloidal silica polymer 'Ludox' produced by Du Pont, primarily developed for use in iron foundries. Ludox has been found to have properties that make it ideal for the extraction of meiofauna. It is available in a range of grades, but Ludox™ is widely used.

As supplied, Ludox™ has a specific gravity of 1.40. The specific gravity needed to extract meiofauna is approximately 1.13, so the stock solution must be diluted. Adding two parts fresh Ludox to three parts freshwater will give a solution of approximately the correct density, but the density should be measured with a hydrometer. For whole sediment samples, some workers prefer to make up a slightly denser solution (specific gravity 1.18) so that the interstitial water in the sediment further dilutes the Ludox to the desired specific gravity. Do not use seawater directly with Ludox, as this can cause the suspended silica to precipitate, rendering the sample useless.

As previously mentioned, Ludox is a colloidal silica solution, and as silica dust is harmful, care must be taken with its use. Sieves, glassware and other equipment such as washbottles, which have been used for Ludox should be soaked in dilute NaOH solution and rinsed in hot water. It is a wise precaution to wear rubber gloves when handling Ludox. The disadvantage of Ludox is that, at the present time, it is generally possible to purchase it only in large quantities (300 kg drums), although some chemical suppliers have been known to supply it in smaller amounts.

A number of centrifugation techniques involving Ludox have been used (e.g. De Jonge & Bouwman, 1977; Pfannkuche & Thiel, 1988), but the following method has been found to be simple and effective. After decantation, or after initial washing, if the sample consists of a small amount of material, carefully wash the extracted portion of the sediment to one side of the sieve, then wash it into a tallform beaker using Ludox. Add at least 10 times the sample volume of Ludox, made up to specific gravity 1.13. Stir vigorously to distribute the sample evenly throughout the volume and leave to settle for at least 40 minutes. Carefully pour the supernatant through a

63 μm sieve into a jug. Return the Ludox to the sample and repeat the process 2–4 times. Wash the extracted fauna thoroughly with fresh water. If the sample is not to be worked on immediately, preserve it with 4% formalin (or 70% alcohol with 5% glycerol) in a suitable container, such as a glass tube with an airtight plastic closure.

Live material: coarse sediment

Where the sediment particles are heavier than the animals living in the interstices, decantation or elutriation techniques may be used, as previously described for preserved material, and these may be simple or elaborate. Seawater, rather than freshwater, should be used. It is important to use filtered seawater to prevent contamination of the sample with organisms which might be present in, for example, the deck water supply of a research vessel. Best results are obtained if the whole sample is first treated with an anaesthetic, since many interstitial animals tend to attach to sand grains when motion occurs. A solution of magnesium chloride isotonic with sea water is widely used, 75.25 g $MgCl_2{\cdot}6H_2O$ dissolved in 1 litre distilled water is approximately isotonic with sea water of 35 psu. (Since this salt is very hygroscopic, a solution is made up more accurately by heating the salt to constant weight at 500–600°C, 35.24 g of the anhydrous salt being used to make 1 litre of solution). This is sufficient to relax the fauna without adverse effects after several minutes' treatment, but the addition of 10% alcohol is also satisfactory.

Elutriation is a more sophisticated, and perhaps more efficient, version of the decanting procedure introduced by Boisseau (1957) for meiofauna work. Hockin (1981) described a simple system for it, constructed from readily available 'Quickfit' laboratory glassware. Tiemann and Betz (1979) used a long, narrow, elutriation vessel, 40 cm in height and 8 cm at its widest part. They emphasise the value of this shape in reducing turbulence and achieving better animal/sediment separation. Elutriation in warm water, described by Uhlig et al. (1973), has improved efficiency for nematodes which stretch out under these conditions, and for ostracods which open their shells thus increasing water resistance.

Behavioural methods for extraction of living animals were originally developed for terrestrial work, and make use of the activity of the organisms in response to changes imposed by the operator on their physical conditions. A technique was developed by Uhlig (1966, 1968) specifically for marine work. The sediment is placed in a tube on a nylon gauze base which just dips into filtered sea water in a collection dish. Crushed sea-water ice is added to a layer of cotton wool on top of the sediment and, as the ice melts, organisms move down into the collecting dish. The method has certain limitations and can be applied only to sandy sediments, which have a capillary structure. Most of the smaller forms leave the deposit and concentrate in the collecting dish but nematodes and some of the larger animals are not extracted quantitatively, and the sediment should later be elutriated or decanted to obtain a complete extraction. The technique is, however, well suited to 'soft' meiofauna (and also to ciliates and flagellates).

Live material: fine sediment

For quantitative extraction of living material from fine muds and silts, hand sorting is generally required. Divide the sample into two size fractions using a fine sieve. A mesh of 63 μm is appropriate since this is usually accepted as defining the upper limit of the silt–clay fraction of the sediment, but even finer meshes (30 or 40 μm) are often used to ensure that most of the fauna is retained in the sieve residue. Most of the silt-clay fraction of the deposit, along with the smallest meiofauna – mainly juvenile stages – passes through the sieve. The residue is hand-sorted under a stereoscopic microscope, attention always being paid to the surface film of the overlying water as some groups, such as Kinorhyncha, tend to be caught by surface tension. Normal hand sorting from fine material passing through the sieve is difficult, and an appropriate subsampling technique can be employed. The volume thus subsampled should be related to the size of the sorting dish so that the settled sub-sample covers the bottom of the dish with a layer only a few grains thick. When viewed by light from below, living animals can easily be seen either directly as they move or by the track they leave. Counts from the sub-sample are adjusted to give an estimate for the total volume of water, and this is added to the sieve residue count to obtain a value for the original sample. The section of the sample awaiting examination must be kept under suitable conditions of temperature and restricted light so that the animals remain in good condition. The whole procedure is time-consuming, and restricts the number of samples that can be dealt with, but it does permit an accurate count of the fauna in fine muddy deposits.

With many organic muds, large quantities of light debris present in the sieve residue severely hamper the sorting of meiofauna. Density separation techniques, used to separate preserved animals from such debris, have not been widely employed for living meiofauna because the flotation media are usually toxic, or cause distortion of the organisms. However, Schwinghamer (1981a) described a method for the extraction of living organisms from such sediments using centrifugation in a buffered mixture of sorbitol and the silica sol 'Percoll' (Pharmacia Fine Chemicals AB, Uppsala, Sweden), which is non-toxic. It may be possible to detoxify the Ludox by dialysis (De Jonge & Bouwman, 1977), and use it with living specimens.

Secondary substrata

Meiofauna may be extracted from secondary substrata by agitating the substrata vigorously in a bucket of seawater, removing the substrata and then pouring the contents of the bucket through a sieve. A narcotising agent is recommended, as many meiofaunal organisms are adapted to grip their substrata tightly. Structurally complex substrata such as algal holdfasts, or ASUs of the nylon-mesh pan-scourer type, should be broken up prior to extracting the meiofauna. If, as is often the case, quantities of coarse or fine sediment or detritus are washed out of the substrata along with the meiofauna, then these can be removed using the appropriate extraction techniques described above for sediments.

6.5 Storage and preservation

For the preservation of specimens, once they are extracted, a concentration of 4% formalin is adequate. To minimise the risks associated with formalin, a solution of 70% ethanol (or industrial methylated spirit) to which 5% glycerol has been added may be preferred. Unless the sample containers are particularly well sealed, the alcohol tends to evaporate slowly. The glycerol does not evaporate and therefore prevents specimens from drying out. It also serves to prevent some organisms such as harpacticoids from becoming brittle. Samples should be checked regularly and topped up with spirit as necessary.

6.6 Sample splitting

Typically, sediment samples contain large numbers of meiofaunal organisms, and it would be impossible to count and identify all of them. Meiobenthic communities are, however, patchy at small spatial scales, so simply taking smaller samples is not the answer. This is why we take larger samples and then routinely identify a proportion, extracted at random from the whole sample, which we refer to as a subsample. As a general rule, a subsample that contains at least 200 specimens is adequate for standard community analyses. It is helpful to keep the subsample size (defined as a fraction of the whole sample) constant within a particular study. Subsampling can be undertaken before extraction of the fauna but it is generally more efficient to split the sieve residue of completely extracted samples of the meiofauna.

Elmgren (1973) built a special sample divider, consisting of a plexiglass cylinder with its bottom divided into eight equal chambers. A preserved sample is poured into the sample divider, a little detergent added (to prevent copepods and ostracods adhering to the water surface), the volume made up to 1 litre and a tightly fitting lid applied. The sample divider is then inverted and vigorously shaken for a short while. The sample is given about one hour to settle to the bottom, during which time a few twists of the sample divider cause material sedimenting onto the dividing walls to fall down into the chambers. The water is slowly drained off through a tap, until it reaches the level of the dividing walls, the rubber stoppers in the bottoms of the chambers are removed, and the subsamples drained into eight small containers. A gentle jet of water washes out remaining sediment. Jensen (1982) described a modified version, with a removable mixing chamber and a central cone-shaped splitter.

Rather more simple techniques have also proved robust and reliable. An extremely simple but effective subsampler can be constructed as follows. Bend the handle of a kitchen ladle so that it hangs with the bowl horizontal. Fill the bowl with water and measure the volume 20 times. Find a container with a flat bottom and vertical sides made from a reasonably strong translucent plastic, with a volume about 30 times that of the ladle. Sawing the top off a chemical reagent container can often make such a container. Measure a volume of water equivalent to 20 times the volume of the ladle into the plastic container, place it on a horizontal surface and allow it to settle. Mark the level of the meniscus with a permanent marker at various points around the container. Carefully remove one ladle full of water, and mark the position of the meniscus again. Repeat this process to give a series of marks on the container denoting volumes equivalent to various numbers of ladles. It is used as follows: if, after extraction, the sample has been stored in formalin or alcohol, wash the sample with freshwater on a 63 μm sieve. Wash the extracted sample into the plastic container with freshwater and add water until the total volume is equivalent to a known number of ladle volumes. The key to efficient subsampling with this apparatus is in agitating the contents of the container in such a way as to distribute the sample throughout the volume homogeneously. Simply stirring the sample with a circular motion concentrates the meiofauna. The best method is to use the ladle to vigorously agitate the sample with a vertical motion for 20–30 seconds, then to carefully remove a single ladle-full to a 63 μm sieve. Carefully rinse the ladle, collecting the washings on the sieve. If further subsamples are required, the remaining volume must be agitated prior to each removal. Once subsamples have been removed, the remaining sample should be returned to formalin or alcohol, and the size of sub-sample removed should be recorded on the container. Note that minor errors in the volume extracted add up, so using this method to remove more than about a quarter of the sample (5 ladle volumes) begins to become unreliable.

Other simple subsampling techniques rely on similar principles. The sample can be washed into a tall-form beaker, made up to a known volume with tap water, vigorously agitated into an even suspension, and a sub-sample rapidly withdrawn using a sampler of known volume such as a Stempel or Hensen pipette (available from Hydro-Bios, Kiel, Germany). It is convenient to adjust the volume in the beaker so that a one-tenth or one-twentieth sub-sample is withdrawn each time. Alternatively, the sample may be washed into a measuring cylinder, made up to a known volume with tap water, agitated by vigorous bubbling induced by blowing through a graduated pipette (with a fairly wide mouth) and then quickly sucking up a measured volume of suspension. This latter technique is not advisable with material that has been in contact with formalin (or other chemicals such as gluteraldehyde for that matter).

6.7 Examination and counting

Sorting

The sorting of meiofaunal samples is time-consuming and labour-intensive. It is necessary to sort samples only if one is interested in identifying the components of the meiofauna which cannot routinely be identified in whole mounts, such as harpacticoid copepods. Picking out is done using fine stainless steel forceps for larger, more robust material or with a fine glass pipette, fine tungsten needle, sharpened quill or similar instrument for smaller more delicate organisms. It may be useful, however, to examine extracted samples, in water, under a binocular microscope with about

250× magnification, in order to check that animals are present, to count individuals of major groups (e.g. nematodes, copepods), or to pick out larger pieces of detritus. A small petri dish with lines scored on the bottom is ideal for the purpose. The spacing between the lines should be slightly less than the field of view of the microscope. The use of a moveable stereo-microscope on an extended arm is also an option. The petri-dish can be fixed in position (with a piece of graph paper beneath it) so that the animals are not disturbed during the scanning of the dish bottom and the surface film (Uhlig et al., 1973).

To aid the detection of animals during sorting, samples may be stained. Rose Bengal has commonly been used for this purpose. It may be added to 10% formalin during sample fixation. 10 ml of a 1% solution, made by dissolving 1 g of the stain in 1 l of 10% formalin, added to each litre of fixative creates a sufficient stain after several days. Samples that have already been extracted may be temporarily immersed in a 1% solution for 15 minutes, although some animals with impenetrable cuticles, such as kinorhynchs and some halacarids will not take up the stain unless they are immersed for 48 hours or more.

Preparation for microscopy

Details of the special requirements of all major meiofaunal groups are given in Table 6.2 (p. 240). The soft meiofauna can usually only be identified with certainty when alive, but it may be helpful to reduce or eliminate mobility by the use of a compression chamber such as the rotary micro-compressor described by Spoon (1978) or the special chamber (available from Hydro-Bios) described by Uhlig and Heimberg (1981), which can also be used with inverted microscopes. With such an instrument it is possible gradually to reduce the gap between the glass base-plate and cover slip (or between two cover slips), and so squeeze the animal until it can no longer move. Alternatively, animals can be placed in a drop of seawater under a standard coverslip, and the water may then be gradually removed using filter paper until the required degree of compression is achieved. Adding an anaesthetic to the water may also be helpful.

For hard meiofauna, it is possible to make permanent slides which can be examined at leisure and retained as a reference collection. There are two approaches to mounting samples of hard meiofauna. For taxonomic work, or for the routine examination of taxa that require dissection, specimens must be picked out and mounted either singly or in small numbers on individual slides, either with or without dissection. For the routine identification of other taxa, bulk mounts have proved to be extremely efficient. A bulk mount is made by mounting a whole sample or subsample on one or more slides. Although a range of mounting media is available and have been recommended for different taxa, the majority of hard meiofauna can be mounted in anhydrous glycerol. The transfer to glycerol needs to be controlled to prevent the collapse of specimens, and this is usually achieved using modifications of the evaporation technique described by Seinhorst (1959).

If individual specimens have been picked out into a smaller container, such as a watch glass, carefully pipette off as much fluid as possible. Add a mixture of dilute ethanol and glycerol. The exact recipe for this mixture is not important, but it should contain approximately 5% glycerol and 10–30% ethanol. This can be mixed up and stored in a labelled wash bottle. For bulk mounts, wash subsamples (or whole samples if subsampling has not been carried out) on a small 63 μm sieve and then wash the material into a cavity block with the same mixture. The watchglass or cavity block is then placed in a desiccator to evaporate off the ethanol and water and leave the specimens or samples in pure glycerol. This process takes several days, but if more rapid transfer is required, the watch or cavity block may be placed on a warm hotplate (20–30°C) or in an oven at 50°C, when evaporation is completed in about 24 hours. If the material is to be left for longer than a day, the cavity block should be partially covered to exclude dust. Although it has been washed several times, formalin-fixed material often produces formalin fumes at this stage, and so it is a good idea to carry out the evaporation in a fume cupboard.

Once evaporation is complete the specimens or samples are ready for mounting. For taxonomic work and for some identifications, it is sometimes advantageous to mount specimens in such a way that they may be viewed from both top and bottom. This may be achieved by mounting specimens between two coverglasses in a special holder such as a Cobb mount. This consists of a metal frame the same size as a normal microscope slide, which holds the double coverglass mount securely in place by means of a pair of plastic or cardboard spacers. Specimens may be viewed from either surface simply by turning the preparation over.

More usually, specimens (and samples) are mounted on standard glass microscope slides. Preparations are made by mounting the specimens in a drop of anhydrous glycerol and, after making sure that they are arranged correctly, carefully covering them with a coverglass of an appropriate size. For temporary mounts, the coverglass may be propped up with small pieces of lens tissue or broken coverglass to prevent flattening of the specimens. Permanent mounts may be prepared by supporting the coverglass with fine glass beads (ballotini) or rods of appropriate diameter, and sealing the edges with an appropriate sealant for fluid mounts such as 'Glyceel' or 'Bioseal No. 2 mountant'. If oil-immersion lenses are to be used, it is important that the sealant compound is resistant to the solvents used to remove the oil, such as xylene.

A useful method for making permanent or semi-permanent slides that is routinely employed in a number of laboratories is the wax ring technique. This involves placing a wax ring on each slide the same shape as, but slightly smaller than, the coverglass intended for use. For bulk mounts, relatively large slides and coverglasses may be appropriate. Wax rings are made by dipping a specially made metal template into molten wax, and then applying it to each slide so as to leave a smooth ring of solid wax on each side. The material to be mounted is then placed in anhydrous glycerol within the ring and the coverglass is placed on top. The wax ring is then carefully melted, slowly lowering the coverglass, by placing the preparation on a hotplate. On removal from the heat it resolidifies, sealing the preparation and supporting the edges of the coverglass at the same time, although one or two coats of an appropriate sealant are usually applied in addition. Bulk mounts made by this process have proved to be useful after 10 years or more, and standard and Cobb mounts made using wax rings are routinely used for taxonomic preparations of marine nematodes.

Counting

Starting at one corner, bulk mounts should be scanned systematically, in such a way that the investigator can be sure that coverage is complete. The use of a mechanical stage with vernier scales on the *X* and *Y* axes is highly recommended. Scanning is normally done using the 10× objective, giving a magnification of 100×. If it is possible to set up the microscope and stage in such a way that one field of view of the 10× objective coincides with a major division of the vernier scale, this facilitates the process greatly. Conventionally, organisms are counted and identified only when the anterior end lies within the field of view. If the slides are examined in a systematic fashion it is then possible to record the position of specimens, for reference purposes or for later closer examination, by reading the coordinates from the vernier scales. If a specimen has come to lie at an awkward angle, or with an important feature hidden under a piece of detritus, applying gentle pressure to the coverglass with a firm pointed implement, such as the point of a pencil, should move it sufficiently to allow identification. Once the slides have been examined they should be stored flat, for example, in cardboard slide trays.

Measurement

Identification often depends on making measurements, which can be achieved from camera-lucida drawings, the scale of which is calibrated with a stage micrometer. The lengths of curved structures can be determined by running an opisthometer (map measurer) along the drawing, and then running it back to zero along a scale drawn from the stage micrometer. Alternatively, a digitizing tablet may be used to measure lengths from drawings or directly through the camera-lucida. By attaching an electronic camera to the microscope, use may be made of sophisticated image analysis software to measure lengths and areas.

6.8 Biomass determination

The small size of most meiofauna poses methodological constraints on direct biomass measurements of individuals. Generally, estimates of individual meio-faunal biomass are not obtained from direct weight or carbon determinations, but from volume estimates. Biomass (wet weight) can be derived from volume by ref-erence to a specific gravity of 1.13 (Wieser, 1960). This nematode-based value is commonly used for other meiofaunal taxa too. For 'hard' taxa, volume estimates are made from measurements of the body length (L) and maximum width (W) of individual specimens. These are converted to volume (V) using the formula

$$V = LW^2C$$

where the value of C relates to the shape of the body (see also Chapter 8). The actual value of C will vary according to the units in which measurements are recorded, and to the shape of each individual animal. As volume estimates made in this way are always somewhat approximate, values of C for individual animals are rarely, if ever, calculated. Instead, approximate values for 'average' animals of various shapes are generally employed (Table 6.3). Volumes of individual animals may also be calculated from scale drawings or photomicrographs. For 'soft' meiofauna, volumes may be calculated by gently squashing specimens to a uniform thickness under a coverslip, and then measuring the area of the specimen. The thickness of the preparation can be determined using the calibration on the fine-focus knob of

Table 6.3 Approximate conversion factors (C) for calculating body volumes (in nl) of different taxa using the equation $V = LW^2C$, where L is the length and W is the maximum body width (both in mm). Values for copepods vary widely according to body form, see Warwick and Gee (1984).

Taxon	Conversion factor (C)		
Acari	399		
Cnidaria	385		
Gastrotricha	550		
Isopoda	230		
Kinorhyncha	295		
Nematoda	530		
Oligochaeta	530		
Ostracoda	450		
Polychaeta	530		
Tanaidacea	400		
Tardigrada	614		
Turbellaria	550		
Copepoda		Body shape	Cross section
'Cylindrical'	750	Parallel sided	Round
'Semi-cylindrical'	560	Narrows evenly towards posterior	Round
'Semi-cylindrical compressed'	630	Narrows evenly towards posterior	Laterally flattened
'Semi-cylindrical depressed'	490	Narrows evenly towards posterior	Dorso-ventrally flattened
'Fusiform'	485	Widest near centre of body	Round
'Pyriform'	400	Narrows markedly behind cephalothorax	Round
'Pyriform depressed'	260	Narrows markedly behind cephalothorax	Dorso-ventrally flattened
'Scutelliform'	230	Broad, almost round	Dorso-ventrally flattened

a good microscope. A microcompression chamber (Uhlig & Heimberg, 1981) can be a useful tool for this type of work.

If direct measurements of carbon or biomass are to be obtained, care should be taken to account for the effects of fixation/preservation (see Chapters 5 and 8). Such direct weight measurements of meiofauna usually require the pooling of a few to many individuals, depending on their size. Electrobalances sensitive to ± 0.1 μg are the most useful device for such measurements. Animals have to be pre-dried to constant weight before measurement. To achieve this, samples should be rinsed on a GF/C glass fibre filter, oven-dried for 24 hours at 60°C, and stored in a desiccator prior to weighing. Ash-free dry weight is obtained by ashing samples in a furnace at 500°C for six hours after drying.

Wet:dry weight ratios have not been determined for most meiofaunal groups. Estimates for dry weight of nematodes vary between 20 and 25% of wet weight (Wieser, 1960; Myers, 1967; Gerlach, 1971), and this value is commonly applied to other meiofaunal taxa as well. Carbon accounts for approximately 40% of dry weight (Feller & Warwick, 1988). The carbon content of meiofaunal wet weight is in the same order as for other invertebrates, the most frequently cited values being 10.6% (Sikora et al., 1977) and 12.4% (Jensen, 1984).

6.9 C/N-stoichiometry and biochemical indicators

There is a general lack of data on the elemental stoichiometry of meiofauna. The small size of these organisms prohibits routine measurements of the C and N content of individuals. Feller and Warwick (1988) proposed a C/N ratio of 4 (with C and N accounting for 40 and 10%, respectively, of animal dry weight) as a rough approximation for meiofauna in general. Generally, CHN analysis requires a minimum of 10 μg of dry material, equivalent to an individual of a large meiofaunal species, but to 20 or more specimens of average-sized nematodes or copepods. Gnaiger and Bitterlich (1984) propose a direct link between CHN stoichiometry and the nature of storage products (protein, lipid and carbohydrate) in animal tissue, but little information is available on the relation between feeding conditions and the nature of storage products in the meiobenthos. A notable exception is the study by Danovaro et al. (1999), who linked seasonal fluctuations in biochemical composition of both coastal and deep-sea nematodes to variations in food supply. The methods used by these authors are not too sophisticated, and will hopefully find broader application in the near future.

6.10 Energy flow measurements

The general principles of energy flow through individuals and populations of benthic organisms, and a detailed overview of the available methods, are given in Chapter 8.

Here we highlight the techniques that are most suitable for use on small organisms, and discuss potential problems and caveats of their application to meiofauna.

Assuming conservation of energy, energy flow through individuals or populations of animals can be described by the equation:

$$Consumption = Production + Respiration\ (and\ metabolism) + Excretion + Egestion$$

where production is the sum of somatic growth and gonad output. Conceptual problems with the use of the above 'static' equation are discussed in Chapter 8. For convenience, in what follows, we discuss each state variable of the above equation separately. Production, however, is treated as a single state variable because of difficulties in determining somatic growth and gonad output separately in field populations of meiofauna.

Consumption

To measure consumption or ingestion rates, the food material ingested must be quantified in some way. Methods aimed at assessing consumption rates by meiofaunal organisms can be roughly divided in four categories: (1) direct observations of ingestion, (2) gut content analyses, (3) measurements of (decreases in) food concentration in a medium containing meiofauna, and (4) tracer techniques. The latter, in turn, can be subdivided into radioactive tracer techniques, stable-isotope tracer techniques and fluorescent tracer techniques.

Direct observations of ingestion

Direct observations of food uptake are largely restricted to the laboratory and are most suitable when food particles are relatively large compared to the size of the consumer. If the organisms under investigation thrive well in a transparent medium such as agar, or in a thin water layer, observations under a good binocular, or, better still, an inverted microscope, can yield detailed information on a variety of feeding interactions. In general, however, the information that can be obtained from observations is more qualitative than quantitative. Quantitative data require prolonged and/or repeated observations. Even when these are available, one has to be careful with their interpretation. Meiofaunal ingestion rates may be strongly affected by a variety of abiotic factors, including light (Buffan-Dubau & Carman, 2000; Moens et al., 2000) and minor temperature changes which are often inevitable when transferring a microcosm from an incubator to the microscope. The patchy distribution of food items and the presence of sand grains may be highly relevant in determining feeding rates in the field, but the former is difficult to mimic accurately in the laboratory, particularly in artificial media or substrata, while the latter hampers direct observations.

Counting pharyngeal pumping rates has been used to quantify ingestion rates by an estuarine nematode (Moens et al., 1996b). Rhabditid nematodes (bacterivores) have a valve apparatus in the pharyngeal metacorpus, the movements of which are relatively easy to observe. They also feed readily when food is presented in thin layers of liquid or on agar media. Most meiofauna, however, do not possess these features, and accurate counts of pharyngeal pulsation rates are only possible when the organism moves slowly and in a plane. The volume of medium ingested per pulsation can be estimated from the volume dilation of the pharynx (De Soyza, 1973; Woombs & Laybourn-Parry, 1984). If food particles are evenly distributed through the medium, the amount of food ingested may be calculated.

Gut content analysis

Observations on gut contents of a number of meiofaunal taxa have proved useful in generating qualitative information on diet. Caution is necessary when interpreting such observations. As an example, Moens et al. (1999b) attributed the high frequency of diatoms in the guts of *Enoploides longispiculosus*, a nematode with known predatory feeding behaviour, to the remnants of prey gut contents.

Pigment analyses of gut contents using high performance liquid chromatography (HPLC) serve specifically for the detection and quantification of microalgal grazing. To our knowledge, the only meiofaunal organisms to which this technique has so far been applied are harpacticoid copepods (Souza-Santos et al., 1995; Santos et al., 1995; Buffan-Dubau et al., 1996; Buffan-Dubau & Carman, 2000). There are, however, no obvious reasons why it could not be applied to other taxa, provided they are sufficiently abundant. The sample size needed for reproducible measurements mainly depends on the targeted pigment(s). As few as 10 copepods may be sufficient for an analysis of chlorophyll *a* (Souza-Santos et al., 1995). If a broader pigment spectrum is taken into account, more detailed information on diet can obviously be obtained, but sample size should be considerably increased (Buffan-Dubau & Carman, 2000). The above-mentioned problems in interpreting gut contents of predators apply here as well. In addition, gut content gives an indication on substrate ingestion rather than utilization. It is not uncommon to find food passing relatively unaltered through the guts of benthic organisms, including meiofauna. In order to estimate ingestion rate from gut content, information on gut passage times is needed. Souza-Santos et al. (1995) thus linked gut pigment analyses to defaecation rate.

Measurement of (decreases in) food concentration

In laboratory microcosm experiments, counting the number of prey organisms before and after adding one or more predatory organisms is, in principle, a straightforward method of measuring food consumption in predacious meiofaunal organisms. Controls without predators have to be run in parallel with the predation trials, to account for, among other things, counting efficiency and the escape of prey to the walls and lids, or out, of experimental chambers. Even in closed containers, and using a transparent medium, prey recovery is seldom 100% efficient. This approach has been used to quantify predation rates of the nematode *Enoploides longispiculosus* on other nematodes and ciliates (Moens et al., 2000; Hamels et al., 2001a). If it is assumed that prey organisms do not multiply over the time course of the incubation, ingestion rate can be readily calculated from the rate of disappearance of prey organisms. If prey items do multiply, as is often the case for protozoans over time scales of more than one hour, the prey growth rate in the absence of predators also has to be determined. If prey growth and predation can be determined at regular, short, time-intervals a model of prey growth rate can be fitted to the data and very accurate calculations of predation rate then become possible. If counts of prey items in the presence of predators are made only after a fixed incubation time, assumptions have to be made about whether prey growth over the experimental interval is linear, exponential, or other, and predation rate can be calculated using a mean prey density. In the latter case, significant bias may be introduced if the predators have a strong functional response over the prey density range of the experiment. This may be avoided by reducing the duration of the experiment, but the time interval must remain long enough to produce a significant reduction in prey density compared to controls, which usually requires incubation times of several hours to one day. Alternatively, predators may be pooled at higher densities, but this increases the risk of interference.

If extrapolations from laboratory experiments to a field situation are to be made, incubations should preferably use prey and predator densities that closely match field densities. Prey density often has a pronounced influence on predation rate (functional response) (Moens et al., 2000; Hamels et al., 2001a), and predators may interfere with one another, thus reducing the *per capita* predation rate (Hamels et al., 2001a). Moreover, depending on the test conditions, prey availability may deviate substantially from natural conditions. If the medium used allows an assessment of prey and predator motility, encounter rate models can be applied to interpret the observed rates of prey disappearance. Similarly, encounter rate models can strongly aid the interpretation of selectivity experiments where multiple prey species are offered simultaneously. Moens et al. (2000) derived a two-dimensional model from the three-dimensional one proposed by Gerritsen and Strickler (1977), for plankton and applied it to predation rates observed in thin agar layers. One of the challenges in the (near) future will be to incorporate substrate heterogeneity and prey refugia introduced by sediment texture and structure into such encounter rate models, and to provide adequate estimates of meiofaunal motility in sediments under natural conditions.

Many experiments aimed at assessing meiofaunal ingestion rates have used starved animals. One should always bear in mind that the responses of starved animals to (often excess) food can substantially differ from normal behaviour. If experiments are designed to determine consumption rates under (near) natural conditions, starvation should be avoided. Experimental organisms should be harvested and pretreated, where necessary, in as short a time as possible.

Direct field assessments of meiofaunal predation rates are difficult to obtain. A few studies have used experimental enrichment with, or removal of, meiofauna in otherwise undisturbed sediment cores to evaluate predation pressure of meiofauna on protozoa (Epstein & Gallagher, 1992; Starink, 1995). However, it is often difficult to interpret the observed changes in food organism densities in terms of direct and/or indirect trophic interactions with meiofauna.

Tracer techniques

Although monitoring changes in the density of bacteria or diatoms before and after the addition of grazers can be used to assess grazing rates under laboratory conditions, most studies of meiofaunal grazing rates on bacteria or microalgae have used *radioactive tracer techniques*, where tracer is added either as free label (inorganic or organic) or as prelabelled food particles. Experiments with additions of labelled food particles under controlled conditions are useful for studying mechanisms of food selectivity and the influence of abiotic factors on food uptake, but less so in determining absolute grazing rates pertinent to field situations. For this to be possible, not only should meiofauna have the same activity level under test conditions as in the field, but the prelabelled food should also be representative of field microbial communities.

Most experiments to date have been performed to determine meiofaunal community grazing rates on bacteria or microalgae by the direct addition of radioactive label ($NaH^{14}CO_3$ for microalgae and [methyl-^{3}H]-thymidine for bacteria) to sediment samples. One should, however, be aware that thymidine is not incorporated by a variety of bacteria, so that rates calculated from trials using this tracer will not accurately reflect grazing on total bacterial standing stock. Alternatives such as ^{3}H-leucine may yield better, yet still not entirely satisfactory, results.

In sediments the administration of label for tracer experiments, so that it is rapidly and evenly distributed without disturbing microbial and meiofaunal organisms, may be problematic. Slurries are frequently used, but these may bias *in situ* grazing rates by disrupting natural microalgal and bacterial patch structures. Alternatives to slurries are the pore water replacement method and the horizontal injection of label into the sediment. The former yields a fairly homogeneous distribution of label but may still affect the benthos adversely, while the latter does not significantly disrupt the sediment but results in poorer label homogenization (Carman et al., 1989; Montagna, 1983, 1984; Jönsson, 1991). A simple water spray can be used to distribute label over the surface of intertidal sediments, hence avoiding sediment disturbance. Surface addition provides label adequately to all photosynthetically active microalgae, while substantially limiting confounding effects from label uptake by chemoautotrophs (Middelburg et al., 2000).

The fixation procedure used to stop experimental incubations may influence results. Formaldehyde has commonly been used in trials with meiofauna but it may permeabilize cuticle to some extent, causing leakage of low-molecular weight metabolites (by far the largest portion of observed losses), and induce egestion or defaecation of (parts of) the grazers' gut contents upon addition of the chemical. In the marine nematode *Pellioditis marina*, fed ^{3}H-labelled bacteria, label loss averaged 40% after one hour and 85% after 24 hours in formaldehyde, irrespective of formaldehyde concentration (Moens et al., 1999a). No further losses occurred beyond 24 hours. Similarly, large label losses occurred with the fixatives glutaraldehyde and ethanol. Cooling samples on ice and fixation with ice-cold formaldehyde, as routinely applied in studies on ciliates and flagellates (Sanders et al., 1989; Sherr & Sherr, 1993) prohibits egestion. If this is followed by immediate freeze-preservation, and sorting of the meiofauna within two hours after thawing, average values for label leakage in *P. marina* are 50%.

Grazing rates can be calculated from the equation:

$$G = 2M/mt$$

where G is the grazing rate, M is the amount of label entering grazers via feeding on bacteria or microalgae, m is the amount of label in the bacteria or microalgae and t is the incubation time (Daro, 1978; Montagna, 1984; Montagna & Bauer, 1988). G is expressed in units of h^{-1}.

For this equation to yield accurate grazing rates, two essential assumptions should be met: (i) label uptake in the grazed compartment should be linear (and label recycling zero), and (ii) label uptake in the grazer compartment should be hyperbolic with time (as grazers are feeding on food which becomes increasingly more labelled) (Daro, 1978; Montagna, 1984, 1993; Montagna & Bauer, 1988). Most studies so far have used relatively long incubation times (in the order of hours) and assumed, rather than proved, that label uptake is hyperbolic. This assumption is often based on preliminary trials where the uptake is measured at hourly or half-hourly intervals. The few available data on gut passage times in actively feeding bacterivorous and herbivorous meiofauna, however, all indicate that the residence time of food in the guts of nematodes and copepods is in the range of 30 seconds to 30 or 40 minutes (see Moens et al., 1999a, for a discussion of this problem). In order to measure ingestion, incubation intervals should be kept as short as possible. The assumption that label recycling from the grazed compartment is negligible is also not so evidently met. Moreover, bacteria and, especially, microalgae produce copious amounts of mucopolysaccharides, which can be readily utilized by other bacteria (Smith & Underwood, 1998; Goto et al., 1999) and perhaps even by some meiofauna (Decho & Moriarty, 1990; Decho & Lopez, 1992). This obviously hampers the interpretation of the pathways of radioactive tracer to the meiobenthos.

The provision of adequate controls for label adsorption to grazers' body surfaces, and for the ingestion of free label (i.e. not assimilated by the grazed compartment) requires careful thought. Poisoned controls, typically using pre-killed (for instance, with formaldehyde) samples, are incubated to account for the former, but living meiofauna often produce mucus and/or carry a variety of micro-organisms on their body surfaces so the adequacy of such controls may be questionable. Parallel dark (in the case of grazing on microalgae) or prokaryotic inhibitor-poisoned (in the case of grazing on bacteria) incubations, where label uptake by the grazed compartment is considered non-existent or minimal, may be used to control for the ingestion of free label. Whereas the use of parallel dark incubations is straightforward, uptake of free organic label together with surface adsorption may account for more than 80% of label entering bacterial grazers (Montagna & Bauer, 1988). The use of prokaryotic inhibitors also requires time-consuming efficiency screening and may adversely affect meiofaunal activity. The determination of experimental and control values is essentially done in two parallel but separate experiments. This has implications for the calculation of means and variances on the obtained grazing rates (Montagna, 1993).

It is unlikely that the rates obtained from relatively short incubations may be extrapolated to 24 hour rates, especially for meiofauna that live in variable environments. In an intertidal environment, for example, it seems plausible that there is a significant tidal impact on feeding. In subtidal environments hydrodynamics, and other abiotic factors, may influence feeding activity. Very few studies have hitherto addressed diel feeding rhythms in meiobenthic organisms (Decho, 1988; Souza-Santos et al., 1995; Buffan-Dubau & Carman, 2000). Clearly, more research is needed on this aspect.

A powerful alternative to radioactive labeling methods is the use of *stable isotope tracers*. Stable isotopes are naturally occurring non-radioactive isotopes of an element, and can be analysed by mass spectrometry. Consumers tend to have stable isotope ratios very similar to those of their food sources, although there is a general tendency towards fractionation resulting in a stepwise increase in stable isotope ratio with trophic level. This increase is very small (on average 1%) for carbon, but substantial enough for nitrogen (3–4%) to discriminate between trophic levels (De Niro & Epstein, 1978, 1981; Rau et al., 1983). Stable isotope ratios generally use the δ notation, where $\delta^{13}C$ is expressed relative to Vienna Pee Dee Belemnite (PDB).

Natural stable isotope ratios of meiofauna can give information on diet, provided the stable isotope ratios of different organic matter sources are sufficiently distinct. In most cases, consumer isotope ratios will reflect a mix of organic matter sources, and the interpretation of natural stable isotope ratios of consumers may not always be straightforward. It is therefore advisable to analyse, where possible, both carbon and nitrogen isotopes. Most mass spectrometric analyses require at least 10 μg of an element (here carbon or nitrogen), which means approximately 100 μg wet weight for carbon analysis and at least five times more for nitrogen. This makes carbon isotope analyses feasible for most meiofauna community studies, sometimes even allowing discrimination between different trophic guilds or dominant genera/species. Currently, biomass is often too limited for nitrogen analysis, but Carman and Fry 2002 propose some modifications to conventional coupled elemental analyzer-mass spectrometer systems which reduce biomass requirements for isotope analyses to as little as 1 μg N and/or 2 μg C.

Of major importance when determining natural stable isotope ratios is the treatment of samples: contact with other organic substances should be kept to a minimum, and sample fixation with ethanol, formaldehyde or other chemicals is to be avoided. It is advisable to isolate organisms alive, or to freeze samples and sort them immediately upon thawing, preferably after simple decanting procedures. If extraction requires some compound like Ludox, preliminary checks should be made to ensure that stable isotope signatures are not affected.

In addition to natural stable isotope analyses, enrichment experiments using either pre-labelled food or free tracer (e.g. $NaH^{13}CO_3$ for labelling microphytobenthos) can be designed analogous to those described for experiments with radioactive tracers (Middelburg et al., 2000). An additional advantage of working with stable isotopes is that mass spectrometric analysis not only gives information on the proportion of the different isotopes of an element, but also allows one to determine the total amount of that element. When working with carbon, this gives a direct and accurate estimate of the biomass of the sample analysed. A major disadvantage is that this is a highly sophisticated technique, requiring expensive equipment.

Finally, *fluorescently labelled food* (bacteria or diatoms) or food analogues (similarly sized but inert microbeads) are now routinely used in studies on protozoan grazing (Epstein & Shiaris, 1992, and references therein). As quantification of ingested particles is through epifluorescence microscopy, this method is perhaps still more laborious than the radiotracer methodology. It may also be hampered by autofluorescence of consumer organisms. In experiments with fluorescently labelled food particles (be they inert microbeads or labelled bacteria or microalgae), the question as to how representative the introduced food (analogue) is for determining grazing rates, and whether it will be selected for or against by members of the meiofauna, remains. An interesting development in this respect is the use of sediment which is stained in its entirety with DTAF, thus allowing both free and attached microbiota to become fluorescently labelled (Starink et al., 1994; Hamels et al., 2001b). If efficient ways of quantifying fluorescent particles inside meiofaunal guts can be found, this may well prove a good method for obtaining realistic estimates of microbial grazing by meiobenthos.

6.11 Production

Meiofaunal biomass in marine sediments roughly spans four orders of magnitude, from 0.01 to 10 g C m^{-2} (Heip et al., 1985). Only a minor fraction ($<1\%$) of the C-input shows up in meiofaunal biomass at any given time (Vranken & Heip, 1986).

Whether or not meiofauna has a significant role in C-fluxes through the benthos, therefore, largely depends on production and respiration rather than biomass.

Since most marine meiofauna reproduce continuously, no distinct cohorts can be recognized in the field, thus rendering direct production estimates from field studies extremely difficult. Some studies of Crustacea (ostracods and copepods) successfully used high temporal resolution sampling and applied production calculation methods for non-cohort populations, but these are exceptional. Laboratory studies have yielded detailed information on a limited number of species. These, however, are almost invariably opportunistic and fast growing and hence not necessarily representative of an entire community. The most commonly used approaches to estimating meiofaunal production are, therefore, indirect methods which use respiration data or a fixed production to biomass ratio (P/B) per generation. For a more conceptual discussion of these and other methods to estimate secondary production, and of the caveats involved in their use, the reader is referred to Chapter 8.

Respiration-based production estimates

McNeil and Lawton (1970) proposed an empirical relationship between production (P, in kcal m^{-2} yr^{-1}) and respiration (R) in short-lived poikilotherm organisms:

$$\log P = 0.8233 \log R - 0.2367$$

The slope of this regression differs significantly from 1, implying that production efficiency $P/(P + R)$ decreases with increasing biomass. Warwick and Price (1979) measured respiration rates in some of the most prominent members of a meiobenthic community, and used this equation to calculate community secondary production. However, as respiration data on meiofauna are rather scarce, this approach cannot routinely be applied.

Alternatively, the following log–log relationship with a slope of 1 between population production (P) and respiration (R) has been used (Humphreys, 1979):

$$\log P = 1.069 \log R - 0.601 \text{ (also in caloric units)}$$

In contrast to the equation proposed by McNeil and Lawton (1970) the slope of this relationship implies that production efficiency is independent of size, a feature that was confirmed by Herman et al. (1984a) in a study on the field production and respiration of one ostracod and two harpacticoid copepod species.

P/B based production estimates

From a laboratory study of two nematode species, Gerlach (1971) derived an annual P/B of 9 for marine meiofauna. This value is composed of two parts: a life-cycle turnover of three (Waters, 1969), and three generations per year. Several studies have since confirmed the value for the life-cycle turnover (see, e.g. Herman et al.,

1984b; Heip et al., 1985), but have demonstrated a large variability in the number of annual generations. In nematodes, the annual number of generations ranges from less than 1 to an astonishing 23 (Vranken & Heip, 1986). Because it is usually impossible to measure the number of generations under field conditions, annual *P*/*B* can only be calculated for species in culture, and it remains to be established how life-cycle characteristics under stable environmental conditions compare to those in the fluctuating natural environment.

An alternative approach based on *P*/*B* uses Banse and Mosher's (1980) regression of body mass upon reaching sexual maturity (M_s, in kcal) to annual *P*/*B* to predict production:

$$P/B = 0.65 M_s^{-0.37}$$

Here again, we have insufficient data to assess how meiofauna fits into this general regression. However, Vranken and Heip (1986) compiled all available development data on marine nematodes (as summarized in Heip et al., 1985, and Vranken & Heip, 1985) to scale *P*/*B* to M_s. The weight dependence of their regression was a little higher than, but not significantly different from Banse and Mosher's regression. Interestingly, the intercept values for meiofauna are on average an order of magnitude lower than in macrofauna, suggesting a low specific production for the former (Heip et al., 1982). This feature remains largely unexplained.

Based on the data obtained from laboratory cultures of brackish-water nematodes, Vranken et al. (1986) proposed a regression for nematodes relating egg-to-egg development time *D* (in days) to temperature *T* (in °C) and adult female wet weight (in μg):

$$\log D = 2.202 - 0.0461T + 0.627 \log W$$

When multiplied by a life-cycle turnover of three, the development rate, 1/*D*, can be used to predict daily *P*/*B*. Again, caution is advised in generalizing from equations based on a limited number of mostly opportunistic species.

Each of the above indirect methods for the calculation of meiobenthic production is subject to a number of conceptual problems. Herman et al. (1984a) discuss the conditions that should be met for meiofauna to comply with a *P*/*B* of three per 'generation time'. It is questionable whether all these assumptions (e.g. that adult body growth is small) are met. In any case, it is advisable to read the paper by Herman et al. (1984a) and related papers, and to be aware of the caveats involved in the application of indirect methods before applying a fixed *P*/*B* ratio or respiration/production regression to your own data.

Nematode production and respiration depend heavily on environmental conditions, such as temperature, salinity and food availability. Of particular interest is the observation that food concentrations above threshold values, while having a moderate and basically linear effect on respiration, appear to impact production profoundly following a hyperbolic pattern (Schiemer, 1982, 1985, 1987). This implies that the production efficiency increases strongly with increasing food supply. Whether this observation is of general relevance, and how production and production efficiency are influenced by other environmental conditions, remain to be established.

Respiration and metabolism

Chapter 8 gives an account of methods for the measurement of aerobic respiration in benthic animals. Studies on meiofauna-sized organisms have mostly used divers and polarographic electrodes. Neither of these techniques is entirely satisfactory. Divers allow extremely sensitive measurements – sometimes at the level of single individuals – but are very laborious and delicate and require relatively long incubation times. Polarographic electrodes are easier to use and accurate readings can be obtained after fairly short incubations, but require several tens or even hundreds of meiofaunal individuals for reliable measurements. Both methods use highly artificial incubation conditions, but results from both appear to be in good agreement and probably give adequate estimates of 'natural' rates (Moens et al., 1996a, see also Chapter 8).

A number of factors may affect meiofaunal respiration rates. The nutritional condition of animals is of major importance. Organisms may show significant decreases in respiration upon starvation for a few days or even a few hours. In the bacterivorous nematode *Pellioditis marina*, for instance, respiration of organisms suspended in water without food decreases within two hours after the onset of starvation, and stabilizes at a maintenance level that is less than one-third of the respiration of well-fed individuals after about eight hours of starvation. This obviously necessitates a standardization of treatments when measuring and comparing respiratory activity of meiofauna.

A major problem when using aerobic respiration rates of meiobenthic organisms, to interpret the magnitude of their contribution to benthic carbon cycles, is the absence of information on their activity under hypoxic or anoxic conditions. Many meiofauna spend much of their lives in hypoxic or anoxic micro-environments. Micro-calorimetry is a tool that integrates all types of metabolic activity, independent of the prevailing oxygen conditions. This is, however, an expensive and highly specialised technique, and it has only seldom been used successfully on meiofauna-sized organisms, using batches of many hundreds of laboratory-reared animals (Butterworth et al., 1989; Braeckman et al., 2002).

Excretion and defaecation

Methods for the quantification of secreted and/or excreted metabolic end products and faeces are discussed in Chapter 8. In principle, they can be applied to members of the meiofauna too, but only if sufficiently high numbers are available. Moreover, except for those organisms which produce clearly distinguishable faecal pellets, it is difficult to distinguish between excretion and defaecation. The difficulties in measuring excretion and defaecation bear strongly upon energy budget calculations and estimates of absorption efficiency in meiofaunal organisms.

6.12 Cultivation of marine and brackish-water meiobenthos

A detailed account of cultivation techniques for marine and brackish-water meiobenthos is well beyond the scope of this chapter, but it is clear that many questions concerning the energetics and role of populations of meiobenthic organisms remain unanswered. Many figures, generalizations, conversion factors etc., which are currently being used for field populations of meiobenthos are, in fact, derived from laboratory-reared species. It is, of course, highly relevant to know about the culture or maintenance conditions under which such figures were produced.

Cultivation may significantly improve our understanding of meiofaunal energetics. Only a limited number of species, most of which belong to the Harpacticoida and Nematoda, have been cultivated in the laboratory. Species amenable to culture are usually opportunistic, have short generation times and high reproductive capacities. Continuous cultures have been established for less than 30 marine and brackish-water nematode species. Of these, 16 belong to a single family (Monhysteridae) typical of detritus-enriched habitats (Moens & Vincx, 1998). Novel cultivation approaches offering suitable conditions for the rearing of a broader range of species would be invaluable, but limited progress in the development of cultivation techniques has been made over the past two decades, and reviews of cultivation attempts and successes published in the seventies and eighties still offer up-to-date information. Among others, the papers by Hicks and Coull (1983) and Chandler (1986) for harpacticoid copepods, by Kinne (1977b) and Moens and Vincx (1998) for nematodes, and by Tietjen (1988) for meiofauna in general, are recommended. Several papers in Kinne's (1977a) book on cultivation techniques of marine organisms also provide highly relevant information.

Most culture systems so far have been small-scale and closed. Examples include petri dishes filled with agar, and conical flasks containing liquid medium. Some harpacticoid copepods, however, have been successfully established in recirculating, continuous, mass cultures (Fukusho, 1980; Bin Sun & Fleeger, 1995).

6.13 Experimental techniques

Community attributes can be correlated with natural and anthropogenic variables in the field. With careful sampling designs, strong evidence can be accumulated as to which environmental variables appear to affect community structure most. Such studies, however, cannot prove cause and effect. For this, experiments are required in which the effects of an individual factor on community structure are investigated, while other factors are held constant or controlled. Field manipulative experiments include, for example, caging experiments to exclude or include predators, or the controlled addition of contaminants to experimental plots. Sediments within the cages or plots are sampled using the appropriate techniques and are treated in the same way as any other samples.

Owing to their small size and life-history characteristics, several meiofaunal taxa, but particularly nematodes and copepods, are useful groups with which to conduct ecological experiments in the laboratory. Two types of experimental set-up are common. In the first, organisms are kept in stock culture and added to experimental treatments as required. This type of approach is generally used to study species in isolation or specific combinations of species. As mentioned above, only certain groups of species are amenable to culture and thus such experiments are of limited use for the study of the majority of ecological questions, although they are very useful for addressing specific hypotheses.

In a more realistic and ecologically relevant set-up, whole meiofaunal communities may be collected in the field and then added to experimental chambers, either within the sediment in which they were collected or else having been extracted alive from their original sediment. Austen et al. (1994) describe a simple 'microcosm' system consisting of 570 ml glass bottles, each stoppered with a rubber bung pierced by two holes, through one of which an air-stone diffuser can aerate the liquid within the bottle. Each bottle contains 80–200 g of sediment (depending on sediment type) with meiofauna and 1 μm-filtered seawater adjusted to an appropriate salinity. Experiments are generally run in the dark, initially at a temperature equivalent to field temperatures on the day of collection, and then raised by 1–2°C per day to a final temperature of 20°C. The experimental temperature is chosen to be higher than field temperatures, stimulating and optimising conditions for growth and reproduction. Experiments are generally run for two months, at the end of which the sediment in each bottle is fixed, removed and treated as a single sample. The system is cheap to build and operate, but it does have some drawbacks. The sediment structure is disturbed while assembling the experimental units. Copepods and meiofaunal groups other than nematodes do not thrive in the system. While many nematode species survive in the system, some groups (deposit feeders) do better than others (microphytic grazers, predators). Thus, the 'naturalness' of the communities within the system is debatable. Despite this, many species do thrive, and presumably interact, feed and reproduce, so it is true to say that meiofaunal communities are maintained in the system. Such a set-up allows comparisons between treatments so that, although nematode communities are different from field communities, differences between groups of samples can be attributed to experimental treatments. This and similar systems have been used for a variety of experiments on the effects of contaminants and dredgings disposal (e.g. Austen & Somerfield, 1997; Schratzberger et al., 2000) and on the effects of physical disturbance, organic enrichment and predation (e.g. Schratzberger & Warwick, 1999a, b).

Larger systems in which meiofaunal communities may be maintained and manipulated are often referred to as 'mesocosms'. Experimental units generally consist of large buckets or boxes containing tens to hundreds of litres of sediment. Various mesocosm designs exist, but the commonest consist of large basins through which natural seawater is circulated, into which the experimental units are placed. Treatments may be allocated to individual containers, or combined within containers, and at the end of the experimental period sediments within the experimental units are sampled using the appropriate techniques and are treated just like any other samples. Mesocosm experiments have the potential advantages that disturbance can be reduced while transferring sediments from the field to the experimental units, the conditions in which meiofauna are maintained are comparatively natural, and the experimental units are large enough to allow interactions between meiofauna and macrofauna to be studied (e.g. Austen et al., 1998).

References

Anderson, R.O. (1959) A modified flotation technique for sorting bottom fauna samples. *Limnology and Oceanography*, **4**, 223–225.

Austen, M.C. & Somerfield, P.J. (1997) A community level sediment bioassay applied to an estuarine heavy metal gradient. *Marine Environmental Research*, **43**, 315–328.

Austen, M.C., McEvoy, A.J. & Warwick, R.M. (1994) The specificity of meiobenthic community responses to different pollutants: results from the microcosm experiments. *Marine Pollution Bulletin*, **28**, 557–563.

Austen, M.C., Widdicombe, S. & Villano, N. (1998) Effect of biological disturbance on diversity and structure of a nematode meiobenthic community. *Marine Ecology – Progress Series*, **174**, 233–246.

Barnett, B.E. (1980) A physico-chemical method for the extraction of marine and estuarine benthos from clays and resistant muds. *Journal of the Marine Biological Association of the United Kingdom*, **60**, 255.

Barnett, P.R.O., Watson, J. & Connelly, D. (1984) A multiple corer for taking virtually undisturbed samples from shelf, bathyal and abyssal sediments. *Oceanologica Acta*, **7**, 399–408.

Banse, K. & Mosher, S. (1980) Adult body mass and annual production/biomass relationships of field populations. *Ecological Monographs*, **50**, 355–379.

Bell, S.S. & Hicks, G.R.F. (1991) Marine landscapes and faunal recruitment: a field test with seagrasses and copepods. *Marine Ecology – Progress Series*, **73**, 61–68.

Bett, B.J., Vanreusel, A., Vincx, M., Soltwedel, T., Pfannkuche, O., Lambshead, P.J.D., Gooday, A.J., Ferrero, T. & Dinet, A. (1994) Sampler bias in the quantitative study of deep-sea meiobenthos. *Marine Ecology – Progress Series*, **104**, 197–203.

Bin Sun, J.W. & Fleeger, J.W. (1995) Sustained mass culture of *Amphiascoides atopus*, a marine harpacticoid copepod, in a recirculating system. *Aquaculture* **136**, 313–321.

Blomqvist, S. (1991) Quantitative sampling of soft-bottom sediments: problems and solutions. *Marine Ecology – Progress Series*, **72**, 295–304.

Boisseau, J.P. (1957) Technique pour l'étude quantitative de la faune interstitielle des sables. *Comptes Rendus du Congrés des Sociétés Savantes de Paris et des Départements*, **1957**, 117–119.

Braeckman, B.P., Houthoofd, K., De Vreese, A. & J.R. Vanfleteren (2002) Assaying metabolic activity in ageing *Caenorhabditis elegans. Mechanisms of Ageing and Development*, **123**, 105–119.

Buffan-Dubau, E. & Carman, K.R. (2000) Diel feeding behaviour of meiofauna and their relationships with microalgal resources. *Limnology and Oceanography*, **45**, 381–395.

Buffan-Dubau, E., de Wit, R. & Castel, J. (1996) Feeding selectivity of the harpacticoid copepod *Canuella perplexa* in benthic muddy environments demonstrated by HPLC analyses of chlorin and carotenoid pigments. *Marine Ecology Progress Series*, **137**, 71–82.

Butterworth, P.E., Perry, R.N. & Barrett, J. (1989) The effects of specific metabolic inhibitors on the energy metabolism of *Globodera rostochiensis* and *Panagrellus redivivus. Revue de Nématologie*, **12**, 63–67.

Carman, K.R., Dobbs, F.C. & Guckert, J.B. (1989) Comparison of three techniques for administering radiolabelled substrates to sediments for trophic studies: uptake of label by harpacticoid copepods. *Marine Biology*, **102**, 119–125.

Carman, K.R. & Fry, B. (2002) Small-sample methods for $\delta^{13}C$ and $\delta^{15}N$ analysis of the diets of marsh meiofaunal species using natural-abundance and tracer-addition isotope techniques. *Marine Ecology – Progress Series*, **240**, 85–92.

Chandler, G.T. (1986) High density culture of meiobenthic harpacticoid copepods within a muddy sediment substrate. *Canadian Journal of Fisheries and Aquatic Sciences*, **43**, 53–59.

Chandler, G.T. & Fleeger, J.W. (1983) Meiofaunal colonization of azoic estuarine sediment in Louisiana: mechanisms of dispersal. *Journal of Experimental Marine Biology and Ecology*, **69**, 175–188.

Cleven, E-J. (1999) An improved method of taking cores in sandy sediments. *Archiv für Hyrdobiologie*, **147**, 65–72.

Craib, J.S. (1965) A sampler for taking short undisturbed marine cores. *Journal du Conseil Permanent International pour l'Exploration de la Mer*, **30**, 34–39.

Cummings, E. & Ruber, E. (1987) Copepod colonization of natural and artificial substrates in a salt marsh pool. *Estuarine, Coastal and Shelf Science*, **25**, 637–645.

Danovaro, R., Dell'Anno, A., Martorano, D., Parodi, P., Marrale, N.D. & Fabiano, M. (1999) Seasonal variation in the biochemical composition of deep-sea nematodes: Bioenergetic and methodological considerations. *Marine Ecology – Progress Series*, **179**, 273–283.

Daro, M.H. (1978) A simplified ^{14}C method for grazing measurements on natural planktonic populations. *Helgoländer wissenschaftliche Meeresuntersuchungen*, **31**, 241–248.

Decho, A.W. (1988) How do harpacticoid grazing rates differ over a tidal cycle? Field verification using chlorophyll-pigment analyses. *Marine Ecology – Progress Series*, **45**, 263–270.

Decho, A.W. & Lopez, G.R. (1992) Exopolymer microenvironments of microbial flora: multiple and interactive effects on trophic relationships. *Limnology and Oceanography*, **38**, 1633–1645.

Decho, A.W. & Moriarty, D.J. (1990) Bacterial exopolymer utilization by a harpacticoid copepod: a methodology and results. *Limnology and Oceanography*, **35**, 1039–1049.

Delamare-Deboutteville, C. (1960) *Biologie des Eaux Souterraines Littorales et Continentales*. Hermann, Paris.
De Jonge, V.N. & Bouwman, L.A. (1977) A simple density separation technique for quantitative isolation of meiobenthos, using colloidal silica Ludox^{TM}. *Marine Biology*, **42**, 143–148.
De Niro, M.J. & Epstein, S. (1978) Influence of diet on distribution of carbon isotopes in animals. *Geochimica et Cosmochimica Acta*, **42**, 495–506.
De Niro, M.J. & Epstein, S. (1981) Influence of diet on the distribution of nitrogen isotopes in animals. *Geochimica et Cosmochimica Acta*, **45**, 341–351.
De Soyza, K. (1973) Energetics of *Aphelenchus avenae* in monoxenic culture. *Proceedings of the Helminthological Society of Washington*, **40**, 1–10.
Dillon, W.P. (1964) Flotation technique for separating fecal pellets and small marine organisms from sand. *Limnology and Oceanography*, **9**, 601–602.
Edgar, G.J. (1991) Artificial algae as habitats for mobile epifauna: factors affecting colonization in a Japanese *Sargassum* bed. *Hydrobiologia*, **226**, 111–118.
Elmgren, R. (1973) Methods of sampling sublittoral soft bottom meiofauna. *Oikos*, Supplement, **15**, 112–120.
Epstein, S.S. & Gallagher, E.D. (1992) Evidence for facilitation and inhibition of ciliate population growth by meiofauna and macrofauna on a temperate zone sandflat. *Journal of Experimental Marine Biology and Ecology*, **155**, 27–39.
Epstein, S.S. & Shiaris, M.P. (1992) Rates of microbenthic and meiobenthic bacterivory in a temperate muddy tidal flat community. *Applied and Environmental Microbiology*, **58**, 2426–2431.
Feller, R.J. & Warwick, R.M. (1988) Energetics. In: *Introduction to the Study of Meiofauna*. (Eds R. P. Higgins & H. Thiel), pp. 181–196. Smithsonian Institution Press, Washington, D.C.
Fenchel, T. (1967) The ecology of marine microbenthos. I. The quantitative importance of ciliates as compared with metazoans in various types of sediments. *Ophelia*, **4**, 121–137.
Fleeger J.W. & Gee, J.M. (1986) Does interference competition determine the vertical distribution of meiobenthic copepods? *Journal of Experimental Marine Biology and Ecology*, **95**, 173–181.
Fleeger, J.W., Thistle, D. & Thiel, H. (1988) Sampling equipment. In: *Introduction to the Study of Meiofauna* (Eds R. P. Higgins & H. Thiel), pp. 115–125. Smithsonian Institution Press, Washington, DC.
Frithsen, J.B., Rudnick, D.T. & Elmgren, R. (1973) A new flow-through corer for the quantitative sampling of surface sediments. *Hydrobiologia*, **99**, 75–79.
Fukusho, K. (1980) Mass production of a copepod *Tigriopus japonicus* in combination culture with a rotifer, *Brachionus olicatilis*, fed ω-yeast as a food source. *Bulletin of the Japanese Society of Scientific Fisheries*, **46**, 625–629.
Gee, J.M. & Warwick, R.M. (1996) A study of global biodiversity patterns in the marine motile fauna of hard substrata. *Journal of the Marine Biological Association of the United Kingdom*, **76**, 177–184.
Gerlach, S.A. (1971) On the importance of marine meiofauna for benthos communities. *Oecologia (Berlin)*, **6**, 176–190.
Gerritsen, J. & Strickler, J.R. (1977) Encounter probabilities and community structure in zooplankton: a mathematical model. *Journal of the Fisheries Research Board of Canada*, **34**, 73–82.

Giere, O. (1993) *Meiobenthology: The Microscopic Fauna in Aquatic Sediments*. Springer-Verlag, Berlin.

Gnaiger, E. & Bitterlich, G. (1984) Proximate biochemical composition and caloric content calculated from elemental CHN analysis: a stoichiometric concept. *Oecologia (Berlin)*, **62**, 289–298.

Goto, N., Kawamura, T., Mitamura, O. & Terai, H. (1999) Importance of extracellular organic carbon production in the total primary production by tidal-flat diatoms in comparison to phytoplankton. *Marine Ecology – Progress Series*, **190**, 289–295.

Gwyther, J. & Fairweather, P.G. (2002) Colonisation by epibionts and meiofauna of real and mimic pneumatophores in a cool temperate mangrove habitat. *Marine Ecology – Progress Series*, **229**, 137–149.

Hamels, I., Moens, T., Muylaert, K. & Vyverman, W. (2001a) Trophic interactions between ciliates and nematodes from an intertidal flat. *Aquatic Microbial Ecology*, **26**, 61–72.

Hamels, I., Muylaert, K., Casteleyn, G. & Vyverman, W. (2001b) Uncoupling of bacterial production and flagellate grazing in aquatic sediments: a case study from an intertidal flat. *Aquatic Microbial Ecology*, **25**, 31–42.

Heip, C., Herman, P.M.J. & Coomans, A. (1982) The productivity of marine meiobenthos. *Academiae Analecta*, **44**, 1–20.

Heip, C., Smol, N. & Hautekiet, W. (1974) A rapid method for extracting meiobenthic nematodes and copepods from mud and detritus. *Marine Biology*, **28**, 79–81.

Heip, C., Vincx, M. & Vranken, G. (1985) The ecology of marine nematodes. *Oceanography and Marine Biology: an Annual Review*, **23**, 399–485.

Heip, C., Willems, K.A. & Goossens, A. (1977) Vertical distribution of meiofauna and the efficiency of the van Veen Grab on sandy bottoms in Lake Grevelingen (The Netherlands). *Hydrobiological Bulletin (Amsterdam)*, **11**, 35–45.

Herman, P.M.J., Vranken, G. & Heip, C. (1984a) Problems in meiofauna energy-flow studies. *Hydrobiologia*, **118**, 21–28.

Herman, P.M.J., Heip, C. & Guillemijn, B. (1984b) Production of *Tachidius discipes* Giesbrecht 1881 (Copepoda: Harpacticoida). *Marine Ecology – Progress Series*, **17**, 271–278.

Hicks, G.R.F. & Coull, B.C. (1983) The ecology of marine meiobenthic harpacticoid copepods. *Oceanography and Marine Biology: an Annual Review*, **21**, 67–175.

Higgins, R.P. (1964) Three new kinorhynchs from the North Carolina Coast. *Bulletin of Marine Science of the Gulf and Caribbean*, **14**, 479–493.

Higgins, R.P. (1977) Two new species of Echinoderes (Kinorhyncha) from South Carolina. *Transactions of the American Microscopical Society*, **96**, 340–354.

Higgins, R.P. & Thiel, H. (Eds) (1988) *Introduction to the Study of Meiofauna*. Smithsonian Institution Press, Washington, DC.

Hockin, D.C. (1981) A simple elutriator for extracting meiofauna from sediment matrices. *Marine Ecology – Progress Series*, **4**, 241–242.

Humphreys, W.F. (1979) Production and respiration in animal populations. *Journal of Animal Ecology*, **48**, 427–453.

Jensen, P. (1982) A new meiofauna sample splitter. *Annales Zoologici Fennici*, **19**, 233–236.

Jensen, P. (1984) Measuring carbon content in nematodes. *Helgoländer Meeresuntersuchungen*, **38**, 83–86.

Joint, I.R., Gee, J.M. & Warwick, R.M. (1982) Determination of fine-scale vertical distribution of microbes and meiofauna in an intertidal sediment. *Marine Biology*, **72**, 157–164.

Jönsson, B. (1991) A ^{14}C-incubation technique for measuring microphytobenthic primary productivity in intact sediment cores. *Limnology and Oceanography*, **36**, 1485–1492.

Kinne, O. (1977a) *Marine Ecology, Volume III, Cultivation, Part 2*. John Wiley & Sons, Chichester, England.

Kinne, O. (1977b) Cultivation of animals: research cultivation, (7) Nematoda. In: *Marine Ecology, Volume III, Cultivation, Part 2* (Ed O. Kinne), pp. 691–709. John Wiley & Sons, Chichester, England.

Kramer, K.J.M., Brockmann, U.H. & Warwick, R.M. (1994) *Tidal Estuaries: Manual of Sampling and Analytical Procedures*. A. A. Balkema, Rotterdam.

Lydell, W.R.S. (1936) A new apparatus for separating insects and other arthropods from soil. *Annals of Applied Biology*, **23**, 862–879.

Lyman, F.E. (1943) A pre-impoundment bottom fauna study of Watts Bar Reservoir area (Tennessee). *Transactions of the American Fisheries Society*, **72**, 52–62.

Mare, M.F. (1942) A study of a marine benthic community with special reference to the micro-organisms. *Journal of the Marine Biological Association of the United Kingdom*, **25**, 517–554.

McIntyre, A.D. (1968) The meiofauna and microfauna of some tropical beaches. *Journal of Zoology*, **156**, 377–392.

McIntyre, A.D. (1971) Deficiency of gravity corers for sampling meiobenthos and sediments. *Nature*, **231**, 60.

McNeil, S. & Lawton, J.H. (1970) Annual production and respiration in animal populations. *Nature*, **225**, 472–474.

McLachlan, A., Dye, A.H. & Van Der Ryst, P. (1979) Vertical gradients in the fauna and oxidation of two exposed sandy beaches. *South African Journal of Zoology*, **14**, 43–47.

Middelburg, J.J., Barranguet, C., Boschker, H.T.S., Herman, P.M.J., Moens, T. & Heip, C.H.R. (2000) The fate of intertidal microphytobenthos carbon: an *in situ* ^{13}C-labeling study. *Limnology and Oceanography*, **45**, 1224–1234.

Moens, T., Herman, P.M.J., Verbeeck, L., Steyaert, M. & Vincx, M. (2000) Predation rates and prey selectivity in two predacious estuarine nematode species. *Marine Ecology – Progress Series*, **205**, 185–193.

Moens, T., Verbeeck, L. & Vincx, M. (1999a) Preservation- and incubation time-induced bias in tracer-aided grazing studies with meiofauna. *Marine Biology*, **133**, 69–77.

Moens, T., Verbeeck, L. & Vincx, M. (1999b) The feeding biology of a predatory and a facultatively predatory marine nematode (*Enoploides longispiculosus* and *Adoncholaimus fuscus*). *Marine Biology*, **134**, 585–593.

Moens, T., Vierstraete, A., Vanhove, S., Verbeke, M. & Vincx, M. (1996a) A handy method for measuring meiobenthic respiration. *Journal of Experimental Marine Biology and Ecology*, **197**, 177–190.

Moens, T., Vierstraete, A. & Vincx, M. (1996b) Life strategies in two bacterivorous marine nematodes: preliminary results. *Marine Ecology – Pubblicazioni della Stazione Zoologica di Napoli I*, **17**, 509–518.

Moens, T. & Vincx, M. (1998) On the cultivation of free-living marine and estuarine nematodes. *Helgoländer Meeresunters*, **52**, 115–139.

Montagna, P.A. (1983) Live controls for radioisotope tracer food chain experiments using meiofauna. *Marine Ecology – Progress Series*, **12**, 43–46.

Montagna, P.A. (1984) *In situ* measurement of meiobenthic grazing rates on sediment bacteria and edaphic diatoms. *Marine Ecology – Progress Series*, **18**, 119–130.

Montagna, P.A. (1993) Radioisotope technique to quantify *in situ* microbivory by meiofauna in sediments. In: *Aquatic Microbial Ecology* (Eds P.F. Kemp, B.F. Sherr, E.B. Sherr & J.J. Cole), pp. 745–753. Lewis Publishers, Boca Raton FL.

Montagna, P.A. & Bauer, J.E. (1988) Partitioning radiolabelled thymidine uptake by bacteria and meiofauna using metabolic blocks and poisons in benthic feeding studies. *Marine Biology*, **98**, 101–110.

Mortensen, R.H. (1925) An apparatus for catching the micro-fauna of the sea bottom. *Videnskabelige Meddelelser fra Dansk naturhistorik Forening i Kjøbenhavn*, **80**, 445–451.

Myers, R.F. (1967) Osmoregulation in *Panagrellus reduvivus* and *Aphelenchus avenae*. *Nematologica*, **12**, 579–586.

Ockelmann, K.W. (1964) An improved detritus-sledge for collecting meiobenthos. *Ophelia*, **1**, 217–222.

Pfannkuche, O. & Thiel, H. (1988) Sample processing. In: *Introduction to the Study of Meiofauna* (Eds R. P. Higgins & H. Thiel), pp. 134–145. Smithsonian Institution Press, Washington, DC.

Pollock, L.W. (1970) Distribution and dynamics of interstitial Tardigrada at Woods Hole, Massachusetts, US.. *Ophelia*, **7**, 145–166.

Rau, G.H., Mearns, A.J., Young, D.R., Olson, R.J., Schafer, H.A. & Kaplan, I.R. (1983) Animal $^{13}C/^{12}C$ correlates with trophic level in pelagic food webs. *Ecology*, **64**, 1314–1318.

Rees, C.B. (1940) A preliminary study of the ecology of a mud-flat. *Journal of the Marine Biological Association of the United Kingdom*, **24**, 185–199.

Renaud-Debyser, J. (1957) Description d'un carrotier adapte aux prelèvements des sables de plate. *Revue de l'Institut Français du Pétrole*, **12**, 501–502.

Renaud-Debyser, J. (1963) Recherches Ecologiques sur la fauna interstitielle des sables (Bassin d'Arcachon, Ile de Bimini, Bahamas). *Vie et Milieu*, Supplement, **15**, 1–157.

Sanders, R.W., Porter, K.G., Bennett, S.J. & DeBiase, A.E. (1989) Seasonal patterns of bacterivory by flagellates, ciliates, rotifers, and cladocerans in a freshwater planktonic community. *Limnology and Oceanography*, **34**, 673–687.

Santos, P.J.P., Castel, J. & Souza-Santos, L.P. (1995) Microphytobenthic patches and their influence on meiofaunal distribution. *Cahiers de Biologie Marine*, **36**, 133–139.

Schiemer, F. (1982) Food dependence and energetics of free-living nematodes. I. Respiration, growth and reproduction of *Caenorhabditis briggsae* (Nematoda) at different levels of food supply. *Oecologia (Berlin)*, **54**, 108–121.

Schiemer, F. (1985) Bioenergetic niche differentiation of aquatic invertebrates. *Verhandlungen der Internationalen Vereinigung für Theoretische und Angewandte Limnologie*, **22**, 3014–3018.

Schiemer, F. (1987) Nematoda. In: *Animal Energetics, Volume 1* (Eds J.F. Vernberg & T.J. Pandian), pp. 185–215. Academic Press, New York.

Schoener, A. (1974) Experimental zoogeography: colonization of marine mini-islands. *American Naturalist*, **108**, 715–738.

Schratzberger, M., Rees, H.L. & Boyd, S.E. (2000) Effects of simulated deposition of dredged material on structure of nematode assemblages – the role of contamination. *Marine Biology*, **137**, 613–622.

Schratzberger, M. & Warwick, R.M. (1999a) Differential effects of various types of disturbances on the structure of nematode assemblages: an experimental approach. *Marine Ecology – Progress Series*, **181**, 227–236.

Schratzberger, M. & Warwick, R.M. (1999b) Impact of predation and sediment disturbance by *Carcinus maenas* (L.) on free-living nematode community structure. *Journal of Experimental Marine Biology and Ecology*, **235**, 255–271.

Schwinghamer, P. (1981a) Extraction of living meiofauna from marine sediments by centrifugation in a silica sol-sorbitol mixture. *Canadian Journal of Fisheries and Aquatic Sciences*, **38**, 476–478.

Schwinghamer, P. (1981b) Characteristic size distributions of integral marine communities. *Canadian Journal of Fisheries and Aquatic Sciences*, **38**, 1255–1263.

Seinhorst, J.W. (1959) A rapid method for the transfer of nematodes from fixative to anhydrous glycerine. *Nematologica*, **4**, 67–69.

Sellmer, G.P. (1956) A method for the separation of small bivalve molluscs from sediments. *Ecology*, **37**, 206.

Sherr, E.B. & Sherr, B.F. (1993) Protistan grazing rates via uptake of fluorescently labeled prey. In: *Handbook of Methods in Aquatic Microbial Ecology* (Eds P.F. Kemp, B.F. Sherr, E.B. Sherr & J.J. Cole), pp. 695–701. Lewis Publishers, Boca Raton, FL.

Sikora, J.P., Sikora, W.B., Erkenbrecher, C.W. & Coull, B.C. (1977) Significance of ATP, carbon and caloric content of meiobentic nematodes in partitioning benthic biomass. *Marine Biology*, **44**, 7–14.

Smith, D.J. & Underwood, G.J.C. (1998) Exopolymer production by intertidal epipelic diatoms. *Limnology and Oceanography*, **43**, 1578–1591.

Somerfield, P.J. & Clarke, K.R. (1997) A comparison of some methods commonly used for the collection of sublittoral sediments and their associated fauna. *Marine Environmental Research*, **43**, 145–156.

Somerfield, P.J. & Jeal, F. (1996) The distribution of Halacaridae (Acari: Prostigmata) among macroalgae on sheltered rocky shores. *Journal of the Marine Biological Association of the United Kingdom*, **76**, 251–254.

Somerfield, P.J., Rees, H.L. & Warwick, R.M. (1995) Inter-relationships in community structure between shallow-water marine meiofauna and macrofauna in relation to dredgings disposal. *Marine Ecology – Progress Series*, **127**, 103–112.

Souza-Santos, L.P., Castel, J. & Santos, P.J.P. (1995) Feeding rate cycle of the epibenthic harpacticoid copepod *Harpacticus flexus*: laboratory experiments using fecal pellet counts. *Vie et Milieu*, **45**, 75–83.

Spoon, D.M. (1978) A new rotary microcompressor. *Transactions of the American Microscopical Society*, **97**, 412–416.

Starink, M. (1995) *Seasonal variation in bacterial production, protozoan grazing, and abundances in sediments of a freshwater littoral zone*. PhD thesis, Universiteit Amsterdam.

Starink, M., Krylova, I.N., Bargilissen, M.J., Bak, R.P.M. & Cappenberg, T.E. (1994) Rates of benthic protozoan grazing on free and attached sediment bacteria measured with fluorescently stained sediment. *Applied and Environmental Microbiology*, **60**, 2259–2264.

Tanner, C., Hawkes, M.W., Lebednik, P.A. & Duffield, E. (1977) Hand-operated suction sampler for collection of subtidal organisms. *Journal of the Fisheries Research Board of Canada*, **34**, 1031–1034.

Thiel, H. & Hessler, R. (1974) Ferngesteuertes Unterwasserfahrzeug erforscht Tiefseeboden. *Umschau in Wissenschaft und Technik*, **74**, 451–453.

Thiel, H. Von, Thistle, D. & Wilson, G.D. (1975) Ultrasonic treatment of sediment samples for more efficient sorting of meiofauna. *Limnology and Oceanography*, **20**, 472–473.

Thistle, D. (1978) Harpacticoid dispersion patterns: Implications for deep-sea diversity maintainence. *Journal of Marine Research*, **36**, 377–397.

Tiemann, H. & Betz, K-M. (1979) Elutriation: Theoretical considerations and methodological improvements. *Marine Ecology*, **1**, 277–281.

Tietjen, J.H. (1988) Culture techniques. In: *Introduction to the Study of Meiofauna* (Eds R.P. Higgins & H. Thiel), pp. 161–168. Smithsonian Institution Press, Washington, D.C.

Uhlig, G. (1966) Untersuchungen zur Extraktion der vagilen Mikrofauna aus marinen Sedimenten. *Zoologischer Anzeiger*, Supplement, **29**, 151–157.

Uhlig, G. (1968) Quantitative methods in the study of interstitial fauna. *Transactions of the American Microscopical Society*, **87**, 226–232.

Uhlig G. & Heimberg, S.H.H. (1981) A new versatile compression chamber for examination of living organisms. *Helgoländer Meeresuntersuchungen*, **34**, 251–256.

Uhlig. G., Thiel, H. & Gray, J.S. (1973) The quantitative separation of meiofauna: A comparison of methods. *Helgoländer wissenschaftliche Meeresuntersuchungen*, **25**, 173–195.

Vranken, G. & Heip, C. (1985) Aspects of the life-cycle of free-living marine nematodes. In: *Progress in Belgian Oceanographic Research* (Eds R. Van Grieken & R. Wollast),. pp. 267–278. Antwerp University Press, Antwerp.

Vranken, G. & Heip, C. (1986) The productivity of marine nematodes. *Ophelia*, **26**, 429–442.

Vranken, G., Herman, P.M.J., Vincx, M. & Heip, C. (1986) A re-evaluation of marine nematode productivity. *Hydrobiologia*, **135**, 193–196.

Warwick, R.M. (1984) Species size distributions in marine benthic communities. *Oecologia (Berlin)*, **61**, 32–41.

Warwick, R.M. (1989) The role of meiofauna in the marine ecosystem: evolutionary considerations. *Zoological Journal of the Linnean Society*, **6**, 229–241.

Warwick, R.M. & Gee, J.M. (1984) Community structure of estuarine meiobenthos. *Marine Ecology – Progress Series*, **18**, 97–111.

Warwick, R.M. & Price, R. (1979) Ecological and metabolic studies on free-living nematodes from an estuarine mudflat. *Estuarine, Coastal and Marine Science*, **9**, 257–271.

Waters, T.F. (1969) The turnover ratio in production ecology of freshwater invertebrates. *The American Naturalist*, **103**, 173–185.

Wieser, W. (1960) Benthic studies in Buzzards Bay. II. The meiofauna. *Limnology and Oceanography*, **5**, 121–137.

Woombs, M. & Laybourn-Parry, J. (1984) Feeding biology of *Diplogasteritus nudicapitatus* and *Rhabditis curvicaudata* (Nematoda) related to food concentration and temperature, in sewage treatment plants. *Oecologia (Berlin)*, **64**, 163–167.

Chapter 7
Deep-Sea Benthic Sampling

J.D. Gage and B.J. Bett

7.1 Introduction

Characteristics of deep-sea sampling gear

From the range of equipment that has been used to sample deep-sea benthos, some samplers have been widely adopted by the research community, while others have been designed and operated only by single workers or institutions. Here we provide an overview of this type of equipment, though this is not an exhaustive review because the development of deep-sea sampling equipment is an active and ongoing process, ranging from simple mechanical improvements to sediment corers to the latest in autonomous underwater vehicle technology.

Most deep-sea benthic sampling is still undertaken from ships. As pointed out by Eleftheriou and Holme (1984) in a previous edition of this handbook, successful sampling in the deep sea using a dredge, trawl or grab positioned at the end of several kilometres of wire is time-consuming and requires special skills, and failure rates can be high. Gear used to sample the deep-sea benthos is usually larger and heavier than equivalent apparatus used in shallow water, not only because larger research vessels are used in deep water beyond the continental shelf edge, but also because faunal densities may be much lower. For this reason, a substantial amount of sediment or area of seabed needs to be targeted in order to obtain a useful sample of the benthic fauna. The necessary winches with the accompanying required steel cable are heavy and can only be accommodated on a sizeable vessel. The layout of a typical medium-sized research ship equipped for benthic sampling is shown in Fig. 7.1.

Sampling programme design and the degree of replication needed in quantitative studies are not dealt with in this chapter, as these are issues which concern all benthic studies. There are, nevertheless, certain problems specific to deep-sea studies: (1) the greater difficulty of replicated sampling within a relatively small area of seabed when using a surface vessel in deep water, (2) the potentially lower densities and smaller body size of the deep-sea macrobenthos and (3) the high species richness and high degree of evenness with large numbers of rare species shown in abundance distributions of deep-sea samples. Consequent on the latter is the relative rarity of individual species. In the deep sea, more and larger samples may be required to

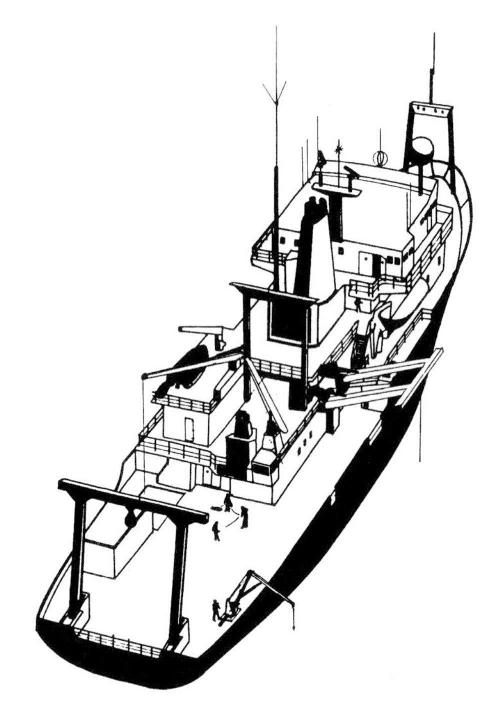

Fig. 7.1 View showing working deck area of a typical ocean-going research ship equipped for sampling the deep-sea benthos. The midships gantry is shown in position for a deployment of a bottom corer. This position is best in order to avoid the sometimes severe heave experienced at the rear end of the ship where the main gantry (or 'A' frame) is usually mounted. A small crane is shown in position for sampling at the surface.

sample the species present to the same degree as is possible in coastal waters. The best indication of this potential problem comes from the analysis of a large number of quantitative macrofauna samples taken off the north western United States along a 180-km transect at 2000 m depth on the continental slope (Grassle & Maciolek, 1992). Collectors' curves, plots of species accumulation against sample accumulation, drawn from these data show no sign of leveling off!

7.2 Benthic sampling from research ships

Deploying sampling gear on wire ropes

Sampling the deepest ocean trench would require at least 11 km of wire (cable or warp) in order to lower the gear vertically to the bottom, and, in addition, if the sampling gear were to be towed along the seabed, the amount of wire necessary to ensure that it reaches the bottom may be substantially greater than the water depth (e.g. ×1.5–3). The ratio of total wire paid-out to water depth, known as the scope, can vary with depth (see Fig. 7.2) as the weight of wire paid-out increasingly dominates the total payload. In practice, however, many deep-sea workers tend to

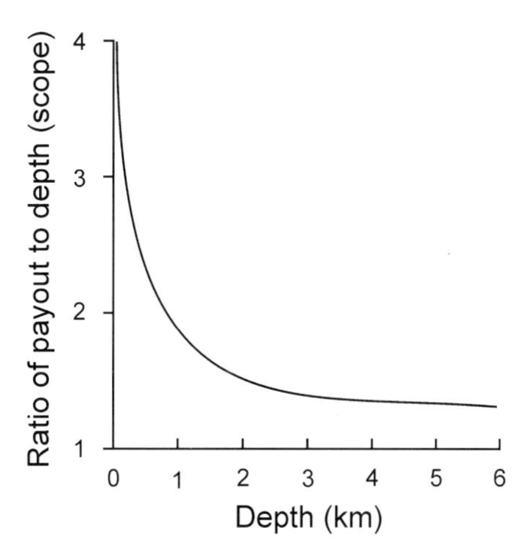

Fig. 7.2 Ratio of wire payout to depth ('scope'), with the ship steaming at 1 knot over the bottom, to ensure an anchor dredge bottoms. Because of its relatively small drag in relation to that from the surface area of the wire and its relatively great weight, the amount paid out needs to be more than twice bottom depth only for depths less than 1000 m.

employ a constant scope for most of their operations. Attempting to fish with a reduced scope usually leads to poor bottom contact and a tendency for the gear to 'skip' across the bottom. As the wire rope has an appreciable weight and drag, it is therefore important to choose wire of the correct gauge. Three types of wire are commonly available on ocean-going research ships equipped for benthic sampling: (1) a small gauge 'hydrographic' wire that may be suitable for the operation of lighter grabs and corers in relatively shallow waters; (2) a heavy gauge 'deep-sea coring' wire suitable for use with almost all vertically deployed deep-sea samplers to great depths (approx. 6000 m); and (3) a tapered trawling wire of mixed gauge and typically of great length (10–15 km) suitable for use with towed samplers to great depth (approx. 6000 m).

It is clearly important to be aware of the mechanical characteristics of any wire in order to avoid approaching the yield point where the wire is damaged by stretching, or the nominal breaking load (NBL) at which it may break. The point where the wire yields to stretching (and is thereafter unusable) is about 70% of the NBL. In current practice, the ratio of NBL to sustained load should not exceed 2.5:1, or 2:1 in the case of transient loads, such as those associated with the heave of the ship. The characteristics of the wires fitted to the British research ships RRS *Discovery* and RRS *Charles Darwin* are summarised in Table 7.1. These mechanical characteristics impose practical limits on the maximum depth of sampling. For example, a 1 tonne sampler may be deployed only to approx. 8000 m on an 18 tonne NBL wire, with allowance for the additional pull-out load and dynamic loading; as the ship heaves this depth needs to be reduced to around 6000–7000 m. The equivalent calculation for a towed sampler is more complex, but a similar limit (approx. 6000 m) is likely. Two other types of wire may be encountered in benthic sampling operations: (1)

Table 7.1 Characteristics of deep trawling and coring wires used for different purposes on research ships operated by the UK Natural Environment Research Council (RRS *Discovery* and RRS *Charles Darwin*).

Typical use	Cable length (m)	Nominal cable breaking load (tonnes)	Cable diameter (mm)	Cable weight in water (kg km^{-1})
Heavy coring	7000	18	16.7	780
Trawling, three part warp ('tapered wire')	8300 (outboard)	13	14.5	609
	4350 (middle)	18	16.4	780
	2350 (inboard)	21	18.0	1133
Hydrographic and light core/grab samplers		2	6.0	130

those made of Kevlar/Aramid-type synthetic materials and (2) armoured conducting or fibre-optic cables. The former may be required for the operation of particularly heavy pieces of equipment or where exceptional water depths are encountered. The latter types allow the use of equipment that requires electrical power to be supplied from the surface vessel and/or to establish direct two-way communication with the deployed instrument. A coaxial conducting cable of 11 km length (18 mm dia.) has been used in German studies on deep-sea fishing with an epibenthic sledge equipped with a television and telemetry systems (Christiansen & Nuppenau, 1997).

Most wires in use for oceanographic sampling are handed, i.e. they have a lay, and so have a natural tendency to twist and spin. It is therefore important to incorporate one or more swivels in the rigging of the equipment to be deployed. Another problem of wire in water is the effect of 'thrumming', i.e. vibrations set up by the passage of water over the wire (and gear). Thrumming can shake and damage equipment, often loosening the best-tightened bolts. When it is likely that the length of a wire will be towed over the seabed it is also common practice to insert a sacrificial pennant between the main wire and the gear. Another common technique employed with towed samplers is the use of a 'pig' or depressor weight. The pig, typically 100–500 kg, is let into the rigging at the inboard end of the pennant, and is used to reduce the scope necessary to fish the gear (and hence the total time required to complete the fishing operation). Whether a pig is used or not, determining the appropriate scope for any new sampler may be a matter of expensive trial and error. The application of real-time telemetry (acoustic or electro-optical) does, however, take much of the guesswork out of fishing operations in deep water.

Use of acoustic and other telemetry systems

Deep-sea sampling can be a very time-consuming business, and a single trawl in the abyss may occupy a research vessel for 16 hours or even more. The cost of a failed attempt is correspondingly high. Failures can be avoided, or aborted early if they do occur, through the use of telemetry from the deployed equipment. The simplest, and most commonly employed form of telemetry is the 'pinger', an acoustic beacon

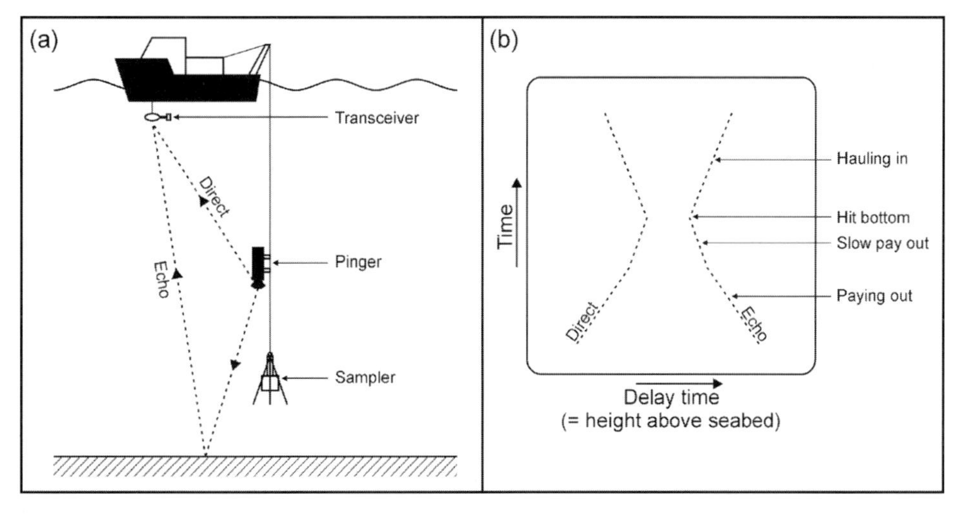

Fig. 7.3 Schematic representation of the operation of a pinger with a deep-sea sampler. (a) Direct and seabed reflected sound paths from pinger to ship-borne transceiver. (b) Visual display ('traces') seen by the operator.

that sends out a constantly repeating, accurately timed, sound pulse termed a 'ping'. The sound pulse is directed towards the seabed, where it reflects and is subsequently detected by the surface vessel. The ping also radiates directly back to the surface vessel. The time difference between the arrival of the direct ping and its echo from the seabed can then be used to calculate the distance of the pinger from the seabed. In practice, the repeating pings and their echoes are displayed visually on a thermal recorder or PC, allowing the operator to continuously monitor the progress of the sampler towards the seabed (see Fig. 7.3). This simple but effective technique is used with most vertically deployed deep-sea samplers.

Simple pingers are not particularly effective when used with towed samplers. Specialised acoustic telemetry systems, often referred to as monitors, are much more suited to such operations. An example of a simple but effective monitor is the otter trawl monitor developed by the Institute of Oceanographic Sciences (now at Southampton Oceanography Centre, SOC, UK). The monitor contains three mercury tilt switches and a calibrated pressure sensor. Data from these sensors are telemetered to the operator as a series of time-delayed pulses. This information is viewed in real time as a series of traces on a thermal recorder or PC display. A reference 'ping' is always present at a fixed repeating rate. Information from the tilt switches is displayed as traces that may be on or off (present or absent) at fixed time delays from the reference. Monitor depth, as pressure, is displayed as a trace of varying delay from the reference (see Fig. 7.4). This system has been successfully used with a semi-balloon otter trawl, a large beam trawl (the French *chalut à perche*) and the SOC epibenthic sledge. The SOC epibenthic sledge (see below) is normally operated with a more sophisticated monitor that also includes telemetry on net opening and closing, camera and video operation etc.

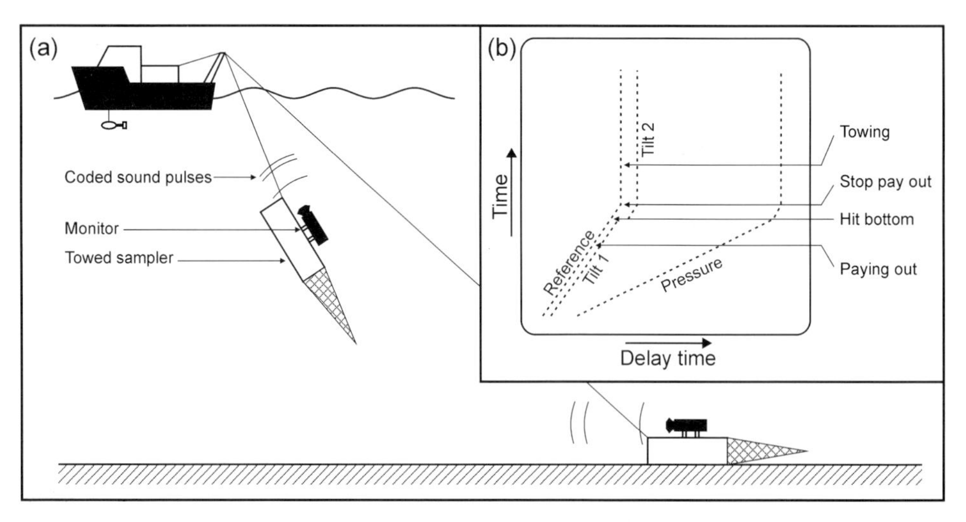

Fig. 7.4 Schematic representation of the operation of an acoustic telemetry system with a towed sampler (based on the SOC OTSB monitor). (a) Net orientation. (b) Visual display ('traces') seen by operator.

Almost any sensor type can be used with an acoustic telemetry system, though with visual display systems there is a limit to the number of traces that an operator can handle. Recent developments in acoustic signal processing now make direct reading of telemetry possible, enabling a greater number of sensors to be monitored simultaneously. With equipment deployed on conducting or fibre optic cables, signals from multiple sensors can be connected directly to ship-borne computer systems for storage and display.

A second major application for acoustic telemetry is the determination of the position of deployed equipment relative to the surface vessel or fixed points on the seabed. Long-baseline acoustic navigation has been available for deep-water operations for some time. Short-baseline systems suitable for operations to 2000 m are now commercially available and could readily be upgraded for deeper work. Long-baseline navigation requires the prior deployment of three or more acoustic transponders on recoverable moorings. Once deployed, the geographic positions of these transponders are determined by making multiple determinations of the relative distances between the surface vessel and the various transponders. Slant range from the ship to the transponder is determined from the time delay between an outgoing coded signal from the ship and the receipt of a reply signal from the transponder. An additional transponder is then fitted to the deployed sampling equipment, which can then range to the moored transponders enabling the position of the sampling equipment to be determined.

Short-baseline navigation operates in a similar manner, but is rather more convenient in that no moored transponders are required. A multi-headed hydrophone is fitted to the surface vessel that enables both slant range and bearing relative to the vessel to be calculated from the returning reply signal of the transponder. If the depth of the deployed equipment is also known, its position relative to the vessel can be calculated. Where the equipment is within a limited horizontal range from the vessel, e.g. in the case of vertically deployed samplers, the system can also calculate depth. In the case of towed samplers, an additional source of depth information is required; for example, this may be by acoustic telemetry from a pressure sensor built in to the transponder.

Trawls

Trawls do not produce quantitative samples despite the use of odometer wheels and other bottom tracking devices in some versions. A recent study involving the intensive use of otter trawling has shown that catches are sufficiently acceptable to enable monitoring of long-term change in the megabenthos (Billett et al., 2001). However, the catches are qualitatively biased and may give an extreme underestimate of the true faunal density (Bett et al., 2001). Further details to those given below may be found in Eleftheriou and Holme (1984). See also Chapter 5 in this volume.

Agassiz and beam trawls (Figs 7.5a and 7.5b) are, nevertheless, still important sampling devices for deep-sea work. The Agassiz, Sigsbee or Blake trawl shown in Fig. 7.5a (named after the naturalist, the commanding officer and the ship *Blake*, respectively) (Agassiz, 1888), was adapted from the gear once universally used by coastal fishermen. It is constructed with D-shaped runners connected across the 2–3 m width (the larger size is usual in deep-sea sampling) of the trawl by heavy rods or beams. The main net, consisting of 20 mm mesh, usually has a cod-end lined with finer meshed 'shrimp' netting (mesh about 10 mm). Protective chafing covers are sometimes rigged on either side to protect the bag from seabed abrasion on rough ground. These devices are principally used for collecting large numbers of benthic mega-epifauna, along with elements of the bentho-pelagic fauna including small demersal fish. Beam trawls also have their origin in coastal fisheries. Generally, such devices can fish only one way up, the buoyant timber beam ensuring correct orientation. French deep-sea biologists from IFREMER (Institut Français de Recherche pour l'Expoitation de la Mer), Brest, have used a commercial pattern, known as the *chalut à perche* (Fig. 7.5b), used in coastal fisheries. This has a beam 6 m wide, with a tapering bag with meshes from 20 mm, then 15 mm and finally 10 mm apertures in the cod-end, and a tickler chain attached along the lower side of the mouth. This gear is normally used with a heavy ball weight (750–1500 kg) secured to the wire 50 m in front of the trawl. Whether this weight will be on the seabed ahead of the trawl opening is arguable, and will certainly depend on its weight and wire payout, and on the drag of the trawl or sled at the end of the wire. Personal experience from watching video footage casts doubt on whether the wire or pig weight are necessarily on the bottom when towing gear with relatively high drag, as the ship's heave can be seen to be transmitted very directly to the gear – even when fishing the SOC epibenthic sledge at 5000 m. These trawls are

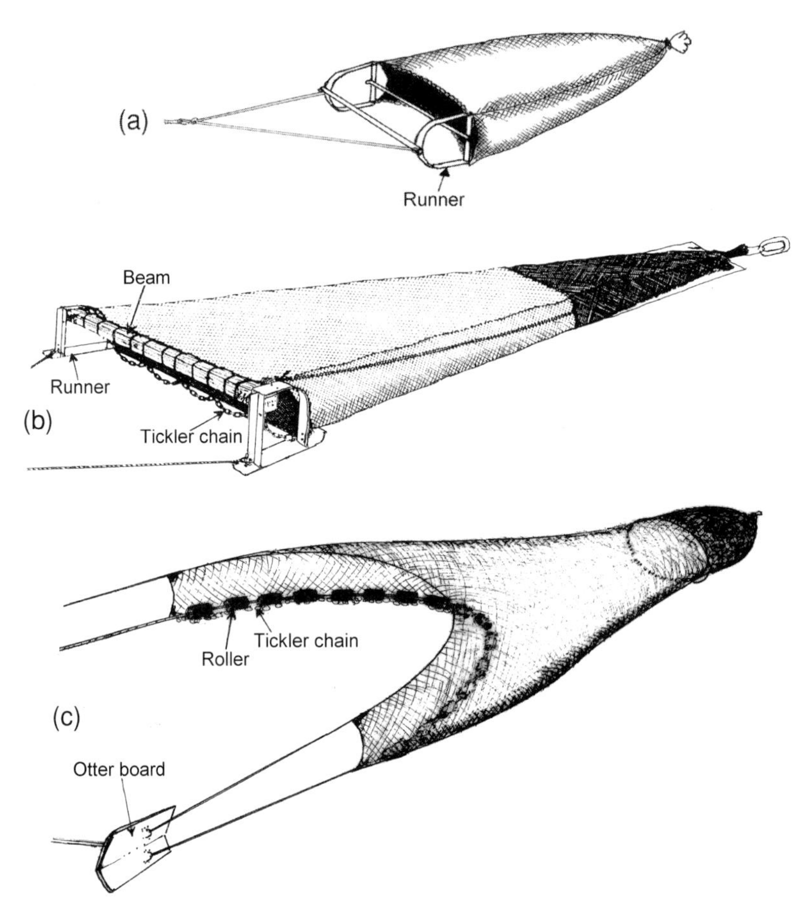

Fig. 7.5 Deep-sea trawls. (a) 3-m wide Agassiz trawl showing D-shaped runners; (b) the 6-m wide French *chalut à peche*, as used by IFREMER, showing timber beam, steel runners, and tickler chain; (c) semi-balloon otter trawl (OTSB) showing otter boards, tickler chain rigged to the footrope, and the rollers to help avoid the trawl biting into the sediment.

normally rigged with a safety strop so that if the trawl comes fast, a rated safety pin is broken between the ends of the wire connecting the paired warps attached to the D-runners, thus transferring the strain to the cod-end. This generally ensures that the gear breaks free and can be recovered intact, even if the catch may be lost.

Other than the large *chalut à perche*, such gears are relatively easy to handle from the ship, and can be operated in relatively heavy weather when the use of other sampling equipment might not be possible. In deep water, neither gear has much drag on descent in relation to that of the wire and the empirical curve provided by Carey and Hancock (1965) for dredging, and may be used to approximate the final payout necessary in relation to depth. The mouth of the trawl should be rigged with chain links lashed along the mouth edges to ensure the leading edge will run over the sediment surface and take up animals. If sufficient payout results in a certain amount of wire moving over the bottom in front of the gear, it would probably create clouds of sediment and mobilise animals that may, or may not, be more easily caught by the trawl opening when it passes later. However, the degree to which this actually occurs is arguable, and needs investigating. Personal experience in the Rockall Trough indicates that repeatable catches are obtained using the Agassiz trawl in deep water when fished over the same ground regardless of weather conditions, so long as the towing speeds, wire payout and distance fished over the ground are roughly equivalent.

Large commercial *otter trawls* are seldom used to sample the megabenthos and demersal fish, because to operate such gear specialist expertise and twin warps are necessary. Most research ships will only have one drum carrying enough wire for such heavy gear. However, as commercial boats fish at even greater depths (depths to 1500 m and deeper are routinely fished in several parts of the world – see Merrett and Haedrich (1997) for a general discussion of deep-sea trawling, including methodology) it is possible that material from these hauls, or even the use of the trawlers themselves, will be available to researchers. Large commercial-type twin-warp otter trawls have been successfully used in deep water by research ships targeting demersal fish (Gordon & Duncan, 1987). Such trawls are usually very successful in obtaining large quantities of benthic megafauna as a by-catch.

Several types of smaller otter trawls have been used to collect the larger bottom fauna along with fish. Those otter trawls that can be fished using a single warp can use more or less the same deck arrangements as might be used for a frame trawl, such as the Agassiz, or epibenthic sled. However, the fish catch from twin warps is different from that obtained when using a single warp (Gordon et al., 1996). The catch from twin warps has been shown to be similar to that obtained from the much larger Granton trawl, both these devices being much more effective than the single warp trawl in catching large, motile species (Gordon & Bergstad, 1992). It is not clear whether this affects the invertebrate catch. However, these authors also show that a single warp trawl is more effective in catching synaphobranchid eels. These normally burrow in sediment and may be disturbed by the single warp if excessive payout causes the towing wire to pass over the ground in front of the trawl. Although not so far demonstrated, it is possible that motile benthos known to shelter in burrows, such as decapod crustaceans, might be similarly affected. However, it is known that the spread of the twin warps of larger trawls acts to drive the more motile fish encountered towards the doors and mouth area. This means that comparisons of fish catch between surveys are often difficult to carry out unless very similar gear is used, and it is fished in the same way.

Of the otter trawls used on a single warp, the small Marinovitch semi-balloon otter trawl (OTSB) (Merrett & Marshall, 1981; Gordon, 1986) has proved to be by far the most popular (Fig. 7.5c). This device is based on a commercial trawl developed for the shrimp fishery in the Gulf of Mexico, and has fine mesh netting throughout the bag. An earlier design, the single warp commercial otter trawl (SWT), used in deep-water programmes in the UK had wide meshes but was fitted with a fine-mesh cod-end (Gordon & Duncan, 1985). The OTSB has a headline width of 14.7 m and has been fished down to depths greater than 5000 m. The progress of the trawl during a tow can be monitored with an acoustic telemetry system (see above) mounted in one of the trawl doors. The OTSB can be fished using either single or paired warps. For the most effective sampling of benthic invertebrates, the single warp seems to be preferable.

Special cod-end devices have been developed which protect the catch by insulating it from thermal shock during recovery through the water column (e.g. Childress, 1985). These are important in physiological studies, as animals will arrive at the surface less stressed than otherwise, and therefore more suitable for physiological experiments, or attempts to culture them in aquaria. Although the temperature range encountered may seem small, most deep-sea animals are very intolerant of temperature changes, usually much more so than changes in hydrostatic pressure (although fishes with swim bladders provide an exception to this).

Epibenthic sleds

The *WHOI Epibenthic Sled* was developed at the Woods Hole Oceanographic Institution (WHOI) in the 1960s, and was the first epibenthic sled to be used in the deep sea (Hessler & Sanders, 1967). Both this sled and its later modifications at the UK Institute of Oceanographic Sciences and now SOC were aimed at catching the smaller megabenthos/larger macrobenthos of the deep-sea floor. The WHOI epibenthic sled (Fig. 7.6a), although aimed more at the fauna associated with the sediment–water interface, is a useful device for sampling the fauna associated with the most superficial layer of sediment. It consists of a fine monofilament nylon mesh bag, resembling a plankton net, mounted inside a steel frame attached to wide runners to allow it to skim over, rather than to sink into, the soft sediment. The runners are symmetrical at the top and bottom so that the sled can work either way up. The collecting net is lashed onto a rectangular, box-like mouth area whose opening measures 81 cm in width and 30 cm height. The upper and lower panels of the box are formed by 'cutting' plates whose frontal edge may be adjusted up or down. The meshes of the main bag are usually 1 mm^2 and the bag is protected within the steel framework by removable panels of steel mesh. It was found that in order to avoid loss of samples from the bag during recovery to the surface, an extension cod end, about 1.3 m long, was required. This is typically fabricated from finer mesh netting (0.3 to 0.5 mm meshes), and is protected by canvas aprons lashed to the afterwidth of the sled framework.

The mouth edges of the WHOI epibenthic sled are designed such that the horizontal cutting blades will cut off a strip of the topmost layer of seabed. Sea-bed observations by Gage (1975) established that this causes clogging of the mouth area (Fig. 7.6b), with the gear obtaining only a small catch that was quickly washed out during recovery. In other words, with the cutting blades set in this way the gear was behaving like a dredge. Gage (1975) showed that the catch obtained from an inshore station was indistinguishable from that obtained from an anchor box dredge (see

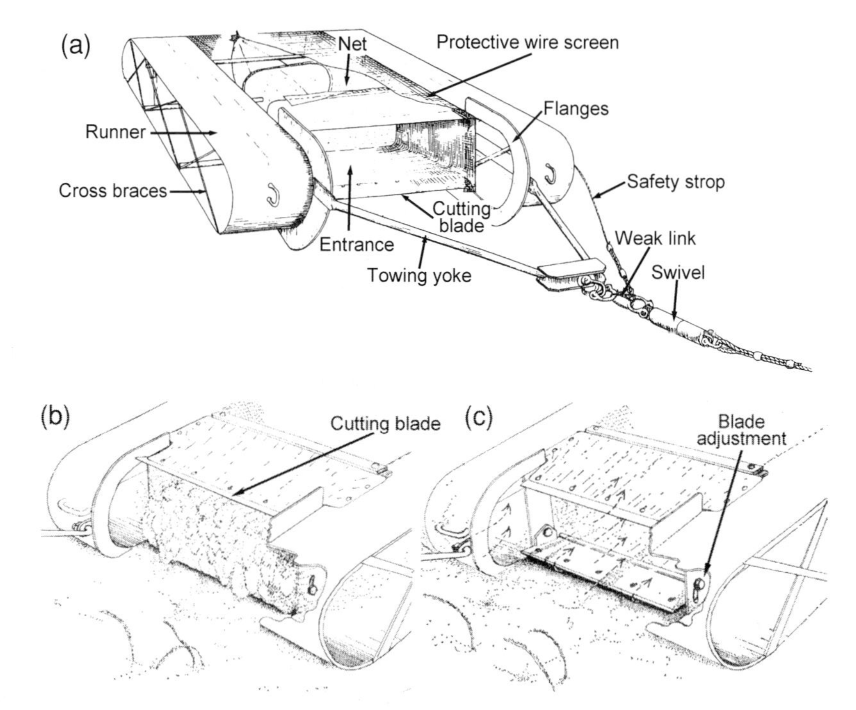

Fig. 7.6 Woods Hole (WHOI) pattern epibenthic sled. (a) Epibenthic sled rigged with main bag extension. (b) Sampling behaviour of sled with mouth 'cutting blades' set as shown in (a). (c) Sampling behaviour when 'cutting blades' are raised.

below) or van Veen grab. If the blade angle is set parallel with, or angled to point slightly above the seabed (Fig. 7.6c), then noticeably different, and much larger, samples are obtained, rich in elements of the smaller macrobenthos associated with the sediment–water interface. Diving observations established that in the latter position, sediment (presumably including fauna) passes up and into the mouth from just below the area of the blade where the sediment is disturbed by a downward pressure wave. Other designs of epibenthic sledge that specifically target the hyper- or supra-benthos, incorporate an angled surface below the mouth opening that has a similar effect (see below). If in the deep sea, part of the wire paid out were to be pulled over the seabed ahead of the gear, clouds of disturbed sediment, along with small benthic animals, would also enter the mouth (although active species, such as natant decapods and fish may well move away). The mesh bag was designed to filter off fine sediment, but, in practice, the cod-end section typically fills with a muddy mixture of fauna and sediment particles that then require careful washing. The WHOI sled has a timer-released metal plate gate that closes the mouth after sufficient time has elapsed for about an hour's tow. A somewhat similar system was developed by IFREMER for use with this gear with the gate being opened by increasing hydrostatic pressure during the lowering to the seabed, and closed by a decrease in pressure during recovery (Guennegan & Martin, 1985). Closure of the gate protects the sample from contamination by planktonic organisms during the slow winching back of the gear (the relatively small drag in relation to its weight ensures that the gear does not fish during lowering to the bottom). This minimises winnowing, and consequent loss of light-bodied fauna, through the open mouth during recovery. The IFREMER version of the epibenthic sled was also fitted with an odometer wheel within one of the runners, so that it would run whichever way up the gear operates on the seabed.

The epibenthic sled ushered in a new era of deep-sea benthic biology by revealing an unexpectedly species-rich deep-water fauna. However, the WHOI sled has operational limitations, such as a tendency for substantial variability between hauls taken at differing speeds over the ground at the same station (Gage et al., 1980).

SOC Epibenthic Sledge was designed in the 1970s by the biology group at the UK Institute of Oceanographic Sciences with the Acoustically Monitored Epibenthic Sledge (Aldred et al., 1976; Rice et al., 1987), in order to determine the speed and distance the sled actually runs over the bottom, and to image the bed photographically before it is sampled. This is a much larger apparatus than the WHOI epibenthic sled, and consists of a steel frame fitted with broad, weighted skids to ensure that the gear arrives at the seabed the correct way up. The mouth measures 2.29 m by 0.61 m and is equipped with a mechanism that can both open and close the mouth. The main net consists of 4.5 mm terylene mesh with a 1.5 m long cod-end of 1 mm monofilament mesh, like that of the WHOI sled. Mounted on the top of the sledge are pressure housings containing a camera and flash that look obliquely downwards and forwards over the bottom, ahead of the mouth. The sledge is fished using an acoustic telemetry system (see above).

As operated today, the SOC sledge system has a number of variants, e.g. (a) replacing the single net with three separate nets (Fig. 7.7), the outer ones having 4.5 mm mesh nets and the centre one a 1 mm mesh net; (b) an additional small 'suprabenthic' net with 0.33 mm meshes for the collection of near-bottom plankton, or hyperbenthos; and most recently, (c) the addition of a self-contained digital video camera system.

Comparison of the bottom area imaged with the final catch shows that the gear often misses quite large epifauna, which nevertheless are clearly visible in the sampling swathe of the sledge. Furthermore, the performance varies from haul to haul (Rice et al., 1979; Rice et al., 1982). Thus, even with this level of sophistication, the sampling performance of the gear is subject to bias, and at best only provides semi-quantitative estimates of the benthic community structure and standing crop. Furthermore, unreliability with reception of the pinger signal monitoring sledge bottom contact and behaviour can sometimes make fishing in deep water uncertain. Real-time visual imagery, using a conducting cable, has been added to a modified version of the SOC sledge as used by the Institut für Hydrobiologie und Fishereiwissenschaft (Hamburg) in order to help overcome such uncertainties, and to provide additional information on density and behaviour of large epibenthic and benthopelagic organisms (Christiansen & Nuppenau, 1997).

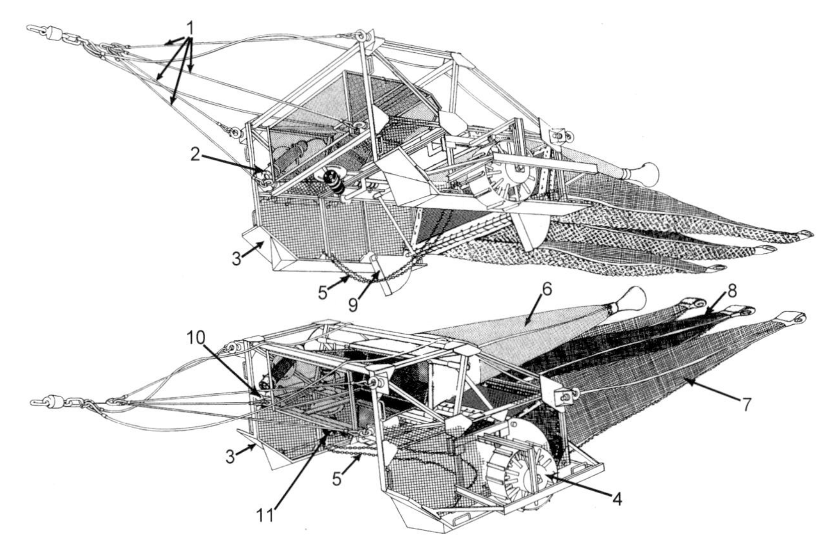

Fig. 7.7 SOC pattern epibenthic sledge. The version illustrated is equipped with three main nets and a 'suprabenthic' net. 1. Towing bridles; 2. Still camera; 3. Skid; 4. Odometer wheel; 5. Tickler chain; 6. Suprabenthic net; 7. Coarse mesh net; 8. Fine mesh net; 9. Quadrant (open and close net); 10. Acoustic monitor; 11. Flash gun.

The deep-sea suprabenthos community consists of motile animals living immediately above the sea bottom. It may include organisms, such as cumaceans, that are able to swim off the bottom, but which normally feed at the sediment interface. The SOC epibenthic sled is designed to catch the more planktonic members of this community in the 'suprabenthic' net mounted above the level of the main nets (see Fig. 7.7). There have been many devices specifically targeting the suprabenthic fauna in shallow water (reviewed in Brattegard & Fosså, 1991), but few have been used in the deep sea.

One device, the *epibenthic sampler* of Rothlisberg and Pearcy (1976), as modified by Torleiv Brattegard (Fig. 7.8) for use on the deep continental shelf and slope (Buhl-Jensen, 1986; Brattegard & Fosså, 1991) provides repeatable samples of the suprabenthos. Brattegard and Fosså (1991) also fitted a reinforced rubber chafing mat, weighted at its rear end, and attached to the rear frame to protect the sampling net (0.5 mm meshes) from damage from rough seabeds. The Rothlisberg and Pearcy (RP) sledge samples about 26–59 cm above the bottom. But because of the upturned runners that extend across the lower width of the frame, the downward pressure creates turbulence as the gear is towed over the bottom so that the sampler, like the WHOI epibenthic sled, actually samples a mixture of superficial fine sediment and near-bottom water. A hinged door equipped with a projecting shoe forces the door open when on the sea bottom, while springs close the door when it is off the bottom in order to prevent contamination of the sample from pelagic fauna.

It is not known to what extent the catch from the RP sampler is similar to that from a WHOI-type epibenthic sled with cutting blades set not to dredge (see above), but

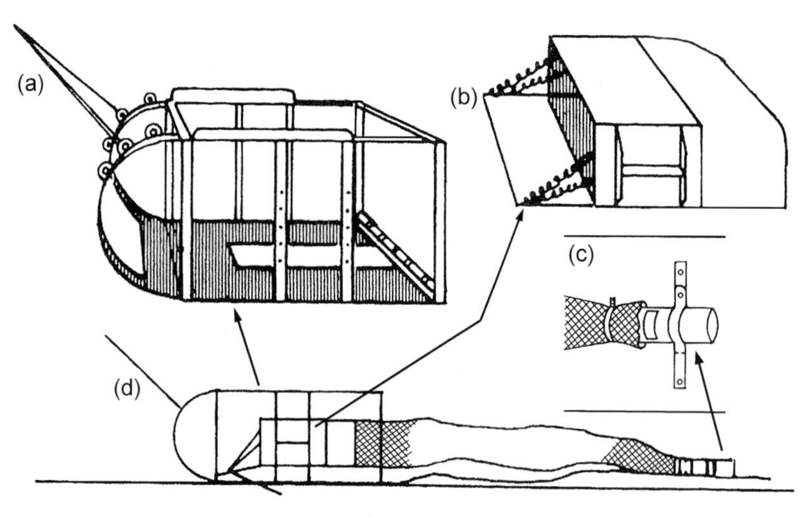

Fig. 7.8 Modified Rothlisberg-Pearcy (RP) epibenthic sled sampler. (a) Sled frame, with rubber chafing apron attached to the rear lower frame; (b) sampling box with door open; (c) cod-end with collector pot strapped to the rubber chafing apron; (d) side view of assembled sled at the bottom, with the door held open by friction with the bottom. After Buhl-Jensen (1986).

it seems likely that they would be quite similar. If this is true, then the catch from the various epibenthic sampling devices reviewed above will differ mainly with respect to the extent they disturb the superficial sediment and avoid any dredging effect.

Comparisons between sledge and trawl samplers

An intercalibration exercise ('INCAL') was undertaken in the north east Atlantic on the French research ship *Jean Charcot* in 1976 involving two versions of the WHOI-type epibenthic sled (one rigged with a 0.5 mm mesh extension bag as used at Oban, and the other with a 1 mm bag as used by IFREMER), the SOC epibenthic sledge rigged with a single net, and the *chalut à perche* beam trawl as used by IFREMER. The intercalibration also used a Reineck box corer, and involved replicate sampling with each of these five sampling gears. The results, expressed as the percentage representation of 15 major taxa in each sample from a station at 4300 m depth on the Meriadzek Terrace are summarised in Fig. 7.9. These, and the results of the ANOVA in Table 7.2, indicate that the catches from all the towed samplers were rather similar, although the macrofaunal catch in replicates from individual gears was more similar than the catch between different gears. Variability within replicates from individual gear, however, did vary, with homogeneity greatest with the epibenthic sledge samples and least with the *chalut à perche*. Differences between gears arising from the degree to which the sediment is disturbed before it enters the net, and perhaps the extent to which the mouth entrance of the gear follows, or slices across, uneven ground might also cause the minor differences between gear observed.

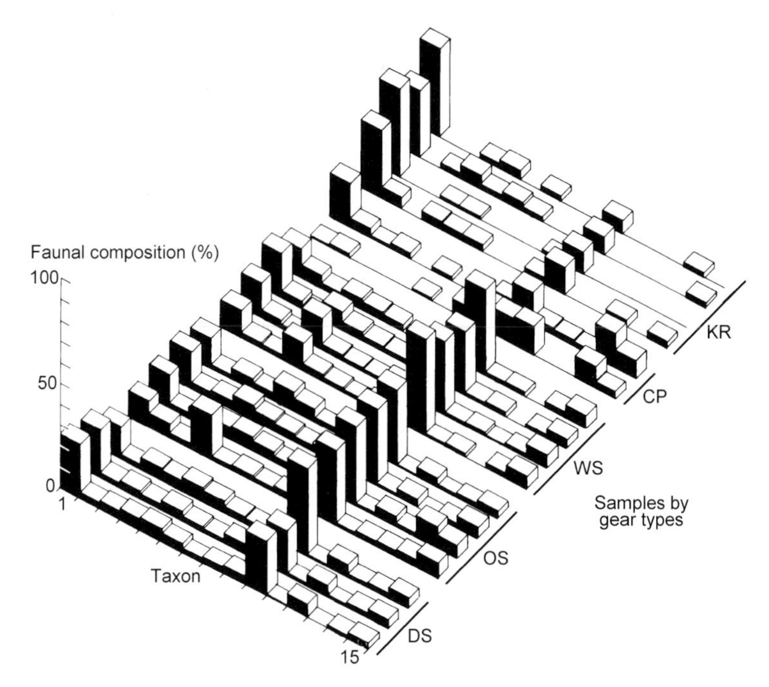

Fig. 7.9 Percentage representation of 15 major macrofaunal taxa in samples from the Meriadzek Terrace. DS, Epibenthic sled with 1-mm meshed bag as operated by IFREMER; OS, epibenthic sled with 0.5 mm extension bag; WS, single-net SOC epibenthic sledge; CP, chalut à perche; KR, Reineck box corer.

Dredges and semi-quantitative sampling

Dredges were used in some of the earliest attempts to sample life on the deep-sea floor (and have provided a starting point for some attempts at quantification – see below), and were simple modifications of seabed sampling gear already in use by fishermen.

Naturalist's dredge

The *naturalist's dredge*, or Ball's dredge, as used on the pioneering deep-sea sampling cruises of HMS *Lightning* and HMS *Porcupine* is an example of a simple modification of existing gear used by fishermen. Dredges were originally towed using hemp ropes, and this practice remained standard up to the voyage of HMS *Challenger*. Wire ropes became standard for deep-sea benthic sampling after the voyages in the Caribbean and Pacific by Alexander Agassiz on the US Coastguard steamer *Blake*. Variants of the Ball's dredge, rigged using canvas protective screens over a frame covering the main bag, and with 'tangles' of long rope tails attached to the after end to catch up surface-living epifauna are illustrated in Figs 7.10a and 7.10b. Long rope tangle tails were also effective on their own attached to an iron bar (Agassiz, 1888), and are sometimes still used to catch epifauna.

Table 7.2 Contingency table analysis of percentage representation of major macrofaunal taxa in the replicates from three different versions of epibenthic sledge deployed on the *INCAL* cruise (see text). Two of these (DS and OS) were based on the Woods Hole Epibenthic Sledge, but had differing lengths of collecting bag, while the other (WS) DS was the larger SOC Epibenthic Sledge (see text for further details). Raw numbers of individuals belonging to 13 of the 15 major taxa (the remaining two were too poorly represented) were standardized to the number expected in a single 1-km haul on the seabed. Data from hauls (identified as numerals) from each gear are set out as contingency tables for numbers of each taxon found in each of the replicates (columns). The null hypothesis (H_0) is that the proportions for each taxon are equivalent.

Gear Taxon (rows)	DS $\Sigma\chi^2$	OS $\Sigma\chi^2$	WS $\Sigma\chi^2$
Polychaeta	123.26	60.88	85.54
Sipuncula	22.36	170.42	27.09
Cumacea	5.12	17.84	3.77
Tanaidacea	5.06	376.57	77.90
Isopoda	1.28	232.37	9.97
Amphipoda	5.13	103.41	2.01
Aplacophora	4.01	7.87	3.42
Gastropoda	16.63	100.87	12.64
Scaphopoda	12.00	73.48	16.69
Bivalvia	75.33	7.01	23.33
Ophiuroidea	28.48	13.98	4.25
Holothuroidea	11.78	974.33	3.14
Ascidiacea	6.05	82.17	10.91

Sample (columns)	$\Sigma\chi^2$	Sample	$\Sigma\chi^2$	Sample	$\Sigma\chi^2$
DS 14	99.37	OS 05	338.36	WS 07	55.51
DS 15	135.71	OS 06	888.70	WS 08	43.04
DS 16	81.40	OS 07	408.32	WS 09	82.66
		OS 08	483.89	WS 10	99.33

	DS	OS	WS
Total χ^2	316.49	2221.20	280.66
df	24	36	36
Discrepancy from H_0	18.38	58.28	15.32

Anchor dredges

A variant of Holme's *anchor dredge* was used by H.L. Sanders for semi-quantitative deep-sea benthic sampling (Sanders et al., 1965). The sediment is collected in a canvas bag supported by an outer protective bag of coarse nylon mesh (Fig. 7.10c). The 'cod-end' of the bag is closed with a cord that is released after recovery to allow the sediment to be emptied out for washing. It is intended to work by cutting a short swathe of the upper layer of the sediment, which passes back into the bag (Fig. 7.10d). In the Anchor Box Dredge of Carey and Hancock (1965), the canvas collecting bag is replaced by a 57 cm wide steel box equipped with removable hardened steel cutting teeth (Fig. 7.10e). A loosely hinged V-shaped throat valve in the mouth of the dredge is designed to prevent contamination by pelagic organisms during lowering and recovery from the seabed. At the bottom, the valve swings up through pressure from entering sediment.

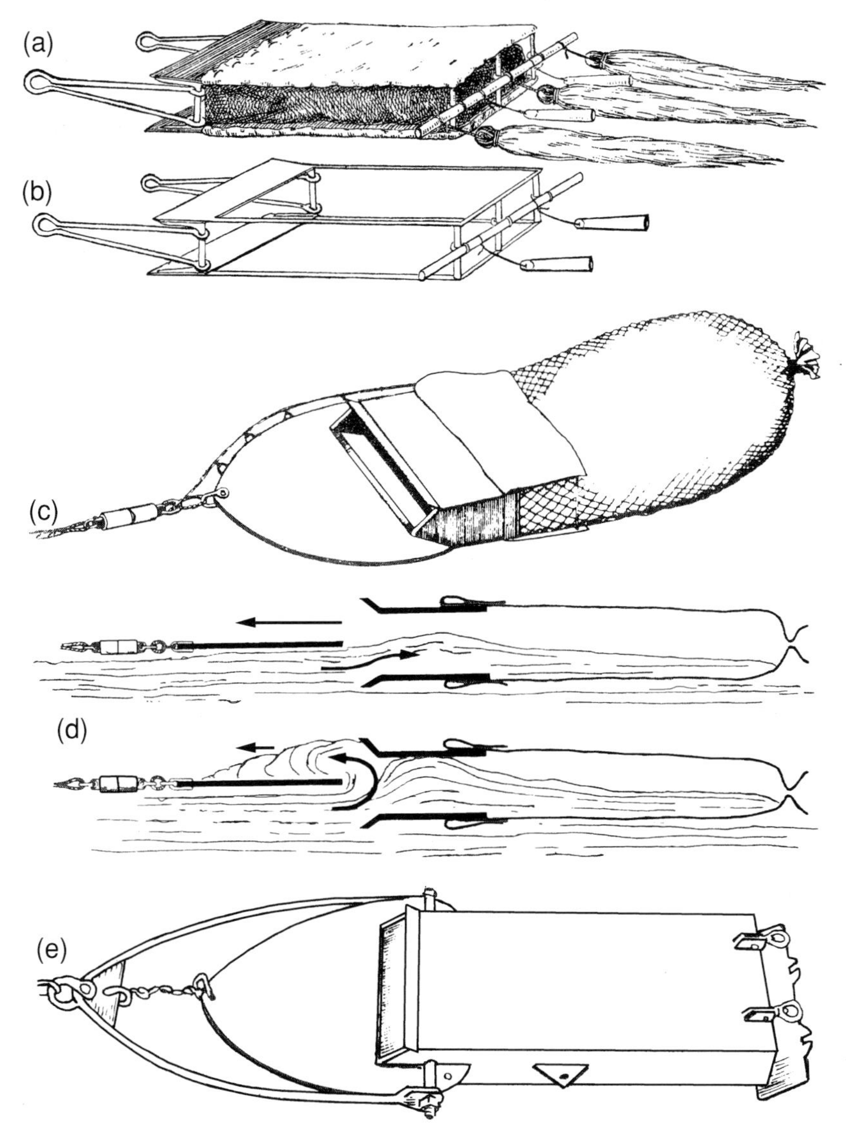

Fig. 7.10 Deep-sea dredges. The early version of the naturalists' dredge shown in (a) and as a frame only in (b), were the types used in the 19th century. The canvas covers protected the mesh bag, while the rope tangles attached to the after end were very successful in collecting spiny organisms in good condition. (c)–(e) Versions of the anchor dredge used in the deep sea. (c) The semi-quantitative anchor dredge used by H.L. Sanders during the 1960s and 1970s, while (d) shows the way the dredge strips off superficial sediment before the opening is blocked preventing more sediment from opening. (e) Anchor box dredge of Carey and Hancock. This dredge is equipped with a free-swivelling V-shaped throat valve; this hangs down blocking the mouth when the gear is raised, but is forced open by incoming sediment during sampling. The sample is removed through the hinged flap at the rear end.

In both Sanders' deep-sea anchor dredge and the anchor box dredge, a planing surface consisting of a sheet of steel plate welded to the front of the dredge prevents it from dipping too deep into the sediment. This, in theory, means that it can strip off the top 10 cm layer of sediment from an area up to 1.3 m^2. Estimates of the area sampled are made by dividing the volume of sediment obtained by the biting depth (10 cm). Diver observations of the Anchor Box Dredge operating on inshore soft mud indicates that only about half of this area is taken up before friction of sediment entering the box clogs the entrance causing further material to be rejected (Gage, 1975). This probably also applies to the operation of the anchor dredge of Sanders et al. (1965), although Gage (1975) observed that the anchor dredge of Sanders et al. would probably collect more sediment because there would be less frictional resistance, causing buckling of the sediment sheet to occur after more sediment had been collected.

Gage (1975) showed that samples taken by the anchor box dredge have fewer species than samples taken from the same sea bottom by careful hand coring by diving. This bias was almost entirely caused by a deficiency in bivalves and particularly crustaceans, which at the station would have been mainly peracarids, such as amphipods and cumaceans. Gage concluded that these animals probably escaped capture by swimming away. The lower-than-expected sampling for bivalves may simply have been the result of the deep burrowing habit of the dominant species. Observations by diving established that the sediment rejection process with the anchor box dredge starts to operate well before the box is completely full.

Quantitative sampling using grabs, box and tube corers

As the emphasis of deep-sea studies changed from descriptive studies of zoogeography and taxonomy to process-related work, there came a need for unbiased, quantitative samples of the composition, natural position and biogenic structures associated with the deep-sea benthos. A good discussion of some of the generic practical problems facing the benthic ecologist in sampling soft sediments is given by Blomqvist (1991). Although quantitative samplers have been available for studies in shallow water since the early part of the 20th century, it has often proved impracticable to use shallow-water gear in the deep sea, without scaling up the size. This is because their small size made them difficult to monitor on long payouts, and they sampled too small an area to collect a large enough sample of fauna at the low densities that often prevail in the deep sea.

Grabs

Petersen's grab and its variants

The first serious attempt at quantitative sampling in the deep sea was made by Spärck during the Galathea Expedition of 1950–1952. He used a version of the Petersen grab sampling 0.2 m^2, which has been used down to 10 120 m depth with a 68% success rate (for further details see Thorson, 1957). The 'Okean' grab used by Russian workers in their worldwide sampling programme during the 1950s and 1960s was a slightly modified version of the Petersen grab. It was equipped with a pendulum trigger weight whose bottom contact released the open jaws to drop into the sediment. The jaws closed as the gear was heaved off the bottom.

The van Veen grab, long used inshore, is available in a 0.2 m^2 version, and has been used in deep water, as have the spring-assisted Smith–McIntyre grab, and the somewhat similar and equally efficient Day grab. However, neither of the latter has been available in a scaled-up size that would make them really suitable for deep-sea use. Those that have been modified in the past for deep-sea use, such as the Campbell grab (an even larger variant of the Petersen grab sampling an area of 0.6 m^2, were equipped with pressure housings for flash and camera, one in each jaw) and the 'Orange-Peel' grab (which used to be commercially available for inshore industrial dredging), a modified version of which was used off southern California, during the 1950s and 1960s, are no longer in use by deep-sea benthic ecologists. These gears are described by Holme (1971), who commented that there appeared to be some risk of loss of material between the closed jaws of the orange peel grab, and also from the top of the grab unless a canvas cover is fitted. The other grabs are better protected from this, but are well known to be associated with a bow wave that can easily blow aside light-bodied animals as the open grab approaches the sediment surface (e.g. Wigley, 1967; Ankar, 1977), and this applies equally to their use in deep water.

LUBS sampler

The Large Undisturbed Bottom Sampler, LUBS (Menzies & Rowe, 1968), consists of a frame supporting a 19-litre ('Mini-LUBS') or 208-litre ('MegaLUBS') removable can or bucket surrounded by hinged weights. A canvas or nylon bag is attached to the bottom, which is pulled shut by tension on throat wires on pullout. It is claimed that this gear performs at least as well as a grab, but possibly is not as robust and efficient as box corers (see below). The largest version ('Megalubs') was used by Menzies to 3200 m, but to our knowledge it does not appear to have been used since then in the deep sea.

Box corers

Other devices designed to take a quantitative sample without the scooping, and sediment-disrupting, action of grab jaws have a long history. But apart from the LUBS sampler (above), those for deep sea use date from the box corers developed originally by sedimentary geologists for recovering, reasonably intact, blocks of the superficial sediment, and the Ekman, Ekman-Lenz and Birge–Ekman grabs developed mainly for limnological use. The latter are small box corers equipped with twin closing jaws that, depending on the model, sample only from 0.031 m^2 (Ekman) to 0.02 m^2 (Ekman-Lenz) and so are unsuited for sampling in deep water.

GOMEX box corer

The Gulf of Mexico (GOMEX) corer is described by Boland and Rowe (1991), and consists of a modification of a large Ekman corer designed by Jonasson and Olausson (1966) for deep-sea use. The stainless steel construction consists of a 25 cm × 25 cm × 50 cm high sample box and exterior fixed-support framework, internal structural linkage that attach weight trays to the closing jaws and the sliding bars that travel inside the support framework, and a release mechanism on the upper crossbar. The weight of the gear keeps it open during descent, and when tension comes off the wire, a release is tripped that allows the spring-loaded jaws to close. This action also allows a hinged lid held open above the corer too close to seal the box containing the sample during recovery. The hinged jaws at the bottom of the box are forced to close by the weights mounted above the box acting on linking slides. This gear samples an area of 0.0625 m^2, which is larger than that of the Reineck corer, but smaller than the USNEL or *Haja* box corers (see below), and perhaps for this reason has been little used elsewhere. Boland and Rowe (1991) found that mean values of macrofaunal organisms were almost twice as high (using a 0.25 mm sieve) from the Norwegian and Greenland Seas compared with results from the Reineck box corer (see below) used by Dahl et al. (1976) from similar depths in the Norwegian and Greenland Sea basins. However, because of the high variability in the samples of Dahl et al. (1976), the means from the two sets of samples are not significantly different.

The Reineck box corer

This device (*Kastengreifer* in German) was developed at the Senckenberg Institute, Germany (Reineck, 1958, 1963) to obtain undisturbed blocks of the sediment for the study of sedimentary stratigraphy and bedforms. It has been commercially available and used by benthic biologists as well as geologists worldwide, along with variants developed, as box corers, in the United States and described by Bouma and Marshall (1964). Large-scale use in the deep sea was pioneered in German studies of the north east Atlantic by Hjalmar Thiel, and France (IFREMER) where the gear was used for quantitative sampling of both macro- and meiobenthos in studies by Alain Dinet. The Reineck box corer samples an area of 20 ×30 cm (0.04 m^2) to a depth of about 45 cm, with the aid of lead ballast mounted above the core box. A pivoted spade closes the box by rotating below it when the heavy gear (over 750 kg) is heaved in.

The INCAL inter-calibration exercise, referred to above, also included samples from a Reineck box corer. As Fig. 7.9 makes clear, the sample composition in terms of representation of 15 major macrobenthic taxa was different from the sled/trawl samples, being proportionally richer in polychaetes and poorer in bivalve molluscs. The result for the box corer probably reflected much more accurately the actual quantitative representation of fauna in the sediment than the epibenthic sled samplers tested. The Reineck sampler is known to cause a bow wave that blows away light-bodied small animals. The proportion of peracarid taxa, which are usually thought to be most susceptible to this, however, is not appreciably lower in these samples than in those from the towed sleds and trawl samplers.

USNEL box corer

The 0.25 m^2 version of the US Naval Electronics Laboratory (USNEL) spade or box corer described by Hessler and Jumars (1974), and well illustrated by Guennegan and Martin (1985), represents a line of development from the Reineck box corer that has been modified for benthic sampling in the deep sea. The deep-sea version is a slightly modified, larger version of smaller corers sampling 0.0625 m^2, previously designed at the Scripps Institution of Oceanography and built at the US Naval Electronics Laboratory as described by Bouma (1969) and later modified at the Texas A&M University by A.M. Rosfelder. These large box corers were equipped with a custom dolly cart that was inserted under the closed spade to support the heavy core while the gear was raised slightly to take its weight and allow a curved plate to be inserted between the bottom of the full box and the upper surface of the spade. The curved plate was then secured to the box so that the spade could be swung back. The box could then be let down on the dolly cart and its fastenings to the box corer released so that it could be wheeled away. For the large, deep-sea benthos version (Hessler & Jumars, 1974), the box was equipped with a line of runoff holes that could be opened sequentially by means of a sliding blanking plate equipped with a spout so that the overlying water could be drained off through a sieve. The large box corer of Hessler and Jumars was equipped with screened vents with flapper valves to allow relatively free passage of water through the box during descent to reduce the bow wave. A pressure-powered safety pin, consisting of a piston partially inserted into an air-filled chamber restrained by rated shear pins (different sizes are used to shear at predetermined depths), prevents the friction release from firing prematurely. Over-pressure finally breaks the shear pins so that the friction release at the top of the column was armed to fire when the load came off the wire when the corer settled on the sea bottom. This device helped protect the corer from premature closure caused by the heave on the wire, although, in practice, pre-firing could easily occur during the latter stage of descent after the safety pins had sheared.

Presently, the most widely used box corer for sampling deep-sea benthos is the large Mk II model (Fig. 7.11). This is sometimes referred to as the USNEL box corer with Sandia modifications – from the Sandia Laboratories in California which were involved in its redesign. This represents significant improvement on the earlier model of Hessler and Jumars (1974). The Mk II version possesses a much more easily removable spade, utilising cam closures and improved venting of the upper box to minimise the ‘bow wave’. There is also a spring-loaded cocking pin that is locked in position until released by a linkage mechanism activated by the corer’s weighted central column passing down through the gimbals as the supporting frame settles on the bottom. The removable spade does away with the need to insert the curved plate to seal the bottom of the box, but does necessitate having a spare

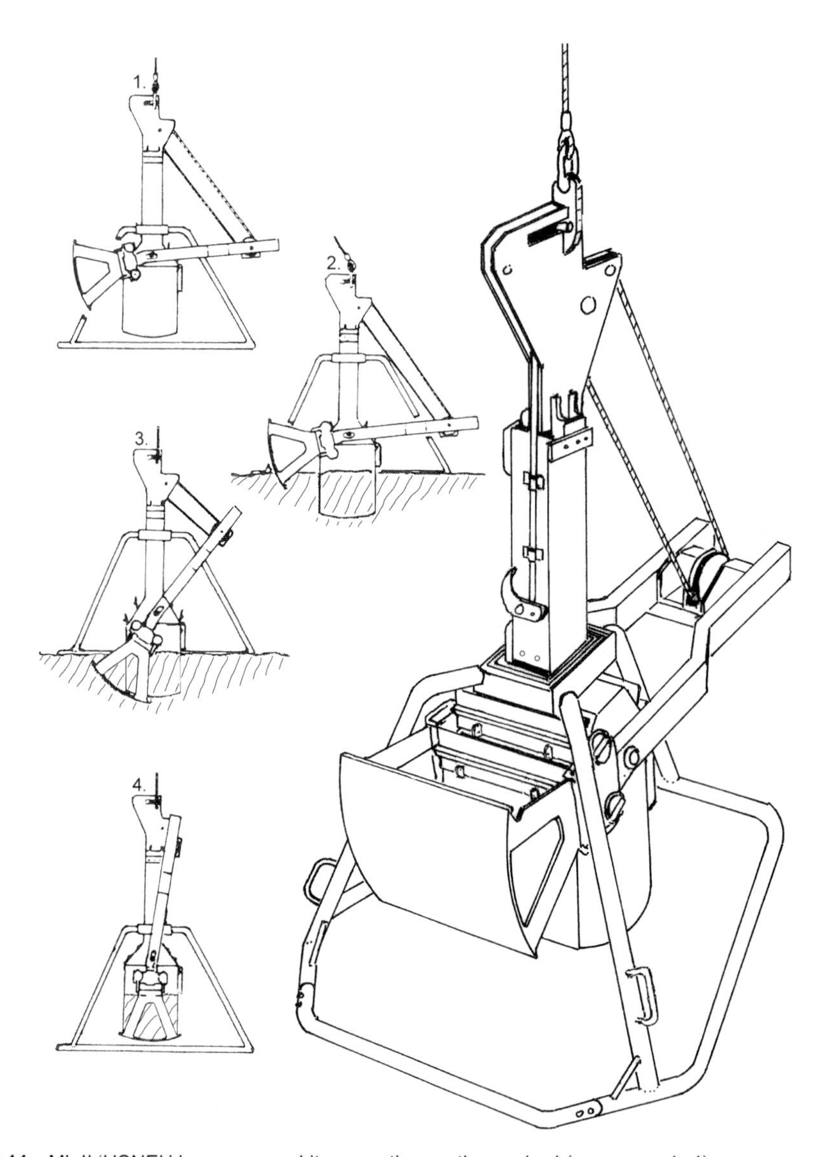

Fig. 7.11 Mk II 'USNEL' box corer and its operation on the seabed (sequence 1–4).

spade available, as well as an empty box, if the corer is to be quickly redeployed for another sample. The sequence of events on the sea bottom while taking the sample is illustrated in Fig. 7.11. Both versions of the USNEL box corer are able to penetrate to 40–50 cm in soft ooze and by means of a rotating spade closure can remove a 0.25 m^2 block of mud intact from the seabed.

Box core 'Vegematic' subcores

A girded insert into the box was used by Jumars (1975) to obtain 25 subsamples, each 10 × 10 cm outside dimension. These subcores are formed from a piece of square extruded aluminium tubing. They are sharpened at the lower end and held together by a frame that is bolted to the inside of the core box. Jumars covered the top of the array of tubes with a mesh screen to prevent animals from moving between them, and named it the 'vegematic' after a kitchen vegetable chopper. After deployment and recovery of the gear and removal of the box, the core tubes are removed individually by unbolting them from the holding frame. Use of the 'vegematic' in the San Diego Trough revealed sampling bias caused by the bow wave of the gear as it approaches the sediment. This was inferred by statistical comparison of the catch of macrobenthos from the 16 outer cores compared to the nine inner ones. Jumars applied a simple 'efficiency of capture' statistic from McIntyre (1971a) defined as the ratio of animals collected to the number actually present. In the present context, this was the ratio from the whole 0.25 m^2 sample against a value, standardised to the same bottom area, reached from the inner nine subcores. It is not clear to what extent the presence of the grid of subcores, by adding to the mass of the gear, increases the 'bow wave' effect compared to the ordinary, open box corer. However, a probably more important consideration is the degree of venting possible through the open core box during descent. This will depend critically on the size of the openings and the degree to which any flaps, used later to cover them, are open. Fig. 7.12 shows a photograph of a partial 'vegematic' sample where only a subset of the subcore tubes had been fitted, and a special plastic subcorer for taking vertical slabs of sediment for examining bioturbation has been fitted.

Fig. 7.12 Vegematic box core sample on deck after recovery from the Arabian Sea. The photograph shows the removed box with 10 subcores in position along with a slab-shaped transparent plastic core (only the top part is visible in front of the vegematics) designed for analysis of biogenic structure in the sediment.

Multi-box corer

A large multiple-box corer was developed by the Alfred Wegener Institute (AWI) and Krupp Maschinenbau GmbH, Kiel, in Germany (Gerdes, 1990). This device (Fig. 7.13) consists of a heavy circular framework to which are attached nine separate box core samplers, each able to sample a 12 × 20 cm area of seabed (total 0.22 m^2). The gear has been successfully used to full ocean depth. The multiple samples may provide some insights into the spatial dispersion of macrobenthic organisms, although like all such devices carrying multiple samplers, their regular arrangement precludes application of statistical tests that assume random placement

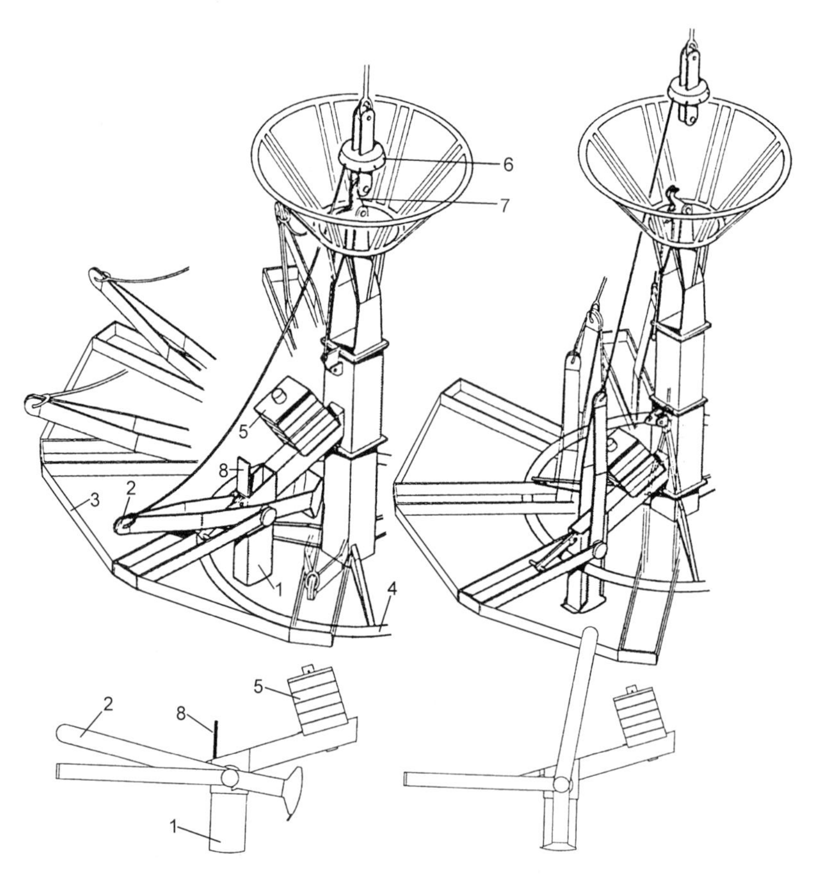

Fig. 7.13 The Alfred Wegener Institute multi-box corer showing circular arrangement of individual box-core samplers, with one of them shown open (left) and closed (right) by drawing its connecting cable up to the terminal bell (6). Each sample box (1) measures 12 × 20 cm by 45 cm deep providing a total sample covering 0.22 m^2 from an area of about 2.3 m^2 of seafloor. Closure of the spade arm (2), which is protected from collision damage with the ship's sides by the fender (3). The Multi-Box Corer lands on a bottom ring (4), and is operated on a single cable with heavy haulage winch on the ship. It weighs up to 2505 kg with maximum ballast added as lead plates (5). On bottoming, the terminal bell (6) on the cable is released from the hook (7) at the top of the corer while the open core boxes slide into the sediment. On heaving the wire, the spade arms are rotated to seal each box, allowing a top flap (8) to close and then the whole gear is heaved out of the seabed.

of sample units. This device has sampled successfully in the deep sea, sometimes incorporating a video camera, although it was originally developed for sampling on the Antarctic continental shelf.

HAJA box corer

This box corer, designed at the Netherlands Institute for Sea Research (NIOZ), is named after the engineers Harry and Jack who developed it there. It is currently obtainable from BV v/h Fa. P. Smit Constructiewerkplaats, Krugerstraat 95–101, 1782 EN Den Helder; Postbox 415, 1780 AK Den Helder, The Netherlands. The device (Fig. 7.14) differs from the USNEL box corer in possessing a double-arm action for effecting closure by rotating the spade into the sediment under the sample box and has better provision for sealing the top edge of the box. The double-arm action seems to avoid the difficulty of a unilateral force experienced with the USNEL box corer as it is heaved out of the bottom, resulting in samples frequently having a tilted surface.

The efficient box sealing (released by the trigger mechanism that closes the box by means of the double spade arms) means that the water overlying the core is effectively sealed in at the time of sampling, and has not been subjected to exchanges

Fig. 7.14 The Netherlands Institute of Sea Research (NIOZ) 'Haja' box corer. See text for details.

during winching up through the water column. It can therefore be used for chemical sampling; tests having demonstrated that oxygen and especially silicate concentrations are in the same range as those taken near the seabed from water bottle samples (Dr Gerard Duineveld, NIOZ, personal communication). The cocking mechanism and pin locking of the HAJA (or 'NIOZ') box corer are somewhat similar to the USNEL Mk II, as is an arrangement for removing the box with the spade in place underneath and replacing it with an empty box for another deployment. However, the HAJA box corer does differ in that it allows both square and circular boxes to be used. The HAJA box corer is available in two sizes, the larger having boxes of 50 cm square, or circular ones of 50 cm diameter, while the smaller model has circular boxes of 30 cm diameter. The base of the frame measures 2.5×1.5 m and it stands 2.7 m high. The frame, including weights, weighs 1400 kg. Although the bottom area of the larger box (0.25 m^2 140 litre capacity) is equivalent to that of the USNEL, that from the circular box is less, approximately 0.196 m^2(110 litre capacity), while that of the 30 cm diameter model is some 0.07 m^2.

Comparing box corers and grabs

Smith and Howard (1972) compared the performance of a modified Reineck box corer sampling 0.0625 m^2 (as illustrated in Bouma, 1969) against the Smith–McIntyre grab. They found that macrofaunal abundance and biomass was higher in samples obtained with the box corer than with the grab, mainly as a result of the box corer recovering older, larger animals from deeper in the sediment. In all cases where there were significant differences in abundance, there was also a difference in the size classes present, with the corer collecting larger individuals.

In work where both the USNEL Mk II and HAJA box corers were deployed, personal experience has shown that the HAJA corer is capable of considerably better success rate in sandy, mixed sediments than a USNEL box corer. This is probably because of the well-sealed sample boxes, but is possibly also helped by the superior mechanical leverage of the spade closure. This is an important advantage on the continental margin where relatively coarse sediments are commonly encountered. However, although as yet unquantified in objective tests, the venting arrangements of the NIOZ corer may result in at least as serious a bow wave effect as the USNEL corer.

Box corers are superior to grabs, which by their scooping action can badly disrupt the sediment. Both are prone to loss of small organisms being blown aside by the 'bow wave' as the gear lands on the sediment, even if venting is provided for within the jaws or above the box. The ability to remove the box and investigate the intact sediment block (after carefully draining away the supernatant water) by removing one of the sides of the box has allowed investigation of subsurface sediment microstructure not previously possible.

Sampling meiofauna using box corers

For meiofauna, samples obtained by the above box corers have been sub-sampled using core tubes of various sizes, but usually covering up to about 25 cm^2. Some investigators (e.g. Thiel, 1983) have noted that this might be reduced to 10 cm^2. Thiel (1966) designed a *meiostecher* (also illustrated in Thiel, 1983) for the purpose of this subsampling. It consists of a shallow rectangular core that can be withdrawn neatly by inserting a sliding shutter down one side and along the bottom. Details of methodology are discussed by Thiel (1983). Thistle et al. (1985) used an array of 5 × 5 cm square tubes as inserts into the nine inner 'vegematic' subcores of a USNEL box corer. However, the bow wave effect produced by descending box corers generally makes this approach unsuitable for truly quantitative meiobenthos studies (see Bett et al., 1994). This is confirmed by the work of Shirayama and Fukushima (1995) who showed that a multiple corer (see below) collected significantly more harpacticoid copepods, meiofaunal organisms that live in the most superficial layer of sediment, than a box core. When they tested the effect among meiofauna as a whole, a significant difference was detected at only one out of the three stations sampled. Nematodes typically dominate the meiofauna and are generally more deeply distributed in the sediment column than harpacticoids. Shirayama and Fukishima (*loc cit.*) concluded that the box corer did not efficiently collect the surface-most flocculent layer of the sediment owing to the bow-wave effect that blows it aside as the corer lands on the bottom.

Further development of box corers

Although it is likely that new devices will soon emerge based on older designs that overcome problems such as the bow wave, none to our knowledge is yet in use for benthic sampling. However, one design developed for geological sampling of soft sediment in the deep sea that is claimed to take undisturbed samples through the important sediment–water interface is the Maxicorer (Gerber et al., 1996). Although essentially similar to the Reineck design, it incorporates controlled penetration of the sample box after bottoming of the supporting frame by means of hydraulic damping, removable acrylic box liners, and sealing of the top of the liner after penetration into the sediment by release of spring-loaded levers.

Tube corers

Barnett–Watson multiple corer

A line of development distinct from that of grabs and box corers has involved either single or multiple arrays of core tubes. It is often more important to the benthic ecologist to sample the flocculent superficial layer of sediment and its contained fauna than to obtain the deep penetration required by geologists that entails the brutally rapid approach of the weighted core tube to the seabed. The required soft landing is achieved by hydraulic damping of the descent of the open core tube after the supporting frame settles onto the seabed. The Barnett–Watson Multiple Corer, or 'Multicorer' (Barnett et al., 1984), a multiple-tube development of the light, inshore single-tube Craib Corer, has become the standard tool for

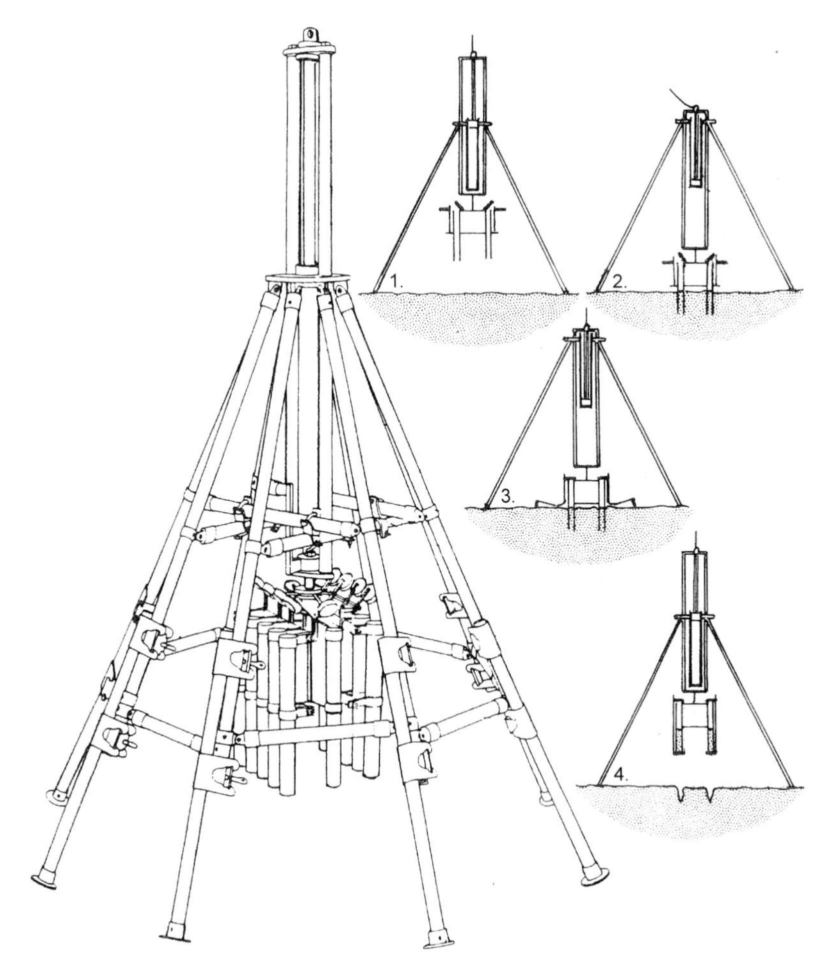

Fig. 7.15 Barnett–Watson, or SMBA-pattern Multiple Corer and its operation on the seabed (sequence 1–4).

obtaining undisturbed sediment samples for meiofauna work, and for the study of sediment biogeochemistry, in deep water (Fig. 7.15). Because of the hydraulically damped, controlled descent of the open core barrels, the gear is capable of taking completely undisturbed samples that retain any flocculent material lying on the sediment surface (e.g. Gooday, 1988). Such material is very easily mobilized by seafloor currents. Rubber hemispheres seal the bottom and top of the tube to prevent loss or disturbance of the core during recovery. The outer framework of the Barnett–Watson Multiple Corer supports a weighted array of 12 plastic core tubes of 56.5 mm internal diameter (25.1 cm^2) hanging from a water-filled dashpot. Figure 7.15 shows the sequence of operation after the frame is lowered on the wire and settles on the seabed. As the wire slackens by paying out a few more metres, the core tube assembly is released and its descent is dampened by the dashpot so that the core tubes enter the sediment as slowly as possible. After a minute or two, the corer is winched back to the ship and the core tubes removed for processing. The core size is ideal for meiofauna in the deep sea, obtaining an excellent sample of manageable size that is free of the bias afflicting subsamples taken from within box cores, as was the practice previously (see Bett et al., 1994; Shirayama & Fukushima, 1995). Comparisons of the chemical characteristics of the overlying water compared to that collected by the USNEL box corer by Shirayama and Fukushima (1995) suggest that the latter may be contaminated during recovery through the water column. However, this failing does not seem to apply to the Haja box corer, as noted above, although in relation to benthic sampling the likely problem of down-wash may be more significant.

The megacorer

This device, the Bowers and Connelly Megacorer (Fig. 7.16), is capable of obtaining up to twelve 10-cm diameter cores penetrating 20–40 cm into the sediment with similar hydraulic damping to minimise the bow wave. McIntyre (1971a) comments on the basis of unpublished work that a 10-cm diameter core tube was more efficient than a much narrower diameter tube (2 cm dia.) used carefully by scuba divers to sample a flocculent surface layer in a Scottish sea loch. The total sample size obtained from a Megacorer deployment (942.5 cm^2, assuming all 12 tubes have worked) is still slightly less than 0.1 m^2 – thought to be the minimum

Fig. 7.16 Bowers and Connelly Megacorer, (left) corer fitted with eight of possible 12 core tubes and carrying an acoustic navigation transponder (upper left of corer frame), (right) a close-up of the top and bottom core closures.

Table 7.3 Sampler bias in the study of deep-sea macrobenthos (>500 μm). Box corer ($n = 5$) and Megacorer ($n = 4$) abundance estimates for the 10 most abundant species and the total macrobenthos from a site at 844 m water depth on the Hebrides Slope, north east Atlantic (58° 49.1′ N 07° 56.6′ W). Note that in all cases the Megacorer estimate is higher than the box core estimate; *t*-test results are shown for comparisons of individual species abundances and an ANOVA test result for the total fauna (all tests based on log-transformed data; ns, not significant).

	Mean abundance (individuals 0.1 m $^{-2}$)		
Taxon	Box corer	Megacorer	*t*-test
Polynoidae indet.	2.25	27.11	$P < 0.005$
Actiniaria indet.	5.91	7.76	ns
Haploops setosa	6.53	6.93	ns
Ampelisca odontoplax	5.33	7.46	ns
Enchytraeidae indet.	2.55	5.87	ns
Amphitritinae indet.	0.64	8.19	$P < 0.001$
Glycera mimica	3.54	4.60	ns
Eriopisa elongata	1.83	3.80	ns
Ophiurae indet.	0.84	4.56	ns
Minuspio cirrifera	2.29	3.46	ns
Total macrobenthos	99.6	213.2	$P < 0.01$ (ANOVA)

for inshore macrobenthos samples. Several more drops of the gear may be necessary to achieve bottom-area coverage equivalent to that from box coring in the deep sea. Furthermore, the regular and almost adjacent arrangement of core tubes means that individual core samples of the set cannot be treated as independent sample units in tests for spatial pattern of organisms. However, quantitative studies show that Megacorer samples contain considerably more fauna. Recent studies on the NE Atlantic margin of Scotland, comparing results from the USNEL Mk II box corer and the Megacorer, indicate surprisingly serious under-sampling (48 to 68% of total abundance) by the box corer relative to the Megacorer (Bett, 2000; Bett & Gage 2000). There is also a measurable qualitative bias, i.e. apparent species composition varies between the two samplers (see Table 7.3). This is interpreted as selective loss of the lighter bodied, more vagile animals that might be expected to be affected by the bow wave caused by the bulky box corer as it approaches the seabed.

7.3 Free-fall equipment

Sediment samples

Free-falling corers and grabs are not in common use, but have been deployed to collect deep-sea samples (principally, for geological purposes). The equipment manufacturer 'Benthos' (Falmouth, MA, USA) produces a 'Boomerang corer' and a 'Boomerang grab'. The free-falling corer is in effect a simple gravity corer (67 mm internal diameter, 1.2 m barrel) fitted with glass buoyancy spheres and an expendable ballast weight that is automatically released during sediment penetration. The Boomerang grab, though somewhat more elaborate, operates on the same principle. The grab was specifically designed for the recovery of manganese nodules, using spring-activated jaws fitted with netting collector bags. As the jaws close, ballast weights are automatically released and the corer begins to ascend.

Traps

This category of sampler includes a wide variety of devices that share the common features of free-fall descent (i.e. they are not lowered on a wire) and buoyant return to the surface for subsequent recovery. Benthic landers come within this category, but given their relative complexity are dealt with in a separate section below.

A range of benthic and bentho-pelagic taxa can be collected using baited traps, as are commonly employed in shallow-water commercial fisheries. Typical examples of trap-caught deep-sea taxa include amphipods and rattail fish; other commonly caught taxa include both natant and reptant decapods, and occasionally motile echinoderms and molluscs. In shallow-water operations, traps are normally set on a surfaced buoyed mooring. This approach is possible in quite deep water, but free-fall systems are generally more convenient for true deep-water work.

A typical trap mooring consists of an upper buoyancy package carrying a strobe light and a radio/satellite beacon to aid relocation at the surface. Suspended below the buoyancy package is a ballasted trap with a release mechanism. Ballast release can be achieved by the use of simple dissolvable magnesium links or timer devices. Acoustically controlled release mechanisms are, however, more convenient as they allow greater control, e.g. release can be delayed to coincide with appropriate surface weather conditions for recovery. Many trap designs are used in shallow water, and there have been several studies of their relative effectiveness (see e.g. Miller, 1990; Whitelaw et al., 1991). Numerous factors may influence catching efficiency, including soak time, trap volume, number of entrances, use of parlours (sub-traps) and other devices to restrict exit, the use of protected or unprotected baits, etc.

In deep-sea applications, traps are generally of simple construction and are relatively small where amphipods are the target taxon (see e.g. Christiansen, 1996) or somewhat larger, where demersal fish are the target (see e.g. Priede et al., 1994). A simple trap design allowing free access to motile epifauna attracted to bait that has been used in French studies at IFREMER is shown in Fig. 7.17a. This contrasts with the much more specialised trap (Fig. 7.17b) designed at the Scripps Institute of Oceanography to catch fast-swimming, giant scavenging amphipods. Such organisms may spend only limited time at the bottom as part of the benthic fauna. Designs to catch mainly benthopelagic fish include a number of more elaborate deep-sea designs. Smith et al. (1979) constructed a single mooring that comprised a multi-hook long line, two traps and a gill net. More ingenious still were two hyperbaric fish traps described by Phleger et al. (1979) and Wilson and Smith (1985).

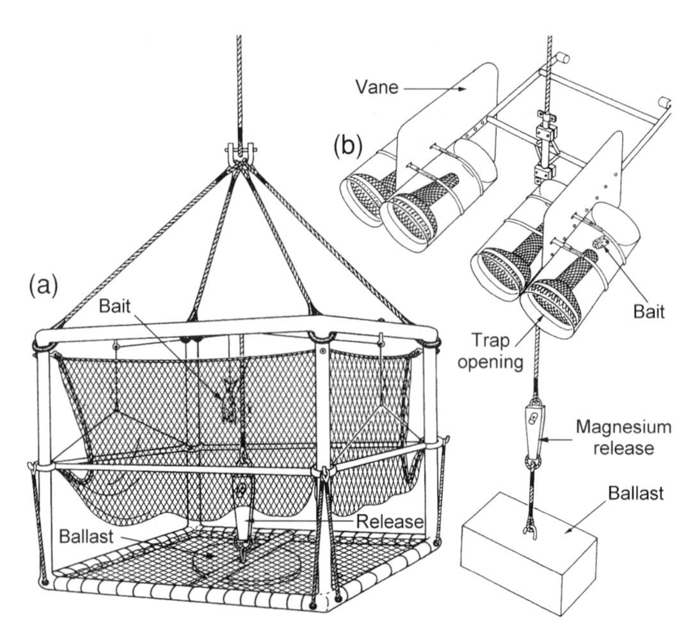

Fig. 7.17 Traps for deep-sea biota. (a) Autonomous design developed at IFREMER to allow free access of carnivores and scavengers, including motile benthic epifauna, to bait. The frame is made of plastic piping and encloses an area of 0.8 m^2 with the bait inside. When the trap is open the netting is gathered up on itself, and when the magnesium wire dissolves in the release mechanism, cords pull the net curtains to close the trap and the ballast weight is released so that the buoyancy (not shown) can lift the trap up to the surface. (b) Small cylindrical trap array moored to the bottom for trapping scavenging amphipods. Each trap contains bait and is able to swivel and orientate with the current by means of vanes and counterweights so that the openings face downstream. Simple conical entrances trap the amphipods inside. The release mechanism allows the traps to lift free from the ballast and rise to the surface for recovery.

Both of these traps are effectively cylindrical pressure vessels deployed with one end open. A baited hook protrudes from the open end, which is automatically withdrawn into the pressure vessel if taken by a fish. Once the hook, and hopefully the fish, is withdrawn, the pressure vessel is closed and the mooring ballast is released.

Other sampling methods

Deep-sea fish and amphipod traps are typically operated for periods ranging from several hours to a few days. If moorings can be left in place for longer periods (e.g. months) then it becomes possible to collect colonising organisms if appropriate substrates are provided. A simple, but effective, example of this technique is given by Harvey (1996) in a study of deep-water wood boring bivalves. Wood panels were opportunistically deployed on an SOC Bathysnap system (Lampitt & Burnham, 1983). When the mooring was recovered some 20 months later the wood panels yielded over 400 specimens of a new species of *Xylophaga*. Other hard substrates could be deployed in a similar manner to study other taxa; for example, Bertram and Cowen (1999) used this approach to study foraminiferal colonisation rates on a seamount (although it should be noted that a submersible was used for emplacement and recovery of the settlement panels). Soft substratum (i.e. sediments) colonisation studies have also been carried out both using submersibles (e.g. Grassle, 1977) and with free-fall systems (e.g. Desbruyères et al., 1980). The design and emplacement of sediment recolonisation trays have been the subject of methodological studies that indicate the need for careful consideration of hydrodynamic effects that may influence the apparent settlement of the fauna (Snelgrove et al., 1995).

7.4 Sampling from manned submersibles and remote operated vehicles (ROVs)

Manned submersibles and ROVs may be used to observe the benthic habitat, to sample, set up, manipulate or otherwise intervene in seabed experiments. There are particular advantages and some disadvantages of sampling using such approaches. The need to control buoyancy and trim means there is a severely limited additional payload capability. Advantages are that it is possible to observe directly the habitat of study, and to monitor any sampling so that there is a degree of control over exactly where and how the sampling gear is positioned to take its sample. Just as for samples taken by a diver, the operator must take special care where random samples are required; ideally a formal randomisation protocol should be employed to avoid biased, deliberate placing of the sampling gear. But, in contrast to sampling using a wire dangled from a ship, samplers can be emplaced precisely in relation to the previous and subsequent samples, or related directly to some feature on the seabed, such as a hydrothermal vent. The latter habitat has provided one of the most important areas of application for manned submersibles in recent years.

With the sediment community, however, an important use for such platforms is for habitat observations that allow the observer to see the larger, epifaunal organisms as living, rather than dead animals, which are usually killed by their journey through perhaps thousands of metres of water column by pressure and thermal changes. Useful observations have been made concerning natural behaviour, providing insights on the interactions of benthic organisms with their environment, which would not have otherwise been possible. Grassle (1980) further suggested that submersibles are the only way to:

- sample small-scale features
- sample repeatedly at a specific site over a period of time
- push sampling devices into the bed without sediment disturbance
- navigate around obstacles in areas of complex topography.

Although the requirement for undisturbed samples to some extent has been overcome using hydraulically damped samplers such as the multi- and megacorers, and perhaps the German maxicorer (see above), the other advantages of submersibles listed above remain largely valid. But perhaps the most important of these is that these direct observations can encourage enthusiastic deep-sea biologists to design experiments to test hypotheses generated by these observations. Although unreplicated and limited in scope, those experiments that have been undertaken to date have provided some very important advances in our understanding of the deep-sea benthos.

Deep-diving manned submersibles

Although several submersibles have been constructed, only a handful possess the capability to dive more than a few hundred metres and are equipped with a manipulator arm, and thus can be useful tools in sampling the deep-sea benthos. One, the DSV (Deep Submergence Vehicle) *Alvin* operated by Woods Hole Oceanographic Institution (USA) stands out as having provided, by far, the greatest body of information on the deep-sea benthic habitat since it was commissioned in 1964. It can carry two scientists and the pilot, and has a maximum operational depth of 4500 m with a normal dive duration of 6–10 hours. But at maximum depth it takes 2 hours to reach the seafloor and another 2 hours to return. It has a maximum payload of 454 kg. Other submersibles include the *Sea Cliff* (originally a sister submersible of the *Alvin*, and also owned by the US Navy) that has a deeper operational depth of 6000 m, while the French *Cyana* and the newer *Nautile* (Fig. 7.18) have maximum

Fig. 7.18 The French manned submersible *Nautile*, equipped with two manipulator arms, shown in an artist's impression approaching a 'black smoker' chimney and the more diffuse flow from 'white smokers' surrounded by alvinellid worms.

working depths of 3000 and 6000 m, respectively, and are owned and operated by IFREMER. The Japanese *Shinkai* 6500, developed by the Japan Marine Science and Technology Centre (JAMSTEC), and the Russian *Mir I* and *Mir II*, also have operational capabilities to 6000 m, and can thus penetrate to all ocean areas except deep trenches. The auxiliary ROV principle (see further below) has already been employed by *Alvin* with the small ROV *Jason Junior* and by the French *Nautile* using the small ROV, *Robin*. By relieving the mother craft from the manoeuvering and manipulations necessary, such systems will provide additional scope to benthic sampling investigations in the deep sea.

Another manned submersible system is the *Johnson-Sea-Link I and II*. These two submersibles (the one capable of serving as a rescue backup for the other in an emergency) have been operating since 1971 and 1975, respectively. Although depth limited to 914 m, they are unique in having a 127 mm thick acrylic sphere allowing panoramic visibility for the pilot and one scientist (an after-sphere made of steel accommodates a less fortunate scientist and one backup crew member, with only normal viewing ports). The *Johnson-Sea-Link* (Fig. 7.19) is equipped

Fig. 7.19 The Johnson-Sea-Link manned submersible operated by Harbor Branch Oceanographic Institution shown suspended from the stern gantry of the mother ship. The pilot and scientific observer can be seen sitting in the acrylic pressure sphere. The manipulator arm (left) can be seen holding the nozzle of a length of corrugated flexible tubing connected to a pump and acting as a 'slurp gun' collector. This discharges into a clear plastic container (visible just above the large, black camera) that is mounted onto a rotating carousel allowing material to be separated and sealed until recovery of the submersible at the end of the dive. Other sampling and experimental equipment is visible on the instrument rack at the front of the craft.

with particularly sophisticated and effective manipulators that are capable, with skilled operation, of undertaking the most delicate tasks. It is well equipped to carry out even quite complicated experimental manipulations on the seabed. The main problem with submersible studies and sampling in the deep ocean is their expense, weather limitation for launch and recovery, and the increasing regulatory framework related to the possible risk to the investigators (Rowe & Sibuet, 1983).

Remote operated vehicles (ROVs)

This is a generic name for a range of platforms that may be towed, tethered or free-moving, but are controlled remotely, usually by means of an 'umbilical' tether and data cable from the surface ship that deployed them. For benthic sampling, only the free-moving ROVs have been used. Because they are usually smaller in size than manned submersibles, payload limitations severely restrict the sampling capability of ROVs. Such limitations can be overcome by the use of auxiliary systems (e.g. free-fall and ascent vehicles) that deliver sampling tools and samples to and from the seabed. For benthic work, the most successful ROVs have been tracked vehicles, or bottom crawlers, which can move over the bottom under control from above. One of the first was the Remote Underwater Manipulator (RUM; Jumars, 1978), which was controlled from a special floating research platform via a conducting umbilical also carrying its power supply. The manipulator arm on RUM was used to take 20 × 20 cm Birge–Ekman box cores from which up to four square subcores, each 10 cm by 10 cm, were taken. Each sampling operation was precisely positioned by triangulation from a bottom-moored transponder net. The exercise yielded 125×0.01 m^2 subcores for analysis of the spatial dispersion of macrobenthic species. Other tracked ROVs have been built (e.g. K.L. Smith's *ROVER*; see Smith et al., 1997), or are planned at the time of writing, with the capability to collect benthic samples.

There are many 'work-class' (designed for industrial use) ROVs in existence, but the majority are tailored for use in underwater engineering applications for the offshore oil industry, and do not usually possess deep-diving capability. There are few ROVs that have a full-ocean depth capability, and these have been designed for research rather than commercial applications. One such is the ROV *Ventana* operated by the Monterey Bay Aquarium Research Institute (MBARI). Another is the French *Victor 6000*, operated by IFREMER. This new vehicle is capable of a wide range of tasks, including sediment sampling (Fig. 7.20).

As noted above, there has been a trend towards the operation of auxiliary ROVs from manned submersibles. There has been a similar trend towards 'two-body' ROV systems, the first body being a deployment and cable management system, the second body being the free swimming ROV. A good example of this is the *Medea–Jason* system of the Woods Hole Oceanographic Institution, USA. When submerged, *Medea* is attached to the surface using a steel-armoured conducting cable by which she is manoeuvered by the support ship (although capable of limited propulsion herself) and serves in a tether management role. From *Medea* can be

Fig. 7.20 IFREMER's *Victor 6000*, a recently built research ROV capable of working at full ocean depth.

deployed *Jason*, a highly manoeuverable and controllable, 2.5 m long mini-ROV with both imaging and bottom sampling capability. The system has an operational limit of 6000 m depth and a dive duration from 6 hours to 1 year. Another deep-diving, two-body system is *Kaiko* (Japanese for 'trench', developed by JAMSTEC). Uniquely, *Kaiko* has a depth capability of 11 000 m and has successfully reached the bottom of the Mariana Trench. The *Kaiko* launcher is fitted with CTD sensors, a sidescan sonar system, and a sub-bottom profiler to make general environmental observations. It also carries simple video cameras for mission monitoring. The *Kaiko* ROV is linked to the launcher via a 250 m umbilical, the ROV having two manipulators and carrying high quality video and still cameras. The French *Victor 6000,* ROV operates in conjunction with a 'tool sled' to which the ROV is tethered by a 100–300 m long cable in order to allow greater flexibility in its work mission.

Autonomous underwater vehicles (AUVs)

A newer concept is to send the unmanned craft to the bottom untethered to undertake a pre-planned work programme that can be repeated over a long period of time. Rather than returning to the surface, the AUV named *ABE* (Autonomous Benthic Explorer), operated by Woods Hole Oceanographic Institution, can undertake a predetermined work programme within an area the size of a city block, and then return to a seabed docking station and hibernate in order to conserve power. It is reactivated on command, by external event, or by an internal clock, to return to its work area for repeat measurements or sampling. This concept promises to provide important information on site-specific variability, which would be very expensive to achieve by consecutive dives using a manned submersible operated by a mother ship, and virtually impossible to achieve by lowering instrumentation to the seabed from a ship.

There are a number of other AUVs, mostly at the design or early proving stage. Those that are operational have been used for upper ocean studies and sidescan sonar seabed surveys. One interesting prospect, soon to be exploited by the *AUTOSUB* vehicle of the Southampton Oceanography Centre, is under-ice missions. *AUTOSUB* is at present not fitted with any tools suitable for benthic sampling but already has the ability to operate very close to the seabed and will soon be fitted with a seabed photography system.

Sampling using submersibles and ROVs

Sampling relies on the manipulator arm fitted to front of the submersible and what is possible depends heavily on the sophistication of this hydraulic arm and the skill of the operator, as well as the maximum lifting capacity of the arm and payload of the submersible. For sampling the seabed sediment, penetrative corers which have been tailored to the particular abilities and constraints of the submersible are used. These range from simple 'Alvin' corers, or similar designs of manipulator-operated tube corers (Fig. 7.21) used in French submersible investigations by IFREMER (Sibuet et al., 1990), equipped with a non-return valve, to prevent loss of sample and retention of the overlying water after the tube is pulled out, to modifications to small sampling gear. An interesting modification of the Barnett–Watson multicorer has been developed by the Monterey Bay Aquarium Research Institute (MBARI) where sample-tube holding assemblies are mounted in individual 'quivers' on hydraulically activated arms and are deployed, fired, and retrieved with the use of the

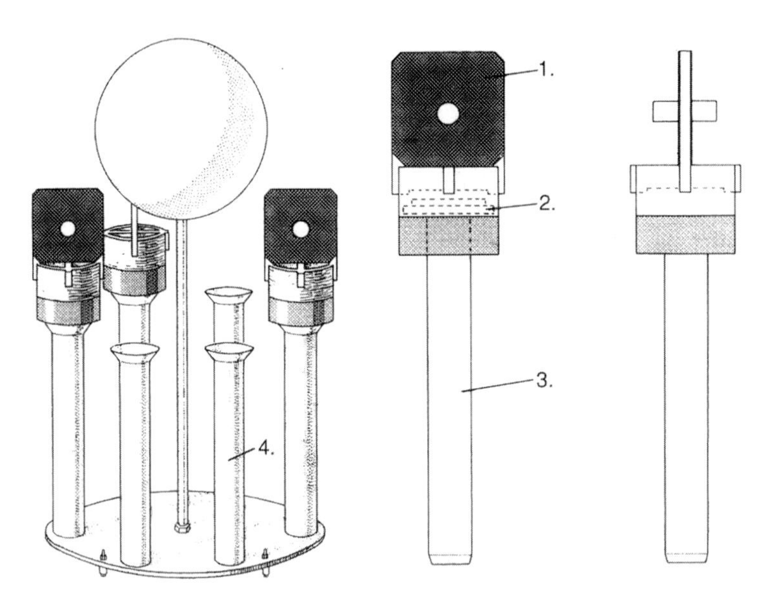

Fig. 7.21 The tube corer developed by IFREMER for use from the submersibles *Cyana and Nautile*. The upper plate (1) is a handle for the submersible arm to grasp. The interchangeable coring tubes (3) fit within the sheaths (4) and are sealed by the one-way valve (2).

manipulator arm on the *Ventana* ROV (see above). The Birge–Ekman box corer (BEBC) is often used, as modified versions of the commercially available models, taking either cores covering 0.023 m^2 or 0.04 m^2 (Rowe & Clifford, 1973). By means of a handle welded to the top of the corer (Fig. 7.22a) it is pushed about 10 cm into the sediment, and the spring-assisted jaws are then triggered by pressing down

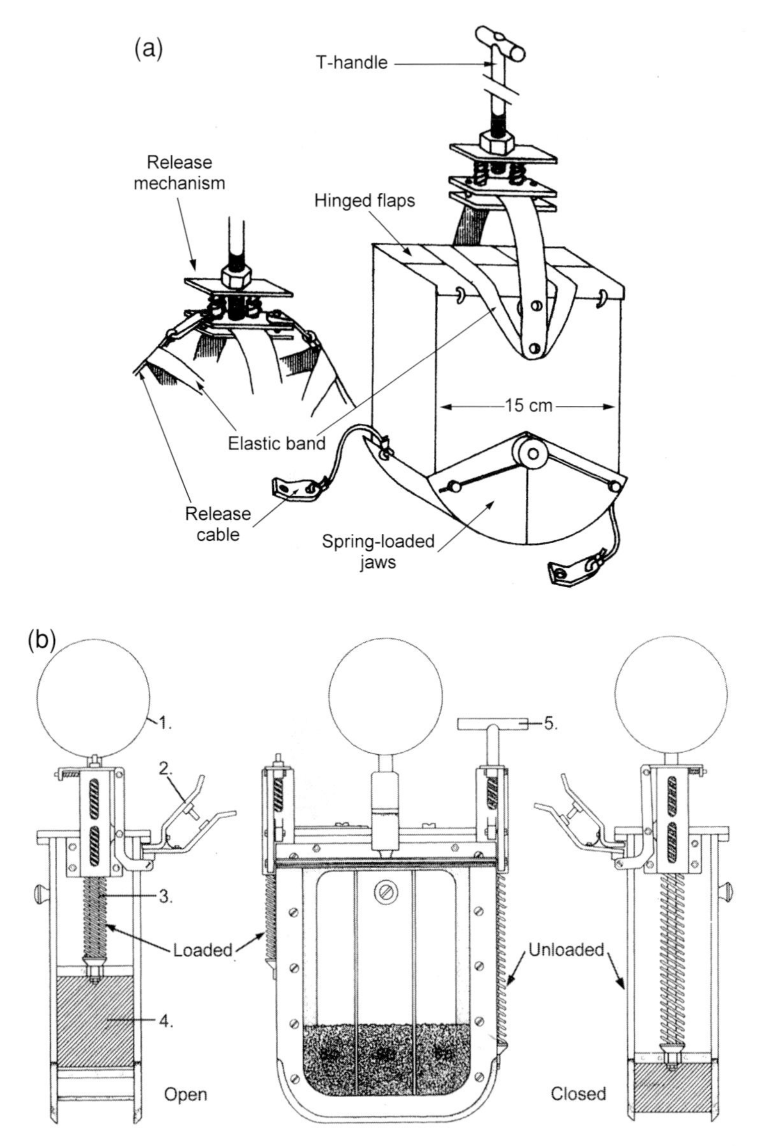

Fig. 7.22 Box core samplers designed for use by submersibles for sediment sampling. (a) The modified Birge–Ekman box core developed for use by *Alvin*. By pressing down on the handle, the spring-loaded jaws are allowed to close, this action also releasing rubber bands, which force down flaps to seal the top space of the core box. (b) Blade box corer designed for use with Cyana and Nautile. The circular handle (1) is grasped by the manipulator arm to emplace the open gear in the sediment; using the manipulator's pincer action squeezing the clips (2) the spring-assisted (3) blades (4) are allowed to close the bottom of the box. The 'T' handle (5) is used to open the corer and load the spring.

on the handle using the submersible's manipulator arm. DSV *Alvin* carried up to 10 on a palette. Each sample requires a separate pre-cocked corer, with each being stored after sampling in the submersible's sample basket. The samples obtained are considered to be of high quality, with careful lowering of the arm avoiding any bow-wave effects. The blade box corer (Fig. 7.22b) is a somewhat similar design to the modified Birge–Ekman, and has been used in French submersible investigations. It is pushed in by the manipulator arm, and the blades are closed by activating the release of powerful springs. This device has also been used as an incubation chamber and can be fitted with injection modules, equipped with a 60 ml commercial syringe, to inject labelled substrates or fixative (formalin) before it is recovered (Sibuet et al., 1990).

The *Johnson-Sea-Link* submersibles are particularly suited to investigations of the rugged topography associated with the upper bathyal, and users and staff at Harbor Branch Oceanographic Institution have developed sophisticated equipment to collect larger megafauna, such as delicate sea lilies and sea urchins, during a dive. A series of removable jars on a rotating carousel allows delicate specimens to be deposited into separate jars, which can be sealed before surfacing. Such specimen collection is often by '*slurp gun*', which is effectively an underwater vacuum cleaner (Fig. 7.19). Slurp guns can also be used to sample epi- and in-faunal benthos in a manner similar to the use of airlifts by divers. By these means, material can be recovered alive in good condition for experimentation or eventual transfer in good condition to special aquaria on land. Biological sampling on hydrothermal vents and deep-sea cold seeps involves similar problems; and manned submersibles (Fig. 7.18) and ROVs are ideally suited for the work as sampling the exotic associated communities remotely from ships is virtually impossible. A wide assortment of specialist devices has been fabricated for sampling and manipulative experiments that are beyond the scope of this chapter to describe. Rowe and Sibuet (1983) summarised some of the other devices that can be operated by DSV *Alvin* including seabed samplers and devices for manipulative experimentation.

Equipment of similar sophistication is becoming available for use with deep-diving ROVs. The 'tool sled' used with IFREMER's *Victor 6000* can be modified to provide flexibility during a mission for various tasks, such as sediment sampling, individual specimen collection and release of passive markers etc. for future investigations.

Benthic landers

These are autonomous free vehicles or platforms that are equipped with buoyancy and ballast release systems (Fig. 7.23) to allow the package to sink slowly to the seabed in order to make measurements (usually a sediment incubation to measure solute uptake, such as oxygen) over a period of time long enough to detect changes due to biological activity. Sealed retrieval of the sediment enclosed by the chamber has been considered desirable. This can be achieved by either a modified Ekman

Fig. 7.23 A modern deep-sea benthic lander designed and operated by the Netherlands Institute of Sea Research (NIOZ) that is capable of recovering sediment samples in addition to carrying out a range of *in situ* incubation experiments.

grab mechanism (e.g. Smith's Free Vehicle Grab Respirometer, FVGR, see Smith & White, 1982), or by a more sophisticated mechanism utilising a spring-driven assembly and hydraulic cylinder (to be pressurised before deployment). A scoop can be rotated through the sediment to activate the lid closure, sealing the chamber top and bottom, as described by Jahnke and Christiansen (1989). A similar lander system capable of retrieving the chamber sediment and its fauna has been used in German studies by GEOMAR, Kiel. These sophisticated instruments are reviewed by Tengberg et al. (1995) and Bagley et al. (2004), but cannot be considered as benthic samplers in the strict sense, and are not considered further, although the samples compare well to those obtained from the same area using larger box core samplers deployed conventionally (based on data from the EU '*BENGAL*' project; see Billett & Rice, 2001).

With larger animals (megafauna), seafloor photographs may provide the best way to describe spatial pattern. Spatial pattern may be much more dynamic with more motile animals, many of which tend to be scavengers. Animals may form highly concentrated, but essentially transient, aggregations around large food parcels. Although such patterns may be difficult to quantify on a large-scale, potential occurrences can be tested by simulating natural food falls, monitored photographically on the seabed (see e.g. Thurston et al., 1995). The elegant approach of getting scavenging fish that are attracted to bait to swallow acoustic tags, and to subsequently track their movement using ranging sonar emplaced at the bottom, is now quite well developed (Smith et al., 1989; Bagley et al., 1990; Priede et al., 1994), and can be seen as one way to sample the community, even if it is restricted to large motile scavengers able to ingest acoustic tags. The utility of emerging technology based on the analysis of acoustic backscatter has also been tested for monitoring the activity of larger animals over the sediment, with promising results (Jumars et al., 1996).

7.5 Processing deep-sea sediment samples for macrobenthos

Processing is generically similar to that usually applied to sampling in shallow water, but there are some characteristics of the deep-sea benthos that require a distinctive approach. Perhaps the main difference is a consequence of the larger bulk of sediment that needs to be washed immediately after recovery in order to preserve the sample of organisms. This must be started quickly, particularly in hot climates, in order to avoid the risk of the fauna starting to decompose. However, the sediment will usually have shielded the organisms from severe temperature change in the upper water column while being winched in. Box core samples in particular, which are recovered more quickly than those from trawls, are usually well insulated from temperature change, and the overlying water, as well as the sediment itself, can feel cold. From a large trawl haul (which may deliver a cod-end full of muddy deposit) the sample size may be in excess of 1000 litres. However, because the fauna may be sparse and small, care needs to be taken when reducing the bulk to a more manageable size for further washing and examination in the laboratory. Therefore, methods commonly applied for washing shallow water samples may be too vigorous and likely to cause avoidable damage to delicate-bodied deep-sea benthic fauna.

Sample washing and sieving

Washing is usually undertaken using seawater delivered at a high flow rate. Where such a rate cannot be supplied from the laboratory areas of the ship, use of one of the ship's fire hose hydrants may be allowed. In either case, unless already installed, a filtration system will need to be interposed to filter off surface plankton sucked in by the pump. The authors have used a filtration capsule containing sintered woven stainless steel wire mesh filters that are capable of taking high liquid flow rates. The filter unit can be removed and cleaned after each deployment. More recently, wound polypropylene 'Schumatex®' filter cartridges have been used, which filter over a

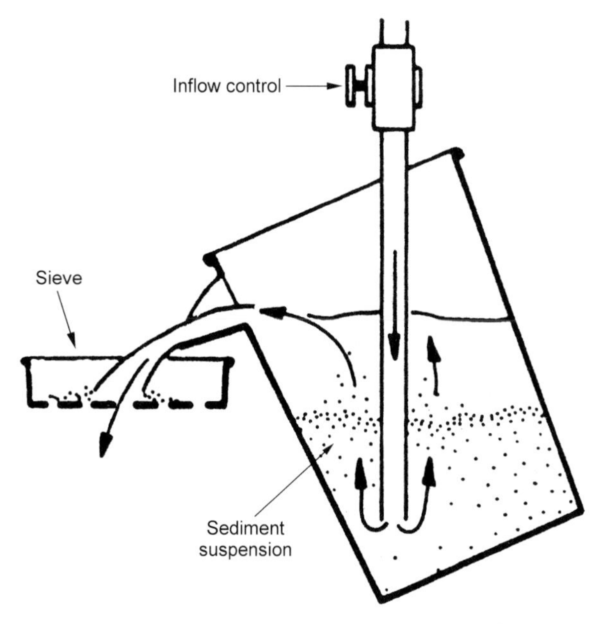

Fig. 7.24 Elutriation method of Sanders et al. (1965) for washing benthic samples from soft sediments in the deep sea. The inclined dustbin into which the sample is gently shovelled is filled with filtered sea water and sufficient flow maintained to allow the bin to overflow through the spout into the sieve. If there is no mechanical mixing this method is very gentle and efficient.

much higher filter surface area and also last for an entire cruise before replacement is needed.

One of the best methods for gently separating the smaller animals from the sediment is still the elutriation technique of Sanders et al. (1965). Aliquots of the muddy sample are placed into a galvanised steel dustbin fitted with a spout (Fig. 7.24). A suspension is created using a large-volume flow of filtered seawater, with the mixture overflowing through the spout into a large-diameter sieve. The latter retains the fauna, but allows through all sediment particles smaller than the sieve aperture. The sample in the dustbin should not be physically stirred, as this may damage delicate animals. The dustbin and sieve are usually used in cradles that can be tilted to optimise flow, with a height adjustment to prevent splashing of the overflow on the sieve. It is very important to prevent the sieve blocking so that it fills with water and overflows. No risk of this occurrence should be allowed, even if it means constant close scrutiny of the process by the operator. Stirring the sample in the dustbin will result in a richer mixture suddenly entering the sieve, and may quickly block it. Periodic removal of washed material from the sieve to the sample jar helps prevent blockages from developing. Some sediments, such as those containing many glass sponge spicules, are difficult to wash as the spicules are retained and fill the sieve. To prevent blockages from starting it may be necessary to agitate the sieve within the cradle; a gentle rotating movement of the sieve by hand backwards and forwards can be useful, and with stubborn sediment directing a flow of water upwards through the sieve from below can rapidly unblock it before it starts overflowing. This action has been semi-automated in the Wilson Auto-Siever© commercially available from Gardline Surveys Limited, Oceanography and Environmental Division, Great Yarmouth NR30 3NG, UK. This consists of a sieve holder mounted in a stainless steel deck table over a rotating system that continually sweeps the underside of the sieve with a low-pressure curtain of water. This maintains sediment fluidity at the sieve interface, thus acting to prevent the sieve apertures from blocking. A lid seals the top of the sieve allowing slightly increased water pressures, but sample loss through overfilling of the sieve with water is prevented by a small, screened escape hole in the centre of the lid. This apparatus has been successfully used with deep-sea samples in the northern Rockall Trough.

Sometimes the uppermost sediment layers of box core (or other) samples are placed directly into pre-cooled seawater formaldehyde solution for fixing while the lower layers are sieved first, thus ensuring that the bulk of the smaller fauna is rapidly fixed. It is usual to process only the upper part of box core samples to some pre-chosen depth (usually 10–20 cm) in order to avoid having to wash the deeper layers which can be unproductive for small animals (but may occasionally harbour some larger, deep-burrowing fauna), and which frequently consist of consolidated clays that are very resistant to elutriation. It is important to avoid direct hosing of specimens on a sieve mesh. Only a gentle flow of water should be used for the final rinse of the sieve residue and its preparation for transfer to a sample container.

Choice of sieve size

The choice of sieve necessary in deep-sea benthic investigations has been somewhat controversial. While it has been recognised that loss of juvenile stages, and even the adult individuals of small-bodied species, may be significant in washing benthic samples through a screen of 1.0 or 0.5 mm apertures, there has been disagreement on the degree of potential loss with deep-sea benthic samples. Eleftheriou and Holme (1984) recalculated the data generated in the benchmark study of Reish (1959), and found that a finer mesh sieve would be required for adequate retention of nematodes and crustaceans, whereas 95% of molluscs were retained on a screen of 0.85 mm. It is not known if the time of sampling would influence the presence of spat of molluscs. In our experience elsewhere, sampling during certain seasons results in much larger numbers of molluscs passing through such a 0.85 mm sieve. With polychaetes, Reish obtained a more variable pattern, with 95% of individuals of one lumbrinereid species being retained by a 1 mm sieve, while it required a 0.27 mm sieve to achieve this level for a cossurid species. The sparse data available where the same or closely related species occur in both deep-sea and shallow water indicate no size difference (Gage, 1978), so that small size in deep-sea sediments may reflect selective colonisation by low-level taxa with small bodies, rather than any more direct dwarfism of individuals. Subsequent studies have concluded that young stages may always be seriously under-represented in benthic samples, whether deep sea or shallow (Grassle et al., 1985; Bachelet, 1990). In practice, some larger-bodied 'meiofaunal' taxa (in the taxonomic sense), such as nematodes and ostracods, may in the deep sea well exceed the median size of 'macrofaunal' taxa (see e.g. Dahl et al., 1976; Gage, 1977; Thiel, 1983). Other workers have maintained that the size-based distinctions, which are now supported by the strongly peaked distribution of numbers of animals present plotted as a function of size in size spectra (Schwinghamer, 1981, 1985; Warwick, 1984), should be maintained as a more meaningful definition of the 'macrofauna'. Against these factors is the argument that it is best to use a sieve size that (i) catches the largest number of target organisms, and (ii) achieves most comparability with results from other studies. In practice, deep-sea workers have found it necessary to wash the sediment through finer meshed screens than those usually applied in benthic sampling in shallow water to collect the same groups of animals familiar in shallow water. As an extreme example, the authors' experience on the oligotrophic abyssal plains of the north east Atlantic (e.g. Madeira and Cape Verde), suggests that to recover one single specimen on a 1 mm mesh may require the total contents of 5–10 box cores!

A 0.42 mm sieve was adopted by H.L. Sanders for deep-water epibenthic sled sampling in the north west. Atlantic in the 1960s (Sanders et al., 1965). Subsequently, R.R. Hessler used a 0.297 mm sieve for abyssal sampling in the central north Pacific (Hessler & Jumars, 1974). Hessler (1974) commented that to catch any animals in such oligotrophic sediments a much finer-meshed sieve was essential, and furthermore that 'macrofaunal' taxa found here were of only meiofaunal size compared to shallow water, a 1 mm sieve catching no macrofauna. A 0.3 mm sieve was used by J.F. Grassle and his co-workers in their benchmark study of the US continental slope (Maciolek et al., 1987a, 1987b; Blake et al., 1987). Finally, a 0.25 mm mesh sieve has been standardised as determining the lower size limit of the macrofauna in French studies on the deep-water benthos dating from the 1970s (Dinet et al., 1985). This sieve size has been adopted in subsequent EU-funded sampling programmes in the deep sea from the 1980s up to the time of writing.

A size-based differentiation of the benthos into micro-, meio- and macrobenthos dates from Mare (1942). But she also recognised that the 'limits of the macro-, meio- and microbenthos will probably vary according to the habitat under consideration'. Mare recognised that the characteristic taxa of each group may well have a size range whose tails extend beyond the lower and upper limits determined by the sieves used. The argument of convenience of retaining the same size limits being more important than adapting to a shift in taxonomic composition caused, for example, by the decreased body size in the deep sea is supported by Thiel (1983). He argued that these benthic subgroups should, by definition, be determined by taxonomic composition rather than strictly by size, even when size as juveniles places them in the next lower category. The pragmatic differentiation based on 'meiofaunal' and 'macrofaunal' taxa now has wide support among deep-sea biologists as a more useful separation (Hessler & Jumars, 1974; Rowe, 1983; Jumars & Gallagher, 1982). This was termed the 'macrofaune, *stricto sensu*' (excluding nematodes, copepods and ostracods) of Dinet et al. (1985). These authors categorise the entire fraction (including any of the above three 'meiofaunal' groups) retained by the 0.25 mm sieve as 'macrofaune, *sensu lato*'.

Analysis of the effect on biomass, and total species richness and abundance, from use of differing sieve sizes has been undertaken using box core samples from the Rockall Trough (Gage et al., 2002). This work is based on sorting of all the metazoan 'macrofauna'. The results show some potential small loss of individuals and species using a 420 μm sieve compared to a 300 or 250 μm sieve, with a more serious loss of individuals, but not of species, if a 500 μm (0.5 mm) sieve is used rather than a 420 μm or smaller sieve (Fig. 7.25). Of the commonly applied statistics to measure species diversity, Shannon's diversity H' is little affected at sieve sizes below 1000 μm (1.0 mm), but evenness, measured as Pielou's J, is more affected, with J declining as sieve size decreases. Comparisons of distribution of species' abundances, with much greater numerical representation of the commoner species in progressively finer sieves, indicate that this is almost certainly the result of adding smaller, young stages to populations of single species, until the 300 μm retention.

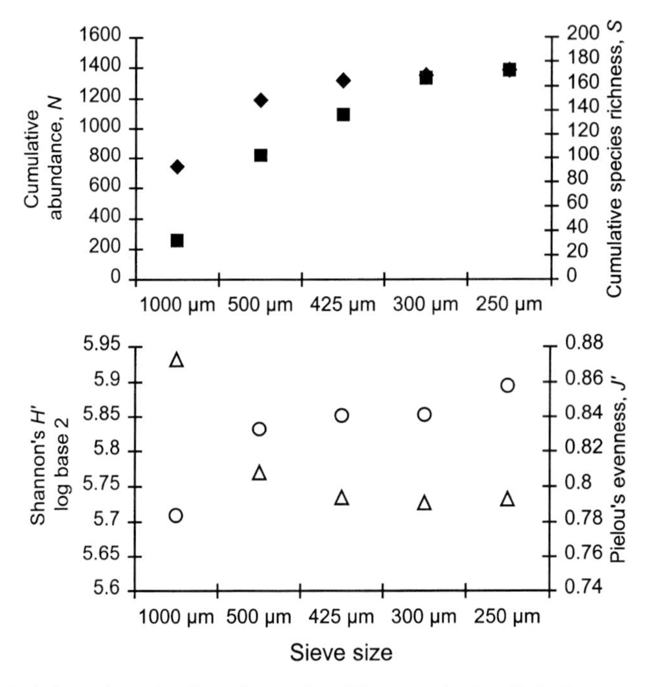

Fig. 7.25 Effect of sieve size retention of macrobenthic invertebrates (Polychaeta excluded), from a deep-sea sample taken from about 1900 m depth in the Rockall Trough, on community characteristics. Upper, total abundance (filled squares) and species richness (filled diamonds); lower, Shannon's diversity index H' ($\log_2$) (open circles) and Pielou's measure of evenness, J (a measure of the degree of heterogeneity in individual species' abundances) (open triangles).

Further processing, fixation and preservation of samples

The use of narcotics or anaesthetising agents to 'relax' organisms is unnecessary as it is unusual to recover deep-sea animals in a viable state. Washed samples are fixed using a buffered seawater solution (10%) of commercial grade formalin on board ship. This is done as soon as possible to avoid deterioration of the sample in warm conditions. The buffering is important in order to avoid decalcification of shells and other skeletal elements caused by the acidity of formaldehyde (formalin is roughly 40% formaldehyde). It can be easily buffered by adding borax powder. Ideally, transfer of the sample to a preservative should take place in about three days (by which time adequate fixation will have occurred). The sample is washed free of formalin, using either filtered sea or fresh water, and transferred to a preservative solution diluted with seawater, if necessary. We use industrial grade ethanol to which is added 10% propylene glycol. The latter is hygroscopic, and will prevent irreparable damage to the organisms if the jar, or open dish, is allowed to dry out. Note that some benthic taxa are better preserved long-term in formalin, and the reader is referred to more specialist works for further details.

We are not concerned with later stages in the analysis of samples, except to mention that large quantities of sediment may be obtained in sledge and trawl hauls, and it may be necessary to subsample the washed sample to manageable units. One of the authors has used an aerated sedimentation column for this purpose that subdivides an epibenthic sled sample into eight equal aliquots (Gage, 1982).

Acknowledgements

Discussions with many individual scientists who have shared their experiences with us has materially improved this chapter in many ways. However, we apologise if we have been unaware of some relevant aspect, development or procedure that should have been included; in mitigation, perhaps some of the most useful, information remains, unpublished and therefore is not widely known. We are grateful to Prof. Lucien Laubier (Institut Océanographique, Paris) and Dr Myriam Sibuet (Institut Français pour l'Exploitation de la Mer, IFREMER) for allowing us to use unpublished data from the *INCAL* cruise organised by IFREMER in 1976, and to Peter Lamont for critically reading through an early draft.

References

Agassiz, A. (1888) *Three cruises of the United States Coast and Geodetic Survey steamer 'Blake'*. Vol. 1. Sampson Low, Marston, Searle & Rivington, London.

Ankar, S. (1977) Digging profiles of and penetration of the van Veen grab in different sediment types. *Contributions from the Askö Laboratory, University of Stockholm, Sweden*, **16**, 22.

Aldred, R.G., Thurston, M.H., Rice, A.L. & Morley, D.R. (1976) An acoustically monitored opening and closing epibenthic sledge. *Deep-Sea Research*, **23**, 167–174.

Bachelet, G. (1990) The choice of a sieving mesh size in the quantitative assessment of marine macrobenthos: a necessary compromise between aims and constraints. *Marine Environmental Research*, **30**, 21–35.

Bagley, P.M., Priede, I.G. & Armstrong, J.D. (1990) An autonomous deep ocean vehicle for acoustic tracking of bottom living fishes. In: *Monitoring the Sea,* IEE Colloquium, Digest no. 182, Ref. E15/E11, pp. 211–213. Institute of Electrical Engineers, London.

Bagley, P.M., Priede, I.G., Jamieson, D., Battle, E.J.V. & Henriques, C. (2004) Lander techniques for deep ocean biological research. *Underwater Technology*, **26**, 3–11.

Barnett, P.R.O., Watson, J. & Connelly, D. (1984) The multiple corer for taking virtually undisturbed samples from shelf, bathyal and abyssal sediments. *Oceanologica Acta*, **7**, 399–408.

Bertram, M.A. & Cowen, J.P. (1999) Temporal variations in the deep-water colonization rates of small benthic foraminifera: The results of an experiment on Cross Seamount. *Deep-Sea Research*, **46**, 1021–1049.

Bett, B.J. (2000) Comparative benthic ecology of the Rockall Trough and Faroe-Shetland Channel, Section 4.3.2 in Environmental Surveys of the Seafloor of the UK Atlantic Margin, Atlantic Frontier Environmental Network [CD-ROM]. Available from Geotek Limited, Daventry, Northants NN11 5EA, UK.

Bett, B.J. & Gage, J.D. (2000) Practical approaches to monitoring the deep-sea environment of the UK Atlantic Margin, Section 6.2 in Environmental Surveys of the Seafloor of the UK Atlantic Margin, Atlantic Frontier Environmental Network [CD-ROM]. Available from Geotek Limited, Daventry, Northants NN11 5EA, UK.

Bett, B.J., Vanreusel, A., Vincx, M., Soltwedel, T., Pfannkuche, O., Lambshead, P.J.D., Gooday, A.J., Ferrero, T. & Dinet, A. (1994) Sampler bias in the quantitative study of deep-sea meiobenthos. *Marine Ecology Progress Series*, **104**, 197–203.

Bett, B.J., Malzone, M.G., Narayanaswamy, B.E. & Wigham, B.D. (2001) Temporal variability in phytodetritus and megabenthic activity at the seabed in the deep northeast Atlantic. *Progress in Oceanography*, **50**, 349–368.

Billett, D.S.M. & Rice, A.L. (2001) The BENGAL programme: introduction and overview. *Progress in Oceanography*, **50**, 13–25.

Billett, D.S.M., Bett, B.J., Rice, A.L., Thurston, M.H., Galeron, J., Sibuet, M. & Wolff, G. (2001) Long-term changes in the megabenthos of the Porcupine Abyssal Plain (NE Atlantic). *Progress in Oceanography*, **50**, 325–348.

Blake, J.A., Hecker, B., Grassle, J.F., Brown, B., Wade, M., Boehm, P.D., Baptiste, E., Hilbig, B., Maciolek, N., Petrecca, R., Ruff, R.E., Starczak, V. & Watling, L. (1987) *Study of biological processes on the U.S. South Atlantic slope and rise. Phase 2.* Final report prepared for US Department of the Interior, Minerals Management Service, Washington, DC, p.414 + appendices.

Blomqvist, S. (1991) Quantitative sampling of soft-bottom sediments: problems and solutions. *Marine Ecology Progress Series*, **72**, 295–304.

Boland, G.S. & Rowe, G.T. (1991) Deep-sea benthos sampling with the Gomex box corer. *Limnology and Oceanography,* **36**, 1015–1020.

Bouma, A.H. (1969) *Methods for the Study of Sedimentary Structures.* John Wiley & Sons, New York.

Bouma, A.H. & Marshall, N.F. (1964) A method for obtaining and analyzing undisturbed oceanic sediment samples. *Marine Geology*, **6**, 231–241.

Brattegard, T. & Fosså, J.H. (1991) Replicability of an epibenthic sampler. *Journal of the Marine Biological Association of the United Kingdom*, **71**, 153–166.

Buhl-Jensen, L. (1986) The benthic amphipod fauna of the west Norwegian continental shelf compared with the fauna of five adjacent fjords. *Sarsia*, **71**, 193–208.

Carey, A.G. & Hancock, D.R. (1965) An anchor-box dredge for deep-sea sampling. *Deep-Sea Research*, **12**, 983–984.

Childress, J.J. (1985) Capture and live recovery of deep-sea crustaceans. *National Geographic Society Research Reports*, **21**, 67–69.

Christiansen, B. (1996) Bait-attending amphipods in the deep-sea: comparison of three localities in the north-eastern Atlantic. *Journal of the Marine Biological Association of the United Kingdom*, **76**, 45–360.

Christiansen, B. & Nuppenau, V. (1997) The IHF Fototrawl: experiences with a television-controlled, deep sea epibenthic sledge. *Deep-Sea Research*, **44**, 533–540.

Dahl, E., Laubier, L., Sibuet, M. & Stromberg, J.-O. (1976) Some quantitative results on benthic communities of the deep Norwegian Sea. *Astarte*, **9**, 61–79.

Desbruyères, D., Bervas, J.Y. & Khripounoff, A. (1980) Rapid recolonisation of deep-sea sediment. *Oceanologica Acta*, **3**, 285–291.

Dinet, A., Desbruyères, D., & Khripounoff, A. (1985) Abondance des peuplement macro- et méiobenthiques: répartition et stratégie d'échantillonage. In: *Peuplements Profonds du Golfe de Gascogne: Campagnes Biogas.* (Eds L. Laubier, L. & C. Monniot), pp. 121–142. Institut Français de Recherche pour l'Exploitation de la Mer (IFREMER), Brest.

Eleftheriou, A. & Holme, N.A. (1984) Macrofauna techniques. In: *Methods for the Study of Marine Benthos* (Eds N.A. Holme & A.D. McIntyre), pp. 140–216. Blackwell Scientific Publications, Oxford.

Gage, J.D. (1975) A comparison of the deep-sea epibenthic sledge and anchor-box dredge samplers with the van Veen grab and hand coring by diver. *Deep-Sea Research*, **22**, 693–702.

Gage, J.D. (1977) Structure of the abyssal macrobenthic community in the Rockall Trough. In: *Biology of Benthic Organisms* (Eds B.F. Keegan, P. O'Ceidigh, & P.S. Boaden), pp. 247–260. Pergamon, Oxford.

Gage, J.D. (1978) Animals in deep-sea sediments. *Proceedings of the Royal Society of Edinburgh*, **76**B, 77–93.

Gage, J.D. (1982) An aerated sedimentation column for subsampling large benthic samples from deep-sea sediments. *Deep-Sea Research*, **29**A, 627–630.

Gage, J.D., Hughes, D.J. & Gonzalez Vecino, J.L. (2002) Sieve size influence in estimating biomass, abundance and diversity in samples of deep-sea macrobenthos. *Marine Ecology Progress Series,* **225**, 97–107.

Gage, J.D., Lightfoot, R.H., Pearson, M. & Tyler, P.A. (1980) An introduction to a sample time series of abyssal macrobenthos: methods and principal sources of variability. *Oceanologica Acta*, **3**, 169–176.

Gerber, H.W., Oebius, H.U. & Grupe, B. (1996) MAXICORER a device for taking undisturbed samples of sediment including the benthic boundary layer. *Sea Technology*, **37**(10), 66–69.

Gerdes, D. (1990) Antarctic trials of the multi-box corer, a new device for benthos sampling. *Polar Record*, **26**(156), 35–38.

Gooday, A.J. (1988) A response by benthic Foraminifera to the deposition of phytodetritus in the deep sea. *Nature*, **332**, 70–73.

Gordon, J.D.M. (1986) The fish populations of the Rockall Trough. *Proceedings of the Royal Society of Edinburgh*, **88**B, 191–204.

Gordon, J.D.M. & Bergstad, O.A. (1992) Species composition of demersal fish in the Rockall Trough, North-eastern Atlantic, as determined by different trawls. *Journal of the Marine Biological Association of the United Kingdom*, **72**, 213–230.

Gordon. J.D.M. & Duncan, J.A.R. (1985) The ecology of the deep-sea benthic and benthopelagic fish on the slopes of the Rockall Trough, Northeastern Atlantic. *Progress in Oceanography*, **15**, 37–69.

Gordon, J.D.M. & Duncan, J.A.R. (1987) Deep-sea bottom-living fishes at two repeat stations at 2200 and 2900 m in the Rockall Trough, northeastern Atlantic Ocean. *Marine Biology*, **96**, 309–325.

Gordon, J.D.M., Merrett, N.R., Bergstad, O.A. & Swan, S.C. (1996) A comparison of the deep-water fish assemblages of the Rockall Trough and Porcupine Seabight, eastern Atlantic: Continental slope to rise. *Journal of Fish Biology*, **49A**, 217–238.

Grassle, J.F. (1977) Slow recolonisation of deep-sea sediment. *Nature*, **265**, 618–619.

Grassle, J.F. (1980) *In situ* studies of deep-sea communities. In: *Advanced Concepts in Ocean Measurements* (Eds F.P. Diemer, F.J. Vernberg, & D.Z. Mirkes), pp. 321–332, University of South Carolina Press, Columbia.

Grassle, J.F. & Maciolek, N. (1992) Deep-sea species richness and local diversity estimates from quantitative bottom samples. *American Naturalist*, **139**, 313–341.

Grassle, J.F., Grassle, J.P., Brown-Leger, L.S., Petereca, R.F. & Coplely, N.J. (1985) Subtidal macrobenthos of Narragansett Bay: field and mesocosm studies of the effects of eutrophication and organic input on benthic populations. In: *Marine Biology of Polar Regions and Effects of Stress on Marine Organisms* (Eds J.S. Gray & M.E. Christiansen), pp. 421–434. John Wiley & Sons, Chichester.

Guennegan, Y. & Martin, V. (1985) Techniques de prélèvement. In: *Peuplement Profonds du Golfe de Gascogne: Campagne BIOGAS* (Eds L. Laubier & C. Monniot), pp. 571–602. Institut Français de Recherche pour l'Exploitation de la Mer (IFREMER), Brest.

Harvey, R. (1996) Deep water Xylophagaidae (Pelecypoda: Pholadacea) from the north Atlantic with descriptions of three new species. *Journal of Conchology,* **35**, 473–481.

Hessler, R.R. (1974) The structure of deep benthic communities from central oceanic waters. In: *The Biology of the Oceanic Pacific* (Ed. C.B. Miller), pp. 79–93. Oregon State University Press, Corvallis.

Hessler, R.R. & Sanders, H.L. (1967) Faunal diversity in the deep sea. *Deep-Sea Research*, **14**, 65–78.

Hessler, R.R. & Jumars, P.A. (1974) Abyssal community analysis from replicate box cores in the central North Pacific. *Deep-Sea Research*, **21**, 185–209.

Holme, N.A. (1971) Macrofauna sampling. In: *Methods for the Study of Marine Benthos* (Eds N.A. Holme & A.D. McIntyre), pp. 80–130. Blackwell Scientific Publications, Oxford.

Jahnke, R.A. & Christiansen, M.B. (1989) A free-vehicle benthic chamber instrument for seafloor studies. *Deep-Sea Research*, **26**, 625–637.

Jonasson, A. & Olausson, E. (1966) New devices for sediment sampling. *Marine Geology*, **4**, 365–372.

Jumars, P.A. (1975) Methods for measurement of community structure in deep-sea macrobenthos. *Marine Biology*, **30**, 245–252.
Jumars, P.A. (1978) Spatial autocorrelation with RUM (Remote Underwater Manipulator): vertical and horizontal structure of a bathyal benthic community. *Deep-Sea Research*, **25**, 589–604.
Jumars, P.A. & Gallagher, E.D. (1982) Deep-sea community structure: three plays on the benthic proscenium. In: *The Environment of the Deep Sea* (Eds W.G. Ernst & J.G. Morin), pp. 217–255. Prentice-Hall, Englewood Cliffs, NJ.
Jumars, P.A., Jackson, D.R., Gross, T.F. & Sherwood, C. (1996) Acoustic remote sensing of benthic activity: a statistical approach. *Limnology and Oceanography*, **41**, 1220–1241.
Lampitt, R.S. & Burnham, M.P. (1983) A free-fall time-lapse camera and current meter system 'Bathysnap' with notes on the foraging behaviour of a bathyal decapod shrimp. *Deep-Sea Research*, **30**, 1009–1017.
Maciolek, N.J., Grassle, J.F., Hecker, B., Boehm, P.D., Brown, B., Dade, B., Steinhauer, W.G., Baptiste, E., Ruff, R.E. & Petrecca, R. (1987a) *Study of biological processes on the U.S. mid-Atlantic slope and rise.* Final report prepared for US Department of the Interior, Minerals Management Service, Washington, DC.
Maciolek, N.J., Grassle, J.F., Hecker, B., Brown, B., Blake, J.A., Boehm, P.D., Petrecca, R., Duffy, S., Baptiste, E. & Ruff, R.E. (1987b) *Study of biological processes on the U.S. North Atlantic slope and rise.* Final report prepared for US Department of the Interior, Minerals Management Service, Washington, DC.
Mare, M.F. (1942) A study of a marine benthic community with special reference to the micro-organisms. *Journal of the Marine Biological Association of the United Kingdom*, **25**, 517–554.
McIntyre, A.D. (1971a) Meiofauna and microfauna sampling. In: *Methods for the Study of Marine Benthos.* (Eds N.A. Holme & A.D. McIntyre), pp. 131–139. Blackwell Scientific Publications, Oxford.
McIntyre, A.D. (1971b) Efficiency of benthos sampling gear. In: *Methods for the Study of Marine Benthos.* (Eds N.A. Holme & A.D. McIntyre), pp. 140–146. Blackwell Scientific Publications, Oxford.
Menzies, R.J. & Rowe, G.T. (1968) The LUBS, a large undisturbed bottom sampler. *Limnology and Oceanography,* **13**, 708–714.
Merrett, N.R. & Marshall, N.B. (1981) Observations on the ecology of deep-sea bottom-living fishes collected off northwest Africa (08°–27°N). *Progress in Oceanography*, **9**, 185–244.
Merrett, N.R & Haedrich, R.L. (1997) *Deep-Sea Demersal Fish and Fisheries.* Chapman & Hall, London.
Miller, R.J. (1990) Effectiveness of crab and lobster traps. *Canadian Journal of Fisheries and Aquatic Science*, **47**, 1228–1251.
Phleger, C.F., McConnaughey, R.R. & Crill, P. (1979) Hyperbaric fish trap operation and deployment in the deep sea. *Deep-Sea Research*, **26**, 1405–1409.
Priede, I.G., Bagley, P.M., Smith, A., Creasey, S. & Merrett, N.R. (1994) Scavenging deep demersal fishes of the Porcupine Seabight, North-East Atlantic: observations by baited camera, trap and trawl. *Journal of the Marine Biological Association of the United Kingdom*, **74**, 481–498.
Reineck, H.E. (1958) Kastengreifer und Lotröhre 'Schnepfe', Gerate zur Entnahme ungestörter, orientierter *Meersgrundproben. Senckenbergiana lethaea*, **39**, 42–48.

Reineck, H.E. (1963) Der Kastengreifer. *Natur und Museum*, **93**, 102–108.

Reish, D.J. (1959) A discussion of the importance of screen size in washing quantitative marine bottom samples. *Ecology*, **4**, 307–309.

Rice, A.L. (1987) Benthic transect photography. In: *Great Meteor East: A Biological Characterisation*, pp. 144–148. Institute of Oceanographic Sciences, Deacon Laboratory, Report No. 248.

Rice, A.L., Aldred, R.G., Billett, D.S.M. & Thurston, M.H. (1979) The combined use of an epibenthic sledge and deep-sea camera to give quantitative relevance to macrobenthos samples. *Ambio Special Report*, **6**, 59–72.

Rice, A.L., Aldred, R.G., Darlington, E. & Wild, R.A. (1982) The quantitative estimation of the deep-sea megabenthos: a new approach to an old problem. *Oceanologica Acta*, **5**, 63–72.

Rothlisberg, P.C. & Pearcy, W.G. (1976) An epibenthic sampler used to study the ontogeny of vertical migration of *Pandalus jordani* (Decapoda, Caridea). *Fishery Bulletin*, **74**, 990–997.

Rowe, G.T. (1983) Biomass and production of the deep-sea macrobenthos. In: *The Sea,* Vol. 8 (Ed. G.T. Rowe), pp. 97–121. Wiley-Interscience, New York.

Rowe, G.T. & Clifford, C.H. (1973) Modifications of the Birge–Ekman box corer for use with SCUBA or deep submergence research vessels. *Limnology and Oceanography*, **18**, 172–175.

Rowe G.T. & Sibuet, M. (1983) Recent advances in instrumentation in deep-sea biological research. In: *The Sea,* Vol. 8 (Ed. G.T. Rowe), pp. 81–95. Wiley-Interscience, New York.

Sanders, H.L., Hessler, R.R. & Hampson, G.R. (1965) An introduction to the study of the deep-sea benthic assemblages along the Gay Head – Bermuda transect. *Deep-Sea Research*, **12**, 845–867.

Schwinghamer, P. (1981) Characteristic size distribution of integral benthic communities. *Canadian Journal of Fisheries and Aquatic Science,* **38**, 1255–1263.

Schwinghamer, P. (1985) Observations on size structure and pelagic coupling of some shelf and abyssal benthic communities. In: *Proceedings of the Nineteenth European Marine Biology Symposium* (Ed. P.E. Gibbs), pp. 347–359. Cambridge University Press, Cambridge.

Shirayama, Y. & Fukushima, T. (1995) Comparisons of deep-sea sediments and overlying water collected using multiple corer and box corer. *Journal of Oceanography*, **51**, 75–82.

Sibuet, M., Floury, L., Alayse-Danet, A.M., Echardour, A., LeMoign, T. & Perron, R. (1990) *In situ* experimentation at the water/sediment interface in the deep sea: 1. Submersible experimental instrumentation developed for sampling and incubation. *Progress in Oceanography*, **24**, 161–167.

Smith, K.L. & Howard, J.D. (1972) Comparison of a grab sampler and large volume corer. *Limnology and Oceanography*, **17**, 142–145.

Smith, K.L. & White, G.A. (1982) Ecological energetics studies in the deep-sea benthic boundary layer: *in situ* respiration studies. In: *The Environment of the Deep Sea.* (Eds W.G. Ernst & J.G. Morin), pp. 279–300, Prentice-Hall, Englewood Cliffs, NJ.

Smith, K.L., Alexandrou, D. & Edelman, J.L. (1989) Acoustic detection and tracking of abyssopelagic animals: description of an autonomous split-beam acoustic array. *Deep-Sea Research*, **36**, 1427–1441.

Smith, K.L., Glatts, R.C., Baldwin, R.J., Beaulieu, S.E., Uhlman, A.H., Horn, R.C. & Reimers, C. (1997) An autonomous, bottom-transecting vehicle for making long time-series measurements of sediment community oxygen consumption to abyssal depths. *Limnology and Oceanography*, **42**, 1601–1612.

Smith, K.L., White, G.A., Laver, M.B., McConnaughey, R.R. & Meador, J.P. (1979) Free vehicle capture of abyssopelagic animals. *Deep-Sea Research*, **26**, 57–64.

Snelgrove, P.V.R., Butman, C.A. & Grassle, J.F. (1995) Potential flow artefacts associated with benthic experimental gear: deep-sea mudbox examples. *Journal of Marine Research*, **53**, 821–845.

Tengberg, A. and 29 others (1995) Benthic chamber and profiling landers in oceanography: a review of design, technical solutions and functioning. *Progress in Oceanography,* **35**, 253–294.

Thiel, H. (1966) Quantitative Untersuchungen über die Meiofauna des Tiefseeboden. *Veröffentlichungen des Instituts für Meeresforschungin Bremerhaven,* **2**, 131–147.

Thiel, H. (1975) The size structure of the deep-sea benthos. *Internationale Revue der Gesamten Hydrobiologie*, **60**, 575–606.

Thiel, H. (1983) Meiobenthos and nanobenthos of the deep sea. In: *The Sea*, Vol. 8 (Ed. G.T. Rowe), pp. 167–230. Wiley-Interscience, New York.

Thistle, D., Yingst, J.Y. & Fauchald, K. (1985) A deep-sea benthic community exposed to strong near-bottom currents on the Scotian Rise (western Atlantic). *Marine Geology*, **66**, 91–112.

Thorson, G. (1957) Sampling the benthos. In: *Treatise on Marine Ecology and Paleoecology*, Vol. 1, (Ed. J. Hedgepeth), pp. 61–86. Geological Society of America, New York.

Thurston, M.H., Bett, B.J. & Rice, A.L. (1995) Abyssal megafaunal necrophages: Latitudinal differences in the eastern North Atlantic Ocean. *Internationale Revue der Gesamten Hydrobiologie,* **80**, 267–286.

Warwick, R.M. (1984) Species size distributions in marine benthic communities. *Oecologia,* **61**, 32–41.

Whitelaw, A.W., Sainsbury, K.J., Dews, G.J. & Campbell, R.A. (1991) Catching characteristics of four fish-trap types on the north west shelf of Australia. *Australian Journal of Marine and Freshwater Research*, **42**, 369–382.

Wigley, R.L. (1967) Comparative efficiencies of van Veen and Smith–McIntyre grab samplers as revealed by motion pictures. *Ecology*, **48**, 168–169.

Wilson, R.R. and Smith, K.L., Jr. (1985) Live capture, maintenance and partial decompression of a deep-sea grenadier fish (*Coryphaenoides acrolepis*) in a hyperbaric trap-aquarium. *Deep-Sea Research*, **32**, 1571–1582.

Chapter 8

Measuring the Flow of Energy and Matter in Marine Benthic Animal Populations

J. van der Meer, C.H. Heip, P.J.M. Herman, T. Moens and D. van Oevelen

8.1 Introduction

It can be said that the flow of energy and matter through an ecosystem may start with the primary producers, who store solar energy into a collection of organic compounds. By using water, carbon dioxide and minerals such as nitrogen and phosphorus, the primary producers (autotrophs) basically produce all organic compounds (such as polysaccharides, lipids and proteins) they need themselves. In contrast, herbivores, carnivores and decomposers (heterotrophs) feed on these energy-rich compounds, which they need both for extracting energy (in order to be able to do 'work') and for obtaining the required 'building blocks' for tissue growth and reproductive material.

Because energy is stored into chemical compounds, the fluxes of energy and mass are closely linked. Yet there is one important difference between these fluxes: ecosystems require the input of (solar) energy from outside, which finally dissipates as heat into outer space. Ecosystems are thus open systems in terms of energy. Material cycles, on the other hand, can be closed, since autotrophs use simple minerals to synthesise complex compounds which are then decomposed again into these simple minerals by heterotrophs. If only heterotrophic organisms or populations are considered, the intake of energy and material are nevertheless closely coupled. Hence the flow of energy may be used as a description of the productivity of herbivores and carnivores. This production is often called secondary production, and in the past much effort has been put in the measurement of secondary production.

The study of energy flows is based on the principle that biological systems obey the laws of thermodynamics. These laws hold at all levels of organisation in biology and thus enable a link between the various levels. The first law, for example, states that energy can neither be created nor destroyed. Hence, energy flows at the population level are directly linked to energy flows at the level of the individual. Indeed, modern work on population dynamics deals with structured populations, in which individuals are no longer treated as identical, but are characterised by their size, their energy stores, and (possibly) other state variables. State-dependent energy fluxes (i.e. energy intake and allocation to maintenance, growth and reproduction) eventually determine the birth and death processes. A similar link exists between the level of the individual and cellular and molecular processes. The study of life-history strategies also profits from an energetics point of view, because the evolutionary demands put on survival and reproduction work only within the constraints on energy fluxes set by the laws of thermodynamics. Hence, while traditionally the rationale for energy flow studies was found in the elucidation of energy transfers within ecosystems or within the practical context of the rational management of resources, its scope embodies almost all biology, including the field of population dynamics and evolutionary studies.

The benthic habitat may contain both the primary producers, the so-called phytobenthos, and herbivores and carnivores, the zoobenthos. In this chapter we focus on the measurement of the flow of energy and matter in the zoobenthos. Measurement of primary production of the phytobenthos and microbial ecology are not discussed here.

State variables and units of measurement

Energy flow studies by different authors are often difficult to compare, mainly for two reasons. First, there is lack of a general theoretical framework and second, the use of different units in which the energy flow or other processes are expressed. With respect to the latter, we strongly recommend the use of the SI system (Taylor, 1995). One should use the metre as the unit of length, kilogram as the unit of mass, newton as the unit of force, joule (not kcal) as the unit of energy and watt (not kJ/day) as the unit of power (or energy expenditure rate). Time is, as far as possible, expressed in seconds, but the use of day and year (notation *a*) is (if really appropriate) allowed.

The state of an animal is usually given by its biomass or its energy content. Biomass is defined here as the mass of an individual animal (or a collection of animals). This mass can consist of living tissue and dead structures (shells, teeth, etc.) that have been built by the animal. When referring to a single organism, biomass is usually called the weight of that organism but it is always expressed in units of mass (kg). Strictly speaking, weight should be given in units of force (newton). Alternatively, biomass is sometimes expressed in terms of the biomass (or the number of molecules) of a specific chemical component (element), e.g. total organic C, total organic N, DNA, ATP, etc. The energy content of animal tissue is often defined as the heat content measured (e.g. in a bomb calorimeter) in relation to the fully oxidised form of the organic material (carbon dioxide, water) as the assumed zero energy state. This definition suffers from one minor source of ambiguity in that it actually refers to the enthalpy instead of to the more useful Gibbs free energy. It thus neglects the entropy, which at constant temperature is proportional to the difference between the enthalpy and the Gibbs free energy, and which is, among other things, dependent upon the actual concentration of the compounds. For practical purposes in animal energetics, however, the entropy can be safely set to zero (Kooijman, 2000).

The lack of a general theoretical framework is, of course, a greater source of concern, and we will come back to this problem. In some modelling approaches, for example, it is essential that the state of an animal should be characterised by both a structural part and energy reserves (Kooijman, 2000; Van der Meer & Piersma, 1994).

8.2 Energy and mass budgets of individual organisms

The flow of energy and matter into, within and out of an individual organism can be divided into a number of separate processes. In the International Biological Programme (IBP) various fluxes were distinguished, the most important ones being:

- *Consumption or ingestion (I)*, total uptake of energy or mass;
- *Absorption (A)*, part of the ingestion that crosses the gut wall;
- *Defaecation (F)*, part of the ingestion that is not absorbed, but leaves the gut as faeces;
- *Growth (dW/dt)*, part of the absorption that is incorporated in the body tissue of the organism;
- *Reproduction (G)*, part of the absorption that is released as reproductive bodies;
- *Excretion (E)*, part of the absorption that is released out of the body in the form of urine, or other exudates (with the exception of reproductive bodies);
- *Respiration (R)*, part of the absorption that is released in association with the oxidation of organic compounds, and thus causes a net loss of CO_2.

Assuming the conservation of energy and mass, the equations, which have played a key role in the IBP, may be written as:

$$A = I - F$$

and

$$dW/dt = A - (G + E + R)$$

where all terms refer to rates, expressed as units of energy or mass (such as the number of carbon atoms) per unit of time. The IBP definition of assimilation as 'physiologically useful energy' has revealed some confusion. Sometimes the term has been regarded as a synonym of absorption, but it also has been circumscribed as absorption minus excretion. Penry (1998) defines assimilation as the anabolic process, i.e. the incorporation of absorbed products into the tissue of the organism. Absorbed products that are not assimilated are used for catabolic processes. We follow that definition in this chapter. In the IBP, relations between various fluxes were usually summarised in terms of so-called coefficients of efficiency, such as the *coefficient of efficiency of absorption*, A/I, and the *coefficient of growth efficiency*, $(dW/dt + G)/I$. When the ingested food contains much inorganic material, which is egested in unchanged form, the *coefficient of efficiency of absorption*, A/I, is often estimated by the Conover ratio $(g_i - g_f)/((1 - g_f)g_i)$, where g_i is fraction organic matter in ingested food (organic matter/(inorganic plus organic matter)) and g_f is the organic fraction in faeces (Conover, 1966). The term coefficient is often misleading, because most of them are far from constant. For example, the growth efficiency will vary strongly with size and food availability (at low food levels, relatively more of the ingested energy is forwarded to body maintenance and thus to respiration).

Hitherto, many studies have followed this IBP recipe of constructing an energy budget for an individual animal. Since the budget must balance, a term particularly hard to measure was often found by difference. In the *scope for growth* approach, for example, all terms apart from growth and reproduction are measured (Bayne & Newell, 1983). The *scope for growth*, which is the difference between absorption and excretion plus respiration, of a 'standard' animal (e.g. a blue mussel of 1 g dry mass) has been frequently used as an indicator of the 'health' of the ecosystem (Smaal & Widdows, 1994; Widdows et al., 1995).

One of the major shortcomings of IBP-type energy budget studies is that the results are only descriptive and very hard to generalise, even towards animals of the same species but which differ in size. Measuring the *scope for growth* of a single individual of a particular size will not elucidate the link between the energy budget and, for example, the age–size relationship. Knowledge of the relations between all energy budget terms and body mass of the individual is required. The terms that were distinguished in the IBP approach have been chosen because they are relatively easy to measure. Unfortunately, their relationship to body mass cannot easily be derived from first principles. Respiration, for example, although it has often been thought to reflect the basal maintenance costs of the body (which are, among other things, due to the maintenance of concentration gradients across membranes and the turnover of structural body proteins), also includes processes that are directly coupled to the ingestion of food and to the growth of body tissue. Hence, while it may be argued from the basic principle of homeostasis that maintenance costs should be directly proportional to body mass (Gurney & Nisbet, 1998; Kooijman, 1986, 2000), a simple mechanistic relationship between respiration and body mass is clearly beyond our reach. A further complicating factor that is not accounted for in the IBP approach, is that not all surplus energy that is available for growth will immediately be used for growth of the structural part of the body. It may be stored temporarily in a reserve tissue buffer. As various processes, such as ingestion and maintenance, will basically be related to the structural part of the body, such distinction between a metabolically active structural part of the body and inactive reserve tissue, seems to be a prerequisite for a proper understanding of energy budgets of individual organisms.

Hence, to establish a link between the energy fluxes into, within and out of an individual and patterns of growth and reproduction, a strategic modelling approach is required. Kooijman (2000) has written the most comprehensive textbook on dynamic energy budget (DEB) models, and we will only briefly, and mainly for illustrative purposes, summarise his DEB model. Various alternative models have been constructed (Gurney et al., 1990; Gurney & Nisbet, 1998; Lika & Nisbet, 2000; Nisbet et al., 1996, 2000), and the major challenge now is to find out what sort of experiments enable the selection of the most appropriate model (Noonburg et al., 1998). A disadvantage of the DEB models compared to the IBP approach (apparently, there is no free lunch) which complicates the process of model selection, is that the state variables and various fluxes (maintenance) are very hard, if not impossible, to measure directly. Ingenuity is called for to link experiments to models.

The kappa-rule DEB model

The kappa-rule DEB model of Kooijman (2000), here restricted to the case of isomorphs (animals that do not alter in shape as they grow in size) and ectotherms (animals that do not heat their body to a constant body temperature), first assumes that organisms basically consist of two parts: a structural body and a reserve pool. Hence, the model uses two state variables: structural body volume and energy density. Structural body volume is related to any practical measure of body length by means of a cubic relationship. Energy density is defined as the amount of energy reserves per unit of structural volume.

The model assumes that absorbed food enters the reserve pool, and that all the energy that subsequently leaves the pool is allocated to either maintenance and growth (where maintenance has priority) or to maturity and reproduction. It is further assumed that (i) ingestion is proportional to the surface area of the organism and is related to food density through a Holling type II curve, (ii) a constant fraction of the ingested energy is absorbed and enters the reserve pool, (iii) the utilisation of the reserve density follows a first-order process, (iv) a fixed proportion of the utilised energy (catabolic rate) is spent on growth plus maintenance, the rest is spent on maturity (in case of juveniles) and reproduction (in case of adults) and (v) maintenance costs are proportional to body volume.

The model predicts that under constant food conditions growth will follow the Von Bertalanffy growth equation.

8.3 Methods for estimating the energy budget of an individual organism

The following section focuses on the practical aspects (the main objective of this handbook) of measuring the energy budget of an individual organism. One should, however, realise that this is only the first step. Modelling the dynamic energy budget of the individual, which includes both the definition of the model structure and the estimation of the model parameters, is the important second step, but will not be discussed here. Dynamic energy budget models have been constructed for a few marine benthic invertebrate species, e.g. the blue mussel (Ross & Nisbet, 1990; Van Haren & Kooijman, 1993) and the Pacific oyster (Ren & Ross, 2001).

Mass, size, chemical composition and energy content

In studies of energy and mass flows, the state of an individual animal is usually expressed in terms of its biomass or its energy content, or alternatively, in terms of some specific chemical component, e.g. total organic carbon content, total organic nitrogen content, DNA, ATP, etc. Overviews are provided in Tables 8.1 and 8.2. Brey et al. (1988), for example, argue for the use of organic carbon content as a biomass indicator. Biomass can be determined either directly on a balance (or with a device based on a string, which actually measures weight) or indirectly through the measurement of body size, i.e. length or volume. A calibration relationship is subsequently used to transform size in mass. Particularly, when a large number of specimens have to be treated, it is much easier to determine individual biomass indirectly.

If the aim is to characterise the animal by a structural part and a pool of reserves, no simple methods are available and many caveats are to be expected (Van der Meer & Piersma, 1994).

Biomass

Three measures of mass are regularly used in benthic studies: wet mass (WM), dry mass (DM) and ash-free dry mass (AFDM). For animals with shells, the shell mass is often determined separately. The procedures to determine mass seem straightforward, but there are many caveats in the methodologies. The outcome therefore depends strongly on the procedure used, which should be described in detail in any benthic study. Unfortunately, despite several efforts there is as yet no standardised procedure (see Brey et al. (1988) for an overview of different methodologies used).

Wet mass is the mass of a specimen, either living or dead, not subjected to drying procedures. There are three main problems with this measure: first, the amount of additional water inside the body or on outer surfaces of the body can

Table 8.1 Direct determination of biomass and chemical composition of individuals.

Variable	Problems	Recommended procedures
Wet mass	Water content Gut content Hard parts (shells, spines, jaws, bones, . . .)	
Dry mass	As with wet mass, in addition temperature and duration of drying	60°C during 24 hours 80°C during 48 hours
	Static electricity for small animals	Degaussing of weighing pan
Ash-free dry mass	Temperature and duration of burning in muffle furnace	550°C during four hours 570–580°C during two hours
Total carbon and nitrogen		Determined directly in an Element Analyser
Carbohydrates		Total carbohydrates according to Gerchacov and Hatcher (1972). Acid soluble carbohydrates extracted in 0.1 M HCl (2 hours at 50°C)
Lipids		Extraction from dried sediments after sonication by direct elution with chloroform and methanol. Analysis according to Bligh and Dyer (1959) and Marsh and Weinstein (1966)
Proteins		Extraction with 0.5 M NaOH during 4 hours. Determination according to Hartree (1972) and Rice (1982)
Energy content	Determination of calorific content requires in the order of 5–10 mg dry mass and is therefore very difficult for small animals	Determination of calorific content of dried tissue with a microbomb calorimeter (Phillipson, 1964) Calorific values can be calculated from CHN analysis (Gnaiger & Bitterlich, 1984), which requires less material

be considerable, e.g. in the gut, on gills, inside the shell, etc. Second, the gut can contain both organic and inorganic material that is not part of the organism. Third is the mass of dead structures, e.g. shells, spines, bones, skeletons, etc. Shells are nearly always removed mechanically.

When the specimens are dried at temperatures below 100°C (usually 60–80 °C) until constant mass (after 24–48 hours), most of the water will be removed but the problem with the gut content remains. In order to determine the ash-free dry mass, specimens are ashed in a muffle furnace at temperatures around 550–580°C for 2 hours. The ash mass is then subtracted from the dry mass in order to obtain the ash-free dry mass. When the carbonate content is large and the ash mass constitutes a considerable fraction of the total dry mass, as in molluscs, barnacles, etc., the loss

Table 8.2 Various conversion factors that have been used to compare carbon content, mass and energy content data of benthic animals.

C/WM (nematodes)	0.124	Jensen (1984)
C/WM (polychaetes, crustaceans, bivalves)	0.043	Rowe (1983)
C/DM (nematodes)	0.4	Feller and Warwick (1988)
C/AFDM (polychaetes, crustaceans, bivalves)	0.375	Rowe (1983)
C/AFDM (aquatic invertebrates)	0.52	Salonen et al. (1976)
Energy/C (aquatic invertebrates)	46 kJ/g	Salonen et al. (1976)
Energy/AFDM (macrofauna)	23 kJ/g	Beukema (1997), Brey et al. (1988)

of mass on ignition could in theory rise to 44% of the ash mass of the calcium carbonate present (Crisp, 1984). When the shell can be removed from the soft tissues, the ash-free dry mass of the shell can be found by decalcifying it in dilute hypochloric acid or by using calcium chelating agents in mildly acid media. When the shell or skeleton cannot be mechanically separated from the tissues, it can often be separated by chemical cleaning with attendant destruction of the soft tissues by boiling the whole animal in 10% by mass of aqueous caustic alkali (see Crisp (1984) for a more detailed description).

AFDM still includes organic material present in the gut, but, in general, AFDM is less variable within and between species than WM or DM. There is, however, a major disadvantage of using AFDM, which is the destruction of the specimens. It is our belief that this should be avoided in most ecological work and that specimens should be conserved whenever possible to allow for later analysis, including molecular analysis (which also has consequences for fixation and preservation; see below).

Determination of biomass is most often done on specimens fixed in some preservative. Formalin (in seawater buffered with 40 g/l borax) and ethyl alcohol are mostly used for this purpose. Formalin (4% solution of formaldehyde in water) is most frequently used but is harmful to human health. When using formalin, good ventilation of the working place is essential. Some substitutes for formalin are dowicil 75 (Dow Chemical) and kohrsolin (Bode Chemie), but they are rarely used (Brey, 1986). Formalin results in loss of dry mass and ash-free dry mass, which is due to leaching of organic compounds. The effect of these three preservatives on dry mass and ash-free dry mass determinations on two mollusc species has been described by Brey (1986). Dry mass did not differ significantly from initial dry mass after 100 days of preservation, but ash-free dry mass stabilised at −22 to −23% of the initial value. The formalin substitute dowicil increased dry mass by 12–14% and ash-free dry mass by 49–54% after 194 days. Kohrsolin did not change dry mass but decreased ash-free dry mass to −14 to −17% of the initial value. Dowicil was not recommended by Brey (1986) because it forms a sticky layer on shells and tissues.

Few parallel data are available on meiofauna. Widbom (1984) found that dry mass of formalin-preserved animals was 44% greater than that of unpreserved animals, and for ash-free dry mass the figure was 29%. Jensen (1984) found carbon losses of 8–24% in nematodes preserved in 4% formaldehyde at room temperature, compared to unpreserved animals or specimens kept frozen in formaldehyde. Effects on specific body components such as low-molecular weight compounds, lipids, proteins, or carbohydrates, may be much more pronounced (Danovaro et al., 1999; Moens et al., 1999).

Body size and size–mass relationships

Determination of length and width are important basic requirements in several methods used in energy budget studies. In hard-bodied species (fish, crustaceans), these linear dimensions are easy to measure but in soft-bodied species they have to be determined from drawings or photographs or calculated using Image Analysis software. This requires calibration. In specific cases, derived measures of width can be used when this is more convenient or when this is the tradition for a certain taxonomic group, for example, measurement of width of a certain segment in polychaetes, the measurement of length or width of jaws, of a certain distance on the carapax of crustaceans, etc. When using such proxies again, calibration is essential. In meiofauna ecology it has long been a tradition to determine volumes. Volumes are calculated from scale drawings made under the camera lucida or from microphotographs, but these are now increasingly calculated using Image Analysis software.

For isomorphic animals, which do not change in shape with the increasing size, such as nematodes in the meiobenthos, a two-step procedure can be used to predict mass from length and width measurements. First, volume V is predicted from length l and width w by the relation $V = alw^2$, where a is a dimensionless body shape factor (Warwick & Price, 1979). Feller and Warwick (1988) made clay models of irregularly shaped meiofauna and measured shape factors for 12 groups of meiofauna (Table 8.3). For copepods, the value of the shape factor varied between 0.23 for depressed and 0.75 for more cylindrical species (Fig. 8.1). Note that a perfect cylinder has a shape coefficient equal to $\pi/6$. Knowledge of the specific density can subsequently be used to transform volume into biomass.

More often, however, the assumption that animals within a species are of the same shape regardless of size is not made. Mass is predicted from length (or some measure of width) using an empirically derived allometric relationship of the form $W = al^b$ (thus, without the requirement that b should be equal to 3). Brey (1999) provided an overview of parameter values that have been used. Usually, the parameters a and b are estimated from a linear regression of $\log W$ versus $\log l$. This procedure however reveals biased parameter estimates and we recommend a non-linear weighted least-squares procedure using untransformed data (Wetherill, 1986).

Mass can be directly calculated from volume when the specific density is known. Density has been determined by flotation in different mixtures of kerosene and bromobenzene (Wieser, 1960) or in a gradient column of bromobenzene and xylene (Low & Richards, 1952). For nematodes, for example, the only datum used is 1.13 (Wieser, 1960).

Table 8.3 Estimated values of the dimensionless shape factor *a* from the equation $V = alw^2$, with *l* length, *w* maximum width and *V* volume, for different meiofauna groups (Feller & Warwick, 1988).

Group	*a*
Nematodes	0.530
Ostracods	0.450
Halacarids	0.399
Kinorhynchs	0.295
Turbellarians	0.550
Gastrotrichs	0.550
Tardigrades	0.614
Hydroids	0.385
Polychaetes	0.530
Oligochaetes	0.530
Tanaids	0.400
Isopods	0.230

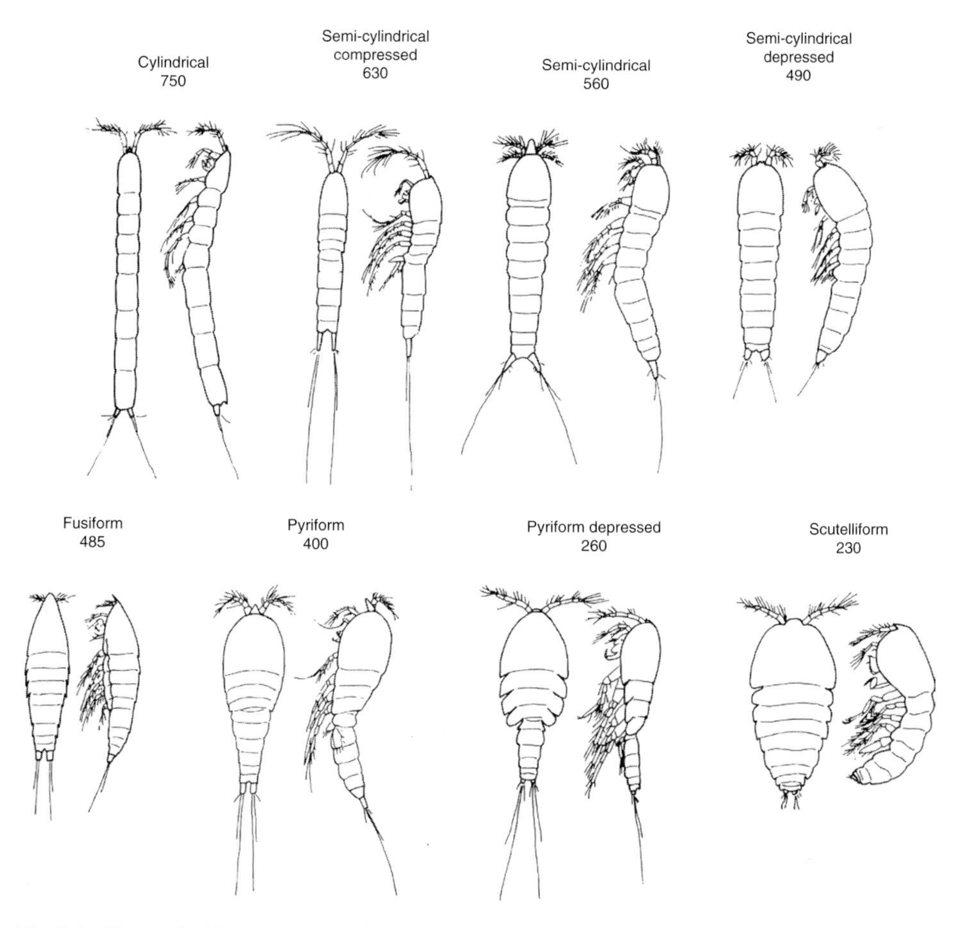

Fig. 8.1 Harpacticoid copepod body forms with the dimensionless conversion factors *a* (×1000) from the equation $V = alw^2$, where *V* is body volume, *l* is length and *w* is width. Reproduced from Warwick and Gee (1984) with permission.

Specific chemical components: organic carbon, organic nitrogen and ATP

There are three groups of methods in use to determine organic carbon (Nieuwenhuize et al., 1994). The first group is based on direct determination by wet oxidation. In the second group of methods, the organic carbon content is obtained by the difference before and after dry combustion at high temperature, analogous to ash-free dry mass. The third group is based on the removal of inorganic carbon by acid leaching before measurement of the carbon content. The method described by Nieuwenhuize et al. (1994) is based on the partitioning of inorganic and organic carbon by acidification with 25% HCl *in situ* and measurement in an element analyser, of which several commercial types exist. In the Carlo Erba NA-1500, the sample is oxidised by combustion at 1010°C using an oxidation catalyst containing Cr_2O_3 and $AgCo_3O_4$. The combustion products subsequently pass a reduction reactor containing elemental copper at 650°C to transform NO_x compounds to N_2 and to remove excess oxygen. After removal of water, N_2 and CO_2 are separated by gas chromatography and detected by thermal conductivity. The minimum amount of material required for carbon is about 30 μg dry mass.

Elemental analysers are now in wide use and have mostly replaced the classical Kjeldahl method for determination of nitrogen. In this method, the tissue is digested in concentrated sulfuric acid containing a catalyst (copper sulphate, selenium oxide, mercuric oxides). After digestion, excess of strong alkali is added and the ammonia released is driven off by passing steam through a condenser unit or in a distillation cell and estimated by titration. Not all the organic nitrogen is converted into ammonia but the difference is small.

ATP content is a measure of instantly available energy. It can be measured on the basis of the principle that when luciferin reacts with oxygen – a reaction that is driven by ATP – in the presence of the enzyme luciferase, light is emitted. This light emission can be monitored luminometrically. The samples need to be stored frozen at −80°C (−20°C can only be considered for short-term storage). Upon analysis, they should immediately be transferred from the freezer to a boiling water bath for 15 minutes in order to destroy ATPase activity, at the same time effectively releasing ATP. Particulate matter is precipitated by centrifugation, and the supernatant containing the ATP is (sub)sampled. Unless detailed information on expected ATP concentrations is already available, it is usually necessary to analyse an ATP dilution series (Braeckman et al., 2001) as well as several dilutions prepared from the sample supernatant. Therefore, the use of a luminometer equipped with a microplate reader greatly facilitates analysis. To avoid bias from slight short-term variability in reaction kinetics, it is advisable to integrate light emission over a time span rather than to perform point measurements (Braeckman et al., 2001). ADP can be completely converted to ATP – and is then measurable as excess ATP – in a coupled reaction, catalysed by pyruvate kinase, where phosphoenolpyruvate is converted into pyruvate.

Energy content

The energy content can be determined directly in a bomb calorimeter (Fraschetti et al., 1994; Phillipson, 1964). Samples of 1 g or less are used. In a micro-bomb calorimeter, samples of 5–10 mg can be used. The material must be thoroughly dried.

The literature on energy content of macrobenthos was summarised by Brey et al. (1988), and Beukema (1997). Brey et al. (1988) provide a table of 376 values of energy conversion based on DM representing 308 species and 255 values based on AFDM, representing 229 species. The general median conversion factor from AFDM to energy content is 23.09 kJ g^{-1}, and the variance for the 255 values is 4.42. The average energy content depends on the amount of carbohydrates, proteins and lipids. The energy values of these compounds are 17.16 kJ g^{-1} DM for carbohydrates, 23.65 kJ g^{-1} DM for proteins, and 39.55 kJ g^{-1} DM for lipids (Crisp, 1984).

Ingestion, absorption and defaecation

Ingestion, absorption and defaecation will be separately discussed for deposit feeders, predators and filter feeders, as each group requires its specific approach.

Ingestion and absorption by deposit feeders

Deposit feeders either ingest whole sediment or selectively choose a particular fraction of the sediment (e.g. based on particle size or density), and digest part of the organic matter incorporated into that sediment. The estimation of their ingestion and absorption rates is fraught with a number of difficulties, related to the nature of the substrate. (1) In a sediment matrix, a number of potential food sources are present: benthic microalgae, bacteria, 'detritus' particles, sorbed organic matter, dissolved organic matter, chemoautotrophs (which may be associated with the animals), and possibly others. Animals may selectively digest only part of these food sources, but are forced to ingest them together with the inedible part of the sediment. (2) Many of the (potential) food sources are poorly specified. Most bacteria in sediments are unknown and not yet cultured and their specific rates of growth, growth efficiency etc., are unknown; the largest part of the sediment organic matter is unspecified, and methods are lacking to chemically characterise all of these; high quality dissolved organic matter, e.g. metabolic intermediates, may have low concentration but high turnover in sediments. When performing feeding experiments on deposit feeders, we are actually offering a black box of potential food to the animals, without the possibility to control exactly what is offered. (3) Deposit feeders typically have low absorption efficiency (see below) and some degree of selectivity in what fraction of the sediment they ingest. As a consequence, it is not uncommon that a higher organic concentration is measured in the faeces compared with the ambient sediment. Direct comparisons of organic content in faeces and sediment are therefore not informative (see Lopez & Levinton (1987) for discussion).

Several methods have been developed to circumvent these problems. Depending on the aims of the study, they have tried either to label selectively and follow the fate of a specific food class (e.g. microphytobenthos) or to label all organic matter as unselectively as possible. In some experiments, artificial 'sediments' have been prepared, with a uniform and well-specified composition, so as to avoid problems associated with selectivity of the animals.

In our discussion, we make a distinction between 'bulk sediment ingestion rate', quantifying the amount of bulk sediment (organic and inorganic fraction) taken into the gut per unit time, 'organic ingestion rate', quantifying the amount of organic matter taken in per unit time. With two different definitions of ingestion rate, the definition of efficiencies (the fraction of ingestion that is digested or absorbed) is ambiguous. Following most authors active in the field, we define efficiencies relative to the organic ingestion rate, thus neglecting the inorganic material ingested. As it can be assumed that inorganic matter leaves the gut relatively unaltered, variability in the ingestion of inorganic matter would introduce undue variability in the estimated values of all efficiencies when total (bulk) ingestion would be used as a yardstick for efficiencies.

Following Penry (1998), three efficiencies related to food uptake can be defined. *Digestion* is defined as the set of enzymatic and chemical processes that break down ingested material into components that can be absorbed across the gut wall. *Absorption* is defined as the uptake, across the gut wall, of digestive products. These absorbed materials can be transported from the gut cells to the rest of the organism. Absorbed products can either be assimilated or metabolised. *Assimilation* is the anabolic process, the incorporation of the products into the tissue of the organism. Absorbed products that are not assimilated are used for catabolic processes. Digestion efficiency is defined as the fraction of ingested food that is broken down to digestive products during gut passage, whether or not these products are absorbed by the organism.

Digestion efficiency is calculated from measurements of the amount of a substance in ingested food (S_i) and the undegraded amount of substance in faeces (S_f): $1 - (S_f/S_i)$. Absorption efficiency is sometimes defined as the fraction of the digestive products absorbed, but this is not a practical definition because it is very difficult to measure the rate of digestion. Similarly, assimilation efficiency can be defined as the fraction of absorbed products that is assimilated, but again the practical definition uses food intake as the denominator. Whenever confusion could arise, it is important to state which definition has been used. The practical definitions are as follows: Absorption efficiency is defined as the fraction of the food taken in that is absorbed: S_a/S_i, where S_a is the amount of the substance that is taken up across the gut wall. Assimilation efficiency is defined as the fraction of the food assimilated, i.e. built into tissues: S_s/S_i, where S_s is the amount of the substance assimilated. Absorption and digestion efficiencies are equal for compounds that are not changed by digestive processes, e.g. easily absorbable components such as glucose, or radioactive or stable isotopes (Penry, 1998). For ill-defined 'food' as is often the case in sediment organic material, the difference between digestion and absorption efficiencies may be caused by egestion of digested molecules, or by bacterial uptake of digestion products in the gut. This difference is often not easy to determine.

Bulk sediment ingestion rate

Since in general organic matter is only a small fraction of the total sediment volume (order 0.1–10% by dry mass, see Berner (1982)), and absorption efficiencies of animals for this organic matter are low, the mass of the ingested bulk sediment changes little during gut passage (typically less than a few per cent, which is below detection limit in most experiments). Consequently, the bulk sediment egestion rate (rate of faeces production) can be taken as a measure for bulk sediment ingestion rate (Taghon, 1988). Bulk sediment ingestion rate is then estimated as the total mass of faeces produced per unit of time per animal.

Several methods have been described for the measurement of faeces egestion rate. For species defaecating at the sediment surface, simple collection of faeces by pipette is possible (Forbes & Lopez, 1990; Lopez & Elmgren, 1989; Taghon, 1988). Alternatives are direct visual inspection of faeces production rate in short experiments (Mayer et al., 1993), or videotaping of worms over longer time intervals (Taghon & Greene, 1998). For animals defaecating below the sediment surface, sieving out faecal pellets may be possible. Wheatcroft et al. (1998) used a 180 μm sieve to remove worms and tubes of *Mediomastus ambiseta*, and a 45 μm sieve for the pellets. Faecal pellets of many species are relatively strongly bound, but it is advisable to reintroduce counted pellets in the sediment, and repeat the procedure to check for losses, as was done by these authors. Sieve mesh sizes may have to be adapted for other species. Collection of faeces does not only allow the estimation of bulk sediment ingestion rates, but also offers the opportunity to study the composition of these faeces.

Cammen (1980a) used fluorescent particles experimentally deposited onto the sediment surface to estimate the bulk sediment ingestion rate of a surface deposit feeder. After a short incubation time, during which the animal ingested sediment coloured by the particles, the animal was killed and its gut contents dissected. In the gut, a distinction could be made between coloured contents in the anterior part of the gut, and uncoloured contents in the posterior part. This allows a quantification of the ingestion during the period when colour was present on the sediment. This type of experiment can be used only if (1) animals feed at the surface where the colour can easily be applied, (2) animals do not select entirely against the coloured particles (note that mild negative selection is not a problem) and (3) gut contents are not vigorously mixed, but rather pass the gut on a first-in-first-out basis. A modification of this method has later been used by Lopez and Elmgren (1989) to determine feeding depth of amphipods.

In dual labelling studies (see below) where the sediment matrix is labelled by adsorbing ^{51}Cr, while the organic matter is ^{14}C labelled by different techniques, the bulk sediment ingestion rate can be estimated by comparing the total amount of ^{51}Cr found back in the faeces to the specific ^{51}Cr activity of the sediment (Forbes & Lopez, 1989).

Organic ingestion rate

Because the major problem for estimating organic ingestion rate from bulk ingestion rate is selectivity of the animals on the particles they take up, a solution may be to offer artificial homogeneous sediment to the animals. As an example, Taghon (1988) used silica sand enriched with baby food (see Tenore (1977) and many later publications – the food value of baby food for worms is better known than that of natural organic matter).

Using radiolabelled food, organic ingestion rate can be estimated in very short incubation experiments. Unlabelled animals are allowed to feed for a short time on labelled food and their gut contents are dissected out. Assuming no substantial digestion has taken place during the short incubation, the radioactivity of the gut contents represents a measure for the ingestion rate of the labelled food type. This approach was followed by Forbes and Lopez (1989) to estimate feeding of *Hydrobia* on radiolabelled benthic algae. Organic ingestion rate is calculated as the amount of ^{14}C ingested per unit of time multiplied by the chl-a to ^{14}C ratio in the sediment multiplied by the C to chl-a ratio in the algae. More elaborate models may be used to model ingestion, absorption and defaecation of radiolabelled food items in animals (Herman & Vranken, 1988; Kofoed et al., 1989).

In dual labelling experiments (see below), it is usually easier to determine the absorption efficiency than the organic ingestion rate. The latter may be calculated from the absorption (digestion) efficiency and the organic content of the faeces (Cammen, 1980b).

Absorption efficiency

The 'dual labelling technique' with ^{51}Cr and ^{14}C deserves special mention as a methodology to measure absorption efficiency. The method was developed by Calow and Fletcher (1972) and adapted for use with deposit feeders by Lopez and Cheng (1983). Lopez et al. (1989) and Charles et al. (1995) give critical accounts of the method. The idea is to apply two tracers to the sediment, one of which is not absorbed by the animal and leaves the gut unaltered, while the other can be absorbed with the food. ^{51}Cr is used as the unabsorbed tracer, as biological membranes are nearly impermeable to the trivalent form of chromium and sediment can easily be labelled with chromium, which readily adsorbs to particles. One or more organic fractions of the sediment mix are ^{14}C labelled (e.g. heterotrophic bacteria by offering a labelled substrate, algae by offering light and inorganic ^{14}C, all organic matter by adsorbing ^{14}C formaldehyde to it).

Homogeneous labelling of food items may require special precautions. Long incubation times are needed, to assure that all organic matter is equally labelled.

When the sediment is taken up, the chromium passes through the gut almost unaltered. In practice, animals do absorb a small amount of ingested chromium, but this can be measured and taken into account. However, part of the ^{14}C is taken up, and as a consequence, the ratio of ^{14}C to ^{51}Cr is different in ingested and egested sediment. This allows various calculations of essential rates.

Lopez and Crenshaw (1982) and Lopez and Cheng (1983) pioneered the use of ^{14}C-formaldehyde to non-selectively label all organic matter in the sediment. Sediment is incubated in 30% NaCl (which reversibly inhibits microbial activity) and ^{14}C-formaldehyde is added. The formaldehyde forms bonds with most organic substrates, and in this way labels the organic matter fairly uniformly. However, desorption may be a problem in some experiments (see Lopez et al. (1989) for details).

Organic absorption efficiency is estimated in dual labelling experiments by the Conover (1966) ratio (see above), which in this case equals one minus $(^{14}C/^{51}Cr)_{faeces}/(^{14}C/^{51}Cr)_{food}$. Note that the dual labelling technique provides a relative estimate of the ratio organic/inorganic matter by means of the $^{14}C/^{51}Cr$ ratio. The ^{51}Cr of the faeces may be corrected for the (slight) uptake (or adsorption) by the animals (Cammen, 1980a). A problem with the application of this formula is that where animals feed selectively on the sediment mix, it is questionable what constitutes the correct ratio in the food. One can easily determine this ratio in the bulk sediment, but this is irrelevant if the animals ingest only a small selection of particles from this bulk. Sampling from the gut contents after very short incubation periods may be a solution (see above).

As an alternative to dual labelling, a mass balance approach with a single labelled food source used in pulse-chase experiments may be used. Animals are allowed to feed on labelled food for a period shorter than the gut residence time, and then allowed to defaecate while being on unlabelled sediment. Faeces are collected quantitatively, and the label activity in faeces and animals at the end of the experiment are determined. This method estimates assimilation, rather than absorption efficiency. Assimilation efficiency is given as $L_a/(L_a + L_f)$, where L_a and L_f are label activity in the animal and faeces, respectively. See Ahrens et al. (2001) for an example, and Penry (1998) for a discussion of advantages and disadvantages of the method. If respiration of the absorbed material in the course of the experiment can be neglected, the assimilation efficiency thus determined will approximate absorption efficiency.

Ingestion and absorption by predators

Many studies on predation in benthic communities focus on the structuring effect of predation on the assemblage structure, community dynamics and the spatial distribution of the prey species (Ambrose, 1991; Beukema et al., 2001; Seitz et al., 2001; Wilson, 1991). While highly relevant, also for the carbon flows through benthic communities, they are outside the scope of the present discussion. Many other experimental or field studies of predation rates are done in the context of a specific predator–prey relation between one or a few species of predator, and species of prey. In these studies, emphasis is seldom on the energy balance of the predator, but rather on its effects on one or more selected prey species. Important aspects for the qualification of a predator–prey relationship are the functional response and the selectivity of the predator. The functional response expresses how the rate of predation (usually expressed as number of prey taken per unit time) varies with the density of prey. Selectivity can be between species, but often is studied between size or age classes of a single species of prey. This aspect is extremely important for the dynamics of the prey species, because many predators select for small sizes and the prey can consequently outgrow predation pressure. However, it is also an important consideration for the study of the predator. Theories of optimal foraging (Stephens & Krebs, 1986) try to predict predator behaviour based on optimisation of energy intake, minimisation of risks, optimisation of time use, etc. Many studies of (epi)benthic predators have focussed on this aspect (see e.g. Cote et al. (2001); Mascaró and Seed (2001); Hiddink et al. (2002) for recent examples).

Estimation of ingestion rate

Most predators are relatively large, and are amenable to visual inspection. In the laboratory, rates of prey intake may be directly observable in relatively short experiments, e.g. experiments on predation by starfish on mussels by Sommer et al. (1999). As an alternative, predators may be encaged with a counted number of prey, and a count of the number of disappeared prey after the experimental incubation allows estimation of the predator ingestion. This approach can also be applied to cages in the field (Dahlhoff et al., 2001). Note that there may be a difference between biomass of prey killed by the predator, and biomass actually ingested, e.g. due to kleptoparasites (Morissette & Himmelman, 2000).

Ejdung et al. (2000) labelled prey species radioactively, and compared the label activity in the predator with the number of prey that had disappeared from the experiments. The favourable comparison showed that recording prey disappearance is a reliable procedure (in the present case, labelling was used to demonstrate that the predator did not only kill, but actually ate the prey). These authors also used fluorescent dye to further confirm the findings.

Gut content analysis can be used both in controlled experiments and as a way to estimate the intake rate of predators caught in the field. Animals are quickly killed, but care is taken that emptying of the gut just before death is avoided. They are subsequently dissected, and hard remains of prey in the digestive tract are enumerated. Estimates of the predator's gut turnover time are needed in order to calculate the effect of the predator on the prey species of interest. Hiddink et al. (2002) provide a recent example of this approach for infaunal predators. If emphasis is placed on the intake rate of the predator itself, a condition for the use of this methodology is that most prey leave measurable and identifiable remains in the predator's digestive tract, which may be difficult to prove. Gut content analysis is an important methodology in fisheries research, and papers discussing the methodology are published.

A critical review with recommendations for standardised methodology is provided by Cortés (1996), where many references to the basic methods used and experimental tests of these methods may be found. Although directed specifically at fish, we believe that many of these methods may also find application in studies of (epi)benthic predators (Cartes & Maynou, 1998). Also, potentially relevant for benthic predators (and even for other benthic groups) is the use of natural radiotracers (in particular ^{137}Cs) to estimate ingestion rates (Forseth et al., 1992; Gingras & Boisclair, 2000; Rowan & Rasmussen, 1994).

For infaunal predators, such as nemerteans or polychaetes, few estimates of ingestion rates are available. Schubert and Reise (1986) provide some rough estimates for *Nephtys hombergii*, based on disappearance of prey in enclosure experiments. They compare them to order-of-magnitude estimates for other (epi-)benthic predators. Ambrose (1991) summarises and averages some scant data from the literature. The general lack of these estimates in the literature is surprising, since the ingestion and absorption by predators is as important for the dynamics of predator–prey dynamics as is the rate of capture of the prey.

Absorption efficiency

We were unable to locate specific studies on absorption efficiency of benthic predators. In principle, the same approaches applicable to other feeding groups could be applied. It may be expected that absorption efficiencies for carnivores are relatively high, in view of the high quality of their food, and this may decrease the importance of an accurate determination of its value, as shown by the following calculation. Suppose absorption efficiency is known to be between 60 and 75%, and absorption rate is to be calculated from ingestion rate I multiplied by absorption efficiency. The estimate will vary between $0.6I$ and $0.75I$, a relative error of approximately 25%. However, if (as in deposit feeders) absorption efficiency is much lower and the same uncertainty range is kept, say a variation between 5% and 20%, then obviously the relative error in the estimated absorption will be 400%. Nevertheless, it remains surprising how profoundly the 'sympathy for the victim' has pervaded the selection of study topics in benthic ecology (note also that fisheries biologists consider predatory fish as prey of fisheries, hence their different (misguided) perception).

Clearance and pumping rates of filter feeders

Pumping rate of filter feeders is defined as the volume of water pumped over the filter per unit of time. Sometimes it is used as a synonym of filtration rate, but the latter term is also used to indicate the mass of solids filtered per unit of time, and we will use it in that sense throughout this text. Operationally, it is usually easier to measure clearance rate, i.e. the volume of water cleared of suspended particles per unit of time. Clearance rate Cl is defined as the product of pumping rate and filtering efficiency. It is equal to pumping rate when the filtering efficiency (the fraction of suspended particles in the water retained by the filter) is 100%. It is lower than pumping rate when filtering efficiency is not perfect.

Riisgård (2001a) recently made a critical review of laboratory methods for the measurement of 'filtration rates' (which, in our use of terms, is actually pumping rates) in bivalves, and this account follows his description of methods. Measurement of pumping rates in other invertebrates may require special technical adaptations, but falls within the same general categories as discussed here (Riisgård & Larsen, 1995, 2000). The critical evaluation of methods and results by Riisgård (2001a) is not generally accepted. See e.g. Cranford (2001) and the reply by Riisgård (2001b). The reader should be aware of the severe controversy in the literature on fundamental aspects of filter feeder physiology (and, as a corollary, on the validity of experimental methods), expressed by the papers of Jørgensen (1996a) and Bayne (1998). Riisgård (2001a) attempts to define minimum requirements for the quality of measurements. In particular, he proposes that any method employed to study physiological regulation of pumping rate always be employed under optimal conditions (where food is a silt-free culture of suitable algal species, with a concentration between lower and upper thresholds where valve opening is reduced, and animals are well acclimated to the laboratory conditions) and that these measurements be used as a methodological check: they should yield pumping rates in the range of published studies summarised in Riisgård (2001a). The latter check seems overly restrictive, as it would automatically reject measurements on populations that for one reason or another would have lower pumping rates. However, it remains a useful suggestion to perform and report a measurement under optimal conditions (if possible, on a well-studied species such as *Mytilus edulis*) as part of all measurement programmes. The following classification of methods was proposed by Riisgård (2001a).

Direct measurement of pumping rate

In these methods, the exhaled water is physically separated from the surrounding water. Separation is possible, e.g. by a rubber apron. The method dates back to the beginning of the twentieth century; older literature is reviewed by Jørgensen (1966b). A more recent implementation by Famme et al. (1986) has demonstrated its extreme sensitivity to the build-up of backpressure on the pumping system. Reliable estimates of 'natural' pumping rates can be obtained only when this backpressure is carefully avoided. Direct measurements seem most useful to study pumping under experimentally manipulated backpressures, allowing the physical characteristics of the animal pump to be defined.

Flow-through chamber method

Animals are placed in a chamber, a suspension of constant and predefined composition is pumped through this chamber, and concentrations of suspended matter are measured at the inflow and outflow of the chamber. The main advantage of the method is that the composition of the feeding suspension can be maintained constant during the experiment. The basic form of the experimental set-up assumes that refiltration by the animal of previously filtered water is excluded. This requires a high flow throughput through the experimental chamber, which moreover should have a proper geometry (Riisgård, 1977). Under these conditions the clearance rate is given by:

$$Cl = Q(1 - C_2/C_1)$$

where Q is the flow rate through the chamber and C_1 and C_2 are the concentrations of inflow and outflow, respectively. The derivation of this equation is as follows: the mass flux flowing into the grazing chamber per unit time is given as $Q{\cdot}C_1$, and the mass flux flowing out is $Q{\cdot}C_2$. The animals take out the difference, i.e. $Q{\cdot}(C_1 - C_2)$. As the animals filter water with a concentration C_1, the volume swept clear is the division of the mass flux by the concentration. It is obvious from the equation that clearance rate calculated according to this formula cannot be larger than the flow rate Q through the chamber. In fact, the validity of the equation above, and, in particular, the assumption of no refiltration, can only be guaranteed if $Q >> C_1$. A compromise must be sought, since the precision of the measurement of C_2/C_1 decreases with Q (smaller concentration differences are more difficult to measure). As a methodological check, it may be useful to plot apparent clearance rate calculated according to the equation versus Q (Riisgård, 2001a). Hildreth and Crisp (1976) used a modified version of the equation that can, under certain conditions, overcome this difficulty. If (and only if) the water in the grazing chamber is perfectly mixed, clearance rate may be estimated from:

$$Cl = Q(C_1/C_2 - 1)$$

which is easily derived by assuming perfect mixing of the water in the grazing chamber, so that the animals filter water with a concentration equal to the outflowing concentration C_2.

Suction method

In this method, described by Møhlenberg and Riisgård (1979) and used by Kiørboe and Møhlenberg (1981), Famme et al. (1986) and Kryger and Riisgård (1988), samples of inhaled and exhaled water are sucked through glass pipes placed a few mm above inhalant and exhalant openings of the filtering animal (Fig. 8.2). The flow rate through the tubes is adjustable, and the clearance rate is calculated with the second equation of the previous paragraph, where C_2 is the concentration in the water collected from the exhalant current, whereas C_1 is the concentration in the inhalant flow. Just as in the flow chamber method using this equation, suction flow must be larger than clearance rate for the method to work, and this should be checked by plotting apparent clearance rate versus suction flow rate. An advantage

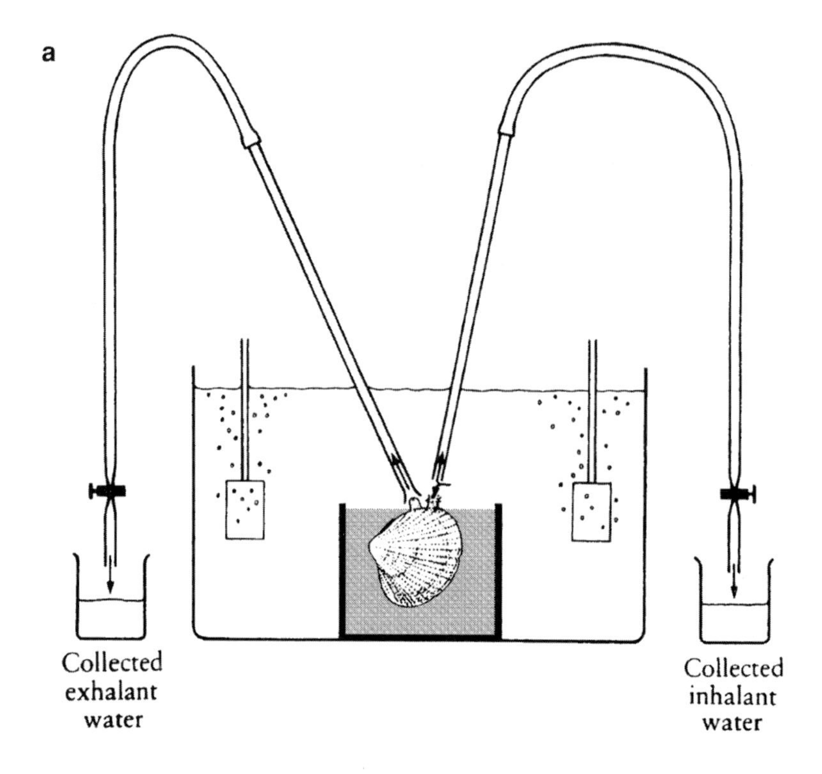

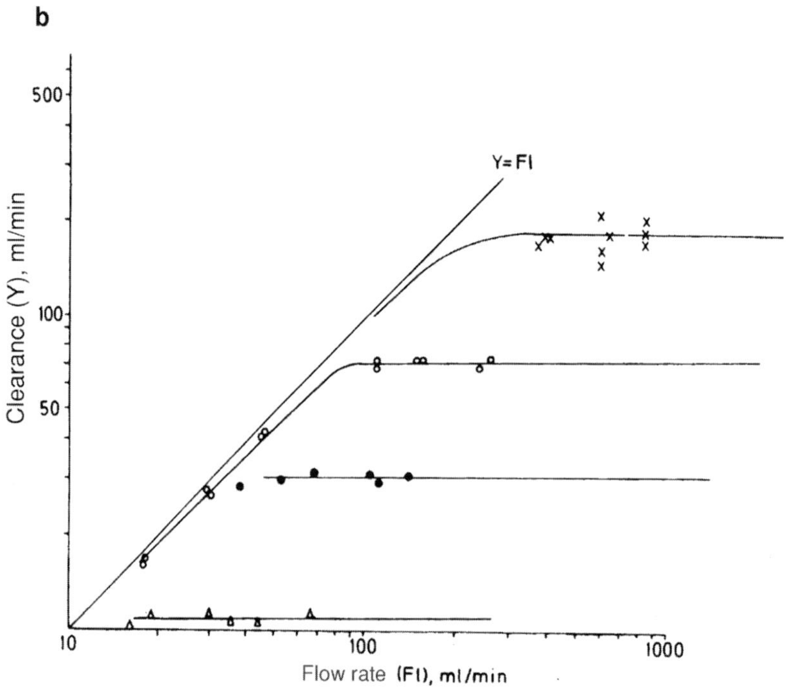

Fig. 8.2 Illustration of the application of the suction method for the measurement of clearance rate. (a) Set-up used for the measurement of filtration rates in suspension feeding bivalves. The glass tubes collect water from the inhalant and exhalant siphons by means of gravity. (b) Clearance rate estimated as a function of suction flow rate. Real clearance rates are reliably estimated at the plateau where they become independent of suction flow rate. The example shows measurements on *Modiolus modiolus* (four individuals of differing size). Reproduced from Riisgård (2001a) with permission.

of the suction method is that it can be applied to animals in a natural position (e.g. buried in sediment) and even, with proper adjustment of the methods, in the field.

Clearance method

This is a basic method, where animals are placed in a container with a food suspension, and the decrease of food concentration is monitored over time. Thorough mixing of the water in the container is needed to avoid the build-up of concentration gradients around the animals, and thus partial re-filtration. A disadvantage of the method is that food concentration decreases during the experiment, but this can be overcome by periodically adding new food supply to the container. The food concentration in the container, at constant clearance rate, follows an exponential decline:

$$C_t = C_0 \exp\left(-\frac{Cl}{V}t\right)$$

where V is the volume of the container and C_0 is the concentration at time 0. It is customary to take several measurements of concentration over time, and estimate clearance as V times the slope of the regression of log(concentration) versus time.

Controlled addition methods

In these methods, controlled additions of food suspension to a (well-mixed) grazing container maintain a constant concentration of food around the animals. Additions of food are measured and clearance rates calculated from them. Winter (1973) and Riisgård and Møhlenberg (1979) used light-based systems to maintain constant concentration of food in an aquarium. When the concentration falls below a set minimum, food is added from a stock solution. The number of food additions is recorded, and clearance calculated from this number. In a slightly different design, Riisgård and Randløv (1981) and Poulsen et al. (1982) used a chemostat algal culture to provide a continuous addition of food in a long-term (45 days) study of mussel filtration. Clearance rate was calculated from the steady-state concentration of algae in the mussel aquarium and the concentration in the water flowing from the algal chemostat.

Measurement of exhalant current velocity

Current velocity in the exhalant stream may be measured with thermistor probes (Vogel, 1994, and references therein). If properly calibrated in a flow from an aperture with the correct geometry, continuous measurements at a single spot within the exhalant current can be sufficient to calculate the volume flux from the exhalant opening. As an alternative, Jones et al. (1992) used a small impeller to measure the exhalant current velocity.

Bio-deposit method

In this method, faeces and pseudofaeçces of the filter feeders are collected, and the proportion of inorganic and organic material in the food suspension and the bio-deposits is determined (see e.g. Hawkins et al., 1996; Cranford et al., 1998). Clearance rate is estimated as:

$$Cl = \frac{f_b P_b}{f_i C_{\mathrm{TPM}}}$$

where f_b and f_i are the fractions of inorganic matter in the bio-deposits and the food suspension, respectively, P_b is the production rate (mass time^{-1}) of bio-deposits and C_{TPM} is the concentration (mass volume^{-1}) of total particulate matter in the food suspension. Riisgård (2001a) suggests that a slightly modified version of this equation, as used by Cranford and Hargrave (1994), does not take into account pseudofaeces production, but this assertion is in error. He correctly points out that, as for the other methods, the clearance rate determined by the bio-deposit method is equal to the pumping rate only if all the particulate matter is filtered with 100% efficiency. However, since the food suspension used in this method is usually naturally occurring suspended matter, some particles may be smaller than the critical size for efficient particle retention. This can lead to estimates of clearance rates lower than pumping rates based on retention of larger and efficiently retained algal suspensions.

Video observation method

A number of authors have used video imaging of particles approaching the gills. From the approach velocity and the surface of the gills an estimate of pumping may be obtained. As with all aspects of filter feeding physiology, this methodology has also been the matter of debate (Beninger, 2000; Riisgård & Larsen, 2000; Silverman et al., 2000; Ward et al., 2000). The utility of video observation is primarily in detecting particle selection mechanisms as well as basic characteristics of the filter, and its use in estimating pumping rates as part of a study of the bioenergetics of the animal is limited. The reader is referred to the references cited above as a useful introduction to the uses, advantages and disadvantages of this type of study.

Absorption efficiencies

Collection of pseudofaeces and faeces, and determination of the organic content both in these and in the food suspension, allows us to calculate a number of rates and ratios relevant to the study of filter feeder physiology (Table 8.4). The table is slightly modified from Hawkins et al. (1998) where references can be found to many studies using this approach. The basis is the formula of Conover (1966), which makes use of the fact that inorganic matter ingested into the digestive system is egested in unchanged form, whereas part of the organic matter ingested is absorbed and, therefore, not egested. The fractions of organic matter in the total dry mass of food suspension, faeces and pseudofaeces, which are relatively easy to determine, are used as the basis for the calculations. The equations are generally applicable; in the absence of pseudofaeces production, its rate can be set equal to zero.

Table 8.4 Rates and ratios relevant for the study of filter feeder physiology.

Feeding parameter	Symbol	Units	Calculation
Clearance rate	Cl	Volume time^{-1}	$Cl = \frac{f_p P_p + f_f P_f}{f_s C_{TPM}}$
Filtration rate	Fi	Mass time^{-1}	$Fi = Cl \cdot C_{TPM} = \frac{f_p P_p + f_f P_f}{f_s}$
Ingestion rate	I	Mass time^{-1}	$I = Fi - P_p$
Net organic ingestion rate	I_{no}	Mass time^{-1}	$I_{no} = Fi \cdot (1 - f_i) - P_p \cdot (1 - f_p)$
Net organic absorption rate	A_{no}	Mass time^{-1}	$A_{no} = I_{no} - P_f \cdot (1 - f_f)$
Net organic selection efficiency		–	$1 - f_p / f_i$
Net absorption efficiency from ingested organics		–	A_{no} / I_{no}
Organic content of ingested matter		–	$I_{no} / (Fi - P_p)$

Note: The net organic ingestion rate and the net organic selection efficiency are influenced by loss of mucus in pseudofaeces; the net organic absorption rate and the net absorption efficiency from ingested organics are influenced by loss of organics in pseudofaeces and by metabolic faecal losses (Hawkins et al., 1996).

The following basic observations are made:

P_p rejection rate is the production rate of pseudofaeces (mass time^{-1})
P_f faeces production rate (mass time^{-1})
f_i fraction inorganic material in the food suspension (–)
f_p fraction inorganic material in pseudofaeces (–)
f_f fraction inorganic material in true faeces (–)

The rejection and faeces production rates may be estimated by collecting (separately) the pseudofaeces and faeces of the filter feeding animal. Organic or inorganic fractions may be determined from CHN analyses or from mass loss on ignition.

Excretion

Urine products

In the deamination step in the catabolisation of proteins, the highly toxic compound ammonium is released into the cells of an animal. Some is transformed into other compounds, such as urea, before it is released into the environment. The release of ammonium and its derivatives requires an effective nitrogen drain. The excretory products have only marginal energy content (Elliot & Davison, 1975). Although of minor importance in terms of energy, energy budgets are traditionally approached using mass balances, in which the products can be significant. Moreover, excretion is studied because the nitrogen compounds can stimulate primary production of N-limited phytoplankton in adjacent waters.

Experimental designs

Excretion processes are measured by monitoring the increase in excretory products in the overlying water of different experimental settings. Closed incubations are the

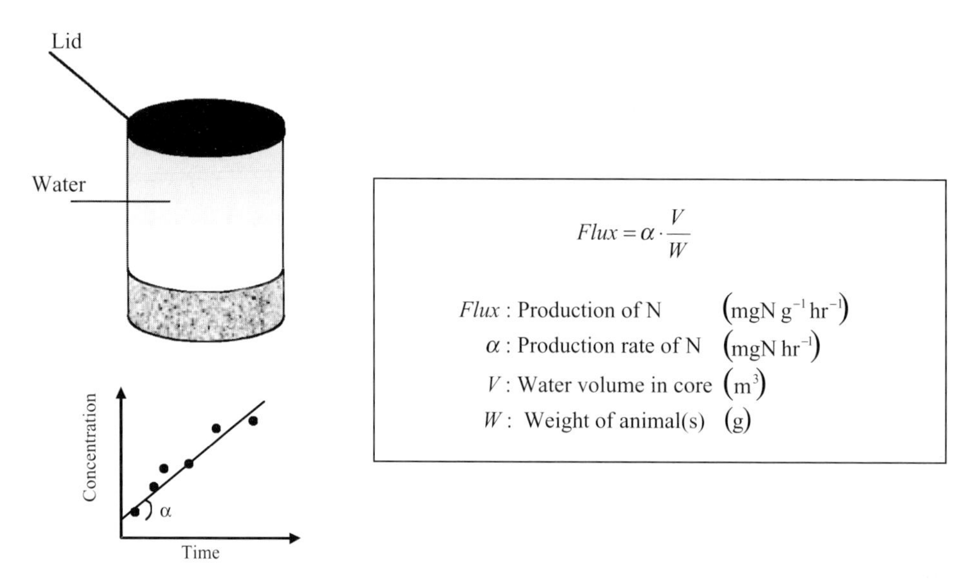

Fig. 8.3 Schematic representation of a closed incubation set-up to monitor increases in excretion products. The simple formulas to transform increases in concentration to fluxes are also presented.

most frequently used setting to measure excretion processes. Animals are placed in a closed system (flasks, cores, chambers or aquaria) with filtered seawater; rotors circulate the water inside the system (Fig. 8.3). Sterilised sand or glass tubes might be used to mimic a more natural environment for polychaetes. During the incubation the increase of excretory products is monitored in the overlying water. The production rate can be estimated by the slope of the linear increase in concentration of the excretory product (Fig. 8.3). The production rate is multiplied by the volume of the core and divided by the desired unit, such as animal mass (Fig. 8.3). Descriptions of experimental settings are given in Migne and Davoult (1997), Hatcher (1994), and Smaal and Vonck (1997). This method is also valid to estimate N production of the whole sediment as benthic chambers or using intact sediment cores (Chapelle et al., 2000, and see Fig. 8.3).

In a flow-through system, the overlying water is continuously refreshed and excretion rates are calculated based on the difference in concentration of excretory products between the outflow and inflow, and are corrected for flow rate and sediment surface (Chapelle et al., 2000; Lavrentyev et al., 2000, see Fig. 8.4). Prior to laboratory incubations, it is useful to allow gut clearance to prevent any leakage of products from ingested food or faeces to prevent over-estimation of true excretion rates (Hatcher, 1994).

It is a common phenomenon that temperature and nutritional condition influence the excretion of ammonium (see Clarke & Prothero-Thomas, 1997, and references therein). For example, ammonium excretion in two benthic cnidarians was found to decrease by 51% after 7 days of starvation and temperature accounted for 44% of the observed variation in the seasonal trend (Migne & Davoult, 1997). Peak

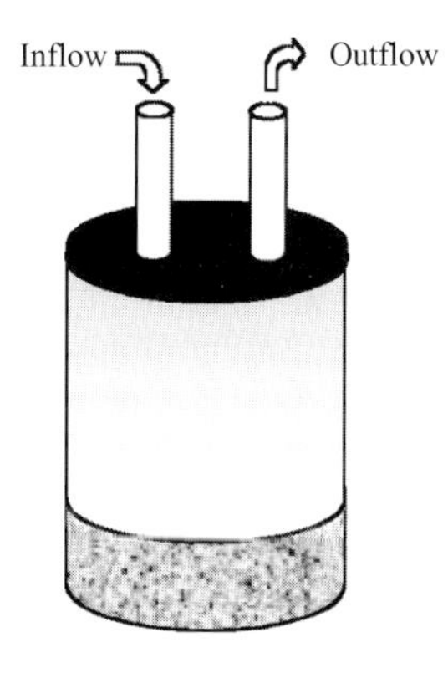

$$Flux = (C_{\text{out}} - C_{\text{in}}) \cdot \frac{Q}{W}$$

$Flux$:	Flux of N product	$(\text{mgN g}^{-1}\ \text{hr}^{-1})$
C_{in} :	Concentration of Inflow	(mgN m^{-3})
C_{out} :	Concentration of Outflow	(mgN m^{-3})
Q :	Flow rate	$(\text{m}^3\ \text{hr}^{-1})$
W :	Weight of the animal(s)	(g)

Fig. 8.4 Schematic representation of a continuous flow set-up to measure excretion fluxes from animals or sediment. The simple formulas to transform concentrations to fluxes are also presented.

excretion rates of mussels (Smaal & Vonck, 1997) and *Monoporeia/Pontoporeia* spp. (Lehtonen, 1995) coincided with blooms of phytoplankton. Brockington and Clarke (2001) estimated that 15–20% of the summer increase in metabolism of polar sea urchins was due to temperature increase, and 80–85% was caused by increased physiological activity. Estimating 'field' excretion rates of collected organisms in laboratory or mesocosm environments thus necessitates short handling and acclimatisation times and maintenance of temperature at ambient level.

Methods to measure excretion products

Possible excretion products comprise ammonium, nitrate/nitrite, amino acids and urea. Most benthic invertebrates are ammonotelic, which means that ammonium dominates the excretory products. Therefore, in the majority of studies only ammonium excretion is monitored, and only certain studies additionally measure other end products. What follows is a short overview of the analytical procedures to determine concentration of the excretion products with proper references.

Ammonium is generally measured with colorimetric method based on Solórzano' (1969) and is explained in most handbooks on sea water analysis (Crompton, 1989; Grasshoff et al., 1983; Strickland & Parsons, 1972). The method is based on the reaction of ammonia with hypochlorite that, with some other additives, gives a blue colour and the intensity of the colour is measured at 625 nm. The precision at which ammonium can be measured is about 0.1 μg N-NH_4^+ dm^{-3} (Strickland & Parsons, 1972). It is most easily applied in an autoanalyser, or flow-segmented analyser. This set-up greatly increases the reproducibility.

Nitrate and nitrite are measured by colorimetric analysis, and the analytical procedure can again be found in all standard handbooks (Crompton, 1989; Grasshoff et al., 1983; Strickland & Parsons, 1972). The method can be handled manually, but an autoanalyser greatly improves reproducibility of the method. The detection of nitrite and nitrate is below 1 μmol N dm^{-3} (Crompton, 1989).

Basically, three different amino acid fractions can be distinguished: (i) free amino acids (FAA), (ii) dissolved combined amino acids (DCAA) and (iii) particulate amino acids (PAA). In fact, only free amino acids can be measured directly and the latter two need hydrolysation prior to determination of the concentration (Cowie & Hedges, 1992). Cowie and Hedges (1992) also provide an accurate description of measuring amino acids using high-performance liquid chromatography (HPLC), as first described by Lindroth and Mopper (1979). The HPLC method has been successfully applied to measure the uptake of dissolved amino acids from natural seawater by the mussel *Mytilus edulis* (Manahan et al., 1983, 1982).

Urea can be measured by (i) enzymatic (with urease as enzyme) hydrolysation to ammonium at elevated temperatures or (ii) by chemically complexing the urea. In the first method, the ureum is hydrolysed and the concentration of ammonium is continuously monitored during the incubation. Strickland and Parsons (1972) describe the method and note that the detection limit is about 0.05 μg N dm^{-3}. The latter method is described by Grasshoff et al. (1983) who report that a detection limit is at 1.4 μg N dm^{-3} with a relative standard deviation of approximately 15% just above the detection limit and 4.5% at higher levels.

Mucus

The production of organic mucus is related to multiple physiological processes ranging from feeding and production of pseudofaeces in bivalves, locomotion and reproduction of molluscs to drifting capability of larvae (Davies and Hawkins, 1998). The comprehensive review by Davies and Hawkins (1998) provides references on different aspects of mucus production in molluscs. Here we will describe only the main methods to measure mucus production in the context of energy budgets.

Gastropods produce mucus on the pedal foot for multiple reasons such as locomotion, attachment, navigation and possibly as a food trap (Davies & Beckwith, 1999; Davies & Hawkins, 1998); some papers report it as being a dominant component of the energy budget. Two methods are described in the literature: either mucus is collected from the pedal foot or from an artificial substrate on which the gastropod has roamed around. Of course, a combination of the two is possible. Peck et al. (1993) estimate mucus production by scraping deposited mucus from a glass plate with a razor blade. Subsequent analysis included C, H, N, protein, lipid, carbohydrate and energy content of mucus. Mucus is collected from the foot by using a rounded end of a pair of forceps (Davies & Williams, 1995) or glass rod (Horn, 1986). On the basis of the difference between a 30-minute and 24-hour 'walk' of limpets on a glass plate, Peck et al. (1993) concluded that reattachment (i.e. the short walk) to the glass plate constituted ~80% of the daily mucus production. This implies that experiments where mucus was scraped from the pedal feet possibly overestimate field rates of mucus excretion.

Bivalve mucus production is primarily related to pre-ingestive selection processes and the production of pseudofaeces (Davies & Hawkins, 1998). For example,

Beninger and St-Jean (1997) use sophisticated techniques to determine the role of mucus in feeding processes and the production location. Quantification of mucus production has proved to be difficult and no single satisfactory method has emerged (see examples in Davies & Hawkins,1998). Yet Urrutia et al. (2001) propose a quantitative method to estimate mucus that is voided with pseudofaeces.

We found no quantitative assessments of mucus excretion by polychaetes in the marine literature. A paper in the field of soil biology describes the quantification of mucus deposition by earthworms (Scheu, 1991). ^{14}C-labelled earthworms were allowed to burrow in an unlabelled environment; the ^{14}C signal in the burrow wall was taken as a measure for mucus excretion by the body surface and the ^{14}C faeces signal as mucus excretion by the intestines. Another paper describes a method to collect mucus qualitatively from earthworms for stable isotope analyses by placing them briefly in a slightly acidified bath (Schmidt et al., 1999). These methods might be starting points for mucus production measurements on polychaetes. Another phenomenon involving mucus is suspension feeding by polychaetes by spinning a mucus net and using ambient of induced flow to capture food particles; the enriched mucus net is subsequently swallowed (Harley, 1950; Riisgård, 1991; Riisgård & Larsen, 1995). In the literature, the focus is on costs of pumping and/or structure of the mucus net (see Rissgård & Larsen, 1995) for references.

Respiration

Adenosine TriPhosphate (ATP) is the energy currency of the cell that supports the cell with energy for all sorts of processes and its production is a continuous process. Aerobic ATP production is driven by the oxidation of organic substrates (e.g. lipids, proteins and carbohydrates) to carbon dioxide and water with oxygen as the final electron acceptor. Under anaerobic conditions bacteria are known to utilise other inorganic electron acceptors (e.g. sulfate, nitrate, iron oxides). This topic is however outside the scope of this chapter (see e.g. Fenchel & Finlay, 1995). In the absence of oxygen, heterotrophic organisms utilise fermentation processes that degrade organic substrates to simpler ones without an external electron acceptor to generate ATP. However, fermentation produces much less ATP than aerobic respiration and is, therefore, energetically less effective. Because hypoxic, dysoxic or anoxic conditions are common, if not prevalent in many marine benthic environments, many species have evolved efficient fermentation pathways (Bryant, 1991; De Zwaan, 1977; Pamatmat, 1980).

Generation and hydrolysis of ATP results in the loss of energy in the form of heat. Moreover, heat production measurement seems a logical step in respiration measurements, as it simultaneously measures aerobic and anaerobic processes. Micro-calorimetry is based on heat production, but has not found many successful applications in research on benthic animals due to methodological difficulties. Other methods for measuring respiration focus on the consumption of oxygen or the formation of carbon dioxide as an end product of aerobic respiration. Thus, in a majority of studies, respiration is considered from the mass balance viewpoint, rather than from that of the energy budget.

Oxygen consumption

Measuring decreases in oxygen concentration or oxygen partial pressure in a closed or (semi-) open system inhabited by organisms is a widespread approach to assess aerobic respiration under both laboratory and field conditions; various methods are currently available such as the Winkler method, Cartesian divers, electrodes and optodes.

Winkler method

The classical Winkler titration, dating back to 1888, is still one of the most accurate techniques for measuring oxygen concentration in water. The chemical background is based on the precipitation of oxygen with manganese as manganese oxide and the subsequent oxidation of iodine to iodide ions in a strongly acidic environment. On the basis of the titration of the iodide ions formed and stoichiometry one can calculate the original oxygen concentration (see e.g. Carpenter, 1965; or Strickland & Parsons,1972, for a full methodological description). An extensive amount of literature is available on automation of the Winkler method (Anderson et al., 1992; Pomeroy et al., 1994, and references therein). Despite this, it remains rather a labour-intensive technique, though it does provide very accurate data in the micromolar range. At concentrations below 3 μM, the analytical precision may be somewhat less (Strickland & Parsons, 1972).

Cartesian diver

Going back as far as the seventeenth century, the principle of the Cartesian diver was developed into an extremely sensitive means for use in cell biology and physiology of small organisms by the Danes Linderstrøm-Lang, Holter and Zeuthen (Lasserre, 1976). It should be stressed that the Cartesian diver is a very labour-intensive and delicate methodology, requiring substantial experience with the methodology before reliable measurements can be obtained. The basic experimental set-up consists of a sealed vessel partially filled with a liquid medium and a small free-floating diver; pressure inside the vessel is adjustable. The diver essentially consists of a glass capillary (0.16–0.5 mm diameter) with a reservoir at one end, into which a sample of living material can be introduced. The reservoir is partially filled with a liquid medium with a headspace above. The diver is sealed at different levels with a CO_2-absorbing NaOH solution, a ring of paraffin oil to prevent water evaporation inside the diver, and a seal containing the same salt solution as the vessel in which the diver is submerged. The headspace in the vessel can be adjusted in such a way that the diver's compound density equals that of the surrounding medium; hence, the diver floats. Oxygen consumption inside the diver reduces its flotation capacity because the respired CO_2 is being absorbed. Consequently, the pressure above the medium must be lowered for the diver to keep floating. The necessary pressure change is directly proportional to the amount of oxygen consumed. Good reviews of this methodology are available (Holter & Zeuthen, 1966; Klekowski, 1971; Lasserre, 1976). In any case, the main merit of the diver methodology is its unsurpassed sensitivity, oxygen consumption values down to 0.01 nM h^{-1} still being accurately measurable.

Interesting variations on the Cartesian diver are the gradient divers described by Hamburger (1981). The gradient diver methodology uses a similar 'stoppered diver' as in the Cartesian diver approach; however, the diver is now introduced into an aqueous density gradient, and is not kept in a fixed position. Its migration in the gradient reflects the consumption of oxygen inside. A variety of other, even more sensitive, diver designs have been used (see Lasserre (1976) for a review). Researchers in the field of ecology/physiology of benthic animals should, however, carefully consider the advantages and disadvantages of extremely sensitive but delicate devices that impose highly artificial conditions upon the experimental animals.

Electrodes

Contrary to the diver technology, polarographic electrodes as originally described by Clark (1956) have a lower sensitivity, approximately 3–6 nM h^{-1} (Holter & Zeuthen, 1966), yet are more suitable for routine use when sufficient biomass (ten to thousands of meiobenthos specimens and one to a few macrobenthos specimens, depending on size and metabolic activity) is available. Polarographic oxygen electrodes allow continuous monitoring of oxygen tension in a solution, enabling the integration of respiration over different intervals of time or oxygen pressure. One should be aware that electrodes have their own background oxygen consumption. Miniaturized Clark-type oxygen electrodes have been used in many different designs, allowing among other things measurements on variable volumes (see Lemmens (1994) for a device with incubation chambers of 10–50 ml; Moens et al. (1996) for one with incubation chambers of 0.3–1 ml and Braeckman et al. (2001) for 1–3 ml chambers), and in closed (Riisgård, 1989) and flow-through chambers (Riisgård & Ivarsson, 1990). The performance and reliability of these systems, especially at low oxygen consumption rates, thus depend on a number of factors, including sensor characteristics, materials constituting the respiration chamber and diffusion in and out of the system via the titration cannula. An in-depth discussion of these factors is given in Haller et al. (1994).

Respiration rates obtained on individual copepods and nematodes via the Cartesian diver methodology have been compared with rates from polarographic electrode measurements on batches of 10–15 copepods (Gyllenberg, 1973) and of 20–10 000 nematodes (Moens et al., 1996), and were found to be in good agreement. Nowadays, several commercially available devices allow parallel, simultaneous measurement of oxygen consumption of a few to many samples, and easy, computer-assisted data processing. Some systems combine polarographic oxygen sensors with pCO_2 and/or pH electrodes, thus enabling direct assessment of the respiratory quotient. The ease with which polarographic electrode measurements can be performed renders them particularly well suited for the study of the influence of varying abiotic factors (e.g. temperature and salinity) on the respiration of laboratory-cultured species or on communities or dominant species extracted live from sediment samples (Moens et al., 1996). In this way, the use of polarographic electrodes is complementary to that of divers. A further advantage of polarographic electrode measurements is that, provided oxygen consumption rates are sufficiently high, short incubation times (less than 30 minutes) yield good results.

Optodes

The use of optodes or optrodes as a tool to measure oxygen concentration was introduced in aquatic ecology by Klimant et al. (1995). The measuring principle of the O_2 optode is based on the ability of oxygen to act as a dynamic fluorescence quencher that decreases the fluorescence quantum yield of an immobilised fluorophore, often a metalloporphyrin complex (Kautsky, 1939; Kohls & Scheper, 2000). In contrast to microelectrodes, they are easy to manufacture, insensitive to stirring, do not consume oxygen and show fairly long-term stability (Klimant et al., 1995). Optodes can be introduced into a variety of closed and flow-through (Sanchez-Pedreno et al., 2000) incubation chambers or even in microtiter plate-format (Kim et al., 1998). They are suited for experimentation under conditions in which conventional chemical analysis or use of polarographic electrodes is difficult, e.g. at high or variable pressure (Stokes & Somero, 1999). They have an acceptably rapid response time (in the order of seconds to minutes). Depending on the type of matrix in which the fluorophore is immobilised, autoclavation of optodes may be possible (Klimant et al., 1999; Voraberger et al., 2001). Similar devices exist for the determination of CO_2 (DeGrandpre et al., 1999; Mills & Eaton, 2000; Weigl & Wolfbeis, 1995).

In ecology, optodes have so far mainly been used for high-resolution mapping of oxygen distributions in sediments and biofilms. See e.g. Glud et al. (1999a, 1996) for the use of planar optodes in determining two-dimensional oxygen distributions; and Glud et al. (1999b) and Wenzhofer et al. (2001) for the application of optodes to deep-sea sediments. Studies at the specimen level involving benthos have been extremely few. Frederich and Portner (2000) used oxygen optodes to determine hemolymph oxygenation levels in the spider crab, *Maja squinado*, under varying temperatures. Holst and Grunwald (2001) applied transparent oxygen optodes to a foraminifer with symbiotic diatoms. Given the properties of optodes, they may well prove to be a fruitful tool in respiration measurements in the future.

Carbon dioxide production

Estimates of metabolic rates based on CO_2 measurements may deviate from those based on O_2 consumption, depending on the type of substrate used (lipid, carbohydrate, protein) and on the prevalence of alternative biochemical pathways (Braeckman et al., 2001).

CO_2 and O_2 concentrations in air can be measured simultaneously by gas chromatography (Abrams & Mitchell, 1978; Mitchell, 1973). Since most incubations with animals will take place in aqueous media, CO_2 concentration is then measured in a head space. A water sample is taken and transferred into a vial (preflushed with nitrogen) that is subsequently sealed with a septum containing cap, providing a head space in the vial. All the aqueous inorganic carbon is transferred to the head space by acidification of the sample through syringe addition of concentrated HCl or H_2PO_4. A sample can then be taken from the head space to measure air CO_2 concentration by gas chromatography. In case the sediment is rich in carbonates, there may be a high background that can strongly decrease the sensitivity of the method.

IRGA (InfraRed Gas Analysis) is an alternative, more sensitive and rapid means of detecting (changes in) CO_2 concentration in air samples. Many commercially available types of IRGA exist. For example, Van Voorhies (2000) and Van Voorhies and Ward (1999) used two different types of IRGA-based devices (the TR2 CO_2 gas respirometry system from Sable Systems and the LiCor 6251 CO_2 analyser) allowing reproducible measurements of CO_2 production of batches of as few as 50 specimens of the nematode *Caenorhabditis elegans*.

^{14}C labelling

The respiration rate of unicellular organisms is so low that it can be measured only at the level of individuals in large species using diver techniques. For smaller species, several to many individuals inevitably need to be lumped, precluding direct assessment of, for example, body size or biomass to respiration allometries. Moreover, stoppered divers are typically incubated for several hours, a period during which effects of starvation and sub-optimal environment may blur true respiration rates at active metabolism. The ^{14}C labelling technique feeds unicellular organisms with a ^{14}C labelled food source during several subsequent generations (Crawford et al., 1994; Stoecker & Michaels, 1991). Protozoan cells are then assumed to have a specific activity that equals that of the radioactive food source (Crawford et al., 1994). Single cells are then incubated – after serial transfer through unlabelled medium – for variable, but mostly short periods of time, and the amount of ^{14}C released into the surrounding medium is determined. Results so obtained on a marine amoeba compared favourably to rates obtained from Cartesian diver measurements (Crawford et al., 1994). A similar method, relying on uniformly labelled organisms, was used to measure the energetic costs of feeding and foraging in a deposit-feeding gastropod (Forbes & Lopez, 1989).

Calorimetry

Facultative anaerobiosis is widespread among the benthos (see, e.g. Newell, 1970). In aquatic sediments, fully aerobic conditions are often restricted to a few millimetres at the sediment surface and the surroundings of oxygenated burrows of deposit feeding macrobenthos. Hence, a first prerequisite for obtaining meaningful energy budgets at the individual or population level is measurement of total metabolic rate, including aerobic and anaerobic. Basically, all catabolic reactions, irrespective of their utilisation of oxygen, evolve heat. The enthalpy change of catabolism approximates that of total metabolism; hence, total metabolic rate can be derived from heat dissipation measures (Kemp & Guan, 1997).

Microcalorimetry can yield accurate measurements of aerobic and anaerobic metabolic heat produced by living tissue. Pamatmat (1978, 1979, 1980, 1983) pioneered the use of microcalorimetry in the ecology of benthic macrofauna, but to our knowledge, only a limited number of other applications have emerged from this work (Shick et al., 1983). In addition to requiring expensive equipment, a major disadvantage of microcalorimetry is the long period needed to attain stable readings, giving rise to possible starvation and abnormal behaviour effects. Another disadvantage lies in the necessity to suppress contaminant growth during measurements.

Growth

Basically, the measurement of growth is straightforward and consists of repeated measurements of the mass of the whole body (or parts of it), for which the methodology has been discussed above. Nevertheless, a complicating factor is that for specific measurements, for example, ash-free dry mass (AFDM) determination, the animal has to be killed. One might then obtain such figures indirectly through non-destructive methods (e.g. by measuring length growth and using a length-mass calibration curve), or alternatively, by starting with a group of similar animals which are subsequently sacrificed over time (see also the next section on population studies).

Animals that produce skeletons may offer the possibility of using growth lines for estimating the age–size relationship over the entire previous lifespan of the collected individual. Such growth lines are formed due to periods of growth cessation, e.g. periods associated with the annual occurrence of colder periods. A crucial prerequisite is of course a reliable aging of the growth lines, and in practice it is not always easy to distinguish between annual rings and so-called disturbance rings (Richardson, 2001). Analysing the growth lines present in the internal structure of, for example, the bivalve shell, after grounding and polishing a sectioned surface, provides a much clearer picture of the growth lines than a visual inspection of the shell surface (Richardson, 1989).

Rhoads and Lutz (1980) provided the classical work on the use of skeletons in (mostly invertebrate) aquatic organisms. Although much attention was paid to molluscs (mainly bivalves and gastropods), growth patterns are also formed in barnacle shells, corals and polychaete jaws. Richardson (2001) reviewed the recent developments in studies of growth line formation and skeletal deposition in molluscs, yet with an emphasis on the use of bivalve shells to reconstruct historical changes in the marine environment.

Reproductive output

Reproductive output is regarded as being part of the production that includes all energy and matter invested into sperm, eggs and material associated with it, such as egg capsules. The two main reasons to investigate energy investment in reproduction can be distinguished: constructing an energy budget of a species under varying environmental conditions or to determine the reproduction strategy. Clarke (1987) hypothesises two alternative strategies of reproductive investment in species living in areas of different latitudes. Brey (1995) reviewed the literature and compiled a dataset to test this hypothesis. In this text we will discuss only the methods of measuring reproductive output and not dwell upon strategies. Both laboratory and field methods are discussed. It should be noted that reproductive output as described below does not include possible 'overhead' costs associated with the production of reproductive material.

Laboratory methods

The majority of papers that estimated fecundity in the laboratory are based on research on the effects of concentration and quality of organic matter in sediment on growth and reproductive effort of deposit feeding polychaetes, with the short-lived opportunistic *Capitella* sp. as a model species (Grémare et al., 1989; Pechenik et al., 2000; Tenore, 1977). Estimating fecundity in the laboratory can be roughly divided into three categories: non-destructive techniques, dissection and biochemical analysis.

The non-destructive techniques include the frequently used method to count eggs under a microscope (Grémare et al., 1989; Honkoop & Van der Meer, 1998) and to determine egg sizes by calliper (Honkoop & Van der Meer, 1998) or image analysis techniques (Clarke, 1993). Other reported methods are collection of embryos from the parent with no apparent adverse effects for the parent (Bridges, 1996) or collection of shed brood from the basin (Pechenik et al., 2000; Prevedelli & Vandini, 1998). Another indirect method, proposed by Crisp (1984), relies on the assumption that the energy content or mass of an animal before spawning minus that after spawning equals the energy content of reproductive output, as applied by Horn (1986). This method introduces some error since all other possible losses, such as respiration, are attributed to reproduction but it is a simple and widely applicable method. The main advantage of the non-destructive techniques is that animals can also be studied after reproduction.

Dissection of the animals to count egg numbers and measure egg sizes is also frequently reported (Linton & Taghon, 2000a; Sola, 1996). This method implies that a suite of chemical analyses on the eggs becomes available such as caloric content (Clarke, 1993; Prevedelli & Vandini, 1998) and C/N analysis (Clarke, 1993), given that sufficient material can be collected.

A biochemical approach to quantify reproductive output based on the animal's lipid composition was proposed by Taghon et al. (1994). Levels of different classes of lipids and glycogen in a deposit feeder were followed for 1 year. Most levels remained stable throughout this period except for levels of the lipid triacylglyceride that were elevated during oogenesis and it 'may represent a complementary method for measuring reproductive effort if these lipids are preferentially used to provision eggs'. Linton and Taghon (2000a) reported the only additional test and they showed a correlation between triacylglyceride and numbers of eggs produced. However, this method has not been subjected to rigorous testing.

Field methods

Estimating reproductive investment in the field requires extensive knowledge on natural spawning period(s), which last for a few weeks and may shift from year to year depending on environmental conditions such as temperature and light. Generally, temperate areas host annual breeders meaning that spawning occurs once a year. In warmer areas, however, reproduction can be a more or less continuous process, complicating the measurement of energy invested in reproduction.

The only method for *in situ* measurement of reproductive output of benthic marine species is described in Qian and Chia (1994), which appears to be labour-intensive. On an intertidal mudflat, experimental trays containing sieved sediment and a known amount of dyed *Capitella* sp. siblings from a single parent were deployed. Growth and death rates were followed during maturation (~2.5 months) by regular collection and redeployment of the experimental trays. Eggs could relatively easily be counted and handpicked since the tube-dwelling *Capitella* sp. deposits eggs within the burrow. Egg numbers, sizes and energy content were determined. Though reproductive output was comparable to the laboratory measurements, the observed variation was very large.

Alternatively, semi-field estimates may be obtained by collecting specimens at the commencement of spawning in the field. Spawning can be awaited or induced shortly after arrival in the laboratory to reduce any bias of keeping animals under laboratory conditions. In this way the previously mentioned laboratory methods of egg counting, dissection or biochemical analysis become available. Spawning is induced by imposing an environmental shock after which eggs and sperm can be collected from the basin for further analysis, for example, by pipetting (Honkoop & Van der Meer, 1998). Reported effective environmental shocks comprise a temperature shock (Honkoop & Van der Meer, 1998), injection of a KCl solution into the mantle of mussels (Honkoop & Van der Meer, 1998) and addition of Prozac to the basin (Honkoop et al., 1999).

Regeneration

Regeneration of body parts, following sublethal predation or other sorts of injuries, is a form of somatic production that should be accounted for when calculating secondary production of a benthic population. Examples of sublethal predation concern siphons in bivalves (De Goeij et al., 2001; Peterson & Quammen, 1982), feeding palps and posterior segments in polychaetes (De Vlas, 1979b; Lindsay & Woodin, 1996; Zajac, 1995), and arms in echinoderms (Lawrence et al., 1999; Lawrence & Vasquez, 1996; Skoeld & Rosenberg, 1996). Injuries due to fights with conspecifics occur in crabs, which may lose chelipeds and walking limbs (Abello et al., 1994; McVean, 1976; McVean & Findlay, 1979). In some species the contribution to overall secondary production can be considerable. In North Inlet, South Carolina, the trophic transfer related to arm regeneration of *Microphiopholis gracillima* ranged from 3.3 up to 9.7 g AFDM m^{-2} a^{-1}, an amount equivalent to total community macrobenthic secondary production in other systems (Pape-Lindstrom et al., 1997). In the Skagerak, south Sweden, about 13% (0.34 g AFDM m^{-2} a^{-1}) of the total production of an *Amphiura filiformis* population was due to arm regeneration (Skoeld et al., 1994).

Predators responsible for the sublethal predation are, for example, shorebirds, flatfish, crabs, and shrimp (De Goeij et al., 2001; Lawrence & Vasquez, 1996; Peterson & Quammen, 1982; Skoeld & Rosenberg, 1996).

The methods to study the importance of regeneration of body parts can be divided into two parts. First, the frequency at which body parts are lost must be estimated. The methods that have been applied are very similar to those that have been used in studying predation rates in general (discussed above): direct observation, encaging or excluding predators and gut content analysis (De Vlas, 1979a). Second, the regeneration rate should be assessed, which can be easily done by taking repeated measurements during regeneration, just as in normal growth studies. A complicating factor is of course that one might have to amputate the body part to enable a proper measurement. In that case one should start with a group of similar animals.

In some cases, however, an approach specific to sublethal predation can be used. Lugworms *Arenicola marina*, which normally live burrowed in the sediment at a depth where they are relatively safe from predators, expose the tip of their tail many times a day when they have to defaecate. During these short moments, the animals are vulnerable to sublethal predation. Predators such as flatfish may crop one or several tail segments. The tail regenerates by lengthening of the remaining segments, and no new segments are ever formed. This observation may imply that the number of segments in a field population can be used to estimate sublethal predation frequency (De Vlas, 1979b).

Finally, one should realise that the damage to body parts may also influence the feeding abilities of the organism, as the browsed tissue is often part of the feeding apparatus (Nilsson, 2000a, 2000b), and this may complicate the description of the energetic effects of sublethal predation (De Goeij & Luttikhuizen, 1998).

8.4 From the individual to the population

Basically, describing the flows of energy through a population is simply a matter of good bookkeeping of the flows of the constituent individuals. In practice, problems arise because not every single individual can be followed. In this section we will discuss what can be said about energy flows through populations when only limited information on individuals is available. The energy flow which has received most attention at the population level is the production of new somatic and reproductive material. The measurement of this so-called secondary production of animal populations is frequently one of the major objectives of descriptive studies of ecosystem energetics.

Secondary production of a population of animals

Secondary production has been defined many times and most definitions date back to the pioneering work of Thienemann (1931), who stated that the production of a population for a known period of time is considered to be the sum of growth increments of all the individuals existing at the start of the investigated period and remaining to the end, as well as the growth of newly born individuals and of those individuals who for various reasons, do not survive to form part of the final population. Two things should be kept in mind with regard to this definition. Firstly, production is apparently considered as a quantity and not as a rate, and although the exact physical dimension depends on the way of expressing the growth increments (this has been done, for example, as dry mass, number of carbon atoms, or nitrogen), the unit time does not occur. Often, the term productivity is specifically reserved for the rate of secondary production (Lindeman, 1942). Some authors, however, characterise production as a (mean) rate or as a rate per unit area. For example, Waters and Crawford (1973) define production as 'that amount of tissue elaborated per unit time per unit area, regardless of its fate'. Macfadyen (1948) gives an early review of these definition problems. We regard the issue to be of minor importance, as long as authors and readers are aware that comparisons between different studies may be seriously hampered when measurements are made over periods of different length or at different times in the year. The second point that we want to draw attention to is the role of mass at birth. Thienemann (1931) referred only to the growth of newly born individuals and did not include their mass at birth. This must then imply that the production of progeny (or, more generally, gonads) should be assigned to the 'growth increment' of the parents.

Thus, if we were to know the exact fate of all individuals in a population, that is, if we were to know their time of birth and death, their growth pattern and their cumulative gonad production, then the calculation of the secondary production over any period of time would be a straightforward and simple exercise. In practice, such detailed information is of course rarely available. In the literature different calculation methods have been proposed for different types of data. Much attention has been paid to data referring to populations with identifiable cohorts, which have been sampled at relatively short time intervals over their entire lifetime. A cohort is a group of individuals who are born at more or less the same time. The reproductive period should be short, relative to the time in between two successive reproductive periods, in order to recognise individuals from those of the preceding or succeeding cohort. The reproductive period should also be short relative to the lifespan in order to obtain relatively accurate age estimates. Methods of separating cohorts are described in the section on demography. Basically, the advantage of such cohort data is that they enable the estimation of both the survival function and the (average) size-age relationship, or growth function. As we will show below, these two functions are the essential ingredients of production estimates. For populations that do not produce identifiable cohorts, estimates of these functions are not easily obtained. Below we will first discuss the case of identifiable cohorts.

Calculation of production of populations with identifiable cohorts

Crisp (1984) mentioned two different approaches in the measurement of the total secondary production of a cohort of animals and stated that these must be clearly distinguished. The first method, which is known as the *increment-summation* method (Winberg, 1971), is to add all the growth increments of all the members of the cohort as they occur during the period under consideration. The second method mentioned by Crisp, and which is known as the *removal-summation* method, is to consider both the matter that leaves the cohort by mortality, and the difference between the total biomass of the cohort at the end of the observation period and at the start of the period. Adding these two terms gives the cohort production. Crisp (1984), as well as many others (Gillespie & Benke, 1979; Rigler & Downing, 1984), argued that the two methods are similar. Below we will first repeat Crisp's argument by using a simple illustrative example.

Consider at time t_1 a cohort with a total number of N_1 animals, each with individual mass w_1. Hence, the total biomass at the start is $N_1 w_1$. A short period later, at time t_2, only N_2 animals are still alive, each with individual mass w_2. All other animals (that is, $N_1 - N_2$) have died. For convenience, we suppose that they all died with a mass of $(w_1 + w_2)/2$. Using the *increment-summation* method, the production equals the sum of growth increments of all the individuals existing at the start of the investigated period and remaining to the end, that is $N_2(w_2 - w_1)$, plus the growth of those individuals which did not survive, i.e. $(N_1 - N_2)((w_1 + w_2)/2 - w_1)$. Alternatively, the *removal-summation method* equals the production to the matter that has left the population by mortality, i.e. $(N_1 - N_2)(w_1 + w_2)/2$, plus the difference between the total biomass of the population at the end of the period and at the start of the period, i.e. $N_2 w_2 - N_1 w_1$. Some simple algebraic manipulation shows that the two approaches do reveal exactly the same result, which also can be written as $(N_1 + N_2)(w_2 - w_1)/2$. An easy interpretation of the latter expression is that all the N_1 animals present at the start of the short period had a growth increment of at least $(w_2 - w_1)/2$, but that only the N_2 animals alive at the end of the period had an additional growth increment of again $(w_2 - w_1)/2$. This formulation forms the basis of the so-called *growth-survivorship method*, or Allen-curve method. The Allen curve gives the relationship between N and w, and the area under the curve equals the production (Fig. 8.5).

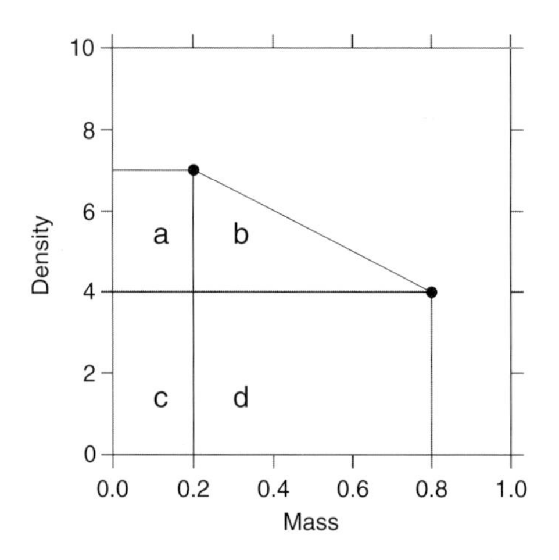

Fig. 8.5 Allen-plot. On the first sampling occasion, seven animals per unit area with a mass of 0.2 units of mass were observed. On the second occasion, four animals per unit area with a mass 0.8 unit of mass were observed. Secondary production over this period equals 3.3. The increment-summation method takes area *b* (which represents the growth increment of those animals that die) plus *d* (the growth increment of the survivors). The removal-summation method first takes area *a* (initial mass of those that die) plus *b*, and then adds *c* (initital mass of the survivors) plus *d* and subtracts *a* plus *c*, which also gives *b* plus *d*.

A more extensive and perhaps more insightful numerical example can be constructed when a cohort has been observed at multiple points in time, e.g. from their birth until all the animals have died (Table 8.5, Figs. 8.6a and 8.6b).

Table 8.5 Production calculations for the 1985 cohort of the bivalve *Macoma balthica* at Balgzand. See text for further explanation.

Year	Density n	Mass u	Δn	Δu	$\bar{n}$	$\bar{u}$	$\Delta n \cdot \bar{u}$	$\bar{n} \cdot \Delta u$
1986	101	0.12						
1987	59	1.43	42	1.31	80	0.775	32.55	104.8
1988	36	2.71	23	1.28	47.5	2.07	47.61	60.8
1989	26	4.35	10	1.64	31	3.53	35.3	50.84
1990	16	6.49	10	2.14	21	5.42	54.2	44.94
1991	9	9	7	2.51	12.5	7.745	54.215	31.375
1992	4	9	5	0	6.5	9	45	0
1993	0	9	4	0	2	9	36	0
Sum							304.875	292.755
Initial mass							−12.12	
Sum							292.755	

One complicating factor that should not be overlooked is the treatment of the mass at birth, or, more generally, the gonad production. Unfortunately, both Crisp (1984) and Rigler and Downing (1984), aiming to show the equivalence between the *removal-summation* and the *increment-summation* method, used examples where

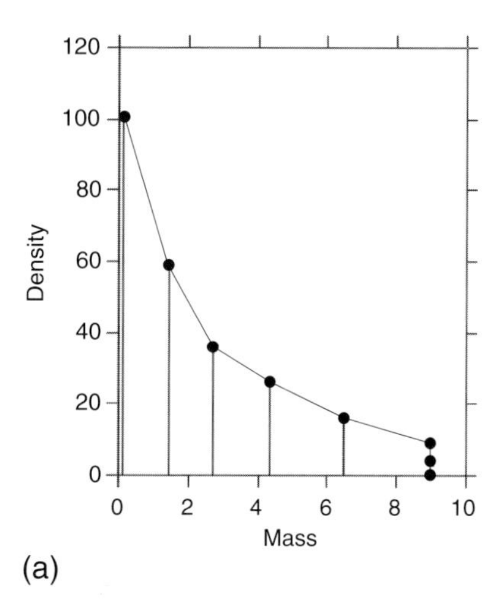

Fig. 8.6a Allen-plot for the 1985 cohort of the bivalve *Macoma balthica* at Balgzand.

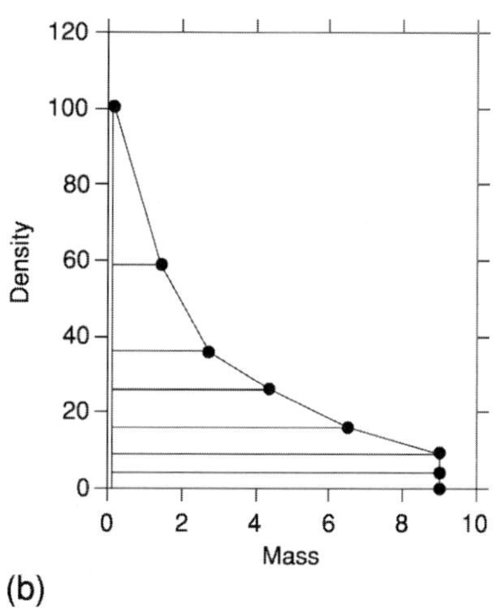

Fig. 8.6b Allen-plot for the 1985 cohort of the bivalve *Macoma balthica* at Balgzand. Illustration of the removal-summation method.

the birth mass was zero and thus did not emphasise this potential source of error. The standard approach is to add the mass of newborn individuals to the production of the parent. This implies that the total initial mass of a cohort should indeed be subtracted when using the *removal-summation* method. However, it also implies that the cumulative gonad production of each individual should be added to the mass at death. Using the *increment-summation* method it implies that the increment should contain the gonad production. The alternative approach is to account the mass at birth to the newborn individual. In that case, the total initial mass of a cohort should not be subtracted when using the *removal-summation* method, and it should be added when using the *increment-summation* or *Allen-curve* method.

Calculation of production when cohorts cannot be identified

Often it is not possible to follow identifiable cohorts through time, perhaps because the population has only been sampled once, or (more frequently), because it is impossible to recognise age classes and the only type of field data available concerns (repeated) observations of the mass distribution of the population. Fortunately, the production rate at any point in time can always be calculated directly from the age distribution or from the mass distribution provided that the age–mass relationship (the growth function) is known. As it is often rather difficult to measure the growth of soft-bodied benthic invertebrates directly in the field, the usual approach is to carry out additional growth measurements in the laboratory. Yet the use of such growth data to estimate production in the field should always be regarded with some suspicion, because it is extremely difficult to mimic natural field conditions in the laboratory, particularly concerning food supply (Crisp, 1984).

Methods of calculating production on the basis of the mass distribution of the population, e.g. the size-frequency method of Hynes and Coleman (1968), later corrected by Hamilton (1969), have caused considerable confusion and debate in the literature (Benke, 1979; Benke & Waide, 1977; Fager, 1969). Similar confusion surrounded the computation of turnover of biomass (as a shortcut to assess production) from turnover of numbers (Rigler & Downing, 1984). Van Straalen (1985) and Aldenberg (1986), who followed a more mathematical approach than usual, showed that such confusions are unnecessary, and below we will base our overview mainly on their work.

Notation

Denote age by x and mass by w. Assume that the mass–age function (or growth function) is a fixed monotonically increasing function $G(x)$, for which it holds that $G(0) = w_0$ and $\lim_{x \to \infty} G(x) = w_\infty$. Hence the maximum mass is assumed finite. Let $g(x) \equiv dG(x)/dx$ denote the growth rate (unit mass per unit time) as a function of age. Furthermore, let $g_w(w) \equiv g\left(G^{-1}(w)\right)$ denote the growth rate as a function of mass. The population's age structure and mass structure at time t are both characterised by a distribution function, $N(x, t)$ and $N_w(w, t)$, respectively (note that N is expressed as the number of individuals per unit of age, respectively mass). That is, the integral

$$\int_{x_1}^{x_2} N(x, t)\, dx \equiv \int_{G(x_1)}^{G(x_2)} N_w(w, t)\, dw$$

denotes the number of individuals present at time t with ages between x_1 and x_2 and, correspondingly, mass between $G(x_1)$ and $G(x_2)$. The age-specific per capita mortality rate at time t, i.e. the number of individuals dying per head per unit of time at age x, is referred to as $Q(x, t)$.

Age-structured data

In accordance with the definitions of Thienemann (1931) and others, the production rate at time t can be defined as the summed individual growth per unit time:

$$P(t) = \int_0^\infty N(x, t)\, g(x)\, dx$$

Hence this definition provides a theoretical justification for application of the *increment-summation* method for calculating production rate on the basis of knowledge on the age structure of the population and the growth function. For each age class

$$\left\{x - \frac{\Delta x}{2}, x + \frac{\Delta x}{2}\right\}$$

the total number, which approximately equals $\Delta x \cdot N(x, t)$, is multiplied by the age-dependent growth rate $g(x)$. Summation over all age classes leads to the following approximation for the production of the population:

$$P(t) \approx \sum N(x, t)\, g(x)\, \Delta x$$

Partial integration of the integral provided above leads to

$$P(t) = -N(0, t)\, G(0) - \int_0^\infty \frac{dN(x, t)}{dx} G(x)\, dx$$

The term $-dN(x, t)/dx$ could be (approximately) interpreted as the change in the numbers within a specific age class as a result of aging, that is, it provides the difference between the numbers that enter and leave a specific age class by aging. A combination with the classical McKendrick–Von Foerster balance equation (see Aldenberg (1986) and Gurney and Nisbet (1998) for an introduction of this balance equation) gives:

$$\frac{\partial N(x, t)}{\partial t} = -\frac{\partial N(x, t)}{\partial x} - N(x, t)\, Q(x, t)$$

which merely formalises that individual's age, but can leave the population only by death. This reveals

$$\begin{aligned} P(t) &= \int_0^\infty N(x, t)\, Q(x, t)\, G(x)\, dx + \int_0^\infty \frac{dN(x, t)}{dt} G(x)\, dx - F(t)\, w_0 \\ &= \int_0^\infty N(x, t)\, Q(x, t)\, G(x)\, dx + \frac{d}{dt} \int_0^\infty N(x, t) G(x)\, dx - F(t)\, w_0 \end{aligned}$$

The individual mass at birth $G(0)$ has been renamed by w_0 and the birth rate $N(0, t)$ by $F(t)$. The latter equation forms the basis of the *removal-summation* method, for calculating production rate on the basis of knowledge on the age structure of the population and the age-specific per capita mortality rate. The three terms give the total mass leaving the population per unit time at time t by death of individuals, plus the rate of increase of biomass at time t, but minus the rate of increase of biomass due to birth at time t, respectively. Note that a condition for the reworking into the second form of the last equation is that $G(x)$ is not a function of time.

Mass-structured data

As was said earlier, it is often impossible to recognise age classes and only information on the mass structure is available. By changing to variable w, the mass-specific analogue of the *increment-summation* method is obtained (Winberg, 1971):

$$P(t) = \int_{w_0}^{w_\infty} N_w(w)\, g_w(w)\, dw$$

Note that we used the relationship between the age and mass distribution functions (in shorthand notation):

$$N(x) = \frac{d \int N(x)\, dx}{dx} = \frac{d \int N_w(w)\, dw}{dw}\frac{dw}{dx} = N_w(w)\, g_w(w)$$

Hence apart from the mass structure, knowledge of the growth rate as a function of mass is required in order to calculate the production rate.

The mass-specific analogue of the *removal-summation* method is provided by changing to variable w,

$$P(t) = -F(t)\, w_0 - \int_{w_0}^{w_\infty} w \cdot d\left[N_w(w)\, g_w(w)\right]$$

This latter equation is the foundation of the Hynes/Hamilton method. The integral from this equation, which in shorthand notation looks as

$$P(t) = -\int_{w_0}^{w_\infty} wd\left[N_w(w)\, dw \cdot \frac{1}{dt}\right]$$

can be approximated by a summation over size classes:

$$P(t) = \sum_i wd\left(\frac{N_w(w)\, \Delta w}{\Delta t}\right) = \sum_i wd\left(\frac{f}{\Delta t}\right)_i$$

where f is the number per size class (the relation between $N_w(w)$ and f was given above), and Δt is the duration of each length class, that is, the time it takes for an individual to pass a length class. Subsequently, the differential is estimated by the difference in $f/\Delta t$ between two adjacent size classes, and, at the same time w is replaced by u (which is a weak point in this discretisation scheme), the mean mass of the two adjacent classes (actually the cubic mean length), resulting in

$$P(t) = \sum_i u\left(\left(\frac{f}{\Delta t}\right)_{i+1} - \left(\frac{f}{\Delta t}\right)_i\right)$$

Hamilton (1969) used size classes of equal width in terms of body length and assumed that the duration of all length classes is the same, hence that $1/\Delta t$ is a constant. Thus, the hidden assumption was made that length growth rate is constant (Aldenberg, 1986). The constant $1/\Delta t$ was called the 'number of times loss occurs' factor i, and was set equal to the number of size classes that are distinguished over the total development. This implies that Δt is expressed in units of total development time. In order to express production in standard units of time, the result of the original Hynes/Hamilton method should therefore be divided by the total development time (Benke, 1979). The important lesson is that the Hynes/Hamilton method does not provide a free lunch! At first sight it might appear that no measurements of the growth function are required, but implicitly the assumption was made that the length growth rate is a constant. It is estimated by the ratio of the total difference in length over the total development time. Alternative discretisation schemes than the one presented here, however, are possible and may lead to slightly different interpretations.

If arbitrary size classes are chosen, an alternative approach suggested by Rigler and Downing (1984), an assumption on the growth function is inevitably made, but it has become rather obscure. So suspicion is always warranted when simplifying assumptions on the growth function are made. A last example of a doubtful assumption, which is nevertheless often applied in the literature as it is so convenient, is the exponential growth assumption (Ricker, 1971). The use of the Hynes/Hamilton method is illustrated in Table 8.6.

In order to obtain a production estimate over a specific period of time, say a year, repeated observations over time of the mass (or age) structure have to be made, and the calculated production rates should subsequently be integrated over time. For stationary populations, which means that the birth rate, the mortality rates and the growth function do not change over time, repeated measurements are of course not needed. If the individuals of a stationary population can be aged (and weighed), no additional measurements of the growth function are required. In this case the calculation of the production entirely resembles the situation of an identifiable cohort followed through its lifetime. This, of course, reflects the well-known fact (which immediately follows from the McKendrick–Von Foerster balance equation) that in a stationary population the age distribution is proportional to the survival function. See Van Straalen (1985) for further details on the link between cohorts and stationary populations.

Table 8.6 Production calculations according the Hynes–Hamilton method. Data from Hamilton (1969), Table 2. See text.

Length l	Volume u	Density n	Δn	$\bar{u} = \left(\frac{l_j + l_{j+1}}{2}\right)^3$	i	$\Delta n \cdot \bar{u}$
0.1	0.001	2682.9				
0.2	0.008	2447.0	235.9	0.003	10	8.0
0.3	0.027	867.9	1579.1	0.016	10	246.7
0.4	0.064	344.0	523.9	0.043	10	224.6
0.5	0.125	148.7	195.3	0.091	10	178.0
0.6	0.216	44.0	104.7	0.166	10	174.2
0.7	0.343	9.8	34.2	0.275	10	93.9
0.8	0.512	2.1	7.7	0.422	10	32.5
0.9	0.729	1.3	0.8	0.614	10	4.9
1.0	1.000	0.9	0.4	0.857	10	3.4
1.1	1.331	0.0	0.9	1.158	10	10.4
Sum						976.6

Production to biomass ratios and the turnover of individuals

Sometimes biologists wish to estimate production for a particular population in a simplified way, which means that they hope to avoid the laborious task of obtaining information on the growth and survival curves. One popular approach is to use published production to biomass ratios (P/B), and combine this information with knowledge of the biomass of the study population. A biomass estimate is of course much easier to obtain than a production figure. The underlying idea of this approach is that populations of the same species or of species with the same ecology or physiology must have similar P/B ratios. Some authors have tried to construct empirical relationships between published P/B ratios and physiological or ecological characteristics of the species: body size (Banse & Mosher, 1980), lifespan (Robertson, 1979), temperature (Tumbiolo & Downing, 1994), food availability, etc. Some studies specifically dealt with marine macrobenthos populations (Brey, 1990; Tumbiolo & Downing, 1994). The unexplained variation that is left over after fitting these empirical relationships is, however, still so large that only rough approximations can be expected when using an externally derived P/B ratio (Banse & Mosher, 1980). Rigler and Downing (1984) conclude that when production estimates are really needed to test specific hypotheses, rough approximations are not very useful. They therefore do not recommend the use of this simplified method. We concur with this point of view.

The ratio of production rate (note that annual production divided by 1 year can be regarded as an average rate) to biomass has the dimension 1 over time. For stationary populations, it can be interpreted as the turnover rate of biomass. Many authors and some textbooks on production (Winberg, 1971) have erroneously stated that for stationary populations the turnover rate of biomass (i.e. the P/B ratio) equals the turnover rate of individuals. The latter rate equals one over the mean lifespan (Bartlett, 1970). Consequently, production has often been calculated as the total biomass divided by mean lifespan. This approach seems to be in agreement with the previously mentioned empirical observation by Robertson (1979) that the P/B ratio was approximately inversely related to maximum lifespan, but Rigler and Downing (1984) showed that the turnover rate of biomass does not always equal the turnover rate of individuals. However, they appear to have missed the critical condition, pointed out by Van Straalen (1985), who showed that the turnover rate of biomass equals the turnover rate of individuals, corrected for the ratio of mean mass at death to mean mass of live individuals. The latter ratio equals 1, if the age-specific per capita mortality rate is constant; in other words, if lifespan is exponentially distributed. Unfortunately, no shorthand rule exists to assess this ratio of mean mass at death to mean mass of live individuals in the case of age-dependent mortality.

Demography: recruitment, growth and survival

Hitherto, we have focussed on the relationship between various mathematical formulations of production rate, and practical methods to calculate production on the basis of various types of data (e.g. data on cohort development, growth rate observations in addition to information on size structure). A more fundamental approach is to model the three underlying processes explicitly (renewal rate, growth rate and survival rate) that together determine population production rate in terms of characteristics at the level of individuals. This challenge is taken up in the following sections.

Estimation of demographic parameters from size- or stage-structured field data: Introduction and principles

Field programmes directed towards estimation of production will typically result in a time series of size-frequency distributions, i.e. plots giving the number of individuals sampled in different size classes. For some populations, in particular of Arthropoda, individuals will usually be classified by developmental stage rather than size; these are called stage-frequency distributions. The basic problem we treat in this section is how to estimate demographic parameters of the population (in particular, patterns of recruitment into the population, growth rates or stage durations and mortality rates) from this type of observation. Although there are cases where the estimation of all these parameters is not needed for a proper estimation of secondary production of the population (McLaren, 1997, and see also the previous discussion on production estimation), in many cases it will be needed for the production estimation. Moreover, the demographic parameters in themselves may be very relevant, e.g. as a means of comparing different populations or in the context of evolutionary studies.

The basic approach to the analysis of size or stage-frequency distributions is to construct a theoretical model of the population dynamics, in which the parameters

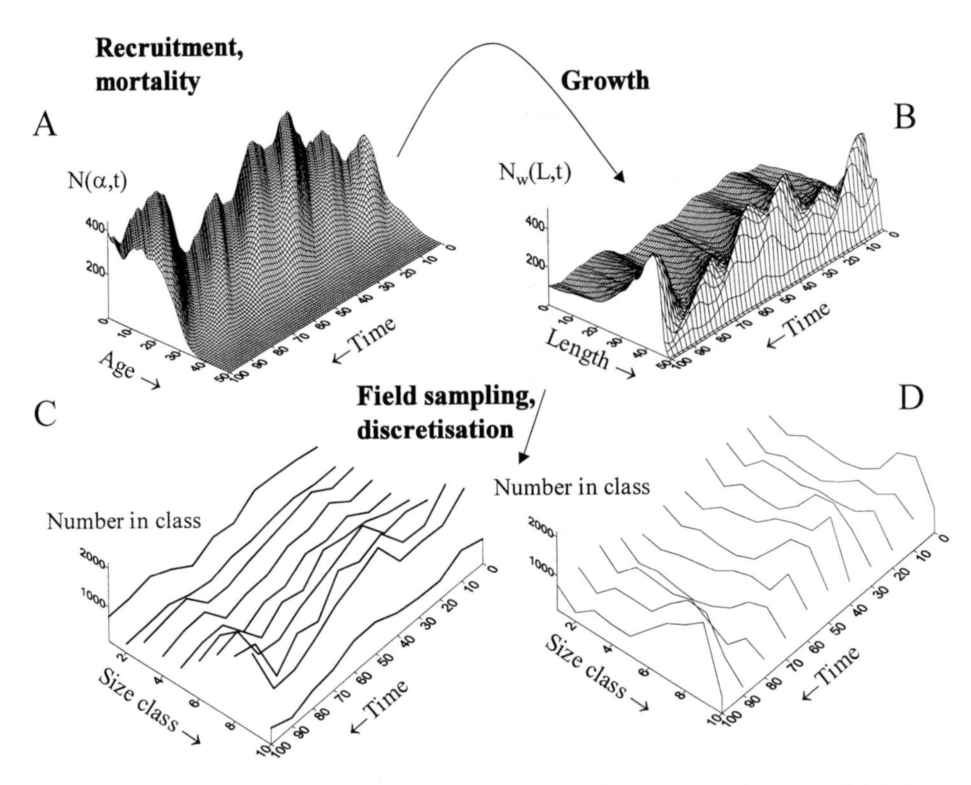

Fig. 8.7 Illustration of the basic approaches in the analysis of size or stage-frequency distributions. The population can be described by a surface in the (age, time) space (A). The form of this curve depends on the recruitment boundary conditions (a moving average function in this example) and on the age- and time-dependent mortality function. Taking into account the (age- and time-dependent) growth function, the population surface in the age–time landscape can be transformed into a surface in a size–time landscape (B). When sampling the population in the field, and using discrete size and time intervals, the observable functions either give a picture of number of individuals in discrete size classes as a function of time (C), or number of individuals in different size classes at different moments in time (D). See text for more details.

are defined, and then apply more or less sophisticated estimation methods to derive the values of these parameters from the available data.

This modelling is illustrated in Fig. 8.7. If $N(x, t)$ is the number of individuals, per unit age interval, of age x at time t, such a population is described by the McKendrick–von Foerster equation:

$$\frac{\partial N(x,t)}{\partial t} + \frac{\partial N(x,t)}{\partial x} + N(x,t)\,Q(x,t) = 0$$

When fully specified, this equation describes a surface in the age–time space (Fig. 8.7a). For the specification, we need to know the mortality function $Q(x, t)$, which is a function of age and time. Furthermore, the initial conditions (i.e. the age structure at time zero) $N(x, 0)$ must be known, as well as the boundary condition $N(0, t)$, which is the birth rate of the population. The boundary condition

$N(\infty, t) = 0$ is of course valid for all animals. A group of animals born at the same moment into the population will describe a trajectory across this surface, which is constrained by the fact that their age will increase by exactly 1 d/d, and that their number can only decline in the process. In Fig. 8.7a, where recruitment was quite variable (it was described by a moving average time series model), these trajectories can easily be recognised, as the peaks in the recruitment function follow an oblique course through the age–time space.

The age of an individual is difficult or, in many cases, impossible to measure directly. Even when it is measurable in principle (e.g. fish otoliths, mollusc bands) it is too time-consuming to measure for a large number of individuals and therefore difficult to apply to population studies. In practice, we have to rely on size information.

The 'population landscape' in the age–time space is projected into a size–time landscape (Fig. 8.7b) by a growth function $G(x, t)$, which, in general, is also a function of age and time. The projection by a non-linear function results in a distortion of the general shape of the landscape. In general, it will tend to collapse the high-age part of the landscape, as growth is slow or absent at older age, while it will extend the range and resolution in the low-age part where growth is fast.

Finally, the field sampling approach will limit our view of the entire landscape to a number of slices at different points in time, where classification of the data in size classes or stages provides a number of discontinuous curves. These can be either time courses of numbers in different classes (Fig. 8.7c), or size-frequency diagrams at different points in time (Fig. 8.7d).

Note that whereas for the sake of clarity Fig. 8.7 uses deterministic functions to model the population, in practice all processes depicted are subject to different sources of natural variability. Added to this are measurement errors, and therefore real data may appear much more scattered than they do in Fig. 8.7. The structure of the natural variability in the recruitment, growth and mortality functions is a particularly complicated problem. Individual variation in the essential parameters of these rates may be considerable, and in addition the different functions can co-vary, when, for instance, fast-growing individuals suffer a different mortality rate than slow-growing ones.

The problem of estimation of demographic parameters from size or stage frequencies is the inverse problem of the construction of the 'field data' as was done in Fig. 8.7. In this construction there were three major functions involved: recruitment, mortality and growth. In addition, initial conditions were needed. In general, when a set of field data is available (which includes the initial conditions up to measurement error), there is more than one set of recruitment, mortality and growth functions that may (within measurement error) reproduce the field data. Even a perfect fit between model predictions and observed data does not prove that the model is a true representation of reality, but merely shows that further improvement or change of the model is not possible until more or different data are tested (Aksnes et al., 1997). Note that this 'perfect fit' is more a theoretical than a real possibility. In order to lead to practical results, the general estimation problem as sketched above must be simplified by one of three approaches: (i) the use of independent auxiliary information, (ii) simplifying assumptions about the essential functions or (iii) both of these.

For the *recruitment* function, independent information is rarely available in benthic species. In zooplankton, and notably in the study of rotifers and daphnids, extensive research has been devoted to the so-called egg-ratio methods, dating back to Edmondson (1960). We do not discuss the approach in detail. Basic publications on the method have been made by Paloheimo (1974), Seitz (1979) and Threlkeld (1979).

In contrast, simplifying assumptions about the recruitment function are very well applicable in benthic research. Many populations have peak recruitment in a short period, a situation which is idealised as 'cohort populations' (see above). Figure 8.8 shows a simulation (with the same functions as used in Fig. 8.7) for a population where recruitment is non-zero only for a short period. The age–time and size–time diagrams clearly show the progression of the peak (mode) of the recruitment function through the size classes with time. *Mode progression* methods use this information to derive an estimate of the growth and mortality functions, or,

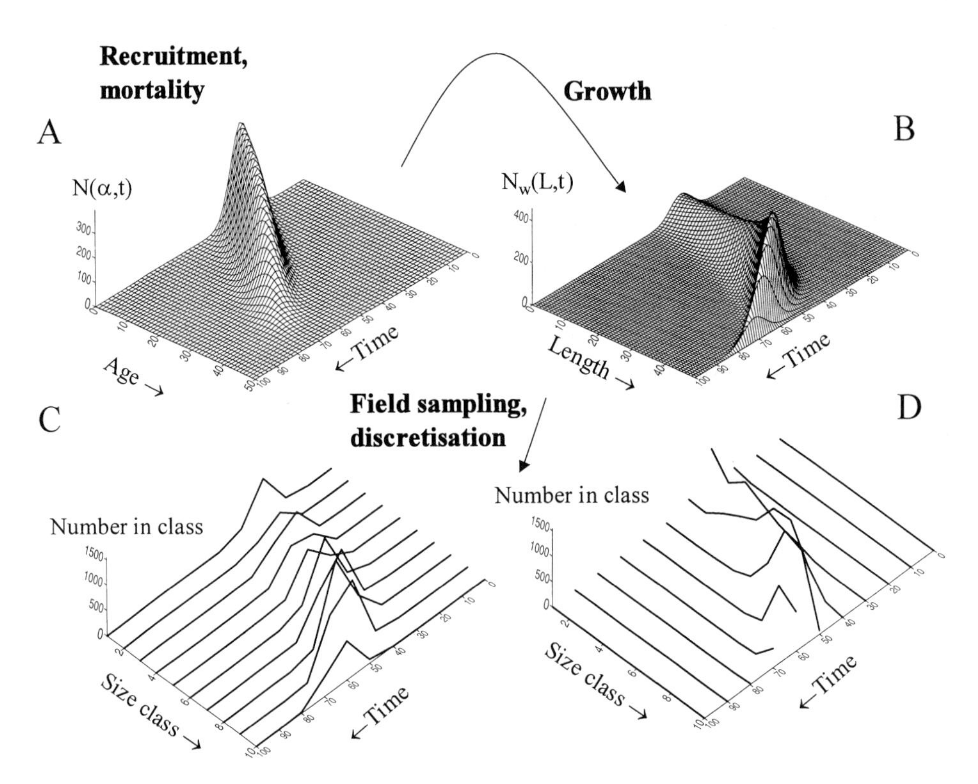

Fig. 8.8 As Fig. 8.7, but illustrating the case of a population with one cohort. Recruitment in this population is restricted to a short period of time compared with the lifetime of the cohort, which greatly simplifies the interpretation of the resulting graphs.

when independent information on the growth function is available, of the mortality function alone (see below).

For the *growth* function, independent auxiliary information is often available from laboratory experiments or from experiments in the field. Studies of stage-structured populations (e.g. copepod populations) can often rely on estimations of stage durations in laboratory experiments. The estimation of development rates from laboratory studies is a complicated statistical problem in itself (Klein Breteler et al., 1994), but because of the well-controlled conditions of these experiments, much less variability is present than in nature, and moreover the experimental set-up can exclude confounding factors, such as continuous new recruitment into the population. Other studies use complementary information from field experiments (see the section on growth of individuals).

As a simplifying assumption, a functional form for the growth function is often applied. In fisheries research, there is a long tradition of using the von Bertalanffy growth equation. This growth function is also derived by the κ-rule DEB model (see the section on energy budgets of individual organisms) and should, with proper scaling, be useful in the study of many other populations. Direct application to stage-structured populations, which are characterised by a decoupling of length growth and growth of structural body mass (the latter being more or less continuous, whereas the former occurs only at stage transitions), is more difficult but should, in principle, be possible. The application of a functional form to the growth function reduces its estimation to the estimation of a few parameters only. For longer-lived populations, a seasonal form of the von Bertalanffy equation is often used, allowing for slower growth in the cold season.

For the *mortality* function, neither external information nor functional representation offers great possibilities. Typically, mortality does not depend on the state of the individual, but on factors outside the population under study, such as the presence of predators, occurrence of diseases, food shortage, etc. An exception is fishery mortality, which is often included as a separate factor in fisheries models, added to 'natural' mortality. It can often be estimated based on fisheries statistics. Another exception is the use of shed ostracod shells conveying information on the time when animals died in the sediment (Herman et al., 1983).

Simplifying assumptions on the mortality function are often included. Some methods (notably theoretical studies investigating stationary populations) assume mortality rate to be constant throughout life. Many methods assume mortality to be constant within a stage or size class. Some methods (see below) impose that mortality rate is a smooth function of time and age (Manly, 1997; Wood, 1994). This may reduce the number of parameters to be estimated, while still allowing the representations of variable mortality functions through life.

Parallel scientific traditions have developed in the estimation of demographic parameters from size-frequency and from stage-frequency data. The former notably developed in fisheries research, where the bulk of the data is in the form of length-frequency and catch number data, and where a (seasonally adjusted) von Bertalanffy growth curve can generally be used to link size to age. Fisheries biologists also have reference datasets based on otoliths, with which they can check the age–length relationships estimated from the field data. Stage-frequency analysis methods have been developed mainly in the context of crustacean zooplankton research. Auxiliary data are generally concerned with the duration of stages as a function of temperature and/or food supply; estimates of egg production may also be available. Although practical procedures may differ for these different types of populations, the underlying population models are very similar. In fact, there is no reason why a length class could not be considered a 'stage', and the growth curve would then be a model for the 'stage durations'. We will therefore combine the different approaches in this discussion, which will be limited to a number of methods that illustrate different possibilities of either external data availability or possibilities for simplifying assumptions. More thorough discussions can be found in Manly (1990), Wood (1994) and Aksnes et al. (1997).

Simple mode progression models

When cohorts are clearly discernible in the population, this simplifies the population surface in the sense of Fig. 8.8a. Population numbers at age 0 will be 0 for most of the time, and only become positive when the cohort is born. Then this peak in numbers will travel through the age–time space along a trajectory like the one depicted in Fig. 8.8a, and the position of this trajectory will be clearly marked as a distinct mode in the age (or size, stage) spectrum of the population. Several methods make use of this characteristic to estimate demographic parameters.

One of the oldest approaches was initiated by Rigler and Cooley (1974) and Gehrs and Robertson (1975). It assumes that all animals that enter a stage or class will spend the whole stage duration in that stage, i.e. will not die during their stay in the stage. Animals can die at the transition between two stages only, and the population trajectory (Fig. 8.8a) is therefore assumed to be a step function with sudden decreases in numbers at stage transitions.

The method calculates a quantity A_j, the total number of 'animal.days' spent in the stage, and a quantity B_j, the 'centre of gravity' of the stage:

$$A_j = \int_0^\infty p_j(x)\,dx \approx \sum_{i=1}^{\pi-1} \frac{1}{2}\{p_{ij} + p_{i+1,j}\}(\tau_{i+1} - \tau_i)$$

and

$$B_j = \frac{\int_0^\infty x p_j(x)\,dx}{\int_0^\infty p_j(x)\,dx} \approx \frac{1}{2A_j}\sum_{i=1}^{\pi-1}\{p_{ij}\tau_i + p_{i+1,j}\tau_{i+1}\}(\tau_{i+1} - \tau_i)$$

These estimates require sampling until all animals of the cohort have left the stage. Manly (1985; 1990) discusses compensations for this requirement. The method then estimates stage durations from the relation:

$$B_{j+1} - B_j = \frac{a_{j+1} + a_j}{2}$$

which necessitates either independent information on one stage duration, or of a relation between the durations of successive stages for solution. The number of animals recruiting into a stage is estimated as:

$$N_j = \frac{A_j}{a_j}$$

again expressing that all animals are assumed to spend a_j units of time in that stage.

An advantage of this method is its simplicity. Its estimation of stage durations can easily lead to oscillations (underestimation in one stage leads to overestimation in the next stage). Most importantly, however, it provides biased estimates in the presence of important mortality within the stages. Mortality has an obvious effect on the time spent by an average individual in the stage, but it also affects the means B_j. This can be seen by the equations, derived by Manly (1977), for a population with (time constant but stage dependent) mortality rates within the stages:

$$A_j = \frac{N_j \mu_j}{1 - \exp(-\mu_j a_j)} \qquad \text{(juvenile stage)}$$

$$B_j = m_j + \frac{1}{\mu_j} - \frac{a_j \exp(-\mu_j a_j)}{1 - \exp(-\mu_j a_j)} \qquad \text{(juvenile stage)}$$

$$A_j = \frac{N_j}{\mu_j} \qquad \text{(adult stage)}$$

$$B_j = m_j + \frac{1}{\mu_j} \qquad \text{(adult stage)}$$

and the equations connecting the stages:

$$N_{j+1} = N_j \exp(-\mu_j a_j)$$
$$m_{j+1} = m_j + a_j$$

In these equations, m_j is a parameter describing the time for peak recruitment into the stage. With appropriate additional assumptions, and making use of the fact that the formulae simplify for the final (absorbing) stage, several operational estimation methods were derived from these equations (Aksnes & Hoisaeter, 1987; Manly, 1977). However, estimations using this method can be unstable and subject to serious errors (Wood, 1994).

Another approach to circumvent the bias present in the basic Rigler and Cooley model, is to use medians, rather than temporal means, in the estimation of stage durations. The median H_j for stage j is defined as the moment in time when 50% of the cohort is observed in stage j or higher, while the other 50% is in earlier stages. The stage duration is then calculated by the difference in successive medians. See Miller (1993) and Miller and Tande (1993) for a discussion of the sensitivity of the method. Bias mainly arises from differences in mortality rates between adjacent stages, but the method behaves reasonably well overall.

A large number of methods have been devised along similar lines to the ones outlined above. We refer to reviews by Manly (1989, 1990). More recently, Manly (1997) published a method which, while still a continuation of the approaches outlined above, also incorporates the feature of 'partially specified ecological model' (Wood, 2001). Recruitment and mortality functions are specified to be functions of time and constrained to be sufficiently smooth, but not given functional forms. Instead, they are specified as exponential functions and exponential polynomials, the parameters of which are estimated in the model estimation procedure. The method gives acceptable results provided stage durations are sufficiently well known.

Mode progression with parametric growth curve

These methods, developed in the fisheries literature, assume that a population is composed of several discernible cohorts, and that growth is governed by a Von Bertalanffy growth equation (Beverton & Holt, 1956). The size-frequency structure of the population at the first sampling occasion is described as a sum of a number of normal distributions, each distribution representing a cohort. The strength of these cohorts can differ from one cohort to another, as is typically the case for fish populations. The size-frequency distribution at the next sampling interval (in fisheries usually the next year) can then be derived as the result of (i) recruitment of one new cohort and (ii) progression in time, with concomitant mortality, of the previous cohorts. Progression in time implies also progression in size, and this is represented by the growth curve, the parameters of which are to be estimated by the method. The essence of the existing methods is therefore the decomposition of subsequent size-frequency plots in their composing cohorts, and the establishment of links between (or 'identities of') the cohorts, so that they show growth compatible with the estimated growth curve.

A widely used computer package used for this analysis is ELEFAN (Pauly, 1987), which is part of a FAO package of programs (Gayanilo et al., 1996). The algorithm identifies modes in the size-frequency distributions, and attaches positive scores to high modes while throughs are given negative scores. In an iterative way, it then identifies cohorts and lets them progress through time and size. The model 'collects' positive and negative scores on its way through the subsequent size-frequency diagrams, and the model with the highest total score is selected. Input requirements, in the form of previous knowledge, are minimal for this method, although it can be forced to accept certain well-known parameter values.

As in ELEFAN, Shepherd length composition analysis SLCA is based on the goodness of fit of the location of modes calculated from a von Bertalanffy growth curve. It uses a different goodness of fit measure, similar to that used in time series analysis, i.e. complex demodulation (Shepherd, 1987). SLCA can be accomplished using the software FiSAT, which is part of the FAO package (Gayanilo et al., 1996). McQuaid and Lindsay (2000) provide a comparison of growth parameters of molluscs estimated by SLCA with direct measurements of growth using tagged individuals. These direct measurements are based on shell marking (Ekaratne & Crisp, 1982) and growth band analysis using acetate peels (Pentilla et al., 1988; Richardson, 1989).

The programs MULTIFAN (Fournier et al., 1990, 1991) and MULTIFAN-CL (Fournier et al., 1997), which are available commercially, are based on the same basic reasoning, but differ from ELEFAN in their estimation procedure. The method is very much tailored to fisheries problems, and starts by specifying 'catch equations' that govern the number of fish in a particular age class in a particular year, as a function of background mortality, fisheries mortality, spatial movement, modelled as a diffusion process in Fournier et al. (1997). Constraints on the mortality parameters (e.g. natural mortality is independent of year and region, but varies with age) can be introduced into the equations. Similarly, assumptions regarding fisheries mortality can be specified. The model specified in terms of age, is transformed to a model as a function of length (most data sets in fisheries have length distributions, not age distributions) by assuming that the mean length of an age class follows a von Bertalanffy growth curve, that the lengths of fish in each age class are normally distributed, and that the standard deviations of these distributions are a linear function of the mean. Fournier et al. (1997) also give an option for density-dependent growth. With the model thus specified (including a specification of the error distributions), estimation of the parameters is done by maximum-likelihood estimation methods, using efficient numerical algorithms. The estimation procedure also yields confidence limits for the parameter estimates. Hypotheses on the processes involved are tested using a Bayesian approach. The method is quite computationally intensive.

Bjorndal and Bolten (1995) compared ELEFAN, SLCA and MULTIFAN on a set of data on green turtles, where the output from the programs could be compared to growth data obtained from tagging. They concluded that MULTIFAN obtained the best output, but that SLCA is useful to conduct prior estimates of the parameters that can be used as input to MULTIFAN.

Populations with continuous recruitment

When recruitment is continuous and (in principle) constant in time, the age distribution in the population, as well as within each of the size, age or length classes is asymptotically stable. Under these assumptions, a fixed proportion of the animals in the class will mature from the class per unit of time. The dynamics of the population can then be represented by matrix models, using discrete time. A population vecter $n(t_i)$ represents abundances of the stages at time t_i. It is updated to its value at t_{i+1} by multiplication with a transition matrix representing growth, mortality and reproduction. In practice, the assumption of stable age distributions within the stages is not too severe when stages are of relatively short duration. The method can then also be extended to make the parameters time-dependent. This method, and its estimation procedures, is discussed in Caswell and Twombly (1989) and Twombly (1994). It is critically compared to other approaches by Wood (1994).

The equivalent of the matrix approach in continuous time representation is a set of ordinary differential equations. Rothschild et al. (1997) discuss a model in the form of a set of ODEs of the general form

$$\frac{dx_i}{dt} = \sum_{j=1}^{q} a_{ij} x_j$$

for which they provide analytical (in simple cases) and numerical estimation techniques. For a similar problem definition, but including explicitly terms related to measurement errors, Ennola et al. (1998) present an estimation method based on Kalman filtering. They show that the method can track reasonably well important (step) changes in the parameters over time.

In the fisheries literature, there is a long tradition of models assuming constant recruitment into a population. In fact, this is the basic assumption of the Beverton and Holt model (Beverton & Holt, 1956). If recruitment is continuous and constant throughout the year and mortality is constant, an exponentially distributed age structure with mortality as the parameter will follow. Under these assumptions, the length-frequency distribution will have declining proportions over size. Beverton and Holt (1956) derived an expression for estimating the mortality from length-frequency data and growth parameters of the von Bertalanffy growth equation. This method has been used as the basis for examining the effect of individual variability on the estimation of demographic parameters (Powell, 1979; Wang & Ellis, 1998). The effect can be substantial, and this aspect has not yet been fully incorporated into models based on more realistic assumptions on the population dynamics.

Smith et al. (1997) estimated growth and mortality parameters of sea urchins both from field size frequency distributions and from laboratory growth experiments. They explicitly incorporated variability in growth parameters of the von Bertalanffy curve, and investigated the effect of this variability on the patterns in size frequency. They conclude that features of a plot of size at $t + 1$ year versus size at t could be explained by this variability. This conclusion contrasts with an earlier publication of Ebert et al. (1999) who used a modified growth equation after Tanaka (1982; 1988) to explain these.

Models using full or partial explicit dynamic equations

An approach called 'systems identification' was introduced by Sonntag and Parslow (1981) and Parslow et al. (1979) and used for copepods by Hay et al. (1988). A dynamic model of the population is formulated as a delay-differential equation and fitted to population data by non-linear fitting routines. Delay-differential models specify the age structure within stages at $t = 0$, and then express the number of individuals leaving a stage as a function of the numbers that have entered the stage one stage duration earlier. In the approach of Hay et al. (1988) several model elements (e.g. recruitment function, mortality) were expressed in functional form with parameters to be estimated from the data. The method was successfully applied to a number of copepod species.

Wood and others (Ohman & Wood, 1996; Wood, 1994) generalise this approach on 'partially specified models'. They assert that it is better not to specify the functional form of unknown characteristics of the population (e.g. the change of mortality rates with time) than to mis-specify them. They, therefore, prescribe only general constraints on the estimated population surface: it should be smooth, positive and it should not imply negative death rates. They use spline functions to represent the population surface and methods of quadratic programming to estimate their parameters. Smoothness, and thus complexity of the solution, is statistically cross-validated on the data. The method requires good estimates of stage durations (which may be function of time), and produces estimates of time- and age-dependent death rates and of time-dependent birth rates with their confidence intervals. The method is tested and compared to other approaches using simulated data. A computer program for the method is available. The approach was later on generalised as a general strategy for 'partially specified ecological modelling' (Wood, 2001).

Conclusions and recommendations

Good estimates of population parameters come at a price. The first, and probably most important price, is to have a good quality data set that approaches the assumptions (e.g. closed population), which is sufficiently detailed in time so as to avoid individuals jumping over classes, and sufficiently long to assure that individuals can reach the final class (especially when cohort methods are used). The second price is that all simulations show that estimations become substantially better when external independent information is available, e.g. on stage (or class) durations, on time courses of mortality or other parameters. The third price is that more sophisticated techniques (e.g. MULTIFAN, Wood's approach) are computationally cumbersome and require sufficient expertise from the user. This effort may be worthwhile only if the data warrant it, and if the problem is sufficiently important. A choice of simpler approaches could be preferable if this is not the case. As a guide for this choice, one should evaluate in particular which external information is available, and which simplifying assumptions or functional forms are most warranted by the data.

8.5 From populations to communities

Due to the complexities of real ecosystems, an understanding of their structure and dynamics cannot easily be obtained by taking into account a detailed description of all the constituent populations and their interactions. Below we will discuss various simplifying approaches that all aim to come to grips with true ecosystems.

Food webs

The food web is a fundamental concept in ecology that pictures the exchanges of energy or matter between living and non-living compartments in a system and across the systems boundaries, and is strongly anchored in a variety of research disciplines. Duplisea (2000) constructed biomass size distributions for the benthos of the Baltic Sea. The sediment characteristics alone did not explain the observed patterns and it was hypothesised that a general applicable theory to explain the different observed patterns would have to include biotic interactions. The effect of contamination on a microcosm soil food web was more pronounced at the higher trophic levels because of contaminant accumulation (Peeters et al., 2001). Consequently, the food web structure influences its susceptibility to disturbance by contaminants. In pelagic ecosystems, trophic transfer efficiency of primary production to higher trophic levels depends, amongst other factors, on food web structure and influences feasible fisheries harvest (Kemp et al., 2001). On the basis of energy patterns in soil food webs, deRuiter et al. (1995) analysed interaction strength patterns in soil food webs and showed that these were organised in such a way that stability was promoted. Estes and Peterson (2000) identified several future research directions in marine benthic environments that crucially depend on food web structure.

Ecological modelling comprises a large research field and it is impossible to go into detail on the different topics. We, therefore, have highlighted the most fundamental concepts and practices and have given readers appropriate references for further exploration. In the next section the food web is captured in a very basic mathematical description, the mass balance, which forms the basis of most modelling exercises. Subsequently, two alternative ways to solve mass balances are presented: inverse analysis and dynamical modelling. Finally, network analysis, which allows the basal properties of the ecosystem to be described in indices, is discussed.

Food web in mass balances

By definition, modelling means simplifying. The real world is very heterogeneous; a sediment core from a mudflat reveals a thin brownish top layer of benthic diatoms, burrows of deposit feeders, hotspots of animals, differently coloured sediment patches and so on. It is impossible to capture this complexity fully in a model, which forces us to make abstractions from nature. One of the arts of modelling is to make these abstractions in such a way that the main topics of interest are well represented in the model. The general approach in ecosystem modelling is to focus on energy or mass flows between compartments using a 0–D approach. This means that spatial heterogeneity is discarded and organisms live in a fully homogeneous world. Organisms are grouped into compartments based on function (e.g. primary producers), size (e.g. macrobenthos) and trophic guild (e.g. herbivores), with non-living compartments (e.g. detritus) completing the model. Furthermore, we apply a fundamental law: conservation of mass. This law is expressed in a mass balance that mathematically describes the processes that influence the concentration of the respective state variable (i.e. the concentration of a compartment) in time. The most general form of a mass balance is a description of the rate of change of a compartment X in terms of the difference between an input and an output term. When the input exceeds the output, the compartment increases in time and vice versa. Steady state occurs when the input equals the output and the concentration does not change in time. A mass balance of a living compartment consists of various terms and could look like (cf. the IBP approach):

$$dX/dt = I - (F + E + R + P)$$

The dX/dt term states the rate of change of the compartment in time, consumption or ingestion I is the total uptake of mass through consumption of different preys, defaecation F is that part of the consumption that is not absorbed in the digestive tract but lost as faeces, excretion E is mass excreted from the body (e.g. urine), respiration R is the loss of mass due to metabolic activity, and predation P refers to predation exerted by higher trophic levels in the food web. Methods to measure each of these processes are given in previous sections of this chapter. When such mass balances are constructed for every compartment, they collectively form a mathematical model of the food web. Choices on how to group different organisms in compartments or which terms make up the mass balance depend on the topic of interest and are always model-specific.

The terms in the general mass balance are based on both the physiology of the individual organism (e.g. defaecation, excretion, respiration) and the interactions between organisms (e.g. consumption and predation). An important aspect in modelling is a realistic mechanistic description of the various mass balance terms, and we will provide some explanation and key references regarding these descriptions. There is no general consensus on the mechanistic description of the physiological terms and as to what complexity is needed for an accurate description (Tett & Wilson, 2000). Some authors use a direct link between respiration and biomass (Pace et al., 1984), whereas others model respiration as a weighted sum of maintenance, absorption and growth processes (Ross et al., 1993; Tett & Wilson, 2000). The most extensive treatment on modelling physiological processes is the Dynamic Energy Budget theory by Kooijman (2000). This theory has not found its way into applications at the ecosystem scale, but given its success at lower levels of organisation, it awaits further exploration (Nisbet et al., 2000).

Interaction between organisms is reflected in the interaction between compartments in the food web. The most commonly used mechanistic description of such interactions is a functional response that describes uptake of a resource as a function of resource density. The Holling types I (equal to the Lotka–Volterra response), II and III are most generally used. The most basic form type I implies a linear uptake of the resource without any saturation. Sometimes this function is used with an upper limit on the uptake to prevent unbounded growth of the consumer. Type II is a hyperbolic function and mathematically similar to Michaelis–Menten enzyme kinetics and Monod growth of nutrient limited autotrophs. At low food concentrations consumption increases almost linearly because practically, it is only prey density that limits consumption. At high resource levels consumption saturates as prey handling eventually becomes limiting. The type III response is a sigmoid function. More functional responses have been described by various authors, all accommodating different mechanisms such as interference competition and different prey handling strategies, but type II is most commonly used (Skalski & Gilliam, 2001).

Papers on the measurements of functional responses in benthic systems are relatively abundant. For example, Balczon and Pratt (1996) measure functional responses of Protozoa, Linton and Taghon(2000b) of deposit feeding polychaetes, Ejdung and Elmgren (1998) of amphipods, Moens et al. (2000) of nematodes, and Sommer (1999) of periwinkles. Trexler et al. (1988) discuss statistical methods to use in the analysis of functional response experiments.

Inverse analysis

The term data assimilation comprises the set of techniques that exist to merge field observations with a mathematical model. Inverse analysis, also termed mass balance inversion, is one such technique that specifically deals with a limited data set consisting only of biomasses and some rate processes and systematically quantifies the different terms in the mass balance description (Vézina & Platt, 1988). A typical food web model has a scarcity of data and this implies an under-determined system, i.e. there are fewer equations (i.e. measurements) than unknowns (i.e. food web flows). This indeterminacy implies that an infinite amount of solutions to the system is available. Therefore, biological constraints and a minimisation criterion are used to select one final solution. The biological constraints mostly have a physiological nature (e.g. bounds imposed on absorption efficiency) and are imposed on unmeasured flows and these ensure that a biologically meaningful solution is selected. The minimisation criterion selects the solution that minimises the sum of squared flows (Vézina & Platt, 1988). This is rather arbitrarily chosen and lacks a sound ecological justification and the effects of these should be kept in mind when evaluating the solution (see Niquil et al. (1998) for a brief discussion).

In some situations, the solution procedure may fail to find a solution to the problem. This means that the data set is inconsistent, or put in another way, there exist conflicts within the data or with the imposed constraints (Vézina & Pace, 1994). Detecting data inconsistencies may not be appealing at first sight but can prove to be valuable. Since all the data on a food web are compiled, inconsistency shows the bottlenecks within the data and this may work directional in future research or point to weaknesses or inaccuracies in the deployed methods.

In summary, three types of input are required: (i) a compartmental representation of the food web where possible links between compartments and across the systems boundaries are defined; (ii) site-specific data such as biomasses of the compartments and rate measurements (e.g. primary production, community respiration); (iii) constraints on (some of) the unmeasured flows. The solution is a static food web with its flow values meeting the rate observations, satisfying the biological constraints and with closed mass balances in a least-squares sense of the flow values.

Though the inverse model output provides insight into the structure of the food web, the sensitivity of the results to changes in the input values and the structure of the model is an important issue. When the model output alters drastically with only moderate changes of input values or model structure, this would severely limit the applicability of inverse analysis since the outcome is too sensitive to the input values. Several types of systematic sensitivity analysis have been applied in various studies. Generally, it was found that mass balance inversion produces robust and stable results. In addition, Vézina & Pahlow (2003) performed numerical twin experiments to examine the accuracy of inverse analysis in reconstructing food web flows and found that (i) food web flows are readily quantified, although in strong recycling systems small flows were over-estimated and large flows were under-estimated, (ii) the accuracy of the solution increases when the food web is solved for both C and N and (iii) the steady state assumption, which is generally applied, does not distort the reconstruction significantly. It is important to realise that the maximal accuracy of any data assimilation technique depends on the quality of the observations.

Inverse analysis has been extensively used in studies on pelagic food webs (Donali et al., 1999; Jackson and Eldridge, 1992; Niquil et al., 1998; Vézina & Platt, 1988) and an almost standardised set of constraints has evolved from the original paper of Vézina and Platt (1988). So far, only three papers appeared on its use in benthic environments. Eldridge and Jackson (1993) quantified elemental flows at two continental slope sites with contrasting oxygen regimes off the coast of California, and showed clear differences in the remineralisation pathways. Chardy et al. (1993) and Leguerrier et al. (2003) inferred food web flows in two French coastal systems. Chardy et al. (1993) showed the potential export of the subtidal community to the adjacent sea and Leguerrier et al. (2003) found that the benthic compartment dominated carbon flows as compared to the pelagic compartment. These examples clearly show that the inverse methodology allows the identification of some of the main features of the systems under study. The inverse solution procedure is discussed in detail in Vézina and Platt (1988) and well demonstrated in Donali et al. (1999) and will not be repeated here. Required algorithms are accessible from the SLATEC library (http://www.netlib.org). A MatLab program is available at http://www-ocean.tamu.edu/~ecomodel/Software/invmodel/invmodel.html that solves the inverse problem. An inverse analysis session is easily started within the modelling environment

Femme {Soetaert, 2002 #277, http://www.nioo.knaw.nl/cemo/femme}, in which a suite of applications becomes available to analyse the results.

In conclusion, inverse analysis is a data analysis tool that is capable of (1) inferring food web flows from a limited set of data and (2) detecting internal inconsistencies in the data and/or our perception of a food web. The food web reconstructions are generally found to be robust to perturbations of input values or the structure of the compartmental food web, demonstrating its applicability. Despite the fact that inverse analysis can provide comprehensive insight in the structure of benthic food webs, its applications have been limited so far. User-friendly interfaces may help us lower the threshold for explorative use.

Dynamical modelling

A shortcoming of inverse analysis is that the solution is static and does not take into account the dynamic state of a food web, nor does it allow making predictions, analysing the system's dynamics or reproducing observed time series of biomasses. These topics are typically approached with dynamical modelling and in this section the basic principles of dynamical modelling are explained.

Dynamical models are driven by a certain forcing function, such as irradiance in pelagic food web models that drives primary production of phytoplankton (Scavia et al., 1988) or sedimentation of organic matter that fuels the food web of benthic subtidal environments (Blackford, 1997; Chardy & Dauvin, 1992). For example, in the latter model, phytoplankton that is not consumed by zooplankton sediments to the benthos where it is subsequently incorporated in the benthic food web.

A very important step concerns the assigning of values to the model parameters. Ideally, these should be estimated experimentally. This is particularly difficult in benthic systems, notably because of access difficulties and small- and large-scale heterogeneity. In order to validate the model, long-term biomass data series is required as, for the same reasons as previously mentioned, these data are difficult to obtain. For example, Pace et al. (1984) analysed the dynamics of a continental food web, including the benthic compartment, with relatively crude assumptions regarding parameter values. Their main objective was to study the general dynamics of such a system, and afterwards the results were compared with benthic production estimates from other systems. Chardy and Dauvin (1992) modelled a sand community in the English Channel based on monthly biomass data. Parameters were taken from the literature and their main objective was to reproduce the observed biomasses and, because of the parameter assumptions, the modelled flows should be regarded with care. Blackford (1997) and Ebenhoh et al. (1995) describe the effort to construct a benthic sub-model of the North Sea within the scope of the generic European Regional Seas Ecosystem Model (ERSEM). They, however, acknowledge that 'most parameterisations are not derived directly from values quoted in the literature but are a synthesis of process measurements, fitting and educated guesswork'.

The models mentioned above are important first steps in dynamically modelling the benthic environment, but as they aim at describing the system's general dynamics based on this relatively scarce field, it clearly shows the current limitations in this area of modelling.

Network analysis

The previous section is devoted to techniques that allow one to infer food web flows. The result is a relatively complex picture that might be difficult to interpret, particularly when one is interested in cycling of matter or comparison of food web properties among systems. Network analysis aims at characterising certain food web properties by the use of indices to facilitate interpretation and comparison. Deriving single indices from an extremely complex system like a food web displays a high level of reductionism. And indeed one has to be careful when applying this analysis. The main problem lies in the lumping of species into compartments, which is severely biased towards the lower trophic levels. The indices should represent a unique property of the ecosystem and be therefore effectively independent of the level of aggregation. But when species are lumped, any exchange between species within a compartment is neglected, making the indices susceptible to the modeller's aggregation level. One may argue that when food webs are aggregated equally and have their topology in common so that any aggregation effect on the indices will be equal. This is not the case. If exchanges within a compartment differ between two systems, a different amount of exchange is not taken into account. Some studies have examined the effects of aggregation on the outcome of the so-called Ascendency index (Abarca-Arenas & Ulanowicz, 2002, and references therein). As expected, the outcome depends on the level and manner of aggregation. Effects on the other indices are as yet unstudied. These unresolved matters raise the question as to how descriptive the indices truly are in an ecological sense. There are no definite rules on how to lump species correctly and systematically within a food web. We maintain that aggregation level and topology should be at least equal among systems, although even this in itself does not guarantee consistent comparison. The indices are grouped into input–output analysis, trophic structure analysis, cycle analysis and topological analysis (Kay et al., 1989; Ulanowicz & Norden, 1990). See also the documentation on the Network Analysis site of R.E. Ulanowicz (see below).

Input–output analysis

Input-output analysis calculates indices that identify the importance of different energy sources for each compartment (dependency coefficient) and the importance of energy sinks from the compartments (contribution coefficient). Therefore, dependency and contribution coefficients contain information on the degree to which the diet of one compartment depends directly and indirectly upon other compartments.

Throughput is the sum of energy that passes through each compartment and allows the ranking of the compartments in order of their ecological importance. Compartments with a high energy throughput are considered more important than compartments that mediate less energy.

Trophic structure analysis

The trophic position of organisms designates its position in the food chain, where primary producers occupy a trophic position of 1, herbivores occupy a position of 2 and so on. However, in nature it is frequently observed that organisms feed on several trophic levels (omnivory), meaning that organisms cannot be assigned to one single trophic position. Rather, a weighted trophic position based on the trophic position of the source compartments and value of the flows is more realistic. For example, an organism that feeds on primary producers and herbivores has a trophic position somewhere between 2 and 3, depending on the degree of consumption of both sources. Within the trophic structure analysis a weighted trophic level is calculated. The Lindeman food chain (Lindeman, 1942) is a food chain that shows the losses and trophic efficiencies when energy is passed from one level to the next. Though seeming to conflict with the former statement that no single trophic positions can be recognised in natural food webs, the concept is nevertheless informative. The trophic positions are not equal to the compartments as specified in the food web, but the compartments are partitioned to the positions based on the degree they contribute to that position. A trophic position of 1.5 means that 50% is partitioned to level 1 and another 50% to level 2. Output of trophic structure analysis allows the construction of the Lindeman food chain.

Cycle analysis

The study of cycling in ecosystems allows the separation of direct energy flow-through and cycling. Cycles are important since these are involved in conservation of material within the system and in feedbacks existing in the food web (Kay et al., 1989). A cycle is a trophic pathway in which the arbitrary starting and ending compartment are the same. Cycle analysis identifies the number of cycles that exist within the food web and calculates the number of transfers of each cycle. It also provides insight into the relative contributions of each compartment to total energy cycling and quantifies the percentage of energy that is involved in cycling, termed the Finn cycling index.

Topological analysis

Topological analysis is a concept borrowed from the information theory and characterises the geometry of the flows by using the ascendancy index (Ulanowicz & Norden, 1990). System ascendancy is defined as the product of the total flow of carbon in the system and the information inherent in the structure of the flows. The more specialised the flows, the more information is required to guide the flows. Maximal ascendancy is achieved when the system is fully specialised and there is no freedom left for the energy to choose its way through the food web. Topological analysis calculates the actual ascendancy of the system as a percentage of the maximal ascendancy, which is 100%. The fraction not covered by the actual ascendancy is the freedom left in the system and is called overhead and is associated with exports from the system, respiration or internal flows.

Network analysis packages

Currently two network analysis packages are freely available: NETWRK (http://www.cbl.cees.edu/~ulan/ntwk/network.html for DOS package and http://www.glerl.noaa.gov/EcoNetwrk/ for Windows package) developed by Ulanowicz and Kay (1991), while ECOPATH (http://www.ecopath.org) is Windows-based and developed by Christensen and Pauly (1992). Recently, Heymans and Baird (2000) compared both packages using data of the Benguela upwelling system. Most important methodological differences between the packages are (i) format of data input, (ii) algorithms of cycle analysis and (iii) inclusion of respiration by primary producers into the calculation procedure. Despite these differences the indices as calculated by both packages fall within the same range, so that both packages can readily be used.

Some applications

Studies so far have focussed on whole ecosystems, such as comparison of three Atlantic US estuaries (Monaco & Ulanowicz, 1997) and characterisation of the upwelling area Benguela off the coast of Namibia (Heymans & Baird, 2000). Heymans and McLachlan (1996) characterised a beach/surf-zone by network analysis and compared it with previously published characterisations, providing an instructive application of network analysis. Note, however, that these studies compared systems that have a different aggregation level and topology.

Conclusions

Network analysis is a tool that characterises food web properties by different types of indices. The degree and way of species aggregation influences the indices and implies that these are not unique to the system. Despite this, the technique is well established in different branches of limnological and marine research, yet relatively under-represented in benthic studies. It might be useful in comparisons between systems, on the premise that at least the compartmentalisation and topology of the food webs is similar.

References

Abarca-Arenas, L.G. & Ulanowicz, R.E. (2002) The effects of taxonomic aggregation on network analysis. *Ecological Modelling*, **149**, 285–96.

Abello, P., Warman, C.G., Reid, D.G. & Naylor, E. (1994) Chela loss in the shore crab *Carcinus maenas* (Crustacea: Brachyura) and its effect on mating success. *Marine Biology*, **121**, 247–52.

Abrams, B.I. & Mitchell, M.J. (1978) Role of oxygen in affecting survival and activity of *Pelodera-punctata* (Rhabditidae) from sewage sludge. *Nematologica*, **24**, 456–62.

Ahrens, M.J., Hertz, J., Lamoureux, E.M., Lopez, G.R., McElroy, A.E. & Brownawell, B.J. (2001) The effect of body size on digestive chemistry and absorption efficiencies of food and sediment-bound organic contaminants in *Nereis succinea* (Polychaeta). *Journal of Experimental Marine Biology and Ecology*, **263**, 185–209.

Aksnes, D.L. & Hoisaeter, T.J. (1987) Obtaining life table data from stage-frequency distributional statistics. *Limnology and Oceanography*, **32**, 514–7.

Aksnes, D.L., Miller, C.B., Ohman, M.D. & Wood, S.N. (1997) Estimation techniques used in studies of copepod population dynamics – a review of underlying assumptions. *Sarsia*, **82**, 279–96.

Aldenberg, T. (1986). Structured population models and methods of calculating secondary production. In: *The Dynamics of Physiologically Structured Populations* (Eds J.A.J. Metz & O. Diekmann), pp. 1–20. Springer-Verlag, Berlin.

Ambrose, W.G. (1991) Are infaunal predators important in structuring marine soft-bottom communities? *American Zoologist*, **31**, 849–60.

Anderson, L.G., Haraldsson, C. & Lindegren, R. (1992) Gran linearization of potentiometric Winkler titration. *Marine Chemistry*, **37**, 179–90.

Balczon, J.M. & Pratt, J.R. (1996) The functional responses of two benthic algivorous ciliated protozoa with differing feeding strategies. *Microbial Ecology*, **31**, 209–24.

Banse, K. & Mosher, S. (1980) Adult body mass and annual production/biomass relationships of field populations. *Ecological Monographs*, **50**, 355–79.

Bartlett, M.S. (1970) Age distributions. *Biometrics*, **26**, 377–83.

Bayne, B.L. (1998) The physiology of suspension feeding bivalve molluscs: An introduction to the Plymouth 'TROPHEE' workshop. *Journal of Experimental Marine Biology and Ecology*, **219**, 1–19.

Bayne, B.L. & Newell, R.I.E. (1983) *Physiological Energetics of Marine Molluscs, the Mollusca. Vol 4: Physiology, Part I*. Academic Press, New York.

Beninger, P.G. (2000) A critique of premises and methods in a recent study of particle capture mechanisms in bivalves. *Limnology and Oceanography*, **45**, 1196–1199.

Beninger, P.G. & St-Jean, S.D. (1997) The role of mucus in particle processing by suspension-feeding marine bivalves: unifying principles. *Marine Biology*, **129**, 389–397.

Benke, A.C. (1979) A modification of the Hynes method for estimating secondary production with particular significance for multivoltine populations. *Limnology and Oceanography*, **24**, 168–71.

Benke, A.C. & Waide, J.B. (1977) In defense of average cohorts. *Freshwater Biology*, **7**, 61–63.

Berner, R.A. (1982) Burial of organic carbon and pytite sulfur in the modern ocean: Its geochemical and environmental significance. *American Journal of Science*, **282**, 451–473.

Beukema, J.J. (1997) Caloric values of marine invertebrates with an emphasis on the soft parts of marine bivalves. *Oceanography and Marine Biology: An Annual Review*, **35**, 387–414.

Beukema, J.J., Dekker, R., Essink, K. & Michaelis, H. (2001) Synchronized reproductive success of the main bivalve species in the Wadden Sea: causes and consequences. *Marine Ecology Progress Series*, **211**, 143–55.

Beverton, R.J.H. & Holt, S.J. (1956) A review of methods for estimating mortality rates in fish populations, with special reference to sources of bias in catch sampling. *Rapports et Procès-Verbaux des Réunions du Conseil International pour l' Exploration de la Mer*, **140**, 67–83.

Bjorndal, K.A. & Bolten, A.B. (1995) Comparison of length-frequency analyses for estimation of growth parameters for a population of green turtles. *Herpetologica*, **51**, 160–167.

Blackford, J.C. (1997) An analysis of benthic biological dynamics in a North Sea ecosystem model. *Journal of Sea Research*, **38**, 213–230.

Bligh, E.G. & Dyer, W. (1959) A rapid method for total lipid extraction and purification. *Canadian Journal of Biochemistry and Physiology*, **37**, 911–917.

Braeckman, B.P., Houthoofd, K. & Vanfleteren, J.R. (2001) Insulin-like signaling, metabolism, stress resistance and aging in *Caenorhabditis elegans*. *Mechanisms of Ageing and Development*, **122**, 673–693.

Brey, T. (1986) Formalin and formaldehyde-depot chemicals: effects on dry weight and ash free dry weight of two marine bivalve species. *Meeresforschung*, **31**, 52–57.

Brey, T. (1990) Estimating productivity of macrobenthic invertebrates from biomass and mean individual weight. *Meeresforschung*, **32**, 329–343.

Brey, T. (1995) Temperature and reproductive metabolism in macrobenthic populations. *Marine Ecology Progress Series*, **125**, 87–93.

Brey, T. (1999) A collection of empirical relations for use in ecological modelling. *Naga*, **22**, 24–28.

Brey, T., Rumohr, H. & Ankar, S. (1988) Energy content of macrobenthic invertebrates: general conversion factors from weight to energy. *Journal of Experimental Marine Biology and Ecology*, **117**, 271–278.

Bridges, T.S. (1996) Effects of organic additions to sediment, and maternal age and size, on patterns of offspring investment and performance in two opportunistic deposit-feeding polychaetes. *Marine Biology*, **125**, 345–357.

Brockington, S. & Clarke, A. (2001) The relative influence of temperature and food on the metabolism of a marine invertebrate. *Journal of Experimental Marine Biology and Ecology*, **258**, 87–99.

Bryant, C. (1991) *Metazoan Life without Oxygen*. Chapman and Hall, London.

Calow, P. & Fletcher, C.R. (1972) A new radiotracer technique involving ^{14}C and ^{51}Cr for estimating the assimilation efficiency of aquatic primary producers. *Oecologia*, **9**, 155–170.

Cammen, L.M. (1980a) A method for measuring ingestion rate of deposit feeders and its use with the polychaete *Nereis succinea*. *Estuaries*, **3**, 55–60.

Cammen, L.M. (1980b) The significance of microbial carbon in the nutrition of the deposit feeding polychaete *Nereis succinea*. *Marine Biology*, **61**, 9–20.

Carpenter, J.H. (1965) The Chesapeake Bay Institute technique for the Winkler dissolved oxygen method. *Limnology and Oceanography*, **10**, 141–143.

Cartes, J.E. & Maynou, F. (1998) Food consumption by bathyal decapod crustacean assemblages in the western Mediterranean: predatory impact of megafauna and the food consumption – food supply balance in a deep-water food web. *Marine Ecology Progress Series*, **171**, 233–246.

Caswell, H. & Twombly, S. (1989). Estimation of stage specific demographic parameters for zooplankton populations: methods based on stage classified matrix projection models. In: *Estimation and Analysis of Insect Populations*. (Eds L.L. McDonald, B.F.J. Manly, J.A. Lockwood & J.A. Logan), pp. 94–107. Springer-Verlag, Berlin.

Chapelle, A., Menesguen, A., Deslous-Paoli, J.M., Souchu, P., Mazouni, N., Vaquer, A. & Millet, B. (2000) Modelling nitrogen, primary production and oxygen in a Mediterranean lagoon. Impact of oysters farming and inputs from the watershed. *Ecological Modelling*, **127**, 161–181.

Chardy, P. & Dauvin, J.-C. (1992) Carbon flows in a subtidal fine sand community from the western English Channel: a simulation analysis. *Marine Ecology Progress Series*, **81**, 147–161.

Chardy, P., Gros, P., Mercier, H. & Monbet, Y. (1993) Benthic carbon budget for the Bay of Saint-Brieuc (Western Channel) – application of inverse method. *Oceanologica Acta*, **16**, 687–694.

Charles, F., Gremare, A. & Amouroux, J.M. (1995) Utilization of C-14 formaldehyde to infer ingestion rates and absorption efficiencies by benthic deposit-feeders. *Marine Ecology Progress Series*, **127**, 121–129.

Christensen, V. & Pauly, D. (1992) Ecopath-II – a software for balancing steady-state ecosystem models and calculating network characteristics. *Ecological Modelling*, **61**, 169–185.

Clark, L.C. (1956) Monitor and control of blood and tissue oxygen tensions. *Transactions American Society for Artificial Internal Organs*, **2**, 41–48.

Clarke, A. (1987) Temperature, latitude and reproductive effort. *Marine Ecology Progress Series*, **38**, 89–99.

Clarke, A. (1993) Egg size and egg composition in polar shrimps (Caridea, Decapoda). *Journal of Experimental Marine Biology and Ecology*, **168**, 189–203.

Clarke, A. & Prothero-Thomas, E. (1997) The influence of feeding on oxygen consumption and nitrogen excretion in the Antarctic nemertean *Parborlasia corrugatus*. *Physiological Zoology*, **70**, 639–649.

Conover, R.J. (1966) Assimilation of organic matter by zooplankton. *Limnology and Oceanography*, **11**, 338–345.

Cortés, E. (1996) A critical review of methods of studying fish feeding based on analysis of stomach contents: application to elasmobranch fishes. *Canadian Journal of Fisheries and Aquatic Sciences*, **54**, 726–738.

Cote, J., Rakocinski, C.F. & Randall, T.A. (2001) Feeding efficiency by juvenile blue crabs on two common species of micrograzer snails. *Journal of Experimental Marine Biology and Ecology*, **264**, 189–208.

Cowie, G.L. & Hedges, J.I. (1992) Improved amino-acid quantification in environmental samples – charge-matched recovery standards and reduced analysis time. *Marine Chemistry*, **37**, 223–238.

Cranford, P.J. (2001) Evaluating the 'reliability' of filtration rate measurements in bivalves. *Marine Ecology Progress Series*, **215**, 303–305.

Cranford, P.J., Emerson, C.W., Hargrave, B.T. & Milligan, T.G. (1998) *In situ* feeding and absorption responses of sea scallops *Placopecten magellanicus* (Gmelin) to storm-induced changes in the quantity and composition of the seston. *Journal of Experimental Marine Biology and Ecology*, **219**, 45–70.

Cranford, P.J. & Hargrave, B.T. (1994) *In situ* time-series measurement of ingestion and absorption rates of suspension-feeding bivalves: *Placopecten magellanicus. Limnology and Oceanography*, **39**, 730–738.

Crawford, D.W., Rogerson, A. & Laybourn-Parry, J. (1994) Respiration of the marine amoeba *Trichosphaerium sieboldi* determined by ^{14}C labelling and Cartesian diver methods. *Marine Ecology Progress Series*, **112**, 135–142.

Crisp, D.J. (1984). Energy flow measurements. In: *Methods for the Study of Marine Benthos* (Eds N.A. Holme & A.D. McIntyre), pp. 284–372. Blackwell, Oxford.

Crompton, T.R. (1989) *Analysis of Seawater.* Butterworths, London.

Dahlhoff, E.P., Buckley, B.A. & Menge, B.A. (2001) Physiology of the rocky intertidal predator *Nucella ostrina* along an environmental stress gradient. *Ecology*, **82**, 2816–2829.

Danovaro, R., Dell'Anno, A., Martorano, D., Parodi, P., Marrale, N.D. & Fabiano, M. (1999) Seasonal variation in biochemical composition of deep-sea nematodes: bioenergetic and methodological considerations. *Marine Ecology Progress Series*, **179**, 273–283.

Davies, M.S. & Beckwith, P. (1999) Role of mucus trails and trail-following in the behaviour and nutrition of the periwinkle *Littorina littorea. Marine Ecology Progress Series*, **179**, 247–257.

Davies, M.S. & Hawkins, S.J. (1998) Mucus from marine molluscs. *Advances in Marine Biology*, **34**, 1–71.

Davies, M.S. & Williams, G.A. (1995) Pedal mucus of a tropical limpet, *Cellana grata* (Gould): energetics, production and fate. *Journal of Experimental Marine Biology and Ecology*, **186**, 77–87.

De Goeij, P. & Luttikhuizen, P.C. (1998) Deep-burying reduces growth in intertidal bivalves: field and mesocosm experiments with *Macoma balthica. Journal of Experimental Marine Biology and Ecology*, **228**, 327–337.

De Goeij, P., Luttikhuizen, P.C., Van der Meer, J. & Piersma, T. (2001) Facilitation on an intertidal mudflat: the effect of siphon nipping by flatfish on burying depth of the bivalve *Macoma balthica. Oecologia*, **126**, 500–506.

De Vlas, J. (1979a) Annual food intake by plaice and flounder in a tidal flat area in the Dutch Wadden Sea, with special reference to consumption of regenerating parts of macrobenthic prey. *Netherlands Journal of Sea Research*, **13**, 117–153.

De Vlas, J. (1979b) Secondary production by tail regeneration in a tidal flat population of lugworms (*Arenicola marina*), cropped by flatfish. *Netherlands Journal of Sea Research*, **13**, 362–393.

De Zwaan, A. (1977) Anaerobic energy metabolism in bivalve molluscs. *Oceanography and Marine Biology: Annual Review*, **15**, 103–87.
DeGrandpre, M.D., Baehr, M.M. & Hammar, T.R. (1999) Calibration-free optical chemical sensors. *Analytical Chemistry*, **71**, 1152–1159.
deRuiter, P.C., Neutel, A.M. & Moore, J.C. (1995) Energetics, patterns of interaction strengths, and stability in real ecosystems. *Science*, **269**, 1257–1260.
Donali, E., Olli, K., Heiskanen, A.S. & Andersen, T. (1999) Carbon flow patterns in the planktonic food web of the Gulf of Riga, the Baltic Sea: a reconstruction by the inverse method. *Journal of Marine Systems*, **23**, 251–268.
Duplisea, D.E. (2000) Benthic organism biomass size-spectra in the Baltic Sea in relation to the sediment environment. *Limnology and Oceanography*, **45**, 558–568.
Ebenhoh, W., Kohlmeier, C. & Radford, P.J. (1995) The benthic biological submodel in the European-Regional-Seas-Ecosystem-Model. *Netherlands Journal of Sea Research*, **33**, 423–452.
Ebert, T.A., Dixon, J.D., Schroeter, S.C., Kalvass, P.E., Richmond, N.T., Bradbure, W.A. & Woodby, D.A. (1999) Growth and mortality of red sea urchins *Strongylocentrotus franciscanus* across a latitudinal gradient. *Marine Ecology Progress Series*, **190**, 189–209.
Edmondson, T.W. (1960) Reproductive rates of rotifers in natural populations. *Memorie dell'Istituto Italiano di Idrobiologia*, **12**, 21–77.
Ejdung, G., Byrén, L. & Elmgren, R. (2000) Benthic predator-prey interactions: Evidence that adult *Monoporeia affinis* (Amphipoda) eat postlarval *Macoma balthica* (Bivalvia). *Journal of Experimental Marine Biology and Ecology*, **253**, 243–251.
Ejdung, G. & Elmgren, R. (1998) Predation on newly settled bivalves by deposit-feeding amphipods: a Baltic Sea case study. *Marine Ecology Progress Series*, **168**, 87–94.
Ekaratne, S.U.K. & Crisp, D.J. (1982) Tidal micro-growth bands in intertidal gastropod shells, with an evaluation of band-dating techniques. *Proceedings of the Royal Society of London, Series B*, **214**, 305–323.
Eldridge, P.M. & Jackson, G.A. (1993) Benthic trophic dynamics in California coastal basin and continental slope communities inferred using inverse analysis. *Marine Ecology Progress Series*, **99**, 115–135.
Elliot, J.M. & Davison, W. (1975) Energy equivalents of oxygen consumption in animal energetics. *Oecologia*, **19**, 195–201.
Ennola, K., Sarvala, J. & Dévai, G. (1998) Modelling zooplankton population dynamics with the extended Kalman filtering technique. *Ecological Modeling*, **110**, 135–149.
Estes, J.A. & Peterson, C.H. (2000) Marine ecological research in seashore and seafloor systems: Accomplishments and future directions. *Marine Ecology Progress Series*, **195**, 281–289.
Fager, E.W. (1969) Production of stream benthos: a critique of the method of assessment proposed by Hynes and Coleman (1968). *Limnology and Oceanography*, **14**, 766–770.
Famme, P., Riisgård, H.U. & Jørgensen, C.B. (1986) On direct measurements of pumping rates in the mussel *Mytilus edulis*. *Marine Biology*, **92**, 323–327.
Feller, R.J. & Warwick, R.M. (1988). Energetics. In: *Introduction to the Study of Meiofauna*. (Eds R.P. Higgins & H. Thiel), pp. 181–196. Smithsonian Institute, Washington, DC.
Fenchel, T. & Finlay, B.J. (1995) *Ecology and Evolution in Anoxic Worlds*. Oxford University Press, Oxford.

Forbes, T.L. & Lopez, G.R. (1990) Ontogenic changes in individual growth and egestion rates in the deposit-feeding polychaete *Capitella* sp 1. *Journal of Experimental Marine Biology and Ecology*, **143**, 209–220.

Forbes, V.E. & Lopez, G.R. (1989) The role of sediment particle-size in the nutritional energetics of a surface deposit-feeder. 1. Ingestion and absorption of sedimentary microalgae by *Hydrobia truncata (vanatta)*. *Journal of Experimental Marine Biology and Ecology*, **126**, 181–192.

Forseth, T., Jonsson, B., Neumann, R. & Ugedal, O. (1992) Radioisotope method for estimating food consumption by brown trout (*Salmo trutta*). *Canadian Journal of Fisheries and Aquatic Sciences*, **49**, 1328–1335.

Fournier, D.A., Hampton, J. & Sibert, J.R. (1997) MULTIFAN-CL: a length-based, age-structured model for fisheries stock assessment, with application to South Pacific albacore, *Thunnus alalunga*. *Canadian Journal of Fisheries and Aquatic Sciences*, **55**, 2105–2016.

Fournier, D.A., Sibert, J.R., Majkowski, J. & Hampton, J. (1990) MULTIFAN: a likelihood-based method for estimating growth parameters and age composition from multiple length frequency data sets illustrated using data for southern bluefin tuna (*Thunnus maccoyii*). *Canadian Journal of Fisheries and Aquatic Sciences*, **47**, 301–317.

Fournier, D.A., Sibert, J.R. & Terceiro, M. (1991) Analysis of length frequency samples with relative abundance data for the Gulf of Maine northern shrimp (*Pandalus borealis*) by the MULTIFAN method. *Canadian Journal of Fisheries and Aquatic Sciences*, **48**, 591–598.

Fraschetti, S., Robertson, M., Albertelli, G., Capelli, R. & Eleftheriou, A. (1994) Calorimetry: use of the Philipson microbomb. *Oebalia*, **20**, 117–127.

Frederich, M. & Portner, H.O. (2000) Oxygen limitation of thermal tolerance defined by cardiac and ventilatory performance in spider crab, *Maja squinado*. *American Journal of Physiology-Regulatory Integrative and Comparative Physiology*, **279**, R1531-R8.

Gayanilo, F.C., Sparre, P. & Pauly, D. (1996) The FAO-ICLARM stock assessment tools (FiSAT) user's guide. *FAO Computerized Information Series*, **6**, 1–186.

Gehrs, C.W. & Robertson, A. (1975) Use of life tables in analyzing the dynamics of copepod populations. *Ecology*, **56**, 665–673.

Gerchacov, S.M. & Hatcher, P.G. (1972) Improved technique for analysis of carbohydrates in sediments. *Limnology and Oceanography*, **17**, 938–943.

Gillespie, D.M. & Benke, A.C. (1979) Methods of calculating cohort production from field data – some relationships. *Limnology and Oceanography*, **24**, 171–176.

Gingras, J. & Boisclair, D. (2000) Comparison between consumption rates of yellow perch (*Perca flavescens*) estimated with a digestive tract model and with a radioisotope approach. *Canadian Journal of Fisheries and Aquatic Sciences*, **57**, 2547–2557.

Glud, R.N., Klimant, I., Holst, G., Kohls, O., Meyer, V., Kühl, M. & Gundersen, J.K. (1999a) Adaption, test and *in situ* measurements with O_2 microopt(r)odes on benthic landers. *Deep-Sea Research I*, **46**, 171–183.

Glud, R.N., Kuhl, M., Kohls, O. & Ramsing, N.B. (1999b) Heterogeneity of oxygen production and consumption in a photosynthetic microbial mat as studied by planar optodes. *Journal of Phycology*, **35**, 270–279.

Glud, R.N., Ramsing, N.B., Gundersen, J.K. & Klimant, I. (1996) Planar optrodes: a new tool for fine scale measurements of two-dimensional O_2 distribution in benthic communities. *Marine Ecology Progress Series*, **140**, 217–226.

Gnaiger, E. & Bitterlich, G. (1984) Proximate biochemical composition and caloric content calculated from elemental CHN analysis: a stoichiometric concept. *Oecologia*, **62**, 289–298.

Grasshoff, K., Ehrhardt, M. & Kremling, K. (1983) *Method of Seawater Analysis*. Second Edition. Verlag Chemie, Weinheim.

Grémare, A., Marsh, A.G. & Tenore, K.R. (1989) Fecundity and energy partitioning in *Capitella capitata* type I (Annelida: Polychaeta). *Marine Biology*, **100**, 365–371.

Gurney, W.S.C., McCauley, E., Nisbet, R.M. & Murdoch, W.W. (1990) The physiological ecology of *Daphnia*: a dynamic model of growth and reproduction. *Ecology*, **71**, 716–732.

Gurney, W.S.C. & Nisbet, R.M. (1998) *Ecological Dynamics*. Oxford University Press, Oxford.

Gyllenberg, G. (1973) Comparison of the Cartesion diver technique and the polarographic method, an open system, for measuring the respiratory rates in three marine copepods. *Commentationes Biologicae*, **60**, 3–13.

Haller, T., Ortner, M. & Gnaiger, E. (1994) A respirometer for investigating oxidative cell-metabolism – toward optimization of respiratory studies. *Analytical Biochemistry*, **218**, 338–342.

Hamburger, K. (1981) A gradient diver for measurement of respiration in individual organisms from the micro- and meiofauna. *Marine Biology*, **61**, 179–183.

Hamilton, A.L. (1969) On estimating annual production. *Limnology and Oceanography*, **14**, 771–782.

Harley, M.B. (1950) Occurrence of a filter-feeding mechanism in the polychaete *Nereis diversicolor*. *Nature*, **165**, 734–735.

Hartree, E.F. (1972) Determination of proteins: a modification of the Lowry method that gives a linear photometric response. *Analytical Biochemistry*, **48**, 422–427.

Hatcher, A. (1994) Nitrogen and phosphorus turnover in some benthic marine invertebrates: implication for the use of C:N ratios to assess food quality. *Marine Biology*, **121**, 161–166.

Hawkins, A.J.S., Bayne, B.L., Bougrier, S., Héral, M., Iglesias, I.P., Navarro, E., Smith, R.F.M. & Urrutia, M.B. (1998) Some general relationships in comparing the feeding physiology of suspension-feeding bivalve molluscs. *Journal of Experimental Marine Biology and Ecology*, **219**, 87–103.

Hawkins, A.J.S., Smith, R.F.M., Bayne, B.L. & Héral, M. (1996) Novel observations underlying the fast growth of suspension-feeding shellfish in turbid environments: *Mytilus edulis*. *Marine Ecology Progress Series*, **131**, 179–190.

Hay, S.J., Evans, G.T. & Gamble, J.C. (1988) Birth, death and growth rates for enclosed populations of calanoid copepods. *Journal of Plankton Research*, **10**, 431–454.

Herman, P.M.J., Heip, C. & Vranken, G. (1983) The production of *Cyprideis torosa* Jones 1850 (Crustacea, Ostracoda). *Oecologia*, **58**, 326–331.

Herman, P.M.J. & Vranken, G. (1988) Studies of the life-history and energetics of marine and brackish-water nematodes. *Oecologia*, **77**, 457–463.

Heymans, J.J. & Baird, D. (2000) Network analysis of the northern Benguela ecosystem by means of NETWRK and ECOPATH. *Ecological Modelling*, **131**, 97–119.

Heymans, J.J. & McLachlan, A. (1996) Carbon budget and network analysis of a high-energy beach/surf-zone ecosystem. *Estuarine Coastal and Shelf Science*, **43**, 485–505.

Hiddink, J.G., Ter Hofstede, R. & Wolff, W.J. (2002) Predation of intertidal infauna on juveniles of the bivalve *Macoma balthica*. *Journal of Sea Research*, **47**, 141–159.

Hildreth, D.I. & Crisp, D.J. (1976) Corrected formula for calculation of filtration-rate of bivalve molluscs in an experimental flowing system. *Journal of the Marine Biological Association of the UK*, **56**, 111–120.

Holst, G. & Grunwald, B. (2001) Luminescence lifetime imaging with transparent oxygen optodes. *Sensors and Actuators B - Chemical*, **74**, 78–90.

Holter, H. & Zeuthen, E. (1966). Manometric techniques for single cells. In: *Physical Techniques in Biological Research* (Eds G. Oster & A.W. Pollister), Vol. 3, pp. 251–317. Academic Press, New York.

Honkoop, P.J.C., Luttikhuizen, P.C. & Piersma, T. (1999) Experimentally extending the spawning season of a marine bivalve using temperature change and fluoxetine as synergistic triggers. *Marine Ecology Progress Series*, **180**, 297–300.

Honkoop, P.J.C. & Van der Meer, J. (1998) Experimentally induced effects of water temperature and immersion time on reproductive output of bivalves in the Wadden Sea. *Journal of Experimental Marine Biology and Ecology*, **220**, 227–246.

Horn, P.L. (1986) Energetics of *Chiton pelliserpentis* (Quoy & Gaimard, 1835) (Mollusca: Polyplacophora) and the importance of mucus on its energy budget. *Journal of Experimental Marine Biology and Ecology*, **101**, 119–141.

Hynes, H.B.N. & Coleman, M.J. (1968) A simple method of assessing the annual production of stream benthos. *Limnology and Oceanography*, **13**, 569–573.

Jackson, G.A. & Eldridge, P.M. (1992) Food web analysis of a planktonic system off Southern California. *Progress in Oceanography*, **30**, 223–251.

Jensen, P. (1984) Measuring carbon content in nematodes. *Helgoländer Meeresuntersuchungen*, **38**, 83–86.

Jones, H.D., Richards, O.G. & Southern, T.A. (1992) Gill dimensions, water pumping rate and body size in the mussel *Mytilus edulis*. *Journal of Experimental Marine Biology and Ecology*, **155**, 213–237.

Jørgensen, C.B. (1966a) *Biology of Suspension Feeding*. Pergamon Press, Oxford.

Jørgensen, C.B. (1996b) Bivalve filter feeding revisited. *Marine Ecology Progress Series*, **142**, 287–302.

Kautsky, H. (1939) Quenching of luminescence by oxygen. *Transactions of the Faraday Society*, **35**, 216–219.

Kay, J.J., Graham, L.A. & Ulanowicz, R.E. (1989). A detailed guide to network analysis. In: *Network Analysis in Marine Ecology. Methods and Applications* (Eds F. Wulff, J.G. Field & K.H. Mann), Vol. 32. Springer-Verlag, Berlin.

Kemp, R.B. & Guan, Y. (1997) Heat flux and the calorimetric-respirometric ratio as measures of catabolic flux in mammalian cells. *Thermochimica Acta*, **300**, 199–211.

Kemp, W.M., Brooks, M.T. & Hood, R.R. (2001) Nutrient enrichment, habitat variability and trophic transfer efficiency in simple models of pelagic ecosystems. *Marine Ecology Progress Series*, **223**, 73–87.

Kim, S.B., Cho, H.C., Cha, G.S. & Nam, H. (1998) Microtiter plate-format optode. *Analytical Chemistry*, **70**, 4860–4863.

Kiørboe, T. & Møhlenberg, F. (1981) Particle selection in suspension-feeding bivalves. *Marine Ecology Progress Series*, **5**, 291–296.

Klein Breteler, W.C.M., Schogt, N. & Van der Meer, J. (1994) The duration of copepod life stages estimated from stage-frequency data. *Journal of Plankton Research*, **16**, 1039–1057.

Klekowski, R.Z. (1971) Cartesian diver microrespirometry for aquatic animals. *Polskie Archiwum Hydrobiologii*, **18**, 93–144.

Klimant, I., Meyer, V. & Kühl, M. (1995) Fiber-optic oxygen microsensors, a new tool in aquatic biology. *Limnology and Oceanography*, **40**, 1159–1165.

Klimant, I., Ruckruh, F., Liebsch, G., Stangelmayer, C. & Wolfbeis, O.S. (1999) Fast response oxygen micro-optodes based on novel soluble ormosil glasses. *Mikrochimica Acta*, **131**, 35–46.

Kofoed, L., Forbes, V.E. & Lopez, G. (1989). Time-dependent absorption in deposit feeders. In: *Ecology of Marine Deposit Feeders*. (Eds G. Lopez, G. Taghon & J. Levinton), pp. 129–148. Springer-Verlag, New York.

Kohls, O. & Scheper, T. (2000) Setup of a fiber optical oxygen multisensor-system and its applications in biotechnology. *Sensors and Actuators B - Chemical*, **70**, 121–130.

Kooijman, S.A.L.M. (1986). Population dynamics on basis of budgets. In: *The Dynamics of Physiologically Structured Populations* (Eds J.A.J. Metz & O. Diekmann), pp. 266–297. Springer-Verlag, Berlin.

Kooijman, S.A.L.M. (2000) *Dynamic Energy and Mass Budgets in Biological Systems*. Second Edition. Cambridge University Press, Cambridge.

Kryger, J. & Riisgård, H.U. (1988) Filtration rate capacities in six species of European freshwater bivalves. *Oecologia*, **77**, 34–38.

Lasserre, P. (1976). Metabolic activities of benthic microfauna and meiofauna: Recent advances and suitable methods of analysis. In: *The Benthic Boundary Layer* (Ed. I.N. McCave), pp. 95–142. Plenum, New York.

Lavrentyev, P.J., Gardner, W.S. & Yang, L.Y. (2000) Effects of the zebra mussel on nitrogen dynamics and the microbial community at the sediment-water interface. *Aquatic Microbial Ecology*, **21**, 187–194.

Lawrence, J.M., Byrne, M., Harris, L., Keegan, B., Freeman, S. & Cowell, B. (1999) Sublethal arm loss in *Asterias amurensis*, *A. rubens*, *A. vulgaris* and *A. forbesi* (Echinodermata: Asterioidea). *Vie et Milieu*, **49**, 69–73.

Lawrence, J.M. & Vasquez, J. (1996) The effect of sublethal predation on the biology of echinoderms. *Oceanologica Acta*, **19**, 431–440.

Leguerrier, D., Niquil, N., Boileau, N., Rzeznik, J., Sauriau, P.G., Le Moine, O. & Bacher, C. (2003) Numerical analysis of the food web of an intertidal mudflat ecosystem on the Atlantic coast of France. *Marine Ecology Progress Series*, **246**, 17–37.

Lehtonen, K.K. (1995) Geographical variability in the bioenergetic characteristics of *Monoporeia*/*Pontoporeia* spp. populations from the northern Baltic Sea, and their potential contribution to benthic nitrogen mineralization. *Marine Biology*, **123**, 555–564.

Lemmens, J.W.T.J. (1994) The western rock lobster *Panulirus cygnus* (George, 1962) (Decapoda, Palinuridae) – the effect of temperature and developmental stage on energy requirements of Pueruli. *Journal of Experimental Marine Biology and Ecology*, **180**, 221–234.

Lika, K. & Nisbet, R.M. (2000) A dynamic energy budget model based on partitioning of net production. *Journal of Mathematical Biology*, **41**, 361–386.

Lindeman, R.L. (1942) The trophic-dynamic aspect of ecology. *Ecology*, **23**, 399–418.
Lindroth, P. & Mopper, K. (1979) High performance liquid chromatographic determination of subpicomole amounts of amino acids by precolumn fluoresence derivatization with *o*-phthaldialdehyde. *Analytical Chemistry*, **51**, 1667–1674.
Lindsay, S.M. & Woodin, S.A. (1996) The effect of palp loss on feeding behavior of two spionid polychaetes: changes in exposure. *Biological Bulletin*, **183**, 440–447.
Linton, D.L. & Taghon, G.L. (2000a) Feeding, growth, and fecundity of *Abarenicola pacifica* in relation to sediment organic concentration. *Journal of Experimental Marine Biology and Ecology*, **254**, 85–107.
Linton, D.L. & Taghon, G.L. (2000b) Feeding, growth, and fecundity of *Capitella* sp. I in relation to sediment organic concentration. *Marine Ecology Progress Series*, **205**, 229–240.
Lopez, G. & Elmgren, R. (1989) Feeding depths and organic absorption for the deposit-feeding benthic amphipods *Pontoporeia affinis* and *Pontoporeia femorata*. *Limnology and Oceanography*, **34**, 982–991.
Lopez, G.R. & Cheng, I.J. (1983) Synoptic measurements of ingestion rate, ingestion selectivity, and absorption efficiency of natural foods in the deposit-feeding molluscs *Nucula annulata* (Bivalvia) and *Hydrobia totteni* (Gastropoda). *Marine Ecology Progress Series*, **11**, 55–62.
Lopez, G.R. & Crenshaw, M.A. (1982) Radiolabelling of sedimentary organic matter with C-14 formaldehyde: preliminary evaluation of a new technique for use in deposit-feeding studies. *Marine Ecology Progress Series*, **8**, 283–289.
Lopez, G.R. & Levinton, J.S. (1987) Ecology of deposit-feeding animals in marine sediments. *Quarterly Review of Biology*, **62**, 235–260.
Lopez, G.R., Tantichodok, P. & Cheng, I.J. (1989). Radiotracer methods for determining utilization of sedimentary organic matter by deposit feeders. In: *Ecology of Marine Deposit Feeders*. (Eds G. Lopez, G. Taghon & J. Levinton), pp. 149–170. Springer-Verlag, Berlin.
Low, B. & Richards, F. (1952) Determination of protein crystal densities. *Nature*, **170**, 412–413.
Macfadyen, A. (1948) The meaning of productivity in biological systems. *Journal of Animal Ecology*, **17**, 75–80.
Manahan, D.T., Wright, S.H. & Stephens, G.C. (1983) Simultaneous determination of net uptake of 16 amino acids by a marine bivalve. *American Journal of Physiology*, **244**, 832–838.
Manahan, D.T., Wright, S.H., Stephens, G.C. & Rice, M.A. (1982) Transport dissolved amino acids by the mussel, *Mytilus edulis*: demonstration of net uptake from natural seawater. *Science*, **215**, 1253–1255.
Manly, B.F.J. (1977) A further note on Kiritani and Nakasuji's model for stage-frequency data including comments on Tukey's jack-knife technique for estimating variances. *Researches on Population Ecology*, **18**, 177–186.
Manly, B.F.J. (1985) Further improvements to a method for analyzing stage-frequency data. *Researches on Population Ecology*, **27**, 325–332.
Manly, B.F.J. (1989). A review of methods for the analysis of stage-frequency data. In: *Estimation and Analysis of Insect Population* (Eds L.L. McDonald, B.F.J. Manly, J.A. Lockwood & J.A. Logan), Vol. 55, pp. 3–69. Springer-Verlag, Berlin.

Manly, B.F.J. (1990) *Stage-Structured Populations: Sampling, Analysis and Simulation.* Chapman and Hall, London.

Manly, B.F.J. (1997) A method for the estimation of parameters for natural stage-structured populations. *Researches on Population Ecology*, **39**, 101–111.

Marsh, J.B. & Weinstein, W.J. (1966) A simple charring method for determination of lipids. *Journal of Lipid Research*, **7**, 574–576.

Mascaró, M. & Seed, R. (2001) Foraging behavior of juvenile *Carcinus maenas* (L.) and *Cancer pagurus* L. *Marine Biology*, **139**, 1135–1145.

Mayer, L.M., Jumars, P.A., Taghon, G.L., Macko, S.A. & Trumbore, S. (1993) Low-density particles as potential nitrogenous foods for benthos. *Journal of Marine Research*, **51**, 373–389.

McLaren, I.A. (1997) Modeling biases in estimating production from copepod cohorts. *Limnology and Oceanography*, **42**, 584–589.

McQuaid, C.D. & Lindsay, T.L. (2000) Effect of wave exposure on growth and mortality rates of the mussel *Perna perna*: Bottom-up regulation of intertidal populations. *Marine Ecology Progress Series*, **206**, 147–154.

McVean, A. (1976) The incidence of autotomy in *Carcinus maenas* (L.). *Journal of Experimental Biology and Ecology*, **24**, 177–187.

McVean, A. & Findlay, I. (1979) The incidence of autotomy in an estuarine population of the crab *Carcinus maenas*. *Journal of the Marine Biological Association of the UK*, **59**, 341–354.

Migne, A. & Davoult, D. (1997) Ammonium excretion in two benthic cnidarians: *Alcyonium digitatum* (Linnaeus, 1758) and *Urticina felina* (Linnaeus, 1767). *Journal of Sea Research*, **37**, 101–117.

Miller, C.B. (1993) Development of large copepods during spring in the Gulf of Alaska. *Progress in Oceanography*, **32**, 295–317.

Miller, C.B. & Tande, K.S. (1993) Stage duration estimation for *Calanus* populations, a modelling study. *Marine Ecology Progress Series*, **102**, 15–34.

Mills, A. & Eaton, K. (2000) Optical sensors for carbon dioxide: an overview of sensing strategies past and present. *Quimica Analitica*, **19**, 75–86.

Mitchell, M.J. (1973) An improved method for microrespirometry using gas chromatography. *Soil Biology and Biochemistry*, **5**, 271–274.

Moens, T., Herman, P.M.J., Verbeeck, L., Steyaert, M. & Vinckx, M. (2000) Predation rates and prey selectivity in two predacious estuarine nematode species. *Marine Ecology Progress Series*, **205**, 185–193.

Moens, T., Verbeeck, L. & Vincx, M. (1999) Preservation and incubation time-induced bias in tracer-aided grazing studies with meiofauna. *Marine Biology*, **133**, 69–77.

Moens, T., Vierstrate, A., Vanhove, S., Verbeke, M. & Vinckx, M. (1996) A handy method for measuring meiobenthic respiration. *Journal of Experimental Marine Biology and Ecology*, **197**, 177–190.

Møhlenberg, F. & Riisgård, H.U. (1979) Filtration rate, using a new indirect technique, in thirteen species of suspension-feeding bivalves. *Ophelia*, **17**, 239–246.

Monaco, M.E. & Ulanowicz, R.E. (1997) Comparative ecosystem trophic structure of three US mid-Atlantic estuaries. *Marine Ecology Progress Series*, **161**, 239–254.

Morissette, S. & Himmelman, J.H. (2000) Subtidal food thieves: interactions of four invertebrate kleptoparasites with the sea star *Leptasterias polaris*. *Animal Behaviour*, **60**, 531–543.

Newell, R.C. (1970) *Biology of Intertidal Animals*. Logos Press, London.

Nieuwenhuize, J., Maas, Y.E.M. & Middelburg, J.J. (1994) Rapid analysis of organic-carbon and nitrogen in particulate materials. *Marine Chemistry*, **45**, 217–224.

Nilsson, H.C. (2000a) Effects of hypoxia and organic enrichment on growth of the brittle star *Amphiura filiformis* (O.F. Mueller) and *Amphiura chiajei* Forbes. *Journal of Experimental Biology and Ecology*, **237**, 11–30.

Nilsson, H.C. (2000b) Interaction between water flow and oxygen deficiency on growth in the infaunal brittle star *Amphiura filiformis* (Echinodermata: Ophiuroidea). *Journal of Sea Research*, **44**, 233–241.

Niquil, N., Jackson, G.A., Legendre, L. & Delesalle, B. (1998) Inverse model analysis of the planktonic food web of Takapoto Atoll (French Polynesia). *Marine Ecology Progress Series*, **165**, 17–29.

Nisbet, R.M., Muller, E.B., Lika, K. & Kooijman, S.A.L.M. (2000) From molecules to ecosystems through dynamic energy budget models. *Journal of Animal Ecology*, **69**, 913–926.

Nisbet, R.M., Ross, A.H. & Brooks, A.J. (1996) Empirically based dynamics energy budget models: theory and application to ecotoxicology. *Nonlinear World*, **3**, 85–106.

Noonburg, E.G. Nisbet, R.M., McCauley, E., Gurney, W.S.C., Murdoch, W.W. & De Roos, A.M. (1998) Experimental testing of dynamic energy budget models. *Functional Ecology*, **12**, 211–222.

Ohman, M.D. & Wood, S.N. (1996) Mortality estimation for planktonic copepods: *Pseudocalanus newmani* in a temperate fjord. *Limnology and Oceanography*, **41**, 126–135.

Pace, M.L., Glasser, J.E. & Pomeroy, L.R. (1984) A simulation analysis of continental shelf food webs. *Marine Biology*, **82**, 47–63.

Paloheimo, J.E. (1974) Calculation of instantaneous birth rate. *Limnology and Oceanography*, **19**, 692–694.

Pamatmat, M.M. (1978) Oxygen-uptake and heat-production in a metabolic conformer (*Littorina irrotata*) and a metabolic regulator (*Uca pugnax*). *Marine Biology*, **48**, 317–325.

Pamatmat, M.M. (1979) Anaerobic heat-production of bivalves (*Polymesoda caroliniana* and *Modiolus demissus*) in relation to temperature, body size, and duration of anoxia. *Marine Biology*, **53**, 223–229.

Pamatmat, M.M. (1980). Facultative anaerobiosis of benthos. In: *Marine Benthic Dynamics* (Eds K.R. Tenore & B.C. Coull), pp. 69–90. University of South Carolina Press, Columbia.

Pamatmat, M.M. (1983) Measuring aerobic and anaerobic metabolism of benthic infauna under natural conditions. *Journal of Experimental Zoology*, **228**, 405–413.

Pape-Lindstrom, P.A., Feller, R.J., Stancyk, S.E. & Woodin, S.A. (1997) Sublethal predation: field measurements of arm tissue loss from the ophiurid *Microphiopholis gracillima* and immunochemical identification of its predators in North Inlet, South Carolina, USA. *Marine Ecology Progress Series*, **156**, 131–140.

Parslow, J., Sonntag, N.C. & Matthews, J.B.L. (1979) Technique of systems identification applied to estimating copepod populations parameters. *Journal of Plankton Research*, **1**, 137–151.

Pauly, D. (1987). A review of the ELEFAN system for analysis of length-frequency data in fish and aquatic invertebrates. In: *Length-Based Methods in Fisheries Research* (Eds D. Pauly & G.R. Morgan), pp. 7–34. International Center for Aquatic Living Resources Management, Manila.

Pechenik, J.A., Berard, R. & Kerr, L. (2000) Effects of reduced salinity on survival, growth, reproductive success, and energetics of the euryhaline polychaete *Capitella* sp I. *Journal of Experimental Marine Biology and Ecology*, **254**, 19–35.

Peck, L.S., Prothero-Thomas, E. & Hough, N. (1993) Pedal mucus production by the Antartic limpet *Nacella concinna* (Strebel, 1908). *Journal of Experimental Marine Biology and Ecology*, **174**, 177–192.

Peeters, E.T.H.M., Dewitte, A., Koelmans, A.A., van der Velden, J.A. & den Besten, P.J. (2001) Evaluation of bioassays versus contaminant concentrations in explaining the macroinvertebrate community structure in the Rhine-Meuse delta, The Netherlands. *Environmental Toxicology and Chemistry*, **20**, 2883–2891.

Penry, D.L. (1998) Applications of efficiency measurements in bioaccumulation studies: Definitions, clarifications, and a critique of methods. *Environmental Toxicology and Chemistry*, **17**, 1633–1639.

Pentilla, J., Jearld, A. & Clark, S. (1988) Age determination methods for Northwest Atlantic species. *NOAA Technical Report of the US Department of Commerce*, **72**, 3–135.

Peterson, C.H. & Quammen, M.L. (1982) Siphon nipping: its importance to small fishes and its impact on the growth of the bivalve *Prothothaca staminea* (Conrad). *Journal of Experimental Marine Biology and Ecology*, **63**, 249–268.

Phillipson, J. (1964) A miniature bomb calorimeter for small biological samples. *Oikos*, **15**, 130–139.

Pomeroy, L.R., Sheldon, J.E. & Sheldon, W.M. (1994) Changes in bacterial numbers and leucine assimilation during estimations of microbial respiratory rates in seawater by the Precision Winkler Method. *Applied and Environmental Microbiology*, **60**, 328–332.

Poulsen, E., Riisgård, H.U. & Mohlenberg, F. (1982) Accumulation of cadmium and bioenergetics in the mussel *Mytilus edulis*. *Marine Biology*, **68**, 25–29.

Powell, D.G. (1979) Estimation of mortality and growth parameters from the length frequency of a catch. *Rapp. P.-v. Réun. Cons. Int. Explor. Mer*, **175**, 167–169.

Prevedelli, D. & Vandini, R.Z. (1998) Effect of diet on reproductive characteristics of *Ophryotrocha labronica* (Polychaeta: Dorvilleidae). *Marine Biology*, **132**, 163–170.

Qian, P.Y. & Chia, F.S. (1994) *In situ* measurement of recruitment, mortality, growth, and fecundity of *Capitella* sp. (Annelida: Polycheata). *Marine Ecology Progress Series*, **111**, 53–62.

Ren, J.S. & Ross, A.H. (2001) A dynamic energy budget model of the Pacific Oyster *Crassostrea gigas*. *Ecological Modelling*, **142**, 105–120.

Rhoads, D.C. & Lutz, R.A. (1980) *Skeletal Growth of Aquatic Organisms*. Plenum Press, New York.

Rice, D.L. (1982) The detritus nitrogen problem: New observations and perspectives from organic geochemistry. *Marine Ecology Progress Series*, **9**, 153–162.

Richardson, C.A. (1989) An analysis of the microgrowth bands in the shell of the common mussel *Mytilus edulis*. *Journal of the Marine Biological Association of the UK*, **69**, 477–491.

Richardson, C.A. (2001) Molluscs as archives of environmental change. *Oceanography and Marine Biology: An Annual Review*, **39**, 103–164.

Ricker, W.E., (Ed) (1971) *Methods for Assessment of Fish Production in Fresh Waters. IBP Handbook No. 3.* Blackwell, Oxford.

Rigler, F.H. & Cooley, J.M. (1974) The use of field data to derive population statistics of multivoltine copepods. *Limnology and Oceanography*, **19**, 636–655.

Rigler, R.H. & Downing, J.A. (1984). The calculation of secondary productivity. In: *A Manual on Methods for the Assessment of Secondary Productivity in Fresh Waters* (Eds J.A. Downing & R.H. Rigler), pp. 19–58. Blackwell, Oxford.

Riisgård, H.U. (1977) On measurements of the filtration rates of suspension feeding bivalves in a flow system. *Ophelia*, **16**, 167–173.

Riisgård, H.U. (1989) Properties and energy-cost of the muscular piston pump in the suspension feeding polychaete *Chaetopterus variopedatus*. *Marine Ecology Progress Series*, **56**, 157–168.

Riisgård, H.U. (1991) Suspension feeding in the polychaete *Nereis diversicolor*. *Marine Ecology Progress Series*, **70**, 29–37.

Riisgård, H.U. (2001a) On measurement of filtration rates in bivalves – the stony road to reliable data: review and interpretation. *Marine Ecology Progress Series*, **211**, 275–291.

Riisgård, H.U. (2001b) Physiological regulation versus autonomous filtration in filter-feeding bivalves: Starting points for progress. *Ophelia*, **54**, 193–209.

Riisgård, H.U. & Ivarsson, N.M. (1990) The crown-filament pump of the suspension-feeding polychaete *Sabella penicillus* – filtration, effects of temperature, and energy-cost. *Marine Ecology Progress Series*, **62**, 249–257.

Riisgård, H.U. & Larsen, P.S. (1995) Filter-feeding in marine macro-invertebrates: pump characteristics, modelling and energy cost. *Biological Reviews*, **70**, 67–106.

Riisgård, H.U. & Larsen, P.S. (2000) A comment on experimental techniques for studying particle capture in filter-feeding bivalves. *Limnology and Oceanography*, **45**, 1192–1195.

Riisgård, H.U. & Mohlenberg, F. (1979) An improved automatic recording apparatus for determining the filtration rate of *Mytilus edulis* as a function of size and algal concentration. *Marine Biology*, **52**, 61–67.

Riisgård, H.U. & Randløv, A. (1981) Energy budgets, growth and filtration rates in *Mytilus edulis* at different algal concentration. *Marine Biology*, **61**, 227–234.

Robertson, A.I. (1979) Relationship between annual production – biomass ratios and lifespans for marine macrobenthos. *Oecologia*, **38**, 193–202.

Ross, A.H., Gurney, W.S.C., Heath, M.R., Hay, S.J. & Henderson, E.W. (1993) A strategic simulation model of a fjord ecosystem. *Limnology and Oceanography*, **38**, 128–153.

Ross, A.H. & Nisbet, R.M. (1990) Dynamic models of growth and reproduction of the mussel *Mytilus edulis* L. *Functional Ecology*, **4**, 777–787.

Rothschild, B.J., Sharov, A.F., Kearsley, A.J. & Bondarenko, A.S. (1997) Estimating growth and mortality in stage-structured populations. *Journal of Plankton Research*, **19**, 1913–1928.

Rowan, D.J. & Rasmussen, J.B. (1994) Bioaccumulation of radiocesium by fish: influence of physico-chemical factors and trophic structure. *Canadian Journal of Fisheries and Aquatic Sciences*, **51**, 2388–2410.

Rowe, G.T. (1983). Biomass and production of the deep-sea macrobenthos. In: *Deep-Sea Biology* (Ed. G.T. Rowe), pp. 97–121. Wiley, New York.

Salonen, K., Sarvala, J., Hakala, I. & Viljanen, M.L. (1976) The relation of energy and organic carbon in aquatic invertebrates. *Limnology and Oceanography*, **21**.

Sanchez-Pedreno, C., Ortuno, J.A., Albero, M.I., Garcia, M.S. & de las Bayonas, J.C.G. (2000) A new procedure for the construction of flow-through optodes. Application to determination of copper (II). *Fresenius Journal of Analytical Chemistry*, **366**, 811–815.

Scavia, D., Lang, G.A. & Kitchell, J.F. (1988) Dynamics of Lake Michigan plankton – a model evaluation of nutrient loading, competition, and predation. *Canadian Journal of Fisheries and Aquatic Sciences*, **45**, 165–177.

Scheu, S. (1991) Mucus excretion and carbon turnover of endogenic earthworms. *Biology and Fertility of Soils*, **12**, 217–220.

Schmidt, O., Scrimgeour, C.M. & Curry, J.P. (1999) Carbon and nitrogen stable isotope ratios in body tissue and mucus of feeding and fasting earthworms (*Lumbricus festivus*). *Oecologia*, **118**, 9–15.

Schubert, A. & Reise, K. (1986) Predatory effects of *Nephtys hombergii* on other polychaetes in tidal flat sediments. *Marine Ecology Progress Series*, **34**, 117–124.

Seitz, A. (1979) On the calculation of birth rates and death rates in fluctuating populations with continuous recruitment. *Oecologia*, **41**, 343–360.

Seitz, R.D., Lipcius, R.N., Hines, A.H. & Eggleston, D.B. (2001) Density-dependent predation, habitat variation, and the persistence of marine bivalve prey. *Ecology*, **82**, 2435–2451.

Shepherd, J.G. (1987). A weekly parametric method for estimating growth parameters from length composition data. In: *Length-Based Methods in Fisheries Research* (Eds D. Pauly & G.R. Morgan), pp. 113–119. International Center for Living Aquatic Resources Management, Manila.

Shick, J.M., De Zwaan, A. & De Bont, A.M.T. (1983) Anoxic metabolic rate in the mussel *Mytilus edulis* L. estimated by simultaneous direct calorimetry and biochemical analysis. *Physiological Zoology*, **56**, 56–63.

Silverman, H., Lynn, J.W. & Dietz, T.H. (2000) *In vitro* studies of particle capture and transport in suspension-feeding bivalves. *Limnology and Oceanography*, **45**, 1199–1203.

Skalski, G. & Gilliam, J.F. (2001) Functional responses with predator interference: viable alternatives to the Holling II model. *Ecology*, **82**, 3083–3092.

Skoeld, M., Loo, L.O. & Rosenberg, R. (1994) Production, dynamics and demography of an *Amphiura filiformis* population. *Marine Ecology Progress Series*, **103**, 81–90.

Skoeld, M. & Rosenberg, R. (1996) Arm regeneration frequency in eight species of Ophiuroidea (Echinodermata) from European sea areas. *Journal of Sea Research*, **35**, 353–362.

Smaal, A.C. & Vonck, A.P.M.A. (1997) Seasonal variation in C, N and P budgets and tissue composition of the mussel *Mytilus edulis*. *Marine Ecology Progress Series*, **153**, 167–179.

Smaal, A.C. & Widdows, J. (1994). The scope for growth of bivalves as an integrated response parameter in biological monitoring. In: *Biomonitoring of Coastal Waters and Estuaries* (Ed. K. Kramer), pp. 247–268. CRC Press, Boca Raton, FL.

Smith, E.B., Williams, F.M. & Fisher, C.R. (1997) Effects of intrapopulation variability on von Bertalanffy growth parameter estimates from equal mark-recapture intervals. *Canadian Journal of Fisheries and Aquatic Sciences*, **54**, 2025–2032.

Sola, J.C. (1996) Population dynamics, reproduction, growth, and secondary production of

the mud-snail *Hydrobia ulvae* (Pennant). *Journal of Experimental Marine Biology and Ecology*, **205**, 49–62.
Solórzano, L. (1969) Determination of ammonia in natural waters by the phenolhypochlorite method. *Limnology and Oceanography*, **14**, 799–800.
Sommer, U. (1999) The susceptibility of benthic microalgae to periwinkle (*Littorina littorea*, Gastropoda) grazing in laboratory experiments. *Aquatic Botany*, **63**, 11–21.
Sommer, U., Meusel, B. & Stielau, C. (1999) An experimental analysis of the importance of body-size in the seastar-mussel predator-prey relationship. *Acta Oecologica*, **20**, 81–86.
Sonntag, N.C. & Parslow, J. (1981) Technique of systems identification applied to estimating copepod production. *Journal of Plankton Research*, **3**, 461–473.
Stephens, D. & Krebs, J.R. (1986) *Foraging Theory*. Princeton University Press, Princeton, NJ.
Stoecker, D.K. & Michaels, A.E. (1991) Respiration, photosynthesis and carbon metabolism in planktonic ciliates. *Marine Biology*, **108**, 441–447.
Stokes, M.D. & Somero, G.N. (1999) An optical oxygen sensor and reaction vessel for high-pressure applications. *Limnology and Oceanography*, **44**, 189–95.
Strickland, J.D.H. & Parsons, T.R. (1972) *A Practical Handbook of Seawater Analysis*. Fisheries Research Board of Canada.
Taghon, G. & Greene, R.R. (1998) Utilization of deposited and suspended particulate matter by benthic 'interface' feeders. *Limnology and Oceanography*, **37**, 1370–1391.
Taghon, G.L. (1988) The benefits and costs of deposit feeding in the polychaete *Abarenicola pacifica*. *Limnology and Oceanography*, **33**, 1166–1175.
Taghon, G.L., Prahl, F.G., Sparrow, M. & Fuller, C.M. (1994) Lipid class and glycogen content of the lugworm *Abarenicola pacifica* in relation to age, growth rate and reproductive condition. *Marine Biology*, **120**, 287–295.
Tanaka, M. (1982) A new growth curve which expresses infinite increase. *Publications of the Amakusa Marine Biological Laboratory*, **6**, 167–177.
Tanaka, M. (1988) Eco-physiological meaning of parameters of ALOG growth curve. *Publications of the Amakusa Marine Biological Laboratory*, **9**, 103–156.
Taylor, B.N. (1995) *Guide for the use of the International System of Units (SI)*. Government Printing Office, Washington, DC.
Tenore, K.R. (1977) Growth of *Capitella capitata* cultured on various levels of detritus derived from different sources. *Limnology and Oceanography*, **22**, 936–941.
Tett, P. & Wilson, H. (2000) From biogeochemical to ecological models of marine microplankton. *Journal of Marine Systems*, **25**, 431–446.
Thienemann, A. (1931) Productionsbegrift in der Biologie. *Archiv für Hydrobiologie*, **22**, 616–621.
Threlkeld, S. (1979) Estimating cladoceran birth rates: the importance of egg mortality and the egg age distribution. *Limnology and Oceanography*, **24**, 601–612.
Trexler, J.C., McCullogh, C.E. & Travis, J. (1988) How can the functional response best be determined? *Oecologia*, **76**, 206–214.
Tumbiolo, M.L. & Downing, J.A. (1994) An empirical model for the prediction of secondary production in marine benthic invertebrate populations. *Marine Ecology Progress Series*, **114**, 165–174.

Twombly, S. (1994) Comparative demography and population dynamics of two coexisting copepods in a Venezuelan flood plain lake. *Limnology and Oceanography*, **39**, 234–247.

Ulanowicz, R.E. & Kay, J.J. (1991) A package for the analysis of ecosystem flow networks. *Environmental Software*, **6**, 131–142.

Ulanowicz, R.E. & Norden, J.S. (1990) Symmetrical overhead in flow networks. *International Journal of Systems Sciences*, **21**, 429–437.

Urrutia, M.B., Navarro, E., Ibarrola, I. & Iglesias, J.I.P. (2001) Preingestive selection processes in the cockle *Cerastoderma edule*: mucus production related to rejection of pseudofaeces. *Marine Ecology Progress Series*, **209**, 177–187.

Van der Meer, J. & Piersma, T. (1994) Invited perspectives in Physiological Zoology: physiologically inspired regression models for estimating and predicting nutrient stores and their composition in birds. *Physiological Zoology*, **67**, 305–329.

Van Haren, R.J.F. & Kooijman, S.A.L.M. (1993) Application of a dynamic energy budget model to *Mytilus edulis* (L.). *Netherlands Journal of Sea Research*, **31**, 119–133.

Van Straalen, N.M. (1985) Production and biomass turnover in stationary stage-structured populations. *Journal of Theoretical Biology*, **113**, 331–352.

Van Voorhies, W.A. (2000) Broad oxygen tolerance in the nematode *Caenorhabditis elegans*. *American Zoologist*, **40**, 1243.

Van Voorhies, W.A. & Ward, S. (1999) Genetic and environmental conditions that increase longevity in *Caenorhabditis elegans* decrease metabolic rate. *Proceedings of the National Academy of Sciences of the United States of America*, **96**, 11399–11403.

Vézina, A.F. & Pace, M.L. (1994) An inverse model analysis of planktonic food webs in experimental lakes. *Canadian Journal of Fisheries and Aquatic Sciences*, **51**, 2034–2044.

Vézina, A.F. & Pahlow, M. (2003) Reconstruction of ecosystem flows using inverse methods: how well do they work? *Journal of Marine Systems*, **40**, 55–77.

Vézina, A.F. & Platt, T. (1988) Food web dynamics in the ocean. I. Best-estimates of flow networks using inverse methods. *Marine Ecology Progress Series*, **42**, 269–287.

Vogel, S. (1994) *Life in moving fluids. The Physical Biology of Flow*. Princeton University Press, Princeton, NJ.

Voraberger, H.S., Kreimaier, H., Biebernik, K. & Kern, W. (2001) Novel oxygen optrode withstanding autoclavation: technical solutions and performance. *Sensors and Actuators B - Chemical*, **74**, 179–185.

Wang, Y.G. & Ellis, N. (1998) Effect of individual variability on estimation of population parameters from length-frequency data. *Canadian Journal of Fisheries and Aquatic Sciences*, **55**, 2393–2401.

Ward, J.E., Sanford, L.P., Newel, R.I.E. & MacDonald, B.A. (2000) The utility of *in vivo* observations for describing particle capture processes in suspension-feeding bivalve molluscs. *Limnology and Oceanography*, **45**, 1203–1210.

Warwick, R.M. & Gee, J.M. (1984) Community structure of estuarine meiobenthos. *Marine Ecology Progress Series*, **18**, 97–111.

Warwick, R.M. & Price, R. (1979) Ecological and metabolic studies on free-living nematodes from an estuarine mudflat. *Estuarine and Coastal Marine Science*, **9**, 257–271.

Waters, T.F. & Crawford, G.W. (1973) Annual production of a stream mayfly population: A comparison of methods. *Limnology and Oceanography*, **18**, 286–296.

Weigl, B.H. & Wolfbeis, O.S. (1995) New hydrophobic materials for optical carbon-dioxide sensors based on ion-pairing. *Analytica Chimica Acta*, **302**, 249–254.

Wenzhofer, F., Holby, O. & Kohls, O. (2001) Deep penetrating benthic oxygen profiles measured *in situ* by oxygen optodes. *Deep-Sea Research I*, **48**, 1741–1755.

Wetherill, G.B. (1986) *Regression Analysis with Applications*. Chapman and Hall, London.

Wheatcroft, R.A., Starczak, V.R. & Butman, C.A. (1998) The impact of population abundance on the deposit-feeding rate of a cosmopolitan polychaete worm. *Limnology and Oceanography*, **43**, 1948–1953.

Widbom, B. (1984) Determination of average individual dry weights and ash-free dry weights in different sieve fractions of marine meiofauna. *Marine Biology*, **84**, 101–188.

Widdows, J., Donkin, P., Brinsley, M.D., Evans, S.V., Salkeld, P.N., Franklin, A., Law, R.J. & Waldock, M.J. (1995) Scope for growth and contaminant levels in North Sea mussels *Mytilus edulis*. *Marine Ecology Progress Series*, **127**, 131–148.

Wieser, W. (1960) Benthic studies in Buzzards Bay. II. The meiofauna. *Limnology and Oceanography*, **5**, 121–137.

Wilson, W.H. (1991) Competition and predation in marine soft-sediment communities. *Annual Review of Ecology and Systematics*, **21**, 221–241.

Winberg, G.G. (1971) *Methods for the Estimation of Production of Aquatic Animals*. Academic Press, London.

Winter, J.E. (1973) The filtration rate of *Mytilus edulis* and its dependence on algal concentration, measured by continuous automatic recording apparatus. *Marine Biology*, **22**, 317–328.

Wood, S.N. (1994) Obtaining birth and mortality patterns from structured population trajectories. *Ecological Monographs*, **64**, 23–44.

Wood, S.N. (2001) Partially specified ecological models. *Ecological Monographs*, **71**, 1–25.

Zajac, R.N. (1995) Sublethal predation on *Polydora cornuta* (Polychaeta: Spionidae): patterns of tissue loss in a field population, predator functional response and potential demographic impacts. *Marine Biology*, **123**, 531–541.

Index